Lecture Notes in Computer Science 15984

Founding Editors

Gerhard Goos
Juris Hartmanis

AF172955

Editorial Board Members

The series Lecture Notes in Computer Science (LNCS), including its subseries Lecture Notes in Artificial Intelligence (LNAI) and Lecture Notes in Bioinformatics (LNBI), has established itself as a medium for the publication of new developments in computer science and information technology research, teaching, and education.

LNCS enjoys close cooperation with the computer science R & D community, the series counts many renowned academics among its volume editors and paper authors, and collaborates with prestigious societies. Its mission is to serve this international community by providing an invaluable service, mainly focused on the publication of conference and workshop proceedings and postproceedings. LNCS commenced publication in 1973.

Fedor V. Fomin · Mingyu Xiao
Editors

Computing and Combinatorics

31st International Computing and Combinatorics Conference
COCOON 2025, Chengdu, China, August 15–17, 2025
Proceedings, Part II

 Springer

Editors
Fedor V. Fomin
Department of Informatics
University of Bergen
Bergen, Norway

Mingyu Xiao
University of Electronic Science
and Technology of China
Chengdu, China

ISSN 0302-9743 ISSN 1611-3349 (electronic)
Lecture Notes in Computer Science
ISBN 978-981-95-0217-2 ISBN 978-981-95-0218-9 (eBook)
https://doi.org/10.1007/978-981-95-0218-9

This Springer imprint is published by the registered company Springer Nature Singapore Pte Ltd.
The registered company address is: 152 Beach Road, #21-01/04 Gateway East, Singapore 189721, Singapore

If disposing of this product, please recycle the paper.

Preface

It is our great pleasure to present the proceedings of the 31st International Computing and Combinatorics Conference (COCOON 2025), held in the vibrant city of Chengdu, China, from August 15 to 17, 2025. As a premier international forum, COCOON 2025 brought together leading researchers and practitioners in algorithms, theory of computation, computational complexity, and combinatorics to share cutting-edge discoveries, exchange ideas, and cultivate collaborative opportunities.

The conference encompassed a broad spectrum of topics, reflecting the dynamic and interdisciplinary nature of the field. Key areas included: Approximation Algorithms, Combinatorial Optimization, Computational Complexity, Computational Geometry, Economics and Computation, Graph Algorithms and Graph Theory, Learning and Data-Related Theory, Parameterized Algorithms, and String Algorithms and Discrete Structures.

This year, COCOON received 191 valid submissions from 24 countries and regions, demonstrating its global reach and academic significance. Each non-withdrawn submission underwent a rigorous peer-review process by the Program Committee, assisted by external referees, with at least three independent double-blind reviews. After careful evaluation, 54 papers were selected for presentation, yielding an acceptance rate of 28.3%.

We were privileged to host four distinguished invited speakers, whose thought-provoking talks enriched the conference:

Andrew Chi-Chih Yao (Tsinghua University, China)
Andrei Bulatov (Simon Fraser University, Canada)
Saket Saurabh (Institute of Mathematical Sciences, India & University of Bergen, Norway)
Shang-Hua Teng (University of Southern California, USA)

Their contributions provided deep insights and stimulated engaging discussions among attendees.

The success of COCOON 2025 would not have been possible without the collective efforts of many individuals and organizations. We extend our deepest gratitude to: The authors for their high-quality submissions; The Program Committee members and external reviewers for their diligent and constructive evaluations; The invited speakers for their enlightening presentations; The local organizers for their meticulous planning and execution; The Steering Committee for their invaluable guidance and support. We also acknowledge the generous sponsorship of: The Algorithms and Logic Group, University of Electronic Science and Technology of China, and The Theoretical Computer Science Committee of the China Computer Federation (CCF). Their support was pivotal in ensuring the conference's success.

Finally, we sincerely thank Springer for publishing the COCOON 2025 proceedings in their prestigious Lecture Notes in Computer Science (LNCS) series, enabling the dissemination of these research contributions to a global audience.

We hope this volume serves as a valuable resource for researchers and inspires further advancements in the fields of computing and combinatorics.

August 2025

Fedor V. Fomin
Mingyu Xiao

Organization

General Chair

Bakh Khoussainov   University of Electronic Science and Technology of China, China

PC Co-chairs

Fedor V. Fomin   University of Bergen, Norway
Mingyu Xiao   University of Electronic Science and Technology of China, China

Program Committee

Faisal Abu-Khzam	Lebanese American University, Lebanon
Xiaohui Bei	Nanyang Technological University, Singapore
René van Bevern	Huawei Technologies Co, Russia
Davide Bilò	University of L'Aquila, Italy
Ivan Bliznets	University of Groningen, Netherlands
Zhipeng Cai	Georgia State University, USA
Karthekeyan Chandrasekaran	University of Illinois, Urbana-Champaign, USA
Xue Chen	University of Science and Technology of China, China
Yong Chen	Hangzhou Dianzi University, China
Pål Grønås Drange	University of Bergen, Norway
Fedor V. Fomin	University of Bergen, Norway
Zhiguo Fu	Northeast Normal University, China
Takuro Fukunaga	Chuo University, Japan
Robert Ganian	Technische Universität Wien, Austria
Serge Gaspers	UNSW Sydney, Australia
Archontia Giannopoulou	National and Kapodistrian University of Athens, Greece
Alexander Grigoriev	Maastricht University, Netherlands
Gregory Gutin	Royal Holloway, University of London, UK
Tanmay Inamdar	IIT Jodhpur, India
Haitao Jiang	Shandong University, China

Bakh Khoussainov — University of Electronic Science and Technology of China, China

Ralf Klasing — CNRS and University of Bordeaux, France

Arie Koster — RWTH Aachen University, Germany

Jan Kratochvil — Charles University, Czechia

Alexander Kulikov — JetBrains, Cyprus

Van Bang Le — University of Rostock, Germany

Yi Li — Nanyang Technological University, Singapore

Chung-Shou Liao — National Tsing Hua University, Taiwan

Guohui Lin — University of Alberta, Canada

Jiamou Liu — University of Auckland, New Zealand

Jingcheng Liu — Nanjing University, China

Shengxin Liu — Harbin Institute of Technology, China

Markus Lohrey — University of Siegen, Germany

Kazuhisa Makino — Kyoto University, Japan

Matthias Mnich — Hamburg University of Technology, Germany

Nicolas Nisse — Inria, France

Yoshio Okamoto — University of Electro-Communications, Japan

Alexander Okhotin — St. Petersburg State University, Russia

Yota Otachi — Nagoya University, Japan

Qi Qi — Renmin University of China, China

Ignasi Sau — Université de Montpellier, France

Dominik Scheder — Chemnitz University of Technology, Germany

Alexander Shen — University of Montpellier, France

Kirill Simonov — University of Bergen, Norway

Frank Stephan — National University of Singapore, Singapore

Toru Takisaka — University of Electronic Science and Technology of China, China

Zhihao Gavin Tang — Shanghai University of Finance and Economics, China

Biaoshuai Tao — Shanghai Jiao Tong University, China

Ioan Todinca — University of Orléans, France

Ryuhei Uehara — Japan Advanced Institute of Science and Technology, Japan

Erik Jan van Leeuwen — Utrecht University, Netherlands

Xiaowei Wu — University of Macau, China

Mingyu Xiao — University of Electronic Science and Technology of China, China

Chao Xu — University of Electronic Science and Technology of China, China

Boting Yang — University of Regina, Canada

Meirav Zehavi — Ben-Gurion University of the Negev, Israel

Jialin Zhang	Chinese Academy of Sciences, China
Louxin Zhang	National University of Singapore, Singapore
Yong Zhang	SIAT, Chinese Academy of Sciences, China
Zhao Zhang	Zhejiang Normal University, China
Binhai Zhu	Montana State University, USA

Local Organizing Committee

Mingyu Xiao	University of Electronic Science and Technology of China, China
Chao Xu	University of Electronic Science and Technology of China, China
Yi Zhou	University of Electronic Science and Technology of China, China
Dong Hao	University of Electronic Science and Technology of China, China
Toru Takisaka	University of Electronic Science and Technology of China, China
Yuting Fang	University of Electronic Science and Technology of China, China
Ting Gou	University of Electronic Science and Technology of China, China
Yiping Liu	University of Electronic Science and Technology of China, China
Junqiang Peng	University of Electronic Science and Technology of China, China
Kangyi Tian	University of Electronic Science and Technology of China, China
Jingyang Zhao	University of Electronic Science and Technology of China, China
Yuxi Liu	University of Electronic Science and Technology of China, China
Binglin Tao	Sichuan Agricultural University, China

Additional Reviewers

Juliette Achdou
Isolde Adler
Tian Bai
Sayan Bandyapadhyay
Aritra Banik

Nikhil Bansal
Sebastian Bielfeldt
Vittorio Bilò
Arthur Braida
Robert Brijder

Anton A. Bukov
Cristian S. Calude
Dipayan Chakraborty
Jou-Ming Chang
Jiejiang Chen
Shengminjie Chen
Xiaoyu Chen
Xiuyang Chen
Yike Chen
Zishang Chen
Eddie Cheng
Kyungjin Cho
Nikolai Chukhin
Yu Cong
Gennaro Cordasco
Rajni Dabas
Renu Dalal
Sebastian Debus
Francesco Diana
Yuejia Dou
Maël Dumas
Michal Dvořák
Ajaykrishnan E. S.
Leah Epstein
Qilong Feng
Yi Feng
Kaito Fujii
Ameet Gadekar
Jiacheng Gao
Claudio Gentile
Valentin Gledel
Petr Golovach
Luciano Grippo
Luciano Gualà
Dong Hao
David Hartman
Klaus Heeger
Milan Hladík
Stepan Holub
Haoqiang Huang
Zengfeng Huang
Lucas Isenmann
Satyabrata Jana
Vít Jelínek
Hua Jiang

Zhile Jiang
Kaspar Kasche
Akinori Kawachi
Marco Kemmerling
Asif Khan
Kei Kimura
Naoki Kitamura
Masashi Kiyomi
Dimitris Kolonelos
Alexander Kononov
Sotiris Kotsiantis
Shubhang Kulkarni
Greg Kuperberg
Jan Matyáš Křišťan
Manuel Lafond
Anissa Lamani
Michael Lampis
Alexandra Lassota
Patrick Lederer
Chia-Wei Lee
Amit Levi
Fei Li
Shuai Li
Weidong Li
Wei Liang
Ya-Chun Liang
Mathieu Liedloff
Bingkai Lin
Cheng-Kuan Lin
Honghao Lin
Hanbin Liu
Maël Luce
Kelin Luo
Zihan Luo
Chensheng Ma
Mengfan Ma
Yuchao Ma
Raghunath Reddy Madireddy
Diego Marcos
Mathurin Massias
Filippos Mavropoulos
Sally Mcclean
Lili Mei
Alexander Melnikov
Ivan Mikhailin

Pedro Montealegre
Junya Nakamura
Anurag Murty Naredla
Daniel Neuen
Meike Neuwohner
André Nichterlein
Maksim Nikolaev
Fedor Noskov
Nacim Oijid
Shmuel Onn
Ioannis Panagiotas
Artem Panin
Anton Paramonov
Bo Peng
Junqiang Peng
Pan Peng
Anthony Perez
Matthias Pfretzschner
Romila Pradhan
Ioannis Psarros
Lianrong Pu
Clemens Puppe
C. Ramya
Adele Rescigno
Andreas Rosowski
Benjamin Rossman
Danil Sagunov
Parikshit Saikia
Alice Sayutina
Christiane Schmidt
Timo Schneider
Sanjay Seetharaman
François t'Serstevens
Carlos Eduardo Silva de Oliveira
M. P. Singh
Alexander Skopalik
Siani Smith
Anastasia Sofronova
Farehe Soheil
Tasuku Soma
Matej Stehlik
Yuichi Sudo
Akira Suzuki
Arman Tadevosian

Arman Tadevosian
Ayman Tajeddine
Kenjiro Takazawa
Karolina Tammemaa
Johannes Tantow
Binglin Tao
Gabor Tardos
İstenç Tarhan
Sergio Thoumi
Alexander Thumm
Kangyi Tian
Ankit Titoriya
Ron Triepels
Dimitra Tsigkari
Freija van Lent
Rob van Stee
Vsevolod Vaskin
Changjun Wang
Chenhao Wang
Dingyu Wang
Hongjie Wang
Qi Wang
Rongquan Wang
Xao Wang
Yanheng Wang
Ye Wang
Zhiqi Wang
Xuan Wu
Kelin Xia
Chenyang Xu
Renzhe Xu
Kuan Yang
Ran Yingli
Peter Zeman
Mengxiao Zhang
Jingyang Zhao
Lei Zhao
Muyang Zhao
Yajie Zhao
Da Wei Zheng
Mingchao Zhou
Yi Zhou
Jiadong Zhu
Weihao Zhu

Contents – Part II

String Algorithms and Discrete Structures

Contents – Part I

Computational Complexity

Computational Geometry

Economics and Computation

Graph Algorithms and Graph Theory

On the Complexity of 2-Club Cluster Editing with Vertex Splitting

Faisal N. Abu-Khzam[1]([✉]), Tom Davot[2], Lucas Isenmann[1], and Sergio Thoumi[1]

[1] Department of Computer Science and Mathematics, Lebanese American University, Beirut, Lebanon
{faisal.abukhzam,lucas.isenmann}@lau.edu.lb, sergio.thoumi@lau.edu
[2] LERIA, University of Angers, 49000 Angers, France
tom.davot@univ-angers.fr

Abstract. Editing a graph to obtain a disjoint union of s-clubs is one of the models for correlation clustering, which seeks a partition of the vertex set of a graph so that elements of each resulting set are close enough according to some given criterion. For example, in the case of editing into s-clubs, the criterion is proximity since any pair of vertices (in an s-club) is within a distance of s from each other. In this work, we consider the vertex splitting operation, which allows a vertex to belong to more than one cluster. This operation was studied as one of the parameters associated with the CLUSTER EDITING problem. We study the complexity and parameterized complexity of the 2-CLUB CLUSTER EDGE DELETION WITH VERTEX SPLITTING and 2-CLUB CLUSTER VERTEX SPLITTING problems. We prove that both problems are NP-complete and APX-hard. On the positive side, we show that the two problems are solvable in polynomial-time on trees and that they are both fixed-parameter tractable with respect to the number of allowed editing operations.

Keywords: Cluster Editing · 2-Club Cluster Edge Deletion · 2-Club Cluster Vertex Splitting · Vertex Splitting · Parameterized Complexity

1 Introduction

Correlation clustering is viewed as a graph modification problem where the objective is to perform a sequence of editing operations (or modifications) to obtain a disjoint union of clusters. Many variants of this problem have been studied in the literature, each with a different definition either of what a cluster is or of the various types of allowed modifications. In the CLUSTER EDITING problem, for example, a cluster was defined to be a clique and the allowed editing operations were edge additions and deletions [9,15,19]. Later, some relaxation models such as s-clubs and s-clans emerged as they were deemed ideal models for clustering

---

This research project was supported by the Lebanese American University under the President's Intramural Research Fund PIRF0056.

F. V. Fomin and M. Xiao (Eds.): COCOON 2025, LNCS 15984, pp. 3–14, 2026.
https://doi.org/10.1007/978-981-95-0218-9_1

biological networks [6, 25]. Subsequent efforts studied overlapping clusters in a graph theoretical context [4, 8, 12]. In this work, we deal with overlapping communities by performing *vertex splitting*, which allows vertices to be cloned and placed in more than one cluster. The CLUSTER EDITING WITH VERTEX SPLITTING (CEVS) problem was first introduced in [4], but relaxation models are yet to be combined with vertex splitting.

The CLUSTER EDITING and CLUSTER DELETION problems were shown to be NP-complete in [19, 26]. Several other variants of the problem have also been proved to be NP-complete. This includes CLUSTER VERTEX DELETION [20], 2-CLUB CLUSTER EDITING [22], 2-CLUB CLUSTER VERTEX DELETION [22], 2-CLUB CLUSTER EDGE DELETION [22], CLUSTER VERTEX SPLITTING [14], and CLUSTER EDITING WITH VERTEX SPLITTING [1]. From a parameterized complexity standpoint, CLUSTER EDITING, CLUSTER (EDGE) DELETION, and CLUSTER VERTEX DELETION are known to be fixed-parameter tractable (FPT) [15, 17]. The same holds for the two club-variants: 2-CLUB CLUSTER EDGE DELETION and 2-CLUB CLUSTER VERTEX DELETION [22], while 2-CLUB CLUSTER EDITING was shown to be W[2]-hard [13]. Furthermore, the CLUSTER EDITING WITH VERTEX SPLITTING problem has also been shown to be FPT [1].

From an approximation standpoint, the CLUSTER EDITING and CLUSTER (EDGE) DELETION problems are APX-hard and have $O(\log n)$ approximation algorithms [10]. On the other hand, CLUSTER VERTEX DELETION has a factor-two approximation algorithm [5]. To the best of our knowledge, problem variants with s-clubs or vertex splitting do not have any known approximation results.

The problems mentioned above are all considered different models of correlation clustering. The s-club models were shown to be effective in some networks where a clique could not capture all information needed to form a better cluster [6, 25]. Vertex splitting proved useful, and in fact essential, when the input data has overlapping clusters, such as in protein networks [23]. So far, vertex splitting has been only used along with cluster editing. In this paper we introduce the operation to the club-clustering variant by introducing two new problems: 2-CLUB CLUSTER VERTEX SPLITTING (2CCVS) and 2-CLUB CLUSTER EDGE DELETION WITH VERTEX SPLITTING (2CCEDVS). These problems seek to modify a graph into a disjoint union of 2-clubs by performing a series of either vertex splitting only (2CCVS) or vertex splitting and edge deletion operations (2CCEDVS).

Our Contribution. We prove that 2CCVS and 2CCEDVS are NP-complete, even when restricted to planar graphs of maximum degree four. Both problems are also shown to be FPT and solvable in polynomial-time on trees. We also show that, unless P = NP, the two problems cannot be approximated in polynomial time with a ratio better than a certain constant greater than one.

2 Preliminaries

We work with simple, undirected, unweighted graphs, and adopt common graph theoretic terminology. Let $G = (V, E)$ be a graph where V is the set of vertices and E is the set of edges. A *path* P_n is a sequence of $n+1$ distinct vertices such that v_i is adjacent to v_{i+1} for each $i \in \{1, 2, \ldots, n\}$. The length of a path in a simple unweighted graph is equal to its number of edges. A *geodesic*, or shortest path, between two vertices in a graph is a path that has the smallest number of edges among all paths connecting these vertices. A *cycle* is a sequence of three or more vertices $v_1, \ldots, v_n$ forming a path such that v_n is adjacent to v_1.

The *distance* between two vertices u and v of G, denoted by $d(u, v)$, is the length of a shortest path between them. An *s-club* is a (sub)graph such that any two vertices are within distance s from each other. Equivalently, the longest path allowed in an s-club is a P_s. A *clique*, or a 1-club, in G is a subgraph whose vertices are pairwise adjacent. An s-club graph is a disjoint union of connected components, such that each component is an s-club. The *open neighborhood* $N(v)$ of a vertex v is the set of vertices adjacent to it. The *degree* of v is the number of edges incident on v, which is $|N(v)|$ since we are considering simple graphs only. A *cut vertex* is a vertex whose deletion disconnects the graph.

A *vertex split* is an operation that replaces a vertex v by two copies v_1 and v_2 such that $N(v) = N(v_1) \cup N(v_2)$. An *exclusive vertex split* requires that $N(v_1) \cap N(v_2) = \emptyset$. In this paper, we do not assume a split is exclusive, but our proofs apply to this restricted version, which is more important in application domains [2]. Therefore, all results that we present also hold for exclusive vertex splitting. In this paper, we introduce the following two problems:

---

s-club Cluster Vertex Splitting

Given: A graph $G = (V, E)$, along with positive integers s and k;

Question: Can we transform G into a disjoint union of s-clubs by performing at most k vertex splitting operations?

---

s-club Cluster Edge Deletion with Vertex Splitting

Given: A graph $G = (V, E)$, along with positive integers s and k;

Question: Can we transform G into a disjoint union of s-clubs by performing a total of at most k edge deletions and/or vertex splitting operations?

We work with the problem variant where s is fixed to 2. For a graph G, we define $2ccvs(G)$ (resp., $2ccedvs(G)$) as the minimum length of a sequence of splits (resp., edge deletions and splits) to turn G into disjoint union of 2-clubs. For example, for a path $P_5 = v_0, \ldots, v_5$ of length 5, we have $2ccvs(P_5) = 2$ by splitting v_2 and v_3 and we have $2ccedvs(P_5) = 1$ by deleting the edge $v_2 - v_3$.

3   The Complexity of 2-Club Cluster Vertex Splitting and 2-Club Cluster Edge Deletion with Vertex Splitting

Since it is obvious that 2CCVS is in NP, we show that it is NP-hard by reduction from 3-SAT [27].

Construction 1. *Consider a 3-CNF formula ϕ, and denote by M the number of clauses and by $\mathcal{V}$ the set of variables. For every variable V, we denote by $a(V)$ the number of clauses where V appears and by $C(V)_1, \ldots, C(V)_{a(V)}$ the clauses where V appears. Our construction proceeds as follows (see Fig. 1):*

- *For each variable V, we create a cycle of length $6a(V)$: $v_1, \ldots, v_{6a(V)}$.*
- *For every clause C where a variable V appears, let j be the index of C in the (above defined) list of the clauses where V appears. We define the vertex v_c as $v_{6(j-1)+1}$ (resp., $v_{6(j-1)+2}$) if its corresponding variable V appears positively (resp., negatively). We denote by v_{c-1} (resp., v_{c+1}) the preceding (resp., following) vertex in the variable cycle. Note that we consider the indices modulo $6a(v)$.*
- *For each clause $C = U \vee V \vee W$, connect the vertices u_C, v_C, w_C into a clique.*

An example is illustrated in Fig. 1. Observe that the obtained graph is of maximum degree 4.

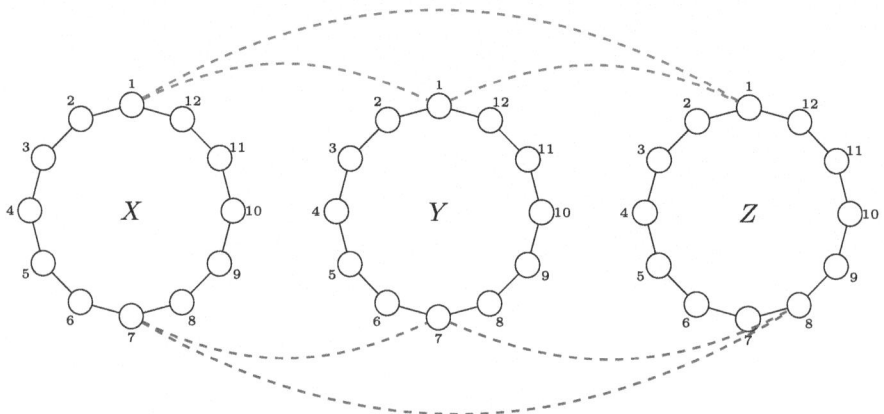

Fig. 1. The graph G for $\phi = (X \vee Y \vee Z) \wedge (X \vee Y \vee \overline{Z})$. The edges of the clause $X \vee Y \vee Z$ are in red and the edges of the clause $X \vee Y \vee \overline{Z}$ are in blue (Color figure online).

Theorem 1. *2CCVS is NP-complete even on graphs with maximum degree four.*

Proof. Let G be the graph obtained by Construction 1. We set $k = 11M$ where M is the number of clauses.

Claim. The graph G has a sequence of at most k splits that turn G into a disjoint union of 2-clubs if and only if ϕ is satisfiable.

($\Leftarrow$) Suppose that ϕ is satisfiable and consider a satisfying assignment σ. Let V be a variable. If V is true (resp., false), then we split v_{2j} (resp., v_{2j+1}) for every $j \in [1, 3a(V)]$ so that it separates the cycles in a disjoint union of paths of length 2. We recall that the indices of the vertices v_j are considered modulo $6a(V)$ in $[1, 6a(V)]$. For each split vertex, one copy will get all the edges coming from outside of the cycle and the other none. Therefore, we have already used $\sum_{i \in |V|} 3a(V_i) = 9M$ splits.

Let C be a clause containing three variables X, Y and Z. As ϕ is satisfied by σ, then there exists a variable which satisfies C. Without loss of generality, we can suppose that X satisfies C. If X is true, then the vertices x_{C-1} and x_{C+1} are split. We split the vertices y_C (resp., z_C) so that one copy is adjacent to vertices of the clause and the other copy is adjacent to vertices of the cycle of Y (resp., Z). For each clause, we make 2 splits. In total, we use $k = 2M + 9M$ splits.

The connected components of the resulting graphs are paths of length 2 (in the variable cycles) and stars centered on x_C vertices where X is the chosen variable satisfying a clause C. We conclude that this sequence of k splits leads to a disjoint union of 2-clubs.

($\Rightarrow$) Conversely, suppose we have a sequence of k splits turning G into a disjoint union of 2-clubs. Let V_i be a variable and suppose that less than $3a(V_i)$ vertices of the cycle are split. As there are $6a(V_i)$ vertices, then there exists two adjacent vertices v_i and v_{i+1}, with $i \in [1, 6a(V_i)]$ which are not split. Thus, the vertices v_{i-1} and v_{i+2} will be at distance at least 3 in the resulting graph, a contradiction. Hence, at least $3a(V_i)$ of the vertices of the cycle must be split.

Let us prove that each clause needs 2 splits. Consider a clause C_j with variables X, Y, and Z. In the resulting graph, there is a copy x'_c of x_c and a copy y'_c of y_c such that x'_c and y'_c are adjacent. Suppose that x'_c is adjacent to x_{c-1} or x_{c+1} and y'_c is adjacent to y_{c-1} or y_{c+1}. This would imply that the resulting graph contains two vertices at distance 3 from each other, a contradiction. Therefore, either x'_c is not adjacent to x_{c-1} and x_{c+1} or y'_c is not adjacent to y_{c-1} and y_{c+1}. Without loss of generality, we can suppose that x'_c is not adjacent to x_{c-1} and x_{c+1}. By considering the edge $y - z$, we prove in the same way that either there is a copy y'_c of y_c which is not adjacent to y_{c-1} and y_{c+1} or there is a copy z'_c of z_c which is not adjacent to z_{c-1} and z_{c+1}. So we need at least 2 splits for every clause which cannot be used to solve the conflicts in the cycle.

Therefore, we need at least $2M + 9M$ splits. As $k = 11M$, we deduce that for each variable V exactly $3a(V)$ vertices of its cycle are split. Let us prove that in every variable cycle, the vertices which are split are not adjacent. Let V be a variable, suppose that two adjacent vertices are split. Then, without loss of generality, we can suppose that v_1 and v_2 are split. We still have to deal with the geodesics $v_1, v_{a(V)}, v_{a(V)-1}, v_{a(V)-2}$ and v_2, v_3, v_4, v_5. Thus, we need $3a(V) - 1$ more splits. It contradicts the fact that we can only use $3a(V)$ splits to resolves all the P_3s of the cycle since we used two splits previously. Therefore, the split vertices disconnecting the cycle of v are not adjacent. As half of the vertices of the cycles must be split, we deduce that there are only 2 ways to split the cycle.

Let us define a truth assignment of the variables. If v_2 is split, then we set V to true; otherwise, we set it to false.

Let us prove that the 3-SAT formula is satisfied by the previous assignment. Let C be a clause with variables X, Y, Z. Without loss of generality we can suppose that $X, Y,$ and Z appear positively in the clause. Suppose that $X, Y,$ and Z are set to false. Therefore, x_1 is split and the edges $x_1 - x_{6a(X)}$ and $x_1 - x_2$ are split. Therefore, the following paths are in the obtained graph: $x_c - x_{c+1} - x_{c+2}$ and $x_c - x_{c-1} - x_{c-2}$. As $y_c - x_c - x_{c+1} - x_{c+2}$ is a path and $z_c - x_c - x_{c+1} - x_{c+2}$ is a path as well, we deduce that we have to split x_1 one more time. In the same way, we prove that y_c and z_c must be split at least 2 times. This contradicts the fact that the sequence of splits is of length at most $k = 2M + 9M$. We conclude that C is satisfied and that we have a yes-instance. This completes the proof.

A similar construction can be used to also show that 2-CLUB CLUSTER EDGE DELETION WITH VERTEX SPLITTING is NP-complete (see [3] for details).

Theorem 2. *The* 2-CLUB CLUSTER EDGE DELETION WITH VERTEX SPLITTING *problem is NP-complete.*

As the two constructions result in planar graphs if the bipartite graph of the 3-SAT instance is planar, we deduce the following from the NP-hardness of PLANAR-3-SAT [21]:

Corollary 1. 2-CLUB CLUSTER VERTEX SPLITTING *and* 2-CLUB CLUSTER EDGE DELETION WITH VERTEX SPLITTING *remains NP-complete on planar graphs with maximum degree four.*

Moreover, knowing that the previous construction is linear in the number of vertices and (resp., Planar) 3-SAT does not admit a $2^{o(n+m)}n^{O(1)}$ (resp., $2^{o(\sqrt{n+m})}n^{O(1)}$) time algorithm [11] unless the Exponential Time Hypothesis (ETH) fails, we conclude the following:

Corollary 2. *Assuming the ETH holds, there is no $2^{o(n+m)}n^{O(1)}$-time (resp., $2^{o(\sqrt{n+m})}n^{O(1)}$)-time algorithm for 2CCVS (resp., on planar graphs) with maximum degree four.*

On the positive side, we show using recursive formulas and dynamic programming that both 2CCVS and 2CCEDVS are solvable in polynomial-time when the input graph is a tree (see [3] for omitted proofs).

Theorem 3. 2CCVS *is solvable in polynomial time on trees.*

Theorem 4. 2CCEDVS *is solvable in polynomial time in trees.*

4 Hardness of Approximation

Our objective in this section is to adapt the construction in Sect. 3 to reduce MAX 3-SAT(4) to 2CCVS. Max 3-SAT(4) is a variant of MAX 3-SAT where each variable appears at most four times in the given formula ϕ.

We also add the following constraint: when a variable V_i appears exactly two times positively and two times negatively, we suppose that the list of clauses in which V_i appears, $C(V_i)_1, C(V_i)_2, C(V_i)_3, C(V_i)_4$, is ordered so that V_i appears positively in $C(V_i)_1$ and $C(V_i)_3$ and negatively in $C(V_i)_2$ and $C(V_i)_4$. This constraint is added to ensure that each unsatisfied clause in ϕ causes an additional split in the construction. In fact, we can observe that if the formula ϕ cannot be satisfied, then we can use an "extra" split in each clause gadget to obtain a solution. However, the inverse does not necessarily hold if there is a variable V_i that occurs two times positively and two times negatively.

Indeed, by using $12 + 1$ splits in the variable cycle, we may be able to satisfy the four clauses where v occurs. In the rest of this section, we show how to obtain a reduction from MAX 3-SAT(4) to 2CCVS to prove the below theorem.

We will make use of linear reductions defined as follows:

Definition 1. (Linear reduction [24]**).** Let A and B be optimization two optimization problems with cost functions $cost$ and $cost'$. We denote by $OPT(I)$ the optimal value of an optimization problem on an instance I. We say that A has a linear reduction to B if there exists two polynomial time algorithms f and g and two positive numbers α and β such that for any instance I of A:

- $f(I)$ is an instance of B.
- $OPT(f(I)) \leq \alpha OPT(I)$.
- For any solution S' of $f(I)$, $g(S')$ is a solution of I and $|cost(g(S')) - OPT(I)| \leq \beta |cost(S') - OPT(f(I))|$.

Theorem 5. 2CCVS *is APX-hard.*

Proof. Recall that in an optimal solution of MAX 3-SAT(4), at least $\frac{7}{8}$ of the clauses are satisfied [16], yielding

$$OPT(\phi) \geq \frac{7M}{8}. \tag{1}$$

Let f be the function transforming any instance ϕ of MAX 3-SAT(4) into a graph G with Construction 1.

We can find a sequence of splits in G turning it into a disjoint union of 2-clubs with $3a(V)$ splits (by splitting one over two vertices of each variable cycle) for every variable cycle and 3 splits for every clause (by splitting the three vertices of the clause triangle) Thus,

$$OPT(G) \leq 3M + 9M = 12M$$

$$\text{As } M \leq \frac{8}{7}OPT(\phi), \text{ we deduce that}$$

$$OPT(G) \leq 12\frac{8}{7}OPT(\phi) = \frac{96}{7}OPT(\phi)$$

Consider an assignment of ϕ maximizing the number of satisfied clauses of ϕ. Let us define a sequence of operations on G_ϕ. For every true (resp., false) variable v, we split the vertices v_{2i} (resp., v_{2i+1}) for every i. Let C be a clause. If C is satisfied, consider v a variable satisfying C. Split the vertices u_C and w_C (like in the proof of Theorem 2) where u and w are the two other variables in C. Otherwise we split the three vertices u_C, v_C and w_C. We denote by SC the set of satisfied clauses of ϕ by σ and UC the set of unsatisfied clauses of ϕ by σ. This sequence is of length $\sum_{v \in V} 3a(v) + \sum_{c \in SC} 2 + \sum_{c \in UC} 3 = 9M + 2M + k$ where k is the number of unsatisfied clauses. Furthermore this sequence of operations turns the graph G_ϕ into a disjoint union of bicliques. We deduce that $OPT(G_\phi) \leq 12M + M - OPT(\phi)$. Thus, $OPT(\phi) + OPT(G_\phi) \leq 12M + M$.

Let X be a sequence of operations turning the graph into a disjoint union 2-clubs. Let us define an assignment $g(X)$. Consider a variable v. According to the proof of Theorem 2, the variable cycle of v needs at least $3a(v) = 12$ operations on its edges and its vertices.

If the variable cycle is using exactly 12 operations, then there are two cases. If the vertices v_{2i} are split for every i, then we set v to true and otherwise to false.

If the variable cycle is using at least $12 + 2$ operations, then we replace these operations by splitting the vertices v_{2i+1} for every i and by splitting the vertices v_1 and v_{13} so that it disconnects the variable cycle and the two positive clauses connected to v. We set v to false.

If the variable cycle is using at $12 + 1$ splits, then we show that it is not possible that it resolves all geodesics of the cycle with the clause edges incident to v_1, v_8, v_{13} and v_{19}. If v is resolving C_{v_1} and C_{v_3}, then we set v to positive. Otherwise, we set v to negative.

As $cost(g(X))$ is the number of satisfied clauses of $g(X)$, we have $13M \leq cost(X) + cost(G(X))$ because for each unsatisfied clause, we can add only one split to satisfy it. Therefore we have:

$$OPT(\phi) + OPT(G_\phi) \leq 13M \leq cost(X) + cost(g(X))$$
$$OPT(\phi) - cost(g(X)) \leq cost(X) - OPT(G_\phi)$$

Thus, $|cost(g(X)) - OPT(\phi)| \leq |cost(X) - OPT(G)|$ because MAX-3SAT(4) is a maximization problem and because 2CCVS is a minimization problem. We have constructed a linear reduction with $\alpha = \frac{96}{7}, \beta = 1$. As $MAX3 - SAT(4)$ is APX-hard, 2CCVS is also APX-hard.

In the same way, we can prove:

Theorem 6. 2CCEDVS *is APX-hard.*

5   The Parameterized Complexity of 2CCEDVS and 2CCVS

As observed for CLUSTER EDITING WITH VERTEX SPLITTING in [1], all edge deletions can be performed before vertex splitting. Thus, we assume that any sequence of operations is equivalent to a sequence of operations where the splitting is performed at the end. Our proof is based on branching on paths of length three whose endpoints are at distance exactly three from each other.

In the case/branch where a vertex is to be split, we simply mark it for splitting and perform this operation at the end, until no such length-three paths exist. This "charge and reduce" approach is explained in more detail in the sequel.

Lemma 1. *Consider a minimal sequence of edge deletions and vertex splittings. Let S be the set of the split vertices. If $v \in S$ and C is a connected component of $G[V(G) \setminus S]$, then each copy of v is either adjacent to all the vertices of $C \cap N(v)$ or to none of them.*

Proof. Suppose there exists a copy u of v and x, y two neighbors of v in C such that u is adjacent to x and not to y. As y is adjacent to v then there exists another copy u' of v such that u' is adjacent to y. As x and y are connected in $G[V(G) \setminus S]$, then u and u' belong to the same connected component of the resulting solution, which is impossible since any two copies of a split vertex must belong to two different clubs, unless the solution is not minimal, a contradiction.

Lemma 2. *Let $G = (V, E)$ be a connected graph and assume a sequence of at most k edge deletions and splits is applied to G. If S is the set of split vertices, then $G[V \setminus S]$ has at most $k + 1$ connected components.*

Proof. The statement simply follows from the fact that the graph is initially connected and a single edge deletion operation or a single split operation can only increase its number of connected components by at most 1.

We now show how to perform the vertex splittings at the end, after marking the vertices to be split and exhausting all the possible edge deletion operations. Given a split set $S = \{v_1, \ldots, v_s\}$, p connected components $C = \{C_1, \ldots, C_p\}$, and a number of possible extra splits e. We consider the algorithm $Aux(S, C, e)$ for trying all sequences of splits on S of length at most $s + e$ such that each vertex of S is split at least once.

At the end of such a sequence, the split set is of size at most $2s + e$ (since we create at most $s + e$ copies). A split is a choice of a vertex in the current split set and one subset of the vertices for each of the two copies. There are at most $2s + e$ choices to select a vertex that will be split. Choosing a set of neighbors for both copies of the split corresponds to selecting two subsets of the current split set and of the connected components (see Lemma 1). There are at most 2^{2s+e+p} choices for the set of neighbors of one copy. Therefore, there are at most $((2s + e)2^{4s+2e+2p})^{s+e}$ sequences of such splits.

Theorem 7. *The complexity of Algorithm* AUX *is in $O((3k2^{8k})^{2k})$ if $p \leq k+1$, $e + s \leq k$.*

We now describe the main algorithm, which returns true if there exists a sequence of length at most k of edge deletions or vertex splittings on a graph which results in a disjoint union of 2-clubs and returns false otherwise. In the

---

Algorithm 1. $f(G, S, k)$

1: **if** $k = 0$ **then**
2:     **return** $Aux(S, G[V(G) \setminus S], 0)$
3: **else if** there exists x, y such that $d(x, y) = 3$ and u and v are interior vertices of a length-3 path between x and y that are not in S **then**
4:     **if** $f(G - xu, S, k - 1)$ **then return** true
5:     **end if**
6:     **if** $f(G - uv, S, k - 1)$ **then return** true
7:     **end if**
8:     **if** $f(G - vy, S, k - 1)$ **then return** true
9:     **end if**
10:     **if** $f(G, S \cup u, k - 1)$ **then return** true
11:     **end if**
12:     **if** $f(G, S \cup v, k - 1)$ **then return** true
13:     **end if**
14: **else**
15:     **return** $Aux(S, G[V(G) \setminus S], k)$
16: **end if**

---

following, $G = (V, E)$ is a graph, $S \subseteq V$ is the set of vertices that were marked for splitting, and k is the number of allowed splits.

The algorithm branches to 5 sub-problems and each step is taking $O(n^3)$ to search for a pair of vertices at distance 3 from each other. Therefore, we deduce:

Theorem 8. *Algorithm 1 solves* 2CCEDVS *in* $O(n^3 5^k (3k2^{8k})^{2k})$. *Hence, the problem is fixed-parameter tractable with respect to total number of edge deletions and vertex splittings.*

Note that a geodesic of length three is also an obstruction in the case of the 2CCVS problem. Therefore we can adapt the previous algorithm by removing the branchings where we delete edges (just remove lines 4–9 of Algorithm 1). We conclude that:

Corollary 3. 2CCVS *is fixed-parameter tractable with respect to solution size.*

6   Concluding Remarks

This paper introduces the 2-CLUB CLUSTER VERTEX SPLITTING (2CCVS) and 2-CLUB CLUSTER EDGE DELETION WITH VERTEX SPLITTING (2CCEDVS) problems. Both problems are shown to be NP-complete in general and that 2CCVS remains NP-complete on planar graphs of maximum degree four. We further considered the polynomial-time approximability of the two problems and showed them to be APX-hard. We believe a constant-factor approximation for 2CCVS is not too difficult to obtain. In fact, this remains an interesting open problem in both cases.

On the positive side, we showed that both 2CCVS and 2CCEDVS are fixed-parameter tractable when parameterized by the number of allowed modifications.

We believe that obtaining an improved algorithm that runs in $O^*(c^k)$ for 2CCVS is not too difficult since a simple branching algorithm would have two cases for each length-three "obstruction path" and the rest consists of performing vertex splits only. However, in the case of 2CCEDVS, obtaining an algorithm with a running time in $O^*(c^k)$ seems much more challenging, and we pose it here as an open problem. Furthermore, we gave polynomial-time algorithms for both 2CCVS and 2CCEDVS when the input is restricted to trees. This suggests considering the parameterized complexity of the two problems when parameterized by treewidth.

Whether the obtained hardness results for 2-clubs hold also for s-clubs (i.e. for any/all $s \geq 3$) remains an open question, to be explored in future work. Other directions that could be explored include the parameterized complexity of the problems with respect to other parameters such as the treewidth of the graph, as well as using additional local parameters such as the number of times a vertex can split, which seems to be a realistic constraint. In fact, the use of local parameters seems to be an effective approach in the context of cluster editing [7,18]. Finally, an interesting open problem is whether the two considered problems admit polynomial-size kernels.

References

1. Abu-Khzam, F.N., et al.: Cluster editing with vertex splitting. Discret. Appl. Math. **371**, 185–195 (2025)
2. Abu-Khzam, F.N., Barr, J.R., Fakhereldine, A., Shaw, P.: A greedy heuristic for cluster editing with vertex splitting. In 4th International Conference on Artificial Intelligence for Industries, AI4I 2021, Laguna Hills, CA, USA, September 20-22, 2021, pages 38–41. IEEE, 2021
3. Abu-Khzam, F.N., Davot, T., Isenmann, L., Thoumi, S.: On the complexity of 2-club cluster editing with vertex splitting. CoRR, abs/2411.04846, 2024
4. Abu-Khzam, F.N., Egan, J., Gaspers, S., Shaw, A., Shaw, P.: Cluster editing with vertex splitting. In: Lee, J., Rinaldi, G., Mahjoub, A.R. (eds.) ISCO 2018. LNCS, vol. 10856, pp. 1–13. Springer, Cham (2018). https://doi.org/10.1007/978-3-319-96151-4_1
5. Aprile, M., Drescher, M., Fiorini, S., Huynh, T.: A tight approximation algorithm for the cluster vertex deletion problem. Math. Program. pages 1–23, 2023
6. Balasundaram, B., Butenko, S., Trukhanov, S.: Novel approaches for analyzing biological networks. J. Comb. Optim. **10**(1), 23–39 (2005)
7. Barr, J.R., Shaw, P., Abu-Khzam, F.N., Chen, J.: Combinatorial text classification: the effect of multi-parameterized correlation clustering. In 2019 First International Conference on Graph Computing (GC), pp. 29–36. IEEE, 2019
8. Baumes, J., Goldberg, M.K., Krishnamoorthy, M.S., Magdon-Ismail, M., Preston, N.: Finding communities by clustering a graph into overlapping subgraphs. IADIS AC **5**, 97–104 (2005)
9. Cai, L.: Fixed-parameter tractability of graph modification problems for hereditary properties. Inf. Process. Lett. **58**(4), 171–176 (1996)
10. Charikar, M., Guruswami, V., Wirth, A.: Clustering with qualitative information. J. Comput. Syst. Sci. **71**(3), 360–383 (2005)

11. Cygan, M., et al.: Lower bounds based on the exponential-time hypothesis. Parameterized Algorithms, pp. 467–521, 2015
12. Fellows, M.R., Guo, J., Komusiewicz, C., Niedermeier, R., Uhlmann, J.: Graph-based data clustering with overlaps. Discret. Optim. **8**(1), 2–17 (2011)
13. Figiel, A., Himmel, A.-S., Nichterlein, A., Niedermeier, R.: On 2-clubs in graph-based data clustering: theory and algorithm engineering. In: Calamoneri, T., Corò, F. (eds.) CIAC 2021. LNCS, vol. 12701, pp. 216–230. Springer, Cham (2021). https://doi.org/10.1007/978-3-030-75242-2_15
14. Firbas, A., Sorge, M.: On the complexity of establishing hereditary graph properties via vertex splitting (2024). arXiv preprint arXiv:2401.16296
15. Gramm, J., Guo, J., Hüffner, F., Niedermeier, R.: Graph-modeled data clustering: exact algorithms for clique generation. Theor. Comput. Syst. **38**, 373–392 (2005)
16. Håstad, J.: Some optimal inapproximability results. J. ACM **48**(4), 798–859 (2001)
17. Hüffner, F., Komusiewicz, C., Moser, H., Niedermeier, R.: Fixed-parameter algorithms for cluster vertex deletion. Theor. Comput. Syst. **47**(1), 196–217 (2010)
18. Komusiewicz, C., Uhlmann, J.: Cluster editing with locally bounded modifications. Discret. Appl. Math. **160**(15), 2259–2270 (2012)
19. Křivánek, M., Morávek, J.: NP-hard problems in hierarchical-tree clustering. Acta Informatica, 23(3):311–323. (1986)
20. Lewis, J.M., Yannakakis, M.: The node-deletion problem for hereditary properties is np-complete. J. Comput. Syst. Sci. **20**(2), 219–230 (1980)
21. Lichtenstein, D.: Planar formulae and their uses. SIAM J. Comput. **11**(2), 329–343 (1982)
22. Liu, H., Zhang, P., Zhu, D.: On editing graphs into 2-club clusters. In Frontiers in Algorithmics and Algorithmic Aspects in Information and Management, pp. 235–246. Springer, 2012
23. Nepusz, T., Yu, H., Paccanaro, A.: Detecting overlapping protein complexes in protein-protein interaction networks. Nat. Methods **9**(5), 471–472 (2012)
24. Papadimitriou, C.H., Yannakakis, M.: Optimization, approximation, and complexity classes. J. Comput. Syst. Sci. **43**(3), 425–440 (1991)
25. Pasupuleti, S.: Detection of protein complexes in protein interaction networks using n-clubs. In European Conference on Evolutionary Computation, Machine Learning and Data Mining in Bioinformatics, pp. 153–164. Springer, 2008
26. Shamir, R., Sharan, R., Tsur, D.: Cluster graph modification problems. Discret. Appl. Math. **144**(1–2), 173–182 (2004)
27. Tovey, C.A.: A simplified np-complete satisfiability problem. Discret. Appl. Math. **8**(1), 85–89 (1984)

A Sufficient Condition for the Existence of Two Completely Independent Spanning Trees

Yang Hu[1,3,4], Bo Ning[1,3,4(✉)], and Xiumin Wang[2,3,4]

[1] College of Computer Science, Nankai University, Tianjin 300350, China
`yang.hu@mail.nankai.edu.cn, bo.ning@nankai.edu.cn`
[2] College of Cryptology and Cyber Science, Nankai University, Tianjin 300350, China
`xiumin.wang@mail.nankai.edu.cn`
[3] Tianjin Key Laboratory of Network and Data Security Technology, Nankai University, Tianjin 300350, China
[4] Key Laboratory of Data and Intelligent System Security, Nankai University, Ministry of Education, Tianjin 300350, China

Abstract. Completely independent spanning trees play an important role in security protection routing and enhancing the robustness of ad hoc networks. In a graph G, a collection of k spanning trees is termed completely independent spanning trees if the paths connecting any pair of vertices x and y in these distinct trees are internally disjoint. Fan (1984) [10] proposed a sufficient condition for the existence of Hamilton cycles. Based on Fan's result, we derive a sufficient condition for the existence of two CISTs and identify the graphs that satisfy Fan's condition but do not contain two CISTs.

Keywords: Completely independent spanning trees · CIST-partition · Hamiltonian sufficient conditions

1 Introduction

Throughout this paper, we consider simple finite graphs. Let G be a graph with vertex set $V(G)$ and edge set $E(G)$. Let $d_G(v)$ denote the degree of a vertex $v \in V(G)$, and let $d_G(u, v)$ denote the distance between two vertices $u, v \in V(G)$. When there is no danger of ambiguity, we use $d(v)$ and $d(u, v)$ instead of $d_G(v)$ and $d_G(u, v)$, respectively. The neighborhood of v in G, denoted by $N_G(v)$, is the set of vertices adjacent to v, excluding v itself. We use $\delta(G)$ to denote the minimum degree of G. For disjoint vertex subsets $S, T \subseteq V(G)$, we define $e(S, T)$ as the number of edges with one endpoint in S and the other in T. A subgraph of G obtained only by deletion of the vertices is called an induced subgraph. We use $G[S]$ to denote the subgraph of G whose vertex set is S and whose edge set consists of all edges of G that have both ends in S. Subgraphs may also be induced by sets of edges. If $E' \subseteq E(G)$, we use $G[E']$ to denote the edge-induced

© The Author(s), under exclusive license to Springer Nature Singapore Pte Ltd. 2026
F. V. Fomin and M. Xiao (Eds.): COCOON 2025, LNCS 15984, pp. 15–28, 2026.
https://doi.org/10.1007/978-981-95-0218-9_2

subgraph of G whose edge set is E' and whose vertex set consists of all ends of edges of E'. Given a partition (V_1, V_2) of $V(G)$, the bipartite graph $G[V_1, V_2]$ has vertex set $V_1 \cup V_2$ and edge set $\{uv \mid uv \in E(G), u \in V_1 \text{ and } v \in V_2\}$. A component of G that is a tree is referred to as a tree component. Specially, an isolated vertex of G also is considered as a tree component. Let $c(G)$ be the number of components of G. A vertex in a tree T is classified as a leaf if its degree in T is 1.

The construction of independent spanning trees in networks has gained significant attention due to their importance in various network tasks, with a notable increase in research over the past decade. Completely independent spanning trees (CISTs for short) are a specific type of independent spanning trees, and this notion refers to two spanning trees T_1 and T_2 of G that are completely independent if, for any two distinct vertices u and v of G, the two u-v paths on T_1 and T_2 are internally disjoint. The notion of completely independent spanning trees was introduced by Hasunuma [12]. The potential of utilizing CISTs to improve the robustness of ad hoc networks has been demonstrated in prior research [20]. Additionally, the critical role of CISTs in enabling secure routing protocols has been further highlighted in [24]. Hasunuma [13] established the $\mathcal{NP}$-hardness of constructing just two CISTs in a general graph G, thereby directing current research towards networks with specific structural characteristics. In addition, research on CISTs has focused on torus networks [15], planar graphs [13], crossed cubes [8, 22, 23], bicububes [7], and DCell data center networks [25], etc.

Increasing the number of completely independent spanning trees can enhance network fault tolerance. Due to cost and resource constraints, practical implementations typically maintain minimal configurations. For two CISTs, this means supporting only two instances as an optimal balance between redundancy and efficiency. The research carried out by Tapolcai [28] demonstrates that networks with two CISTs can achieve full protection against random single-element failures.

Hamilton cycles are important objects in structural and extremal graph theory. Recent results have shown that some sufficient conditions for the existence of Hamilton cycles can also be used as sufficient conditions for the existence of two CISTs. Recall that Dirac's condition [9], Ore's condition [21] and Fan's condition [10] are three classical sufficient conditions for Hamilton cycles in graphs.

Theorem 1 (Dirac, [9]). *If G is a graph on $n \geq 3$ vertices such that $\delta(G) \geq \frac{n}{2}$, then G has a Hamilton cycle.*

Theorem 2 (Ore, [21]). *Let G be a graph on $n \geq 3$ vertices. If $d(u) + d(v) \geq n$ for every pair of non-adjacent vertices $u, v \in V(G)$, then G has a Hamilton cycle.*

Theorem 3 (Fan, [10]). *Let G be a 2-connected graph on $n \geq 3$ vertices. If for every pair of vertices $u, v \in V(G)$ with $d(u, v) = 2$, it holds that $\max\{d(u), d(v)\} \geq \frac{n}{2}$, then G has a Hamiltonian cycle.*

Interestingly, Araki [1] affirmed that the existence of two CISTs is implied by Dirac's condition. Fan, Hong, and Liu [11] proved that Ore's condition indicates the existence of two CISTs.

Theorem 4 (Araki, [1]). *Let G be a graph on $n \geq 7$ vertices. If $\delta(G) \geq \frac{n}{2}$ then G has two completely independent spanning trees.*

Theorem 5 (Fan, Hong, Liu, [11]). *Let G be a graph on $n > 8$ vertices. If $d(u) + d(v) \geq n$ for every pair of non-adjacent vertices u and v, then G has two completely independent spanning trees.*

Hong and Liu [16] further extended the result of Araki [1] by demonstrating that the generalization of Dirac's condition implies the existence of k CISTs. Other research has demonstrated that the neighborhood union condition for Hamiltonian graphs also ensures the existence of two CISTs, see [17,18,26]. Recently, Ma and Cai [19] proved that for a connected graph where every pair of vertices at distance 2 has a degree sum of at least the total number of vertices, there exist two CISTs. For additional sufficient conditions for graphs containing CISTs, we refer the reader to [2–6,14,27,29].

Inspired by Theorem 3, we focus on investigating a new sufficient condition for the existence of two CISTs.

Theorem 6. *Let G be a 2-connected graph on $n \geq 7$ vertices. Let $U = \{u \mid u \in V(G)$ and $d(u) < n/2\}$. If G satisfies the following conditions:*

(i) $e(v, U) \leq \frac{|U| - c(G[U])}{2}$ for any vertex $v \notin U$;
(ii) For any component H of $G[U]$ which is K_2 or K_3, we have $e(V(H), V(G) \setminus V(H)) \geq 3$;
(iii) For any two vertices x and y with $d(x, y) = 2$, we have $\max\{d(x), d(y)\} \geq \frac{n}{2}$,

then G has two CISTs.

The lower bound $n \geq 7$ in Theorem 6 is essential. In general, a graph G with n vertices must have at least $2(n-1)$ edges to admit two completely independent spanning trees. For instance, although $K_{3,3}$ satisfies the conditions of Theorem 6, it has only 9 edges and thus does not contain two CISTs. Similarly, $K_{2,2}$, with just 4 edges, also fails to admit two CISTs despite satisfying the conditions of Theorem 6.

Remark 1. The graph shown in Fig. 1 consists of a complete graph K_{2r} ($r \geq 2$), a set of r independent edges denoted by M_r, and a matching between the vertex sets of K_{2r} and M_r. This graph contains $4r$ vertices with a degree sequence of $(2, 2, \ldots, 2, 2r, \ldots, 2r)$, where $2r$ vertices have degree 2 and $2r$ vertices have degree $2r$. Although the graph satisfies Fan's condition, it does not have two CISTs.

Remark 2. The following examples also show that Fan's condition is not strong enough to guarantee the existence of two CISTs. Consider the graphs G_1, G_2, and G_3 shown in Fig. 2, where $|V(G_1)| = |V(G_2)| = n$. These graphs satisfy Fan's condition, but do not have two CISTs. Moreover, these graphs show that (i), (ii), and (iii) are necessary for two CISTs in Theorem 6. Specifically,

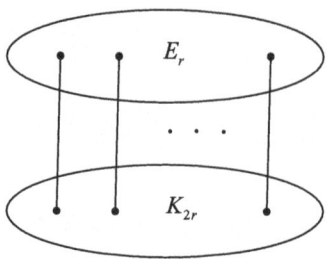

Fig. 1. A graph satisfying Fan's condition but containing no two CISTs.

- Graphs G_1 and G_2 satisfy conditions (i) and (iii) of Theorem 6 but not (ii). By the definition of completely independent spanning trees, they do not have two CISTs.
- The graph G_3 satisfies conditions (ii) and (iii) of Theorem 6 but not (i), and it does not have two CISTs.

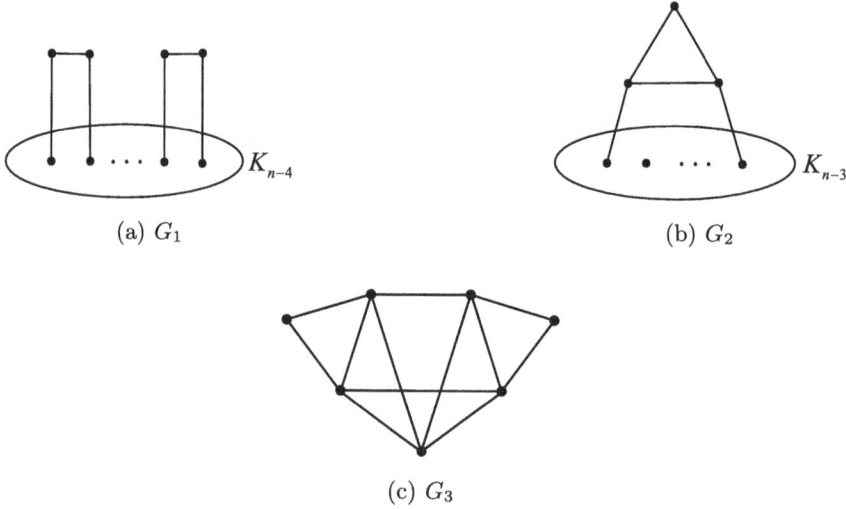

(a) G_1

(b) G_2

(c) G_3

Fig. 2. Examples of graphs satisfying Fan's condition but containing no two CISTs.

From Theorem 6, we obtain the following corollary.

Corollary 1. *Let G be a 2-connected graph on $n \geq 7$ vertices with $\delta(G) \geq 3$. Let $U = \{u \mid u \in G \text{ and } d(u) < n/2\}$. If G satisfies the following conditions:*

(i) $e(v, U) \leq \frac{|U| - c(G[U])}{2}$ for any vertex $v \notin U$;
(ii) For any two vertices x and y with $d(x, y) = 2$, we have $\max\{d(x), d(y)\} \geq \frac{n}{2}$,

then G has two CISTs.

To prove Theorem 6, we need the following two definitions and a new theorem.

Definition 1 (Araki, [1]). *Let G be a 2-connected graph and $\{V_1, V_2\}$ be a partition of $V(G)$. Then $\{V_1, V_2\}$ is a CIST-partition of G if we have:*

(i) $G[V_1]$ and $G[V_2]$ are connected; and
(ii) $G[V_1, V_2]$ has no tree components.

Definition 2. *Let $\{S, T\}$ be a CIST-partition of G, and let $u \in S$ and $v \in T$. Then $\{u, v\}$ is called a pair of CIST-vertices of $\{S, T\}$ in G.*

Theorem 7. *Let G be a 2-connected graph on $n \geq 4$ vertices. Let $t \leq \frac{n}{2}$ be a positive integer. Suppose that $W_1, W_2, \ldots, W_t \subseteq V(G)$ are disjoint such that $|W_i| = 2$ for $1 \leq i \leq t$. If $\delta(G) \geq \frac{n+t}{2}$ then G has a CIST-partition such that W_i is a pair of CIST-vertices for any $1 \leq i \leq t$.*

Our paper is organized as follows. In Sect. 1, we present the research background related to CISTs and outline the research objectives of this paper. In Sect. 2, we present some preliminaries and prove several new lemmas to be used in this paper. In Sect. 3, we present the sketch of proofs of Theorem 6 and Theorem 7. In Sect. 4, we construct a graph G with $\delta(G) \geq 4$ which satisfies Fan's condition but contains no two CISTs.

2 Preliminaries

In this section, we introduce some essential preliminaries and prove several new lemmas which are fundamental to our main results.

Lemma 1. *(Theorem 2.2 in [1]) A connected graph G has two completely independent spanning trees if and only if G has a CIST-partition.*

Lemmas 2, 3 and 4 are established by applying specific graph operations to derive essential properties of the graph under study.

Lemma 2. *Let G_i be a connected graph, and let $\{V_1^i, V_2^i\}$ be a CIST-partition of G_i for $i = 1, 2$, where G_1 and G_2 are vertex-disjoint. Let $u_i \in V_1^i$ and $v_i \in V_2^i$ for $i = 1, 2$. Let G be a graph with $V(G) = V(G_1) \cup V(G_2)$ and $E(G) = E(G_1) \cup E(G_2) \cup \{u_1 u_2, v_1 v_2\}$. Then $\{V_1^1 \cup V_1^2, V_2^1 \cup V_2^2\}$ is a CIST-partition of G.*

Proof. Since G_i is connected and $\{V_1^i, V_2^i\}$ is a CIST-partition of G_i, by Definition 1, we have $G_i[V_j^i]$ is connected for $i, j = 1, 2$. As $u_1 u_2, v_1 v_2 \in E(G)$, where $u_i \in V_1^i$ and $v_i \in V_2^i$ for $i = 1, 2$, we know $G[V_1^1 \cup V_1^2]$ and $G[V_2^1 \cup V_2^2]$ are connected. Since $\{V_1^i, V_2^i\}$ is a CIST-partition of G_i, $G_i[V_1^i, V_2^i]$ has no tree component for $i = 1, 2$. Then $G[V_1^1 \cup V_1^2, V_2^1 \cup V_2^2]$ has no tree component. By Lemma 1, $\{V_1^1 \cup V_1^2, V_2^1 \cup V_2^2\}$ is a CIST-partition of G. The proof is complete. $\square$

Lemma 3. *Let G_1 and G_2 be vertex-disjoint and connected, where $|V(G_1)| \geq 4$ and $|V(G_2)| \geq 4$. Let u_i and v_i be distinct vertices of G_i for $i = 1, 2$. Then $G_1 \cup G_2 \cup \{u_1u_2, v_1v_2\}$ has two CISTs if and only if G_i has a CIST-partition such that $\{u_i, v_i\}$ is a pair of CIST-vertices for $i = 1, 2$.*

Proof. Let $G := G_1 \cup G_2 \cup \{u_1u_2, v_1v_2\}$. We assume that G has two CISTs. By Lemma 1, G has a CIST-partition. Let $\{S, T\}$ be a CIST-partition of G. Obviously, we have $S \cap V(G_i) \neq \emptyset$ and $T \cap V(G_i) \neq \emptyset$ for $i = 1, 2$. Suppose not. Then there exists some $j \in \{1, 2\}$, say $j = 1$, such that $S \cap V(G_1) = \emptyset$ or $T \cap V(G_1) = \emptyset$. Without loss of generality, assume the first case holds, and we have $S \subseteq V(G_2)$. Since $|V(G_1)| \geq 4$, $G[S, T]$ has an isolated vertex, which contradicts the fact that $G[S, T]$ has no tree component. Let $S_i := S \cap V(G_i)$ and $T_i := T \cap V(G_i)$ for $i = 1, 2$. Since $\{S, T\}$ is a CIST-partition of G, $G[S]$ and $G[T]$ are connected.

We shall show that $\{u_1, u_2\} \subseteq S$ or $\{u_1, u_2\} \subseteq T$. Suppose not. Without loss of generality, assume that $u_1 \in S$ and $u_2 \in T$. Since $u_1 \in V(G_1)$ and $u_2 \in V(G_2)$, $u_1 \in S_1$ and $u_2 \in T_2$. There exist four cases to discuss whether $G[S]$ and $G[T]$ are connected:

Case 1: $v_1 \in S_1$ and $v_2 \in S_2$. Since $e(G_1, G_2) = 2$, $G[T]$ is disconnected.

Case 2: $v_1 \in T_1$ and $v_2 \in T_2$. Since $e(G_1, G_2) = 2$, $G[S]$ is disconnected.

Case 3: $v_1 \in S_1$ and $v_2 \in T_2$. Since $e(G_1, G_2) = 2$, $G[S]$ and $G[T]$ are disconnected.

Case 4: $v_1 \in T_1$ and $v_2 \in S_2$. Since $e(G_1, G_2) = 2$, $G[S]$ and $G[T]$ are disconnected.

Similarly, by symmetry, we can show if $u_1 \in T$ and $u_2 \in S$, $G[S]$ or $G[T]$ is disconnected, which contradicts the fact that $G[S]$ and $G[T]$ are connected. Thus, $\{u_1, u_2\} \subseteq S$ or $\{u_1, u_2\} \subseteq T$. By a similar argument as above, we have $\{v_1, v_2\} \subseteq S$ or $\{v_1, v_2\} \subseteq T$. Without loss of generality, assume that $\{u_1, u_2\} \subseteq S$. In the following, we assume $\{v_1, v_2\} \subseteq S$. Since $e(G_1, G_2) = 2$, $G[T]$ is not connected, which contradicts the fact that $G[T]$ is connected. So $\{v_1, v_2\} \subseteq T$.

Finally, we shall prove that G_i has a CIST-partition and $\{u_i, v_i\}$ is a pair of CIST-vertices. Since $G[S]$ is connected and $u_i \in S_i$ for $i = 1, 2$, we have $G[S_1]$ and $G[S_2]$ are connected. As $v_i \in T$ and $v_i \in V(G_i)$ for $i = 1, 2$, we have $v_i \in T_i$ for $i = 1, 2$. Similarly, since $G[T]$ is connected and $v_i \in T_i$ for $i = 1, 2$, $G[T_1]$ and $G[T_2]$ are connected. Furthermore, as G_1 and G_2 are vertex-disjoint and $G[S, T]$ has no tree component, $G_i[S_i, T_i]$ has no tree component for $i = 1, 2$. Thus, $\{S_i, T_i\}$ is a CIST-partition of G_i, where $\{u_i, v_i\}$ is a pair of CIST-vertices for $i = 1, 2$.

Now, we suppose that G_i has a CIST-partition such that $\{u_i, v_i\}$ is a pair of CIST-vertices for $i = 1, 2$. By Lemma 2, G has a CIST-partition. By Lemma 1, G has two CISTs. The proof is complete. □

Lemma 4. *Let G be a connected graph. Let $u, v, w \in V(G)$ and $N_G(u) = \{v, w\}$. Then G has a CIST-partition if and only if $G - u$ has a CIST-partition such that $\{v, w\}$ is a pair of CIST-vertices.*

Proof. Assume that G has a CIST-partition. Let $\{S, T\}$ be a CIST-partition of G. Then $G[S, T]$ has no tree component. Without loss of generality, we suppose that $u \in S$. If $\{v, w\} \subseteq S$, then u is an isolated vertex of $G[S, T]$. So $\{v, w\} \cap T \neq \emptyset$. Next, we shall show $\{v, w\} \cap S \neq \emptyset$. If not, then $\{v, w\} \subseteq T$, and as $G[S]$ is disconnected, a contradiction. Thus, $\{v, w\} \cap S \neq \emptyset$. Let $G' := G - u$. Since $G[S, T]$ has no tree component and u is a pendent vertex of $G[S, T]$, $G'[S \backslash \{u\}, T]$ has no tree component. Since $\{S, T\}$ is a CIST-partition of G, $G[S]$ and $G[T]$ are connected. Then $G[S] - u$ and $G[T]$ are connected. Hence $G - u$ has a CIST-partition such that $\{v, w\}$ is a pair of CIST-vertices.

Assume that $G - u$ has a CIST-partition $\{S', T'\}$ such that $\{v, w\}$ is a pair of CIST-vertices in $G - u$. By Definition 2, $(G - u)[S']$ and $(G - u)[T']$ are connected, and $(G - u)[S', T']$ has no tree component. Since $N_G(u) = \{v, w\}$, $G[S' \cup \{u\}]$ is connected and $(G - u)[S' \cup \{u\}, T]$ has no tree component. Then $\{S' \cup \{u\}, T'\}$ is a CIST-partition of G. The proof is complete. $\qquad \square$

Lemmas 5 and 6 are constructed by defining specific graph structures and analyzing their properties, which serve as essential tools for the proof of Theorem 7.

Lemma 5. *Let t and n be two positive integers such that $t \leq \frac{n}{2}$. Let G be a connected graph on n vertices with $\delta(G) \geq \frac{n+t}{2}$. Suppose that $V_1 \subseteq V(G)$ and $V_2 := V(G) \backslash V_1$ are non-empty vertex subsets such that $t \leq |V_1| < \frac{n+3t+4}{4}$. Then $c(G[V_2]) \leq 2$; furthermore, if equality holds, then every component of $G[V_2]$ is 2-connected and has at most $\lfloor \frac{n-t}{2} \rfloor - 1$ vertices.*

Proof. Suppose to the contrary that $G[V_2]$ has at least three components. Let u be a vertex of a component with the fewest vertices in $G[V_2]$. Then

$$d(u) \leq (\frac{|V_2|}{3} - 1) + |V_1| < \frac{n+t}{2},$$

which contradicts the condition $\delta(G) \geq \frac{n+t}{2}$. Thus $G[V_2]$ has at most two components.

Assume that $G[V_2]$ has two components, say H_1 and H_2. Since $\delta(G) \geq \frac{n+t}{2}$, $|V(H_i)| \geq \frac{n+t}{2} - |V_1| + 1$ for $i = 1, 2$. For $i = 1, 2$, we have

$$|V(H_i)| \leq |V_2| - (\frac{n+t}{2} - |V_1| + 1) = \frac{n-t}{2} - 1.$$

Next, we shall show H_1 is 2-connected. Suppose that u is a cut vertex of H_1. Let H' be a component with the fewest vertices in $H_1 - u$. Then each vertex of $V(H')$ has degree in G at most

$$|V(H')| - 1 + 1 + |V_1| \leq \frac{|V(H_1)| - 1}{2} - 1 + 1 + |V_1| < \frac{n+t}{2},$$

which contradicts the condition $\delta(G) \geq \frac{n+t}{2}$. Then $\kappa(H_1) \geq 2$. Similarly, we have $\kappa(H_2) \geq 2$. Then every component of $G[V_2]$ is 2-connected if $G[V_2]$ has two connected components. The proof is complete. □

Lemma 6. *Let t and n be two positive integers such that $1 \leq t < \frac{n+2}{5}$. Let G be a connected graph on n vertices with $\delta(G) \geq \frac{n+t}{2}$. Suppose that V_1^1 and $V_2^1 := V(G) \setminus V_1^1$ are non-empty such that $|V_1^1| = 2t - 1$ and $G[V_1^1]$ is connected. Let $U_i \subseteq V_1^1$ with $|U_i| = t$ for $i = 1, 2$. If $G[V_1^1, V_2^1]$ has at least one isolated vertex, then G has a CIST-partition $\{V_1', V_2'\}$ such that $U_i \subseteq V_i'$ for $i = 1, 2$.*

Proof. Let $w \in V_1^1$. For each $i \geq 1$, we use the following algorithm to construct some required vertex subsets.

```
1 begin
2     i ← 1;
3     while G[V₁ⁱ, V₂ⁱ] has an isolated vertex do
4         Select an isolated vertex xᵢ of G[V₁ⁱ, V₂ⁱ];
5         Select a vertex yᵢ ∈ (V₂ⁱ\U₂) ∩ N_G(xᵢ) ∩ N_G(w);
6         V₁^{i+1} ← V₁ⁱ ∪ {yᵢ};
7         V₂^{i+1} ← V₂ⁱ\{yᵢ};
8         i ← i + 1;
9     end
10 end
```

First, we shall show that the above algorithm is feasible, that is, there is a least positive integer s such that $G[V_1^s, V_2^s]$ has no isolated vertex and x_{s-1} is an isolated vertex in $G[V_1^{s-1}, V_2^{s-1}]$, where (V_1^{s-1}, V_2^{s-1}) and (V_1^s, V_2^s) are two partitions of G.

We first claim that all isolated vertices of $G[V_1^1, V_2^1]$ belong to V_2^1. Suppose to the contrary that V_1^1 has an isolated vertex of $G[V_1^1, V_2^1]$. Since $|V_1^1| = 2t - 1$, we choose a vertex $u \in V_1^1$ as such an isolated vertex. Then

$$d_G(u) \leq |V_1^1| - 1 = 2t - 2 < \frac{n+t}{2},$$

which contradicts the condition $\delta(G) \geq \frac{n+t}{2}$. Thus, all isolated vertex of $G[V_1^1, V_2^1]$ belong to V_2^1.

We also claim that for all $2 \leq j \leq s - 1$, there exists a vertex $y_j \in (V_2^j \setminus U_2) \cap N_G(x_j) \cap N_G(w)$. Since $\delta(G) \geq \frac{n+t}{2}$, $|N_G(x_j) \cap N_G(w) \cap V_2^j| \geq t + 2$ for $1 \leq i \leq s - 1$. Then V_2^j has a vertex $y_j \in V(G) \setminus U_2$ which is adjacent to x_j and w for $1 \leq j \leq s - 1$.

Then, we claim that for all $2 \leq j \leq s - 1$, all isolated vertices of $G[V_1^j, V_2^j]$ belong to V_2^j. Since $x_{j-1} \in V_2^j$, we have

$$|V_2^j| \geq d_G(x_{j-1}) - 1 + 1 \geq \frac{n+t}{2},$$

where "-1" counts $|\{y_{j-1}\}|$ and "$+1$" counts $|\{x_{j-1}\}|$. Thus $|V_1^j| \leq \frac{n-t}{2}$. Then all isolated vertices of $G[V_1^j, V_2^j]$ belong to V_2^j for $2 \leq j \leq s - 1$. Since $|V_2^i| =$

$|V_2^{i-1}| - 1$ and G is a finite graph, there is a least positive integer s such that $G[V_1^s, V_2^s]$ has no isolated vertex and x_{s-1} is an isolated vertex in $G[V_1^{s-1}, V_2^{s-1}]$. This proves the claim.

In the following, we shall show $G[V_1^s]$ and $G[V_2^s]$ are connected. Suppose to contrary that $G[V_2^s]$ is not connected. Let G_1 be a connected component of $G[V_2^s]$ such that $x_{s-1} \in V(G_1)$. Then we have $|V(G_1)| \geq d_G(x_{s-1}) - 1 + 1 \geq \frac{n+t}{2}$. Let $u \in V_2^s \setminus V(G_1)$. Then we have

$$d_G(u) \leq |V_2^s \setminus V(G_1)| + |V_1^s| = |V_2^s| - |V(G_1)| + |V_1^s| = \frac{n-t}{2},$$

which contradicts the fact $\delta(G) \geq \frac{n+t}{2}$. Thus $G[V_2^s]$ is connected. Since y_i is adjacent to w for $1 \leq i \leq s - 1$, $G[V_1^s]$ is connected. This proves what we want.

Finally, we shall prove that: (i) if $G[V_1^s, V_2^s]$ contains no tree component, then $\{V_1^s, V_2^s\}$ is a CIST-partition of G with $U_i \subseteq V_i^s$ for each $i \in \{1, 2\}$; (ii) If $G[V_1^s, V_2^s]$ contains one tree component, then $\{V_1', V_2'\}$ is a CIST-partition of G with $U_i \subseteq V_i'$ for each $i \in \{1, 2\}$. Note that the first case is true by Definition 1. In the following, we suppose that $G[V_1^s, V_2^s]$ has a tree component.

Since $x_{s-1} \in V_2^s$ and $e(x_{s-1}, V_1^s) = 1$, we have $|V_2^s| \geq d_G(x_{s-1}) - 1 + 1 \geq \frac{n+t}{2}$. So $|V_1^s| \leq \frac{n-t}{2}$. We shall also show that all leaves of $G[V_1^s, V_2^s]$ belong to V_2^s. Suppose to the contrary that V_1^s has a vertex u which is a leaf in $G[V_1^s, V_2^s]$. Then $d_G(u) \leq |V_1^s| - 1 + e(u, V_2^s) \leq \frac{n-t}{2}$, which contradicts the fact $\delta(G) \geq \frac{n+t}{2}$. So all leaves of $G[V_1^s, V_2^s]$ belong to V_2^s.

Let $v \in V_1^s$. Since $|V_1^s| \leq \frac{n-t}{2}$,

$$e(v, V_2^s) = d_G(v) - e(v, V_1^s) \geq \frac{n+t}{2} - (|V_1^s| - 1) \geq t + 1.$$

Hence, each tree component has at least $t + 1$ leaves of V_2^s in $G[V_1^s, V_2^s]$. Let $u \in V_2^s \setminus U_2$ be a leaf of $G[V_1^s, V_2^s]$. Let $V_1' := V_1^s \cup \{u\}$ and $V_2' := V_2^s \setminus \{u\}$. Then

$$|N_G(u) \cap V_2'| \geq d_G(u) - 1 \geq \frac{n+t-2}{2}.$$

Let u' be a vertex of $V_1' \setminus \{u\}$. Then

$$|(N_G(u') \cap V_2') \cap (N_G(u) \cap V_2')| \geq (\frac{n+t}{2} - |V_1'| + 1) + \frac{n+t-2}{2} - |V_2'| = t.$$

Since $G[V_1^s, V_2^s]$ has no isolated vertex, each vertex of V_2^s is adjacent to some vertices of V_1^s. Since $V_1' = V_1^s \cup \{u\}$ and $V_2' = V_2^s \setminus \{u\}$, V_2' has no isolated vertex in $G[V_1', V_2']$. Since u and each vertex of $V_1' \setminus \{u\}$ has at least t common neighbors in V_2', $G[V_1', V_2']$ is connected. Let v be a vertex of $V_1' \setminus N_G(u)$. Then

$$|N_{G[V_1', V_2']}(u) \cap N_{G[V_1', V_2']}(v)| \geq t + 2 \geq 3.$$

It follows that there is a $K_{2,3}$ in $G[V_1', V_2']$, and also a cycle. Thus, $G[V_1', V_2']$ is not a tree component.

At last, we shall prove that $G[V_2']$ is connected. Suppose to the contrary that $G[V_2']$ is disconnected. Let $x \in V_2^s$ be a leaf of $G[V_1^s, V_2^s]$. Then $e(x, V_2') = d_G(x) - e(x, V_1') \geq \frac{n+t}{2} - 2$. Let G' be a component of $G[V_2']$ such that $x \in V(G')$, and let $y \in V_2' \setminus V(G')$. Then

$$|V(G')| \geq d_G(x) - e(x, V_1') + 1 \geq \frac{n+t}{2} - 1.$$

So $d_G(y) \leq |V(G)| - 1 - |V(G')| \leq \frac{n-t}{2}$, which contradicts the condition $\delta(G) \geq \frac{n+t}{2}$. Thus, $G[V_2']$ is connected. Obviously, $G[V_1']$ is connected.

By Definition 1, $\{V_1', V_2'\}$ is a CIST-partition such that $U_i \subseteq V_i'$ for $i = 1, 2$. The proof is complete.   □

3 Proof of Theorem 6 and Theorem 7

Proofs of Theorem 7 (Sketch). The proof is divided into four cases based on the values of t:

Case 1: $t = 1$.

We analyze the bipartition $G[\{w_1^1\}, V(G) \setminus \{w_1^1\}]$. If this bipartition has no isolated vertex, we directly construct a CIST-partition by setting $V_1 = \{w_1^1, u\}$, where $u \in N_G(w_1^1) \setminus \{w_2^1\}$, and $V_2 = V(G) \setminus V_1$. The connectivity of $G[V_1]$ and $G[V_2]$ is guaranteed by the minimum degree condition and Lemma 5. If the bipartition has an isolated vertex, we apply Lemma 6 to ensure the existence of a CIST-partition.

Case 2: $1 < t < \frac{n+2}{5}$.

We construct a connected subgraph $G[E_{q+1}]$ that contains exactly one vertex from each W_i. The construction relies on the minimum degree condition to ensure that sufficient vertices are available for selection. Then we use some new methods to construct two vertex subsets V_1 and $V_2 := V(G) \setminus V_1$ such that $G[V_1]$ is connected and $|V_1| = 2t - 1$. If $G[V_2]$ is disconnected, we refine the partition by carefully adding selected vertices from $N(w_1^1)$ to restore connectivity. The key point is to ensure that $G[V_1, V_2]$ has no tree component, which is achieved by using the high connectivity of G.

Case 3: $\frac{n+2}{5} \leq t \leq \frac{n}{3}$.

We extend the construction from Case 2. By carefully selecting vertices to construct V_1 and V_2, we ensure that $G[V_2]$ remains connected and that $G[V_1, V_2]$ has no tree component. If $G[V_2]$ is disconnected, we adjust the partition using vertices from V_1 to maintain the required properties. The proof relies on the fact that the minimum degree condition guarantees the existence of sufficient common neighbors between vertices in V_1 and V_2.

Case 4: $\frac{n}{3} < t \leq \frac{n}{2}$.

We use the high connectivity of G to directly construct a CIST-partition. By ensuring that each W_i contributes exactly one vertex to V_1, we guarantee that $G[V_1]$ and $G[V_2]$ are connected, and $G[V_1, V_2]$ has no tree component. The proof relies on the fact that the minimum degree condition ensures that each vertex in V_1 and V_2 has sufficiently many neighbors, which prevents the formation of tree components in the bipartition.

In all cases, the proof relies on the minimum degree condition $\delta(G) \geq \frac{n+t}{2}$ and the application of key lemmas (Lemmas 1, 5, and 6) to ensure the existence of a CIST-partition where each W_i is a pair of CIST-vertices for $1 \leq i \leq t$.   □

Proofs of Theorem 6. (Sketch) We first show each component of $G[U]$ is a clique. Let H be a component of $G[U]$. If H consists of only one vertex, it is also a clique. Hence H has at least two vertices. Suppose that H is not a clique. Choose a shortest path in H, and there is a pair of vertices of distance 2 in G. By (iii), there is a heavy vertex, which is in $V \setminus U$, a contradiction. This proves the claim.

Let $G_1, G_2, \ldots, G_t$ be all connected components of $G[U]$, where $|V(G_i)| \geq 4$ for $1 \leq i \leq q$ and $|V(G_i)| \leq 3$ for $q+1 \leq i \leq t$. Let $G' = G - U$ with n_1 vertices. Since $e_G(v, U) \leq \frac{|U| - c(G[U])}{2}$ for any $v \notin U$, we have

$$d_{G'}(v) \geq \frac{n_1 + |U|}{2} - \frac{|U| - c(G[U])}{2} = \frac{n_1 + c(G[U])}{2} = \frac{n_1 + t}{2}.$$

Since G is 2-connected, there are at least two disjoint paths between G' and G_i for $1 \leq i \leq t$. Let w_1^i, w_2^i be two vertices of $V(G) \setminus U$ which are adjacent to two vertices v_1^i and v_2^i of G_i, respectively, where $1 \leq i \leq q$. For each $q + 1 \leq j \leq t$, let w_1^j and w_2^j be two vertices of $V(G')$ which are adjacent to one vertex of G_i if G_i has a vertex adjacent to two vertices of G'; otherwise, let w_1^i and w_2^i be two vertices of $V(G')$ adjacent to two vertices of G', respectively.

By Theorem 7, G' has a CIST-partition $\{V_1', V_2'\}$ such that $\{w_1^i, w_2^i\}$ is a pair of CIST-vertices for $1 \leq i \leq t$. By Theorem 7, G_i has a CIST-partition $\{V_1^i, V_2^i\}$ such that $\{v_1^i, v_2^i\}$ is a pair of CIST-vertices of G_i, where $|V(G_i)| \geq 4$ and $1 \leq i \leq q$.

Let $S = \bigcup_{q+1 \leq i \leq t} V(G_i)$. Let $V_1 := V_1' \cup \bigcup_{1 \leq i \leq q} V_1^i$ and $V_2 := V_2' \cup \bigcup_{1 \leq i \leq q} V_2^i$. If $U = \emptyset$, G has a CIST-partition by Lemma 2. By Lemma 1, G has two CISTs.

If $S \neq \phi$, then $G[V(G) \setminus S]$ has a CIST-partition $\{V_1, V_2\}$. The cases for $2 \leq |V(G_i)| \leq 3$ and $t = q + 1$ are illustrated in Fig. 3.

In (a), let $Z_1 := V_1 \cup \{v_1\}$ and $Z_2 := V_2 \cup \{v_2\}$; In (b), let $Z_1 := V_1 \cup \{u_1\}$ and $Z_2 := V_2 \cup \{u_2, u_3\}$; In (c), let $Z_1 := V_1 \cup \{w_1\}$ and $Z_2 := V_2 \cup \{w_2, w_3\}$. Clearly, $G[Z_1, Z_2]$ has no tree component and $G[Z_i]$ is connected for $i = 1, 2$. By Definition 1, $\{Z_1, Z_2\}$ is a CIST-partition of G.

If $|V(G_{q+1})| = 1$ and $t = q + 1$, then G has a CIST-partition by Lemma 4.

Similarly, G has two CISTs if $t \geq q + 2$ by Lemma 3 and Lemma 4. The proof is complete.   □

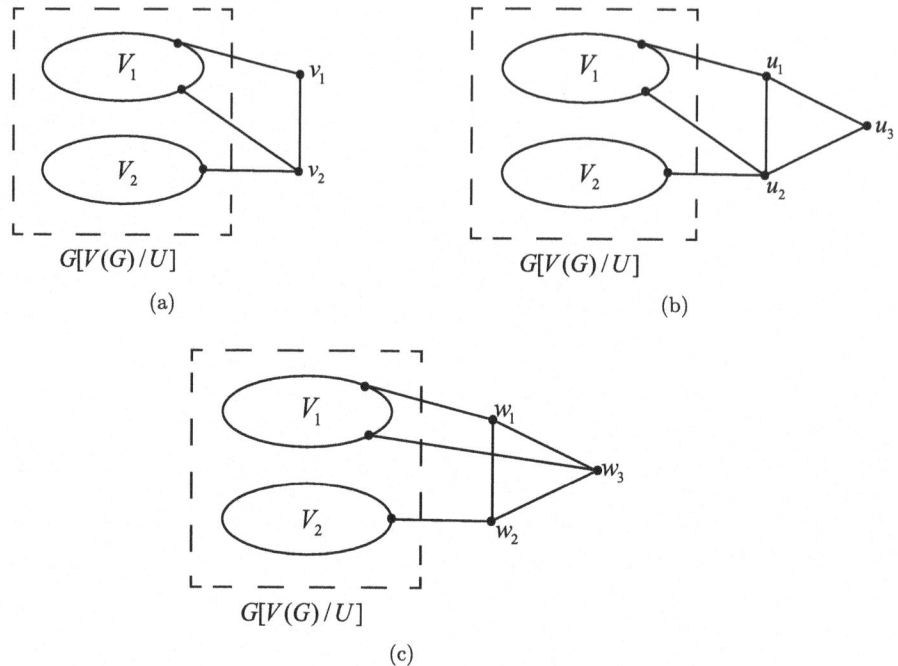

$$G[V(G)/U]$$

(a)

$$G[V(G)/U]$$

(b)

$$G[V(G)/U]$$

(c)

Fig. 3. xxx

4   One Concluding Remark

A natural question arises from Fan's condition and the research presented: Does there exist a graph with minimum degree at least 4 that satisfies Fan's condition but contains no two CISTs? The following example provides an affirmative answer. (See Fig. 4.)

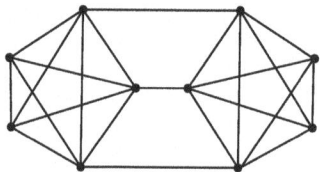

Fig. 4. A graph with minimum degree 4 that satisfies Fan's condition but contains no two CISTs.

References

1. Araki, T.: Dirac's condition for completely independent spanning trees. J. Graph Theory **77**(3), 171–179 (2014)
2. Chang, H.Y., Wang, H.L., Yang, J.S., Chang, J.M.: A note on the degree condition of completely independent spanning trees. IEICE Trans. Fundam. Electron. Commun. Comput. Sci. **98**(10), 2191–2193 (2015)
3. Chen, X., Li, J., Lu, F.: Two completely independent spanning trees of P_4-free graphs. Graphs Comb. **39**(2), 30 (2023)
4. Chen, X., Li, M., Xiong, L.: Two completely independent spanning trees of claw-free graphs. Disc. Math. **345**(12), 113080 (2022)
5. Chen, X., Liu, Q., Yang, X.: Two completely independent spanning trees of split graphs. Disc. Appl. Math. **340**, 76–78 (2023)
6. Chen, X., Zhang, J., Xiong, L., Su, G.: Every 2-connected $\{claw, Z_2\}$-free graph with minimum degree at least 4 contains two cists. Disc. Appl. Math. **360**, 51–64 (2025)
7. Chen, Y.H., Tang, S.M., Pai, K.J., Chang, J.M.: Constructing dual-cists with short diameters using a generic adjustment scheme on bicubes. Theor. Comput. Sci. **878**, 102–112 (2021)
8. Cheng, B., Wang, D., Fan, J.: Constructing completely independent spanning trees in crossed cubes. Disc. Appl. Math. **219**, 100–109 (2017)
9. Dirac, G.A.: Some theorems on abstract graphs. Proc. Lond. Math. Soc. **3**(1), 69–81 (1952)
10. Fan, G.: New sufficient conditions for cycles in graphs. J. Comb. Theory Ser. B **37**(3), 221–227 (1984)
11. Fan, G., Hong, Y., Liu, Q.: Ore's condition for completely independent spanning trees. Disc. Appl. Math. **177**, 95–100 (2014)
12. Hasunuma, T.: Completely independent spanning trees in the underlying graph of a line digraph. Disc. Math. **234**(1–3), 149–157 (2001)
13. Hasunuma, T.: Completely independent spanning trees in maximal planar graphs. In: Goos, G., Hartmanis, J., van Leeuwen, J., Kučera, L. (eds.) WG 2002. LNCS, vol. 2573, pp. 235–245. Springer, Heidelberg (2002). https://doi.org/10.1007/3-540-36379-3_21
14. Hasunuma, T.: Minimum degree conditions and optimal graphs for completely independent spanning trees. In: Lipták, Z., Smyth, W.F. (eds.) IWOCA 2015. LNCS, vol. 9538, pp. 260–273. Springer, Cham (2016). https://doi.org/10.1007/978-3-319-29516-9_22
15. Hasunuma, T., Morisaka, C.: Completely independent spanning trees in torus networks. Networks **60**(1), 59–69 (2012)
16. Hong, X., Liu, Q.: Degree condition for completely independent spanning trees. Inf. Process. Lett. **116**(10), 644–648 (2016)
17. Hong, X., Zhang, H.: A hamilton sufficient condition for completely independent spanning tree. Disc. Appl. Math. **279**, 183–187 (2020)
18. Li, J., Su, G., Song, G.: New comments on "a hamilton sufficient condition for completely independent spanning tree". Disc. Appl. Math. **305**, 10–15 (2021)
19. Ma, J., Cai, J.: Fan's condition for completely independent spanning trees. arXiv preprint arxiv:2502.11522 (2025)
20. Moinet, A., Darties, B., Gastineau, N., Baril, J.L., Togni, O.: Completely independent spanning trees for enhancing the robustness in ad-hoc networks. In: 2017 IEEE 13th International Conference on Wireless and Mobile Computing, Networking and Communications (WiMob), pp. 63–70. IEEE (2017)

21. Ore, O.: A note on hamiltonian circuits. Am. Math. Monthly **67**, 55 (1960)
22. Pai, K.J., Chang, J.M.: Constructing two completely independent spanning trees in hypercube-variant networks. Theor. Comput. Sci. **652**, 28–37 (2016)
23. Pai, K.J., Chang, R.S., Wu, R.Y., Chang, J.M.: A two-stages tree-searching algorithm for finding three completely independent spanning trees. Theor. Comput. Sci. **784**, 65–74 (2019)
24. Pai, K.J., Chang, R.S., Wu, R.Y., Chang, J.M.: Three completely independent spanning trees of crossed cubes with application to secure-protection routing. Inf. Sci. **541**, 516–530 (2020)
25. Qin, X.W., Chang, J.M., Hao, R.X.: Constructing dual-cists of dcell data center networks. Appl. Math. Comput. **362**, 124546 (2019)
26. Qin, X.W., Hao, R.X., Pai, K.J., Chang, J.M.: Comments on "a hamilton sufficient condition for completely independent spanning tree". Disc. Appl. Math. **283**, 730–733 (2020)
27. Qin, X.W., Hao, R.X., Wu, J.: Construction of dual-cists on an infinite class of networks. IEEE Trans. Parallel Distrib. Syst. **33**(8), 1902–1910 (2021)
28. Tapolcai, J.: Sufficient conditions for protection routing in ip networks. Optim. Lett. **7**, 723–730 (2013)
29. Yuan, J., Zhang, R., Liu, A.: Degree conditions for completely independent spanning trees of bipartite graphs. Graphs Comb. **38**(6), 179 (2022)

Undecidability of Polynomial Inequalities in Subset Densities and Additive Energies

Yaqiao Li[✉]📛

Shenzhen University of Advanced Technology, Shenzhen, Guangdong, China
liyaqiao@suat-sz.edu.cn

Abstract. Many results in extremal graph theory can be formulated as certain polynomial inequalities in graph homomorphism densities. Answering fundamental questions raised by Lovász, Szegedy and Razborov, Hatami and Norine proved that determining the validity of an arbitrary such polynomial inequality in graph homomorphism densities is undecidable. We observe that many results in additive combinatorics can also be formulated as polynomial inequalities in subset's density and its variants. Based on techniques introduced in Hatami and Norine, together with algebraic and graph construction and Fourier analysis, we prove similarly two theorems of undecidability, thus showing that establishing such polynomial inequalities in additive combinatorics are inherently difficult in their full generality.

Keywords: Undecidable · Subset density · Homomorphism density · Additive energy

1 Introduction

A central theme in additive combinatorics [20] is the interplay between subset density and additive structures, the most studied being arithmetic progressions. A long line of work in this direction culminated in Szemerédi's theorem, which states that any subset of natural numbers with sufficiently high density contains a long arithmetic progression. A variety of remarkable proofs of Szemerédi's theorem have since been discovered, and have been called by Tao [19] a "Rosetta stone" connecting various fields ranging from ergodic theory to Fourier analysis.

Let $A \subseteq G$ be a subset of a finite abelian group G, and $\alpha(A) = |A|/|G|$ denotes its density. One way to measure additive structure is to consider the size of sumset $A + B = \{a + b : a \in A, b \in B\}$. The ratio $\sigma(A) = \frac{|A+A|}{|A|}$ is called the *doubling constant* of A. For example, $\sigma(A) = O(1)$ when A is an arithmetic progression, while $\sigma(A)$ is large when A is random. Another key notion to quantify the additive structure of A is its *additive energy* $\mathrm{E}(A)$ defined to be the number of additive quadruples in A,

$$\mathrm{E}(A) = \left| \{(a_1, a_2, a_3, a_4) \in A^4 : a_1 + a_2 = a_3 + a_4\} \right|. \tag{1}$$

© The Author(s), under exclusive license to Springer Nature Singapore Pte Ltd. 2026
F. V. Fomin and M. Xiao (Eds.): COCOON 2025, LNCS 15984, pp. 29–40, 2026.
https://doi.org/10.1007/978-981-95-0218-9_3

It is easy to see that $|A|^2 \le \mathrm{E}(A) \le |A|^3$. The higher additive energy, the "more" additive structures there are. Additive energy has been studied in various settings [3,5,7,17]. Below are some general results that concern with subset densities and additive energies.

- The Kneser's theorem [16], generalizing the Cauchy-Davenport theorem, states that $|A + B| \ge |A| + |B| - |H|$, where $A, B \subseteq G$ and H is the stabilizer of $A + B$. This can be rephrased as

$$\alpha(A + B) - \alpha(A) - \alpha(B) + \alpha(H) \ge 0, \quad \forall\, A, B \subseteq G, \quad \forall\, G. \qquad (2)$$

- The Plünnecke-Ruzsa inequality states that $|rB - sB| \le c^{r+s}|A|$ where $c = |A + B|/|A|$, and $rB - sB$ denotes the set $B + \cdots + B - B - \cdots - B$ with r B's in the sum and s B's in the difference. This can also be equivalently reformulated as

$$\alpha(A + B)^{r+s} - \alpha(A)^{r+s-1} \cdot \alpha(rB - sB) \ge 0, \quad \forall\, A, B \subseteq G, \quad \forall\, G. \qquad (3)$$

- The doubling constant and additive energy are closely related. Let

$$r_A(x) = |\{(a_1, a_2) \in A^2 : a_1 + a_2 = x\}|, \qquad (4)$$

which counts how many pairs from A sum to x. By double counting we have $|A|^2 = \sum_{x \in A+A} r_A(x)$. By Cauchy-Schwarz,

$$|A|^4 = \left(\sum_{x \in A+A} r_A(x) \right)^2 \le \left(\sum_{x \in A+A} r_A(x)^2 \right) \cdot \left(\sum_{x \in A+A} 1^2 \right) = \mathrm{E}(A) \cdot |A + A|.$$

Rewriting gives $\mathrm{E}(A)/|A|^3 \ge 1/\sigma(A)$, often described as small doubling implies high additive energy, see a related inequality in [13]. Dividing both sides with $|G|^4$, we get

$$\frac{\mathrm{E}(A)}{|G|^3} \cdot \alpha(A + A) - \alpha(A)^4 \ge 0, \quad \forall\, A \subseteq G, \quad \forall\, G. \qquad (5)$$

The formulation (2), (3) and (5) demonstrate that many results in additive combinatorics can be stated as the non-negativity of certain polynomials in subset's densities and additive energies of related subsets, and suitable (sometimes repeated) application of Cauchy-Schwarz inequality is often useful for deriving these results, see examples in [10]. Other deep results in additive combinatorics such as Freiman's theorem, Balog-Szemerédi-Gowers theorem, and the polynomial Freiman-Ruzsa conjecture are also in similar spirit, see a recent breakthrough in [9,14].

Szemerédi originally proved his theorem for arithmetic progressions by developing a far-reaching theorem called Szemerédi regularity lemma, which is a major step toward the modern study of dense graph limits developed by Lovász et al. and the flag algebra by Razborov, in which the graph homomorphism density plays a key role. The homomorphism density of a (small) graph H in a graph G, denoted by $t(H, G)$, is the probability that a random mapping (not necessarily injective) from the vertices of H to the vertices of G maps every edge of H to an edge of G. Many important results in extremal graph theory can be described using polynomial inequalities in homomorphism densities. For example, Goodman's theorem [8], which generalizes the classical Mantel-Turán theorem, says that $t(K_3, G) - 2t(K_2, G)^2 + t(K_2, G) \geq 0$ holds for every graph G, where K_2 and K_3 denote the edge graph and the triangle graph, respectively. As another example, Sidorenko conjectures that $t(H, G) - t(K_2, G)^{|E(H)|} \geq 0$ holds for every bipartite graph H and every graph G, where $E(H)$ denotes the edge set of graph H.

The above instances show that understanding what polynomial inequalities hold is of central importance in either additive combinatorics or extremal graph theory. From the perspective of computational complexity, it is therefore a natural and fundamental question to understand the difficulty of establishing such polynomial inequalities involving the homomorphism density in extremal graph theory, or involving the subset densities and its variants in additive combinatorics. The first question in different forms concerning extremal graph theory has been formally asked by Lovász, Szegedy and Razborov, and was answered in negative in a breakthrough by Hatami and Norine [11]. Specifically, they showed that the problem of determining the validity of a polynomial inequality between homomorphism densities is undecidable. Recently, their results have been extended in [1,2,4]. In this paper, we consider the second question concerning additive combinatorics. Adapting the techniques in [11], together with algebraic and graph construction and Fourier analysis, we prove similarly two theorems of undecidability, thus showing that establishing such inequalities in additive combinatorics are inherently difficult in their full generality.

We need some notation to state the theorems. Let

$$L = \{f_1(g_1, \ldots, g_k), \ldots, f_d(g_1, \ldots, g_k)\}$$

be a formal system of linear forms of variables $g_1, \ldots, g_k$. Recall $A \subseteq G$ where G denotes a finite abelian group. We use $L \in A$ to denote that $f_i(g_1, \ldots, g_k) \in A$ for every i. By an abuse of notation, we define

$$t(L, A) := \Pr[L \in A] = \Pr[f_1(g_1, \ldots, g_k) \in A, \cdots, f_d(g_1, \ldots, g_k) \in A], \quad (6)$$

where $(g_1, \ldots, g_k) \in G^k$ is sampled uniformly at random. Note that we have used the notation $t(\cdot, \cdot)$ to denote both the graph homomorphism density and the probability of $L \in A$. The specific meaning of $t(\cdot, \cdot)$ should be clear from the context. Below are some examples of $t(L, A)$.

(i) $L = \{f(g) = g\}$. Then, $t(L, A) = \Pr[g \in A] = |A|/|G| = \alpha(A)$ is the density of A.

(ii) $L = \{f_1(g, h) = g, f_2(g, h) = h, f_3(g, h) = g + h\}$. Then, $t(L, A) = \Pr[g \in A, h \in A, g + h \in A] = \left(\sum_{x \in A} r_A(x)\right)/|G|^2$, where $r_A(x)$ is as defined in (4).

(iii) $L = \{f_1(g, h, k) = g, f_2(g, h, k) = h, f_3(g, h, k) = k, f_4(g, h, k) = g + h - k\}$.
Then, $t(L, A) = \Pr[g \in A, h \in A, k \in A, g + h - k \in A] = \mathrm{E}(A)/|G|^3$, i.e., it
is a normalized version of the additive energy of A appearing in (5).

These examples show that, depending on the system of linear forms L, the
probability $t(L, A)$ can express both subset density and additive energy, and
other more complicated variants.

Let L, L' be two systems of linear forms. We interpret the formal product
$L \cdot L'$ as $t(L \cdot L', A) = t(L, A) \cdot t(L', A)$. In this way, we may consider general
quantum system of linear forms where we allow formal product of systems of
linear forms.

Theorem 1. *The following problem is undecidable.*

- *INSTANCE: Two positive integers m, k, quantum systems of linear forms
  $L_1, \ldots, L_m$ on k variables $g_1, \ldots, g_k$, and integers $a_1, \ldots, a_m$.*
- *QUESTION: Does the inequality $a_1 t(L_1, A) + \cdots + a_m t(L_m, A) \geq 0$ hold for
  every subset $A \subseteq G$ and every finite abelian group G?*

Note that the polynomial $\sum_i a_i t(L_i, A)$ is linear in the variables $t(L_i, A)$,
which includes not only $\alpha(A)$ and $\mathrm{E}(A)$, but also other quantities as we discussed
earlier. The second theorem concerns more specifically to only subsets densities
and additive energies.

Theorem 2. *The following problem is undecidable.*

- *INSTANCE: A positive integer k $\geq$ 2, and a polynomial
  $q(x_1, \ldots, x_k, y_1, \ldots, y_k) \in \mathbb{Z}[x_1, \ldots, x_k, y_1, \ldots, y_k]$.*
- *QUESTION: Does the inequality $q(\alpha(A_1), \ldots, \alpha(A_k), \mathrm{E}(A_1), \ldots, \mathrm{E}(A_k)) \geq
  0$ hold for all $A_i \subseteq G_i$ and all finite abelian groups G_i, for $i = 1, \ldots, k$?*

These theorems are ultimately based on the classical Matiyasevich's theorem of undecidability [15], that is, given a multivariate polynomial with integer
coefficients, the problem of determining whether it is non-negative for every
assignment of natural numbers to its variables is undecidable. The undecidability already occurs for a finite number of variables in Theorem 1 and Theorem
2, in fact $k = 9$ suffices by Jones [12], see also [6,18]. Technically, we need to
construct certain maps that build correspondence between densities and natural
numbers, with respect to related polynomials in question.

2   Proof of The First Theorem

2.1   Two Lemmas

Let $I_t = [1 - \frac{1}{t}, 1 - \frac{1}{t+1})$ be sub-intervals of $[0, 1]$ for $t = 1, 2, 3, \ldots$. Define

$$h(x) := \frac{3t^2 - t - 2}{t(t + 1)} x - \frac{2(t - 1)}{t + 1}, \qquad x \in I_t.$$

Let $h(1) = 1$. Note that $h(x)$ is piecewise linear on $[0,1]$, and $h(1 - \frac{1}{t}) = \frac{(t-1)(t-2)}{t^2}$. The function $h(x)$ comes from Bollobás lower bound of the triangle homomorphism density in terms of the edge density, specifically, the following lower bound between homomorphism densities holds for every graph G,

$$t(K_3, G) \geq h(t(K_2, G)). \tag{7}$$

see detail in [11, Section 5]. Let

$$R := \{(x, y) \in [0, 1]^2 : y \geq h(x)\}. \tag{8}$$

Then, Bollobás lower bound implies that $(t(K_2, G), t(K_3, G)) \in R$.

The first lemma is a consequence of Lemma 5.1 and Lemma 5.4 in [11].

Lemma 1. *The following problem is undecidable.*

- *INSTANCE: A positive integer $k \geq 6$, and a polynomial $q(x_1, \ldots, x_k, y_1, \ldots, y_k) \in \mathbb{Z}[x_1, \ldots, x_k, y_1, \ldots, y_k]$.*
- *QUESTION: Does the inequality $q(x_1, \ldots, x_k, y_1, \ldots, y_k) \geq 0$ hold for all $(x_1, \ldots, x_k, y_1, \ldots, y_k)$ with $(x_i, y_i) \in R$ for every $1 \leq i \leq k$?*

Let $\mathfrak{g} = (g_1, \ldots, g_k)$ denote a k-tuple of formal variables. Define a system of linear forms

$$L(\mathfrak{g}) := \{\neg(k+1)g_1\} \cup \left(\bigcup_{p=1}^{k+2} \bigcup_{j=2}^{k} \{p(g_j - jg_1)\} \right). \tag{9}$$

Here, the notation $\neg(k+1)g_1$ means that when we consider $L \in A$ we are asking for $(k+1)g_1 \notin A$. Define two related systems

$$M(\mathfrak{g}) := L(\mathfrak{g}) \cup \{g_1, \ldots, g_k\}, \tag{10}$$
$$L(\mathfrak{g}, j, z) := L(g_1, \ldots, g_{j-1}, z, g_{j+1}, \ldots, g_k), \quad j = 1, \ldots, k. \tag{11}$$

Lemma 2. *Let $S = \{0, 1, \ldots, k\} \subseteq G = \mathbb{Z}_{(k+1)^2}$, let $\mathfrak{g} = (g_1, \ldots, g_k) \in G^k$, then*

- *if $L(\mathfrak{g}) \in S$, then $g_j = jg_1$ and $g_j \neq 0$ for every $j = 1, \ldots, k$;*
- *if $M(\mathfrak{g}) \in S$, then $g_j = j$ for every $j = 1, \ldots, k$.*

Proof. The condition $(k+1)g_1 \notin S$ implies $g_1 \neq 0$. For each $\delta_j = g_j - jg_1$, since we require $\delta_j, 2\delta_j, \ldots, (k+2)\delta_j$ all lie in S, but S contains only $k+1$ distinct elements, by the pigeonhole principle two of them must be equal and hence we have $q_j\delta_j = 0$ for some $1 \leq q_j \leq k+1$. Since $\delta_j \in S = \{0, 1, \ldots, k\}$, we then must have $\delta_j = 0$, i.e., $g_j = jg_1$. The second claim for $M(\mathfrak{g}) \in S$ follows from the extra constraints $g_j = jg_1 \in S$ for every j.

2.2   Proof of Theorem 1

Let G denote a finite abelian group. Given two subsets $B, C \subseteq G$, define an associated *directed* graph U where $V(U) = B$, and for $b_1, b_2 \in B$, there is a directed edge $(b_1, b_2) \in E(U)$ if and only if $b_1 - b_2 \in C$.

Let $A \subseteq G$, let $\mathfrak{g} = (g_1, \ldots, g_k) \in G^k$. Define

$$\mathcal{V}_j(\mathfrak{g}, z) := M(\mathfrak{g}) \cup L(\mathfrak{g}, j, z), \quad j = 1, \ldots, k.$$

Define

$$B_j(\mathfrak{g}) = \{z \in G : \mathcal{V}_j(\mathfrak{g}, z) \in A\} \subseteq G, \tag{12}$$

$$C_j(\mathfrak{g}) = (B_j(\mathfrak{g}) \cap A) - g_j \subseteq G. \tag{13}$$

Then construct the associated directed graph $U_j(\mathfrak{g})$ from $B_j(\mathfrak{g})$ and $C_j(\mathfrak{g})$ accordingly. For $j = 1, \ldots, k$, define

$$\mathcal{E}_j(\mathfrak{g}, z, z') := M(\mathfrak{g}) \cup L(\mathfrak{g}, j, z) \cup L(\mathfrak{g}, j, z') \cup L(\mathfrak{g}, j, g_j + z - z') \cup \{g_j + z - z'\}.$$

Intuitively, the system of linear forms $\mathcal{E}_j$ is to count the edges in the directed graph $U_j(\mathfrak{g})$. Define $\mathcal{T}_j$ similar to $\mathcal{E}_j$ but to count the triangles, i.e.,

$$\begin{aligned}
\mathcal{T}_j(\mathfrak{g}, z, z', z'') := {} & M(\mathfrak{g}) \cup L(\mathfrak{g}, j, z) \cup L(\mathfrak{g}, j, z') \cup L(\mathfrak{g}, j, z'') \\
& \cup L(\mathfrak{g}, j, g_j + z - z') \cup \{g_j + z - z'\} \\
& \cup L(\mathfrak{g}, j, g_j + z' - z'') \cup \{g_j + z' - z''\} \\
& \cup L(\mathfrak{g}, j, g_j + z'' - z) \cup \{g_j + z'' - z\}.
\end{aligned}$$

For the polynomial q as in Lemma 1, define a related polynomial q^* with integral coefficients as follows

$$q^*(v_1, \ldots, v_k, e_1, \ldots, e_k, t_1, \ldots, t_k) := q\left(\frac{e_1}{v_1^2}, \ldots, \frac{e_k}{v_k^2}, \frac{t_1}{v_1^3}, \ldots, \frac{t_k}{v_k^3}\right) \prod_{j=1}^{k} v_j^{3 \deg(q)}.$$

Finally, define the quantum system of linear forms

$$\psi(q^*) := q^*(\mathcal{V}_1, \ldots, \mathcal{V}_k, \mathcal{E}_1, \ldots, \mathcal{E}_k, \mathcal{T}_1, \ldots, \mathcal{T}_k).$$

Proof (Proof of Theorem 1). By Lemma 1, it suffices to show that $q(x, y) \geq 0$ for $(x, y) \in R^k$ is equivalent to $t(\psi(q^*), A) \geq 0$ for every $A \subseteq G$ and for every G.

We first show $q(x, y) \geq 0$ implies $t(\psi(q^*), A) \geq 0$. Note that for those $\mathfrak{g} \in G^k$ such that $M(\mathfrak{g}) \notin A$, we have $t(\mathcal{V}_j(\mathfrak{g}, z), A) = t(\mathcal{E}_j(\mathfrak{g}, z, z'), A) = t(\mathcal{T}_j(\mathfrak{g}, z, z', z''), A) = 0$. On the other hand, for those $\mathfrak{g}$ such that $M(\mathfrak{g}) \in A$, we have

$$t(K_2, U_j(\mathfrak{g})) = \frac{t(\mathcal{E}_j(\mathfrak{g}, z, z'), A)}{t(\mathcal{V}_j(\mathfrak{g}, z), A)^2}, \quad t(K_3, U_j(\mathfrak{g})) = \frac{t(\mathcal{T}_j(\mathfrak{g}, z, z', z''), A)}{t(\mathcal{V}_j(\mathfrak{g}, z), A)^3}. \tag{14}$$

To verify the equality, as an example, we have

$$
\begin{aligned}
t(\mathcal{E}_j(\mathfrak{g}, z, z'), A) &= \Pr[g_j + z - z' \in B_j(\mathfrak{g}) \cap A, z \in B_j(\mathfrak{g}), z' \in B_j(\mathfrak{g})] \\
&= \Pr[z - z' \in C_j(\mathfrak{g}) \mid z, z' \in B_j(\mathfrak{g})] \cdot \Pr[z \in B_j(\mathfrak{g})]^2 \\
&= t(K_2, U_j(\mathfrak{g})) \cdot t(\mathcal{V}_j(\mathfrak{g}, z), A)^2.
\end{aligned}
$$

One can verify the other equality similarly. This implies,

$$
t(\psi(q^*), A) = \begin{cases}
q(t(K_2, U_1(\mathfrak{g})), \ldots, t(K_2, U_k(\mathfrak{g})), \\
\quad t(K_3, U_1(\mathfrak{g})), \ldots, t(K_3, U_k(\mathfrak{g}))) \\
\quad \cdot \prod_{j=1}^{k} t(\mathcal{V}_j(\mathfrak{g}, z), A)^{3 \deg(q)}, & M(\mathfrak{g}) \in A; \\
0, & M(\mathfrak{g}) \notin A.
\end{cases} \tag{15}
$$

Therefore, since by Bollobás lower bound we have $(t(K_2, G), t(K_3, G)) \in R$ holds for every graph G, the claim follows.

Next, we show $q(x, y) < 0$ for some $(x, y) \in R^k$ implies $t(\psi(q^*), A) < 0$ for some $A \subseteq G$ for some finite abelian group G. By the analysis in [11, Lemma 5.4], we know that there in fact exists some $(x_j^* = 1 - \frac{1}{n_j}, y_j^* = 2x_j^2 - x_j) \in R$ in which $n_j \in \mathbb{N}$ for every $j \in [k]$, such that $q(x^*, y^*) < 0$.

Let H be a finite abelian group with subgroups $H_1, \ldots, H_k$ satisfying $\frac{|H_j|}{|H|} = \frac{1}{n_j}$ for every $j \in [k]$. Note that such group and subgroups exist by, e.g., considering direct product of cyclic groups. In addition, let $H_0 = \emptyset$. Let $G = \mathbb{Z}_{(k+1)^2} \times H$, and $S = \{0, 1, \ldots, k\} \subseteq \mathbb{Z}_{(k+1)^2}$. Consider a subset $A \subseteq G$ defined as follows

$$
A := \bigcup_{j \in S} (j \times (H - H_j)).
$$

By Lemma 2, if $\mathfrak{g} = (g_1, \ldots, g_k) \in G$ satisfies $M(\mathfrak{g}) \in A$, then, $g_j = (j, h) \in j \times (H - H_j)$. By the definition (12), we have for $j = 1, \ldots, k$,

$$
B_j(\mathfrak{g}) = j \times H, \quad C_j(\mathfrak{g}) = (j \times (H - H_j)) - g_j = 0 \times ((H - H_j) - h). \tag{16}
$$

As an example, we show $B_1(\mathfrak{g}) = 1 \times H$. We already know for every j, the first component in $g_j \in G$ equals to j. Since for $B_1(\mathfrak{g})$, its element $z = (z_1, z_2) \in B_1(\mathfrak{g}) \subseteq G$ plays the role of g_1 in the constraints given by the linear forms L, we will have $((k+1)z_1, (k+1)z_2) \notin A$. Since $H - H_0 = H$, $(k+1)z_2 \in H - H_0$ always holds, hence, we must have $(k+1)z_1 \neq 0$, which implies $z_1 \neq 0$. Furthermore, the other constraints says that $p(g_j - jz) \in A$ for every $j \geq 2$ and for $p = 1, \ldots, k+2$. If we focus on z_1, we would have $p(j - jz_1) \in S$ for every $j \geq 2$ and for $p = 1, \ldots, k+2$. This implies $z_1 = 1$. But now we have $(k+1)z_1 = k+1 \notin S$, hence, the value z_2 could be arbitrary. In other words, $B_1(\mathfrak{g}) = 1 \times H$. The rest can be verified similarly.

The condition (16) implies that if $z, z' \in B_j(\mathfrak{g})$, then we have $z = (j, a)$ and $z' = (j, b)$ for some $a, b \in H$, and $z - z' = (0, a - b)$. Hence,

$$
\frac{t(\mathcal{E}_j(\mathfrak{g}, z, z'), A)}{t(\mathcal{V}_j(\mathfrak{g}, z), A)^2} = t(K_2, U_j(\mathfrak{g})) = \Pr[z - z' \in C_j(\mathfrak{g}) \mid z, z' \in B_j(\mathfrak{g})]
$$

$$
= \Pr[a - b \in H - H_j - h \mid a, b \in H]
$$

$$
= \frac{|H - H_j|}{|H|} = 1 - \frac{1}{n_j} = x_j^*.
$$

Similarly, we have $\frac{t(\mathcal{T}_j(\mathfrak{g}, z, z', z''), A)}{t(\mathcal{V}_j(\mathfrak{g}, z), A)^3} = t(K_3, U_j(\mathfrak{g})) = 2(1 - \frac{1}{n_j})^2 - (1 - \frac{1}{n_j}) = y_j^*$. Hence, by (15)

$$
t(\psi(q^*), A) = q(x^*, y^*) \cdot \prod_{j=1}^{k} t(\mathcal{V}_j(\mathfrak{g}, z), A)^{3 \deg(q)} < 0.
$$

3   Proof of The Second Theorem

The proof of Theorem 1 relies on the Bollobás lower bound (7) of the triangle homomorphism density in terms of the edge density. Now, to study polynomial inequalities involving subset's density and additive energy, it would be useful to establish a similar bound.

3.1   An Upper Bound of Additive Energy

We review some notation of Fourier analysis on a finite abelian group G. For every $\xi \in G$, let $\chi_\xi : G \to \mathbb{C}$ be the characters on G defined as $\chi_\xi(x) = e^{2\pi i \xi x}$. Given two functions $f, g \in L^2(G)$, their convolution $f * g$ is defined as usual $(f * g)(x) := \mathbb{E}_{y \in G} f(x - y) g(y)$ where the expectation is taken over y uniformly distributed in G. The Fourier transform $\hat{f} : G \to \mathbb{C}$ of f is defined as $\hat{f}(\xi) := \mathbb{E}_{x \in G} f(x) \overline{\chi_\xi(x)}$. The L_2 norm of f is defined as $\|f\| = (\mathbb{E}_{x \in G} |f(x)|^2)^{1/2}$.

For a subset $A \subseteq G$, let $A(x) := 1_A(x)$ be the indicator function for subset A. Recall the notation $r_A(x)$ as defined in (4). By definition, we have $\frac{r_A(x)}{|G|} = (A * A)(x)$. For the additive energy, we have $\mathrm{E}(A) = \sum_{x \in G} r_A(x)^2$. Hence,

$$
\mathrm{E}(A) = \sum_{x \in G} r_A(x)^2 = |G|^2 \sum_{x \in G} (A * A)(x)^2 = |G|^3 \mathbb{E}_{x \in G} (A * A)(x)^2
$$

$$
= |G|^3 \|(A * A)(x)\|^2.
$$

Therefore, if we normalize $\mathrm{E}(A)$ we get $\frac{\mathrm{E}(A)}{|G|^3} = \|(A * A)(x)\|^2$. From now on, when we use $\mathrm{E}(A)$ we mean this normalized version. We now establish the following lemma which bounds the additive energy of a subset $A \subseteq G$ by its density.

Lemma 3. *Let G be a finite abelian group, let $A \subseteq G$ be a non-empty subset with density $\alpha = \frac{|A|}{|G|}$, then the (normalized) additive energy $\mathrm{E}(A)$ satisfies*

$$\mathrm{E}(A) \leq \alpha^3 - \alpha^4 \left(\left\{ \frac{1}{\alpha} \right\} - \left\{ \frac{1}{\alpha} \right\}^2 \right), \tag{17}$$

where $\left\{ \frac{1}{\alpha} \right\} = \frac{1}{\alpha} - \lfloor \frac{1}{\alpha} \rfloor$ denotes the fractional part of $\frac{1}{\alpha}$. In particular, $\mathrm{E}(A) \leq \alpha^3$ when $\frac{1}{\alpha} \in \mathbb{N}$, i.e., when $|A|$ divides $|G|$.

Proof. Using $A(x) = 1_A(x)$, we have $\|A(x)\|^2 = \alpha$, hence by Parseval's identity,

$$\sum_{\xi \in G} |\hat{A}(\xi)|^2 = \|A(x)\|^2 = \alpha, \tag{18}$$

in which $|\hat{A}(\xi)| = \frac{1}{|G|} \left| \sum_{x \in G} A(x) \overline{\chi_\xi(x)} \right| \leq \frac{|A|}{|G|} = \alpha$, hence,

$$|\hat{A}(\xi)|^2 \leq \alpha^2, \quad \forall \, \xi \in G. \tag{19}$$

Since $\mathrm{E}(A) = \|(A * A)(x)\|^2$, applying Parseval's identity another time we have

$$\mathrm{E}(A) = \|(A * A)(x)\|^2 = \sum_{\xi \in G} |\widehat{A * A}(\xi)|^2 = \sum_{\xi \in G} |\hat{A}(\xi) \cdot \hat{A}(\xi)|^2 = \sum_{\xi \in G} |\hat{A}(\xi)|^4. \tag{20}$$

To get an upper bound of $\mathrm{E}(A)$ is equivalent to maximize $\mathrm{E}(A)$ in the form of (20) under the constraints (18) and (19). The Karush–Kuhn–Tucker conditions imply that $\mathrm{E}(A)$ will be maximized when there are as many $|\hat{A}(\xi)| = \alpha$ as possible, which gives

$$\mathrm{E}(A) = \sum_{\xi \in G} |\hat{A}(\xi)|^4 \leq \left\lfloor \frac{1}{\alpha} \right\rfloor \cdot \alpha^4 + \left(\alpha - \left\lfloor \frac{1}{\alpha} \right\rfloor \cdot \alpha^2 \right)^2$$

$$= \left\lfloor \frac{1}{\alpha} \right\rfloor \cdot \alpha^4 + \left(\frac{1}{\alpha} - \left\lfloor \frac{1}{\alpha} \right\rfloor \right)^2 \cdot \alpha^4$$

$$= \left\lfloor \frac{1}{\alpha} \right\rfloor \cdot \alpha^4 + \left\{ \frac{1}{\alpha} \right\}^2 \cdot \alpha^4.$$

This simplifies to (17) after a direct calculation.

Consider $A = \{0\} \subseteq \mathbb{Z}_n$, we find that $\alpha(A) = \frac{1}{n}$ and $\mathrm{E}(A) = \frac{1}{n^3}$, hence the upper bound is tight when $\frac{1}{\alpha} \in \mathbb{N}$.

3.2   Proof of Theorem 2

We first prove a simple lemma. Let $S = \{\frac{1}{n} : n \in \mathbb{N}\} \subseteq (0, 1]$.

Lemma 4. *The following problem is undecidable.*

- *INSTANCE: A positive integer $k \geq 2$, and a polynomial $p(x) = p(x_1, \ldots, x_k) \in \mathbb{Z}[x_1, \ldots, x_k]$.*
- *QUESTION: Does $p(x_1, \ldots, x_k) \geq 0$ hold for every $(x_1, \ldots, x_k) \in S^k$?*

Proof. Let $q(x_1, \ldots, x_k) \in \mathbb{Z}[x_1, \ldots, x_k]$, and let

$$p(x_1, \ldots, x_k) = \left(\prod_{i=1}^{k} x_i^{\deg q} \right) q(\frac{1}{x_1}, \ldots, \frac{1}{x_k}).$$

Then $p(x_1, \ldots, x_k) \in \mathbb{Z}[x_1, \ldots, x_k]$, and clearly $q(x_1, \ldots, x_k) \geq 0$ for $(x_1, \ldots, x_k) \in \mathbb{N}^k$ is equivalent to $p(\frac{1}{x_1}, \ldots, \frac{1}{x_k}) \geq 0$ for $(\frac{1}{x_1}, \ldots, \frac{1}{x_k}) \in S^k$. But the former is undecidable by Matiyasevich's theorem, hence the latter is undecidable too.

We are now ready to prove Theorem 2.

Proof (Proof of Theorem 2). Let $p(x_1, \ldots, x_k) \in \mathbb{Z}[x_1, \ldots, x_k]$. Choose

$$M = \max \left\{ 1, 30 \max \left\{ \left\| \left(\frac{\partial p}{\partial x_1}, \ldots, \frac{\partial p}{\partial x_k} \right) \right\|_{\infty} : x \in [0,1]^k \right\}, \right.$$
$$\left. 3 \max \left\{ \left\| \left(\frac{\partial^2 p}{\partial x_1^2}, \ldots, \frac{\partial^2 p}{\partial x_k^2} \right) \right\|_{\infty} : x \in [0,1]^k \right\} \right\} > 0.$$

Let $g(\alpha) = \alpha^3$. Let $h(\alpha) = \alpha^3 - \alpha^4(\{\frac{1}{\alpha}\} - \{\frac{1}{\alpha}\}^2)$ for $\alpha \in (0,1]$, where $\{\frac{1}{\alpha}\} = \frac{1}{\alpha} - \lfloor \frac{1}{\alpha} \rfloor$ denotes the fractional part of $\frac{1}{\alpha}$. Let $h(0) = 0$. Let $\delta(\alpha) = g(\alpha) - h(\alpha) \geq 0$. Note that $\delta(\alpha) = 0$ or equivalently $h(\alpha) = g(\alpha)$ if $\alpha \in S$. Define a region

$$R := \{(\alpha, \beta) \in [0,1]^2 : \beta \leq h(\alpha)\}. \tag{21}$$

Define

$$q(x_1, \ldots, x_k, y_1, \ldots, y_k) := p(x_1, \ldots, x_k) + M \sum_{i=1}^{k} (g(x_i) - y_i).$$

We also use the abbreviation $(x, y) = (x_1, \ldots, x_k, y_1, \ldots, y_k)$.

We first show that the following are equivalent: (i) $p(x) \geq 0$ for all $x \in S^k$, (ii) $q(x, y) \geq 0$ for all $(x, y) \in R^k$.

To show (ii) implies (i), note that if $x \in S^k$, i.e., if $x_i \in S$ we have $(x_i, g(x_i)) \in R$, hence letting $y_i = g(x_i)$, we have $p(x) = q(x, y) \geq 0$ as desired.

Next we show (i) implies (ii). Consider an auxiliary function

$$\tilde{q}(x_1, \ldots, x_k) := p(x_1, \ldots, x_k) + M \sum_{i=1}^{k} \delta(x_i).$$

Since $M > 0$ and when $(x_i, y_i) \in R$ we have $g(x_i) - y_i \geq \delta(x_i)$, hence $q(x, y) \geq \tilde{q}(x)$ holds for $(x, y) \in R^k$. Hence, it suffices to show $\tilde{q}(x) \geq 0$ for $(x, y) \in R^k$. Let $I_t = [\frac{1}{1+t}, \frac{1}{t}]$, $t \in \mathbb{N}$. For $\alpha \in I_t$, we have $\{\frac{1}{\alpha}\} = \frac{1}{\alpha} - t$. Hence,

$$\delta(\alpha) = -t(t+1)\alpha^4 + (2t+1)\alpha^3 - \alpha^2, \quad \alpha \in I_t.$$

We have $\delta'(\alpha) = -4t(t+1)\alpha^3 + 3(2t+1)\alpha^2 - 2\alpha$ and $\delta''(\alpha) = -12t(t+1)\alpha^2 + 6(2t+1)\alpha - 2$ for $\alpha \in I_t$. With some calculation using calculus, we have

$$\delta'(\alpha) \geq \frac{1}{20} > 0, \quad \alpha \in \left[\frac{1}{3}, \frac{2}{5}\right] \cup \left[\frac{1}{2}, \frac{7}{10}\right],$$

and

$$\delta''(\alpha) \leq -\frac{1}{2} < 0, \quad \alpha \in \cup_{t \geq 3} I_t \cup \left[\frac{2}{5}, 1/2\right] \cup \left[\frac{7}{10}, 1\right].$$

By our choice of M, we have

$$\frac{\partial \tilde{q}}{\partial x_i} = \frac{\partial p}{\partial x_i} + M\delta'(x_i) > 0, \quad x_i \in \left[\frac{1}{3}, \frac{2}{5}\right] \cup \left[\frac{1}{2}, \frac{7}{10}\right],$$

and

$$\frac{\partial^2 \tilde{q}}{\partial x_i^2} = \frac{\partial^2 p}{\partial x_i^2} + M\delta''(x_i) < 0, \quad x_i \in \cup_{t \geq 3} I_t \cup \left[\frac{2}{5}, 1/2\right] \cup \left[\frac{7}{10}, 1\right].$$

Hence, we see that $\tilde{q}(x)$ achieves its locally minimal value when $x \in S^k$, in which case $\tilde{q}(x) = p(x) \geq 0$ as needed.

By Lemma 3, $(\alpha(A_i), \mathrm{E}(A_i)) \in R$. Hence, if there exists $A_i \subseteq G_i$ for $i = 1, \ldots, k$ such that $q(\alpha(A_1), \ldots, \alpha(A_k), \mathrm{E}(A_1), \ldots, \mathrm{E}(A_k)) < 0$ this would imply $q(x, y) < 0$ for some $(x, y) \in R^k$. On the other hand, note that the proof of the equivalence also shows that the local minima of $q(x, y)$ are achieved at $x \in S^k$. Hence, if $q(x, y) < 0$ for some $(x, y) \in R^k$ there will be $x \in S^k$ (and $y_i = x_i^3$) to certify this. The remark after the proof of Lemma 3 shows that there are $A_i \subseteq G_i$ such that $(\alpha(A_i), \mathrm{E}(A_i))$ achieves these values, hence we will have $q(\alpha(A_1), \ldots, \alpha(A_k), \mathrm{E}(A_1), \ldots, \mathrm{E}(A_k)) < 0$. Therefore, the undecidability result of Theorem 2 follows from Lemma 4.

Acknowledgement. The author thanks Hamed Hatami for generously sharing his insights, and referees for useful comments.

References

1. Blekherman, G., Raymond, A., Singh, M., Thomas, R.R.: Simple graph density inequalities with no sum of squares proofs. Combinatorica **40**(4), 455–471 (2020)
2. Blekherman, G., Raymond, A., Wei, F.: Undecidability of polynomial inequalities in weighted graph homomorphism densities. In: Forum of Mathematics, Sigma, vol. 12, p. e40. Cambridge University Press (2024)

3. Bloom, T.F., Chow, S., Gafni, A., Walker, A.: Additive energy and the metric Poissonian property. Mathematika **64**(3), 679–700 (2018)
4. Chen, H., Lin, Y., Ma, J., Wei, F.: Undecidability of polynomial inequalities in tournaments. arXiv preprint arXiv:2412.04972 (2024)
5. de Dios Pont, J., Greenfeld, R., Ivanisvili, P., Madrid, J.: Additive energies on discrete cubes. Discrete analysis (2023)
6. Gasarch, W.: Hilbert's tenth problem: refinements and variants. ACM SIGACT News **52**(2), 36–44 (2021)
7. Goh, M.K.: On an entropic analogue of additive energy. arXiv preprint arXiv:2406.18798 (2024)
8. Goodman, A.W.: On sets of acquaintances and strangers at any party. Am. Math. Mon. **66**(9), 778–783 (1959)
9. Gowers, W., Green, B., Manners, F., Tao, T.: On a conjecture of Marton. Ann. Math. (2025)
10. Gowers, W., Milićević, L.: A quantitative inverse theorem for the u^4 norm over finite fields. arXiv preprint arXiv:1712.00241 (2017)
11. Hatami, H., Norine, S.: Undecidability of linear inequalities in graph homomorphism densities. J. Am. Math. Soc. **24**(2), 547–565 (2011)
12. Jones, J.P.: Universal diophantine equation. J. Symbolic Logic **47**(3), 549–571 (1982)
13. Katz, N.H., Koester, P.: On additive doubling and energy. SIAM J. Discret. Math. **24**(4), 1684–1693 (2010)
14. Liao, J.J.: Improved exponent for Marton's conjecture in f_2^n. arXiv preprint arXiv:2404.09639 (2024)
15. Matiyasevič, Y.: Enumerable sets are diophantine. Mathematical logic in the 20th century, pp. 269–273 (2003)
16. Nathanson, M.B.: Additive Number Theory: Inverse Problems and the Geometry of Sumsets, vol. 165. Springer, New York (1996)
17. Shao, X.: Additive energies of subsets of discrete cubes. In: Proceedings of the Royal Society of Edinburgh Section A: Mathematics, pp. 1–22 (2024)
18. Sun, Z.W.: Further results on Hilbert's tenth problem. Sci China Math **64**, 281–306 (2021)
19. Tao, T.: The dichotomy between structure and randomness, arithmetic progressions, and the primes. In: International Congress of Mathematicians (Madrid, 2006), vol. 1, pp. 581–608 (2007)
20. Tao, T., Vu, V.H.: Additive Combinatorics, vol. 105. Cambridge University Press, Cambridge (2006)

Approximation Algorithm
for Prize-Collecting Hypergraph Vertex
Cover with Fairness Constraints

Xiaofei Liu[1][(✉)] and Weidong Li[2]

[1] School of Information Science and Engineering, Yunnan University,
Kunming, China
lxfjl2016@163.com
[2] School of Mathematics and Statistics, Yunnan University, Kunming, China

Abstract. We are given a hypergraph $H = (V, E)$ and m groups $E_1, E_2,$ $\ldots, E_m$ of E, where each vertex in V has a cost, each (hyper-)edge in E has a profit and a penalty, and each group has a profit requirement. The prize-collecting hypergraph vertex cover with fairness constraints is to select a vertex set to minimize the total cost of the selected vertices and the total penalty of the uncovered edges subject to the fairness constraints, *i.e.*, for each group E_i, the total profit of covered edges in E_i exceeds the profit requirement. In this paper, based on the LP-rounding technique and the guessing method, we design an f-approximation algorithm in $n^{O(m/f)}$ time, where $f = \max_{e: e \in E} |e|$. Thus, when m is a constant, our algorithm is a polynomial-time f-approximation algorithm, which coincides with the lower bound of the hypergraph vertex cover problem if the unique game conjecture holds.

Keywords: Hypergraph vertex cover · fairness constraints · prize-collecting · approximation algorithm

1 Introduction

The vertex cover problem is one of the earliest NP-hard problems studied in combinatorial optimization [1]. In the vertex cover problem, we are given a graph $G = (V, E)$ with vertex set V and edge set E. For each vertex, there is an associated cost. The vertex cover problem is to select a vertex set to cover all the edges such that the total cost of the selected vertices is minimized, where an edge is said to be covered if at least one of its incident vertices is in this set. Karp [1] proved that the vertex cover problem is NP-hard. Then, Khot and Regev [2] improved that this problem cannot be approximated within $2 - \epsilon$ for any $\epsilon > 0$ under the unique game conjecture (UGC). Several polynomial-time 2-approximation algorithms are known for this problem via the LP-rounding technique and the primal-dual framework [3,4].

Under many conditions, covering all edges may not be a good strategy due to resource limitations or expensive costs, and it is natural to consider the partial version and the prize-collecting version of the vertex cover problem. In the

F. V. Fomin and M. Xiao (Eds.): COCOON 2025, LNCS 15984, pp. 41–52, 2026.
https://doi.org/10.1007/978-981-95-0218-9_4

partial vertex cover problem, the objective is to find a minimum cost vertex set that covers at least k edges, where $k \leq n$ is a given constant. Several polynomial-time 2-approximations are known in this problem via the LP-rounding technique and the local-ratio technique [5,6]. In the prize-collecting vertex cover problem, edges in E may not be covered while the uncovered edges should be paid penalties, and the objective is to minimize the total cost of selected vertices plus the total penalty of uncovered edges. Several polynomial-time 2-approximations are known in this problem via the LP-rounding technique and the primal-dual framework [7,8]. Combining the partial and the prize-collecting versions, we obtain the k-prize-collecting vertex cover problem. Based on the guessing technique and the primal-dual framework, Liu et al. [9] presented a 2-approximation algorithm with a running time of $O(|E| \cdot |V|^2)$.

The hypergraph vertex cover problem is a natural generalization of the vertex cover problem, where a (hyper-)edge in a hypergraph is a vertex set. Feige [10] proved that it cannot be approximated within a factor $(1 - o(1)) \ln |V|$ unless $NP \subseteq DTIME(|V|^{O(\log \log |V|)})$. Dinur and Steurer [11] proved the same lower bound under the assumption that $P \neq NP$. Bansal and Khot [12] proved that an approximation guarantee asymptotically better than f cannot be achieved in polynomial-time under UGC, where $f = \max_{e:e \in E} |e|$. For this problem and its generalizations, such as the partial, prize-collecting and k-prize-collecting versions, the f-approximation exists via the LP-rounding method [3], the local ratio method [6] and the primal-dual framework [13,14]. Könemann et al. [13] considered the generalized partial versions, in which each edge has a profit, and the objective is to find a minimum cost vertex set in which the total profit of the covered edges is at least K, where K is a given profit requirement. They presented a $(\frac{4}{3} + \epsilon) \cdot f$-approximation algorithm via the Lagrangian relaxation technique. Liu and Li [15] considered the generalized partial multicut problem with penalty, and presented a $(\frac{8}{3} + \epsilon)$-approximation algorithm for the problem in trees. Guruswami et al. [16] considered the k-uniform k-partite hypergraph vertex cover problem, and proved that an approximation guarantee asymptotically better $\frac{f}{2}$ cannot be achieved in polynomial-time under UGC. Lovász [17] presented a $\frac{f}{2}$-approximation algorithm for the k-uniform k-partite hypergraph vertex cover problem based on the standard LP relaxation.

Fairness is a significant societal concern that prevents resource monopolization and waste, and enhances the acceptability and legitimacy of decisions [18]. Bera et al. [19] considered the vertex cover with fairness constraints, which is a generalization of the partial vertex cover problem, where a partition $E_1, E_2, \ldots,$ E_m of the edges along with covering requirements $k_1, k_2, \ldots, k_m$ is given, and the objective is to find a minimum cost vertex set that covers at least k_i edges of E_i for each $i \in [m]$. Bera et al. [19] showed that it is NP-hard to obtain an approximation guarantee asymptotically better than $O(\log m)$, and presented an $O(\log m)$-approximation algorithm by using the Monte-Carlo randomization method. Bandyapadhyay et al. [20] considered the unweighted version of the vertex cover with fairness constraints, i.e., $c_v = 1$ for any $v \in V$, and presented a $(2 + \epsilon)$-approximation algorithm with a running time of $O(|V|^{\frac{m}{\epsilon}})$ by using the LP-rounding technique, note that, this algorithm is polynomial-time if m is a constant. Inamdar and Varadarajan [21] considered the hypergraph vertex cover with fairness constraints, and presented an $O(f + \log m)$-approximation algorithm by using the Monte-Carlo randomization method. Then, Hung and Kao [22] presented an $(f \cdot H(m) + H(m))$-approximation algorithm by using the primal-dual technique for the some problem, where $H(m)$ is the m-th harmonic number. Recently, Wang et al. [23] considered the prize-collecting hypergraph vertex cover with fairness constraints, and presented an $O(f + \log m)$-approximation algorithm by using the Monte-Carlo randomization method, and an $(m \cdot f)$-approximation algorithm by using the primal-dual technique.

Motivated by the above work, in this paper, we consider the prize-collecting hypergraph vertex cover with fairness constraints, which is to select a vertex set such that the total profit of covered edges by this set for each $i \in [m]$ is at least P_i and the total cost of selected vertices plus the total penalty of uncovered edges is minimized, where P_i is the profit requirement for each group E_i, and $[m] = \{1, 2, \ldots, m\}$. Based on LP-rounding technique and the guessing method, we design an f-approximation algorithm in $n^{O(m/f)}$ time. Note that, when m is a constant, our algorithm is a polynomial-time f-approximation algorithm, which coincides with the lower bound of the hypergraph vertex cover problem if the unique game conjecture holds.

The remainder of this paper is structured as follows. In Sect. 2, we first provide the formal definition of the prize-collecting hypergraph vertex cover with fairness constraints, and then present a preprocessing step to obtain the expected approximation factor. In Sect. 3, we present the f-approximation algorithm. In Sect. 4, we provide a brief conclusion.

2 Preliminaries

2.1 Formal Definition and Integer Programming

We are given a hypergraph $H = (V, E)$, where V is a vertex set, and E is a (hyper-)edge set. Each vertex $v \in V$ has a cost c_v, and each edge e, which is a subset of V, has a profit p_e and a penalty π_e, where we define $f = \max_{e:e \in E} |e|$. Given m groups $E_1, E_2, \cdots, E_m$, each group E_i has a profit requirement P_i, where each group E_i is a subset of E. For any $S \subseteq V$, we define $E(S) = \{e \in E \mid e \cap S \neq \emptyset\}$. The prize-collecting hypergraph vertex cover with fairness constraints (PCHVCF) is to select a subset of vertices $S \subseteq V$ that minimizes $\sum_{v:v \in S} c_v + \sum_{e \in E \setminus E(S)} \pi_e$ subject to fairness constraints, that is, $\sum_{e \in E(S) \cap E_i} p_e \geq P_i$ for each $i \in [m]$.

Similar to the integer programming formulation in [13], x_v is a binary variable for each vertex $v \in V$, where

$$x_v = \begin{cases} 1, & \text{if } v \text{ is selected to cover some edge,} \\ 0, & \text{otherwise.} \end{cases}$$

z_e is a binary variable for each edge $e \in E$, where

$$z_e = \begin{cases} 1, & \text{if } e \text{ is the uncovered edge,} \\ 0, & \text{otherwise.} \end{cases}$$

The integer programming of the PCHVCF can be formulated as follows.

$$\min \sum_{v:v \in V} c_v x_v + \sum_{e:e \in E} \pi_e z_e$$

$$s.t. \quad \sum_{v:v \in e} x_v + z_e \geq 1, \; \forall e \in E, \tag{1}$$

$$\sum_{e:e \in E_i} p_e z_e \leq p(E_i) - P_i, \; \forall i \in [m],$$

$$x_v, z_e \in \{0,1\}, \; \forall v \in V, \; \forall e \in E,$$

where we define $p(E') = \sum_{e:e \in E'} p_e$ for any $E' \subseteq E$, the first set of constraints of (1) guarantees that each edge $e \in E$ is either covered by at least one of the selected vertices or an uncovered edge, and the second set of constraints of (1) guarantees that the total profit of the uncovered elements of any feasible solution is at most $p(E_i) - P_i$ for each $i \in [m]$, i.e., the total profit of the covered elements of any feasible solution is at least P_i for each $i \in [m]$, which satisfies all fairness constraints.

2.2 Preprocessing Step

To obtain the expected approximation factor, we need a preprocessing step: guessing the l largest cost vertices $v^*_{\max,1}, \ldots, v^*_{\max,l}$ in S^* for instance $\mathcal{H} = (H, c, p, \pi, E_i, P_i)$ of the PCHVCF, where

$$\begin{cases} l = \left\lceil \dfrac{m}{f-1} \right\rceil; \\ v^*_{\max,l'} = \arg \max\limits_{v \in S^* \setminus \{v^*_{\max,1}, \ldots, v^*_{\max,l'-1}\}} c_v, \; \forall l' \in [l]; \end{cases}$$

and S^* is an optimal solution of the PCHVCF. Let

$$S_{\max} = \{v^*_{\max,1}, \ldots, v^*_{\max,l}\}.$$

We have that any vertex $v \in V \setminus S_{\max}$ with $c_v > c_{v^*_{\max,l}}$ is not in S^*, i.e., we only need to select vertices from the vertex set $\{v \in V \mid c_v \leq c_{v^*_{\max,l}}\}$ (excluding $S_{\max}$) for the PCHVCF. Therefore, we construct an auxiliary instance $\mathcal{H}_{\setminus S_{\max}} = (H', c', p, \pi, E'_i, P'_i)$ by removing vertex set $S_{\max}$, where $H' = (V \setminus S_{\max}, E(V \setminus S_{\max}))$, in which $E(V \setminus S_{\max}) = E \setminus \{e \in E \mid e \cap S_{\max} \neq \emptyset\}$. We define

$$\begin{cases} c'_v = c_v, & \forall\, v \in V \setminus S_{\max} \ \text{with}\ c_v \le c_{v^*_{\max,l}}; \\ c'_v = +\infty, & \forall\, v \in V \setminus S_{\max} \ \text{with}\ c_v > c_{v^*_{\max,l}}; \\ E'_i = E_i \cap E(V \setminus S_{\max}), & \forall\, i \in [m]; \\ P'_i = \max\Big\{0, P_i - \displaystyle\sum_{e:e \in E_i \cap E(V \setminus S_{\max})} p_e\Big\}, & \forall\, i \in [m], \end{cases}$$

where the second set of equality guarantees that vertices in $\{v \in V \mid c_v > c_{v^*_{\max,l}}\}$ are not selected. Similar to the proof of Lemma 2.1 in [9], the following lemma is not hard to obtain.

Lemma 1. *For any instance $\mathcal{H}$ of the PCHVCF, let $S_{\max}$ be the set of the l largest cost vertices in an optimal solution, and the auxiliary instance $\mathcal{H}_{\setminus S_{\max}}$ can be constructed in polynomial-time. We have*

$$OPT_{\setminus S_{\max}} + \sum_{v:v \in S_{\max}} c_v = OPT,$$

where OPT and $OPT_{\setminus S_{\max}}$ are the optimal values of instances $\mathcal{H}$ and $\mathcal{H}_{\setminus S_{\max}}$, respectively.

3 An f-Approximation Algorithm for the PCHVCF

In this section, we present a polynomial-time LP-rounding algorithm for the PCHVCF on instance $\mathcal{H}_{\setminus S_{\max}} = (H', c', p, \pi, E'_i, P'_i)$. Then, we show how to make use of it to find the expected feasible solution for the PCHVCF.

3.1 Algorithm After Preprocessing

For simplicity of notation in this section, we still use (H, c, p, π, E_i, P_i) to denote the auxiliary instance $\mathcal{H}_{\setminus S_{\max}}$.

Before presenting the LP-rounding algorithm, we state the LP-relaxation of integer programming (1) for instance $\mathcal{H}_{\setminus S_{\max}}$.

$$\min \sum_{v:v \in V} c_v x_v + \sum_{e:e \in E} \pi_e z_e$$

$$s.t. \quad \sum_{v:v \in e} x_v + z_e \ge 1, \ \forall e \in E, \tag{2}$$

$$\sum_{e:e \in E_i} p_e \cdot z_e \le p(E_i) - P_i, \ \forall i \in [m],$$

$$x_v \ge 0, z_e \ge 0, \ \forall v \in V, \ \forall e \in E.$$

Note that, we need not to add the constraints $x_v \le 1$ and $z_e \le 1$ since they are automatically satisfied in an optimal solution.

Our LP-rounding algorithm consists of two phases inspired by [20].

First Phase. We compute an optimal solution $(\{x_v^*\}_{v\in V}, \{z_e^*\}_{e\in E})$ of the LP-relaxation (2) and modify it to obtain another feasible solution $(\{\tilde{x}_v\}_{v\in V}, \{\tilde{z}_e\}_{e\in E})$, which has a special structure introduced in Lemma 2.

Second Phase. Based on $(\{\tilde{x}_v\}_{v\in V}, \{\tilde{z}_e\}_{e\in E})$, we construct a sparse linear programming that does not contain too many constraints. Finding a basic optimal solution of this programming, we obtain the feasible solution by rounding all fractional variables to 1.

Lemma 2. *We can find a feasible solution* $(\{\tilde{x}_v\}_{v\in V}, \{\tilde{z}_e\}_{e\in E})$ *of relaxed programming (2) in polynomial-time with the following properties:*

(i) There is a function $\Phi : E \to V$ *such that for each edge e, $\Phi(e)$ is one vertex in e, and $\tilde{z}_e$ is equal to the $1 - \tilde{x}_{\Phi(e)}$.*

(ii) $\sum_{v:v\in V} c_v\tilde{x}_v + \sum_{e:e\in E} \pi_e \cdot (1 - \tilde{x}_{\Phi(e)}) \leq f \cdot OPT_{LP}$, *where* $f = \max_{e:e\in E} |e|$, *and OPT_{LP} is the optimal value of the LP-relaxation (2).*

Proof. Using the Ellipsoid algorithm, we can find an optimal solution $(\{x_v^*\}_{v\in V}, \{z_e^*\}_{e\in E})$ of the LP-relaxation (2) in polynomial-time. For each edge $e \in E$, we set $\Phi(e)$ to the vertex v_e with $v_e = \arg\max_{v:v\in e} x_v^*$, and have

$$\sum_{v:v\in e} x_v^* \leq f \cdot x_{\Phi(e)}^*, \ \forall e \in E, \tag{3}$$

by $f = \max_{e:e\in E} |e|$. We construct the solution $(\{\tilde{x}_v\}_{v\in V}, \{\tilde{z}_e\}_{e\in E})$ by modifying the solution $(\{x_v^*\}_{v\in V}, \{z_e^*\}_{e\in E})$, where

$$\begin{cases} \tilde{x}_v = \min\{1, f \cdot x_v^*\}, & \forall \, v \in V; \\ \tilde{z}_e = \max\{0, 1 - f \cdot x_{\Phi(e)}^*\}, & \forall \, e \in E. \end{cases}$$

It is obvious that $(\{\tilde{x}_v\}_{v\in V}, \{\tilde{z}_e\}_{e\in E})$ can be found in polynomial-time, and

$$\tilde{z}_e = 1 - \tilde{x}_{\Phi(e)}, \quad \forall \, e \in E,$$

which implies that property **(i)** follows.

Furthermore, for any $e \in E$, we have

$$\begin{aligned} 1 - \tilde{x}_{\Phi(e)} &= 1 - \min\{1, f \cdot x_{\Phi(e)}^*\} = \max\{0, 1 - f \cdot x_{\Phi(e)}^*\} \\ &\leq \max\{0, 1 - \sum_{v:v\in e} x_v^*\} \\ &\leq \max\{0, z_e^*\} \\ &= z_e^*, \end{aligned} \tag{4}$$

where the first inequality follows from inequality (3), and the second inequality follows from $\sum_{v:v\in e} x_v^* + z_e^* \geq 1$ for any $e \in E$ by the first set of constraints of LP-relaxation (2). Then,

$$\sum_{v:v\in V} c_v\tilde{x}_v + \sum_{e:e\in E} \pi_e \cdot (1 - \tilde{x}_{\Phi(e)}) \leq f \cdot \sum_{v:v\in V} c_v x_v^* + \sum_{e:e\in E} \pi_e z_e^* \leq f \cdot OPT_{LP},$$

where the first inequality follows from inequality (4), and the second inequality follows from $f \geq 2$. This statement implies that property **(ii)** follows.

Next, we prove that $(\{\tilde{x}_v\}_{v \in V}, \{\tilde{z}_e\}_{e \in E})$ is a feasible solution of relaxed programming (2). Based on property **(i)**, we have

$$\tilde{z}_e = 1 - \tilde{x}_{\Phi(e)} \geq 1 - \sum_{v:v \in e} \tilde{x}_v, \quad \forall\, e \in E,$$

where the inequality follows from $\tilde{x}_v = \min\{1, f \cdot x_v^*\} \geq 0$. Rearranging this inequality, we have that $\sum_{v:v \in e} \tilde{x}_v + \tilde{z}_e \geq 1$ for any $e \in E$, which implies that $(\{\tilde{x}_v\}_{v \in V}, \{\tilde{z}_e\}_{e \in E})$ satisfies the first set of constraints of LP-relaxation (2). Based on inequality (4), for any $e \in E$, we have $\tilde{z}_e = 1 - \tilde{x}_{\Phi(e)} \leq z_e^*$ by inequality (4), and

$$\sum_{e:e \in E_i} p_e \cdot \tilde{z}_e \leq \sum_{e:e \in E_i} p_e \cdot z_e^* \leq p(E_i) - P_i, \quad \forall\, i \in [m],$$

where the second inequality follows from $\sum_{e:e \in E_i} p_e \cdot z_e^* \leq p(E_i) - P_i$ for any $i \in [m]$ by the second set of constraints of LP-relaxation (2). This statement implies that $(\{\tilde{x}_v\}_{v \in V}, \{\tilde{z}_e\}_{e \in E})$ satisfies the second set of constraints of LP-relaxation (2). Thus, $(\{\tilde{x}_v\}_{v \in V}, \{\tilde{z}_e\}_{e \in E})$ is a feasible solution of the LP-relaxation (2).

Therefore, the lemma holds.   □

Then, we introduce the following definition based on $(\{\tilde{x}_v\}_{v \in V}, \{\tilde{z}_e\}_{e \in E})$ and the function Φ. For each $i \in [m]$ and each vertex $v \in V$, let

$$E_{i,v} = \{\, e \in E_i \mid \Phi(e) = v \,\}$$

be the set of edges $e \in E_i$ with $\Phi(e) = v$. Since Φ is a function from E to V, $\{E_{i,v}\}_{v \in V}$ is a partition of E_i, i.e.,

$$\bigcup_{v:v \in V} E_{i,v} = E_i, \text{ and } E_{i,v} \cap E_{i,v'} = \emptyset, \ \forall v, v' \in V \text{ and } v \neq v.$$

Using the function Φ and $\{E_{i,v}\}_{v \in V}$ for each $i \in [m]$, we can construct the following sparse linear programming, which is in **Second phase**,

$$\max \sum_{v:v \in V} p(E_{1,v}) \cdot y_v$$

$$s.t. \quad \sum_{v:v \in V} p(E_{i,v}) \cdot y_v \geq P_i, \ \forall i \in [m] \setminus \{1\}, \tag{5}$$

$$\sum_{v:v \in V} c_v \cdot y_v + \sum_{e:e \in E} \pi_e \cdot (1 - y_{\Phi(e)}) \leq f \cdot OPT_{LP}$$

$$y_v \in [0, 1], \ \forall v \in V.$$

where variable y_v denotes whether or not the vertex $v \in V$ is selected in the solution, and $p(E_{i,v}) = \sum_{e:e \in E_{i,v}} p_e$.

Lemma 3. *There is a feasible solution to sparse linear programming (5) whose objective function value is at least P_1.*

Proof. For any vertex $v \in V$, we set $y_v = \tilde{x}_v$. Then, $\{y_v\}_{v \in V}$ satisfies the second constraint of (5) by property **(ii)** of Lemma 2. For any $i \in [m]$, we have

$$
\sum_{v:v \in V} p(E_{i,v}) \cdot y_v = \sum_{v:v \in V} p(E_{i,v}) \cdot \tilde{x}_v
$$

$$
= \sum_{v:v \in V} \sum_{e:e \in E_{i,v}} p_e \cdot \tilde{x}_v,
$$

$$
= \sum_{v:v \in V} \sum_{e:e \in E_{i,v}} p_e \cdot \tilde{x}_{\Phi(e)},
$$

$$
= \sum_{v:v \in V} \sum_{e:e \in E_{i,v}} p_e \cdot (1 - \tilde{z}_e)
$$

$$
= \sum_{e:e \in E_i} p_e \cdot (1 - \tilde{z}_e)
$$

$$
= p(E_i) - \sum_{e:e \in E_i} p_e \cdot \tilde{z}_e
$$

$$
\geq P_i,
$$

where the third equality follows from $v = \Phi(e)$ for any $e \in E_{i,v}$ by $E_{i,v} = \{e \in E_i | \Phi(e) = v\}$, the fourth equality follows from property **(i)** of Lemma 2, the fifth equality follows from the fact that $E_{i,v}$ is a partition of E_i, and the inequality follows from the fact that $(\{\tilde{x}_v\}_{v \in V}, \{\tilde{z}_e\}_{e \in E})$ is a feasible solution of relaxed programming (2) and the second set of constraints of relaxed programming (2). Thus, we have that $\{y_v\}_{v \in V}$ is a feasible solution whose objective value is no less than P_1. Therefore, this lemma follows. □

Lemma 4. *Let $\{y_v^*\}_{v \in V}$ be a basic optimal solution of sparse linear programming (5), then the number of fractional variables in $\{y_v^*\}_{v \in V}$ is at most m.*

Proof. There are $2n+m$ constraints and n variables in sparse linear programming (5). Since $\{y_v^*\}_{v \in V}$ is a basic optimal solution, the number of linearly independent tight constraints in $\{y_v^*\}_{v \in V}$ is n [24]. There are only m constraints that are neither $y_v \geq 0$ nor $y_v \leq 1$, *i.e.*, the number of constraints, which is either $y_v \geq 0$ or $y_v \leq 1$ in $\{y_v^*\}_{v \in V}$ is at least $n - m$. Thus, we can assert that at least $n - m$ of the y_v^* in $\{y_v^*\}_{v \in V}$ must be set to $\{0, 1\}$, and the number of fractional variables in $\{y_v^*\}_{v \in V}$ is at most m. □

Lemma 5. *Let $S' = \{v \in V | y_v^* > 0\}$, then S' is a feasible solution for the auxiliary instance $\mathcal{H}_{\backslash S_{\max}}$ and its objective value is*

$$
OUT_{\backslash S_{\max}} \leq f \cdot OPT_{\backslash S_{\max}} + m \cdot c_{v_{\max,l}},
$$

where $\{y_v^*\}_{v \in V}$ is a basic optimal solution of sparse linear programming (5), $OUT_{\backslash S_{\max}} = \sum_{v \in S'} c_v + \sum_{e:e \in E \backslash E(S')} \pi_e$ and $OPT_{\backslash S_{\max}}$ is the optimal value of the PCHVCF on $\mathcal{H}_{\backslash S_{\max}}$.

Proof. Assume that $\{y_v^*\}_{v \in V}$ is a basic optimal solution of sparse linear programming (5), then for any $i \in [m]$, we have $\sum_{v:v \in V} p(E_{i,v}) \cdot y_v^* \geq P_i$, where the inequality for $i = 1$ follows from Lemma 3, and the inequality for $i \in [m] \setminus \{1\}$ from the first set of constraints of sparse linear programming (5). Based on the definition of $E_{i,v}$, $S' = \{v \in V | y_v^* > 0\}$ can satisfy all fairness constraints of instance $\mathcal{H}_{\setminus S_{\max}}$, which is a feasible solution for instance $\mathcal{H}_{\setminus S_{\max}}$.

Let $z_e^* = \max\{0, 1 - \sum_{v:v \in e} y_v^*\}$ for each $e \in E$, then we have

$$z_e^* = \max\{0, 1 - \sum_{v:v \in e} y_v^*\} \leq \max\{0, 1 - y_{\Phi(e)}^*\} \leq 1 - y_{\Phi(e)}^*, \ \forall e \in E, \quad (6)$$

where the first inequalities follow from $y_v^* \geq 0$ for any $v \in V$, and the second inequalities follow from $y_{\Phi(e)}^* \leq 1$. Furthermore, for any $e \in E \setminus E(S')$, we have $e \cap S' = \emptyset$ and $\sum_{v:v \in e} y_v^* = 0$, *i.e.*,

$$z_e^* = \max\{0, 1 - \sum_{v:v \in e} y_v^*\} = 1, \ \forall e \in E \setminus E(S').$$

Since $S' = \{v \in V | y_v^* > 0\}$, the objective value of S' is

$$OUT_{\setminus S_{\max}} = \sum_{v:v \in S'} c_v + \sum_{e:e \in E \setminus E(S')} \pi_e$$

$$= \sum_{v:y_v^*=1} c_v + \sum_{v:0<y_v^*<1} c_v + \sum_{e:e \in E \setminus E(S')} \pi_e$$

$$\leq \sum_{v:v \in V} c_v \cdot y_v^* + \sum_{v:0<y_v^*<1} c_v + \sum_{e:e \in E \setminus E(S')} \pi_e \cdot z_e^*$$

$$\leq \sum_{v:v \in V} c_v \cdot y_v^* + \sum_{v:0<y_v^*<1} c_v + \sum_{e:e \in E} \pi_e \cdot z_e^*$$

$$\leq \sum_{v:v \in V} c_v \cdot y_v^* + \sum_{v:0<y_v^*<1} c_v + \sum_{e:e \in E} \pi_e \cdot (1 - y_{\Phi(e)}^*)$$

$$\leq f \cdot OPT_{LP} + m \cdot c_{v_{\max,l}}$$

$$\leq f \cdot OPT_{\setminus S_{\max}} + m \cdot c_{v_{\max,l}},$$

where the first inequality follows from $y_v^* \geq 0$ for any $v \in V$ and $z_e^* = 1$ for any $e \in E \setminus E(S')$, the second inequality follows from $z_e^* = \max\{0, 1 - \sum_{v:v \in e} y_v^*\} \geq 0$ for any $e \in E$, the third inequality follows from inequality (6), the fourth inequality follows from the second constraint of sparse linear programming (5) and $|\{v \in V | 1 > y_v^* > 0\}| \leq m$ by Lemma 4, and the last inequality follows from the fact that the LP-relaxation (2) is a relaxation of integer programming (1) for instance $\mathcal{H}_{\setminus S_{\max}}$. Therefore, the lemma holds. $\qquad \square$

3.2   The Whole Algorithm

We present the whole algorithm for instance $\mathcal{H}$. In fact, we cannot know $S_{\max}$ in advance; however, we can assume that $S_{\max}$ is known using the guessing technique. We guess that each subset $\hat{S}$ with at most l vertices of V is $S_{\max}$, where

$$l = \left\lceil \frac{m}{f-1} \right\rceil.$$

For a guessed subset $\hat{S}$, we construct the auxiliary instance $\mathcal{H}_{\backslash \hat{S}}$ mentioned in Sect. 2.2, and find a feasible solution $S'_{\backslash \hat{S}}$ generated by the algorithm in Sect. 3.1 for instance $\mathcal{H}_{\backslash \hat{S}}$. By the definition of $\mathcal{H}_{\backslash \hat{S}}$, $S'_{\backslash \hat{S}} \cup \hat{S}$ is a feasible solution for instance $\mathcal{H}$, and we define the objective value of $S'_{\backslash \hat{S}} \cup \hat{S}$ as

$$OUT_{\hat{S}} = \begin{cases} OUT_{\backslash \hat{S}} + \sum\limits_{v:v\in\hat{S}} c_v, & \text{if } \max\limits_{v:v\in S'_{\backslash \hat{S}}} c_v \leq \min\limits_{v:v\in\hat{S}} c_v; \\ +\infty, & \text{otherwise,} \end{cases}$$

where $OUT_{\backslash \hat{S}} = \sum_{v\in S'_{\backslash \hat{S}}} c_v + \sum_{e:e\in E(V\backslash \hat{S})\backslash E(S'_{\backslash \hat{S}})} \pi_e$, and $E(V \backslash \hat{S}) = E \backslash \{e \in E | e \cap \hat{S} \neq \emptyset\}$. We output the solution S' with the minimum objective value among all the guesses, i.e., $S' = S'_{\backslash \hat{S}^*} \cup \hat{S}^*$, where

$$\hat{S}^* = \arg \min_{\hat{S}:\hat{S}\subseteq V \text{ and } |\hat{S}|\leq l} OUT_{\hat{S}}.$$

Theorem 1. *For the PCHVCF, we can find the feasible solution S' in $n^{O(m/f)}$ time, and its objective value is $OUT \leq f \cdot OPT$, where OPT is the optimal value of instance $\mathcal{H}$.*

Proof. Let S^* be an optimal solution for instance $\mathcal{H}$, if $|S^*| \leq l$, we can find the optimal solution by the guessing technique. Thus, we assume that $|S^*| > l$ and let $S_{\max}$ be the set of the l largest cost vertices in S^*. On the auxiliary instance $\mathcal{H}_{\backslash S_{\max}}$, we can find a feasible solution $S'_{\backslash S_{\max}}$, and the objective value of $S'_{\backslash S_{\max}} \cup S_{\max}$ is

$$OUT_{S_{\max}} = OUT_{\backslash S_{\max}} + \sum_{v\in S_{\max}} c_v$$

$$\leq f \cdot OPT_{\backslash S_{\max}} + m \cdot c_{v_{\max,l}^*} + \sum_{v\in S_{\max}} c_v$$

$$\leq f \cdot OPT_{\backslash S_{\max}} + m \cdot \frac{1}{l} \cdot \sum_{v:v\in S_{\max}} c_v + \sum_{v\in S_{\max}} c_v$$

$$\leq f \cdot OPT_{\backslash S_{\max}} + m \cdot \frac{f-1}{m} \cdot \sum_{v:v\in S_{\max}} c_v + \sum_{v\in S_{\max}} c_v$$

$$= f \cdot \left(OPT_{\backslash S_{\max}} + \sum_{v:v\in S_{\max}} c_v \right),$$

where the first inequality follows from Lemma 5, the second inequality follows from $c_{v_{\max,l}} = \min_{v:v\in S_{\max}} c_v$ and $|S_{\max}| = l$, and the last inequality follows from $l = \lceil \frac{m}{f-1} \rceil \geq \frac{m}{f-1}$.

Since $\hat{S}^* = \arg\min_{\hat{S}:\hat{S}\subseteq V \text{ and } |\hat{S}|\leq l} OUT_{\hat{S}}$, the objective value of S' is

$$OUT = OUT_{\hat{S}^*} \leq OUTS_{\max} \leq f \cdot (OPT_{\backslash S_{\max}} + \sum_{v:v\in S_{\max}} c_v) = f \cdot OPT,$$

where the last equality follows from Lemma 1.

For each guessed subset $\hat{S}$, we can construct the auxiliary instance $\mathcal{H}_{\backslash \hat{S}}$ in $O(|V|)$ time. On instance $\mathcal{H}_{\backslash \hat{S}}$, we need to find a feasible solution $S'_{\backslash \hat{S}}$ by solving two linear programmings, which implies that $S'_{\backslash \hat{S}}$ can be found in polynomial-time of n. Since $|\hat{S}| \leq l$ for any guessed subset $\hat{S}$, the total number of different guess subsets is at most n^l. Based on $l = \lceil \frac{m}{f-1} \rceil$, we can find the expected feasible solution in $n^{O(m/f)}$ time. □

Corollary 1. *When m is a constant, we can find a feasible solution S' in polynomial-time and its objective value is $OUT \leq f \cdot OPT$.*

4   Conclusions

In this paper, we consider the prize-collecting hypergraph vertex cover with fairness constraints, and design an f-approximation algorithm in $n^{O(m/f)}$ time based on the LP-rounding technique and the guessing method. When m is a constant, our algorithm is a polynomial-time f-approximation algorithm, which coincides with the lower bound of the hypergraph vertex cover problem if the unique game conjecture holds.

Acknowledgments. The work is supported in part by the National Natural Science Foundation of China [Grant No. 12461059] and Yunnan Fundamental Research Projects [Grants No. 202501AS070076, No. 202501AS070170, No. 202401AT070442].

Disclosure of Interests. The authors declare that they have no known competing financial interests.

References

1. Karp, R.M.: Reducibility among combinatorial problems. In: Complexity of Computer Computations, pp. 85–103. Springer, New York (1972)
2. Khot, S., Regev, O.: Vertex cover might be hard to approximate to within $2 - \epsilon$. J. Comput. Syst. Sci. **74**(3), 335–349 (2008)
3. Hochbaum, D.S.: Approximation algorithms for the set covering and vertex cover problems. SIAM J. Comput. **11**(3), 555–556 (1982)
4. Bar-Yehuda, R., Even, S.: A linear-time approximation algorithm for the weighted vertex cover problem. J. Algorithms **2**(2), 198–203 (1981)
5. Bshouty, N.H., Burroughs, L.: Massaging a linear programming solution to give a 2-approximation for a generalization of the vertex cover problem. In: Proceedings of the 15th Annual Symposium on Theoretical Aspects of Computer Science, STACS 98, pp. 298–308. Springer, Berlin (1998)

6. Bar-Yehuda, R.: Using homogeneous weights for approximating the partial cover problem. J. Algorithms **39**(2), 137–144 (2001)
7. Hochbaum, D.S.: Solving integer programs over monotone inequalities in three variables: a framework for half integrality and good approximations. Eur. J. Oper. Res. **140**(2), 291–321 (2002)
8. Bar-Yehuda, R., Dror, R.: On the equivalence between the primal-dual schema and the local ratio technique. SIAM J. Discret. Math. **19**(3), 762–797 (2005)
9. Liu, X., Li, W., Yang, J.: A primal-dual approximation algorithm for the k-prize-collecting minimum vertex cover problem with submodular penalties. Front. Comp. Sci. **17**(3), 173404 (2023)
10. Feige, U.: A threshold of $\ln n$ for approximating set cover. J. ACM **45**(4), 634–652 (1998)
11. Dinur, I., Steurer, D.: Analytical approach to parallel repetition. In: Proceedings of the Forty-Sixth Annual ACM Symposium on Theory of Computing, STOC 2014, pp. 624–633. Association for Computing Machinery, New York (2014)
12. Bansal, N., Khot, S.: Inapproximability of hypergraph vertex cover and applications to scheduling problems. In: Abramsky, S., Gavoille, C., Kirchner, C., Meyer auf der Heide, F., Spirakis, P.G. (eds.) ICALP 2010. LNCS, vol. 6198, pp. 250–261. Springer, Heidelberg (2010). https://doi.org/10.1007/978-3-642-14165-2_22
13. Könemann, J., Parekh, O., Segev, D.: A unified approach to approximating partial covering problems. Algorithmica **59**, 489–509 (2011)
14. Gandhi, R., Khuller, S., Srinivasan, A.: Approximation algorithms for partial covering problems. J. Algorithms **53**(1), 55–84 (2004)
15. Liu, X., Li, W.: An approximation algorithm for the K-prize-collecting multicut problem in trees with submodular penalties. Math. Struct. Comput. Sci. **34**(3), 193–210 (2024)
16. Guruswami, V., Sachdeva, S., Saket, R.: Inapproximability of minimum vertex cover on k-uniform k-partite hypergraphs. SIAM J. Discret. Math. **29**(1), 36–58 (2015)
17. Lovász, L.: On minimax theorems of combinatorics. Doctoral thesis, Mathematiki Lapok, **26**, 209–264 (1975)
18. Hung, E., Kao, M.J.: Approximation algorithm for prize-collecting vertex cover with fairness constraints. Algorithmica **84**, 1–12 (2022)
19. Bera, S.K., Gupta, S., Kumar, A., Roy, S.: Approximation algorithms for the partition vertex cover problem. Theoret. Comput. Sci. **555**, 2–8 (2014)
20. Bandyapadhyay, S., Banik, A., Bhore, S.: On colorful vertex and edge cover problems. Algorithmica **85**, 3816–3827 (2023)
21. Inamdar, T., Varadarajan, K.:. On the partition set cover problem. CoRR. arxiv:1809.06506 (2018)
22. Zhou, M., Zhang, Z., Ding-Zhu, D.: Approximation algorithm for vertex cover with multiple covering constraints. J. Comb. Optim. **48**, 20 (2024)
23. Wang, Q., Hou, B., Zhang, G., Liu, W.: Approximation algorithms for the partition set cover problem with penalties. SSRN. https://doi.org/10.2139/ssrn.4903712 (2024)
24. Bertsimas, D., Tsitsiklis, J.N.: Introduction to Linear Programming. Athena-Scientific (1997)

Sum-of-Max Chain Partition of a Tree

Ruixi Luo, Taikun Zhu, and Kai Jin[✉]

School of Intelligent Systems Engineering, Shenzhen Campus of Sun Yat-Sen
University, No. 66, Gongchang Road, Shenzhen 518107, Guangdong,
People's Republic of China
jink8@mail.sysu.edu.cn

Abstract. Path partition problems on trees have found various applications. In this paper, we present an $O(n \log n)$ time algorithm for solving
the following variant of path partition problem: given a rooted tree of n
nodes $1, \ldots, n$, where vertex i is associated with a weight w_i and a cost
s_i, partition the tree into several disjoint chains $C_1, \ldots, C_k$, so that the
weight of each chain is no more than a threshold w_0 and the sum of the
largest s_i in each chain is minimized. We also generalize the algorithm
to the case where the cost of a chain is determined by the s_i of the vertex with the highest rank in the chain, which can be determined by an
arbitrary total order defined on all nodes instead of the value of s_i.

Keywords: Chain Partition · Tree Partition · Binomial Heap ·
Dynamic Programming · Heap-over-Heap

1 Introduction

Sequence and tree partition problems have been extensively studied in the past
decades [2,3,5,23]. In this paper, we study a variant of the tree partition problems, namely, the sum-of-max chain partition of a tree.

Above all, it is necessary to clarify the definition of a chain in the context
of tree partitioning. We adopt the chain partition definition as in Misra and
Tarjan's work [21]. A chain in this context is defined as follows: A path of a
rooted tree is a *chain* if for each node pair (u, v) on the path, either u is the
ancestor of v, or v is the ancestor of u.

Sum-of-max chain partition problem. Given a rooted tree T of n
nodes $1, \ldots, n$, where each vertex i is associated with a weight w_i and a
cost s_i, partition the tree into several disjoint chains $C_1, \ldots, C_k$, where k
is arbitrary, so that the total weight of each chain is no more than a given
threshold w_0 and the sum of the largest s_i in each chain is minimized.
Without loss of generality, we assume that $w_i \leq w_0$ for all i.

This research is supported by Department of Science and Technology of Guangdong
Province (Project No. 2021QN02X239) and Shenzhen Science and Technology Program
(Grant No. 202206193000001, 20220817175048002).

Let the largest s_i in each chain be referred to as the **cost** of the chain. The total cost of a chain partition $P = \bigcup_{i=1}^{k} C_i$ is the sum of the largest parameter s in each chain, as defined above. Formally, $\text{cost}(P) = \sum_{1 \leq i \leq k} \max_{j \in C_i} s_j$.

Partitioning a tree into paths has been studied by a number of researchers. In [27], the problem of partitioning a tree into paths with each path having no more than k nodes was solved with a linear time algorithm. The sum-of-max chain partition problem on trees can be seen as a generalization of the problem with variable weights and costs on nodes to some extent. The main contribution of this paper is an $O(n \log n)$ time algorithm for solving sum-of-max chain partition problem on trees mentioned above.

1.1   Related Works

Misra and Tarjan [21] investigated a type of chain partition problem where a tree is divided into several chains, each of which does not exceed the weight limit w_0, and the objective is to minimize the cost of edges contained in no chains. An $O(n \log n)$ algorithm for this problem was presented in their paper.

The *heavy-light decomposition* technique [25] introduces a type of chain partition of trees, which finds numerous applications [1,7,11,15]. Similarly, the *long-path decomposition* technique [4] defines another chain partition of trees, which can be used to speed up several dynamic programming algorithms on trees. Both of the mentioned chain partitions can be computed in linear time.

Path partition problems have drawn more attention in the literature [6,9,16, 22]. For example, partitioning a graph G into the minimum number of paths has been studied extensively [5,17,19]. Boesch et al. [5] provided a simple polynomial algorithm for the special case where G is a tree. This type of partition problem on trees is related to the Hamiltonian completion of a tree [10,24].

Some path partition problems of trees with costs related to the number of paths have also been studied in the past. Cai et al. [8] studied ω-path partition of a edge-weighted tree and gave a linear time algorithm for the problem. Jin and Li [14] presented a linear path partition algorithm for edge-colored trees. Yan et al. [27] presented a linear-time algorithm for the k-path partition problem in trees, which considers partitioning a tree into the smallest number of paths such that each path has at most k vertices. The path partitions problems mentioned above all aim for minimizing the number of the resulting paths. This kind of objective can be seen as a special case of sum-of-max objective with uniform costs for all elements. In other words, by setting uniform costs for all elements, the sum-of-max objective degenerates into minimizing the number of paths in the partition. As a sequence can be treated as a tree without branches, the problem in this paper can be seen as an extension of the sum-of-max sequence partition problem studied in [13], where an $O(n)$ time algorithm is given.

Partition a tree into subtrees has also been investigated extensively. A few references about such problems can be found in the introduction of [13].

The decision version of our chain partition problem is NP-complete for a directed graph G. This is because the chain partition problem can be reduced to the minimum path cover problem when all weights and costs are set to be 1.

Then, the minimum path cover problem can further be reduced to a classic NP-complete problem, namely the Hamiltonian path problem [12], since a minimum path cover consists of one path if and only if there is a Hamiltonian path in G.

1.2 Applications

The partition problem on trees can be applied in real-life, such as water environment protection, industrial investment planning, power grid maintenance, biological sample supply and preservation, etc.

In a river environmental protection scenario, we need to send several staff members to clean the river. Each staff member is responsible for a continuous section of the river. Since the river forks, each unbranched river segment can be seen as a vertex of the tree.

Assume that each person can clean a river of length w_0 at most, each segment i is associated with length w_i and equipment requirement s_i. The equipment that each staff member needs is determined by the largest s_i of the river segments he is responsible for. An effective sum-of-max chain partition guarantees a good staff allocation plan with minimum total equipment requirement.

Our sum-of-max chain partition problem can also be applied to industrial investment. Assume a cooperation focuses on oil related industries. Then, the supply chains based on oil can be seen as a tree, with oil as the root, various consumer goods as leaf vertices and intermediate products between them. Now, the cooperation intends to establish a number of factories, each produces several products on the tree.

In order to ensure the continuity of production, only continuous products on the tree, i.e. a chain, can serve as the supply chain segment for a factory's production. Suppose a product i on the tree requires w_i area of land, and s_i level of pollution. Each factory can occupy no more than w_0 area of land and its pollution is determined by the largest s_i of the products it produces. A good sum-of-max chain partition can reduce the total pollution of the factories.

2 Preliminaries

Denote the given tree and its subtree rooted at vertex v by T and T_v, respectively. For each vertex u in T_v, denote by $T[v, u]$ the chain from v to u, by $T[v, u)$ the chain from v to the parent of u, and by $T(v, u]$ the chain $T[v, u]$ without v. By this definition, $T[v, v)$ is empty. The set of children of v is denoted by $\mathsf{Child}(v)$.

Definition 1. *The* window *of v, denoted by* win_v, *refers to the set of vertices u in T_v for which the chain $T[v, u]$ weights no more than w_0.*

A vertex u in T_v is s-maximal *if each vertex p in $T[v, u)$ admits that $s_p < s_u$. For $u \in T_v$ and $u \neq v$, let* $\mathsf{next}_v(u)$ *be the closest proper s-maximal ancestor of u in T_v. Note that $\mathsf{next}_v(u)$ is well-defined since the root v of T_v is always s-maximal in T_v.*

Let win_v^* *denote the s-maximal vertices in* win_v *and* win_v^- *denote vertices in* $\mathsf{win}_v \setminus \mathsf{win}_v^*$.

Our algorithm follows a dynamic programming paradigm.

Let $F[v]$ denote the minimum cost to partition T_v into chains. Obviously, our goal is to compute $F[r]$, where r denotes the root of T.

Let us consider how to find the optimal partition of T_v. Obviously, v must belong to some chain, e.g. $T[v,i]$. Define $\mathsf{cost}(v,i)$ to be the minimum cost to partition T_v into chains, one of which equals $T[v,i]$. Then, we have

$$F[v] = \min_{T[v,i]\text{weights no more than}w_0} \mathsf{cost}(v,i). \tag{1}$$

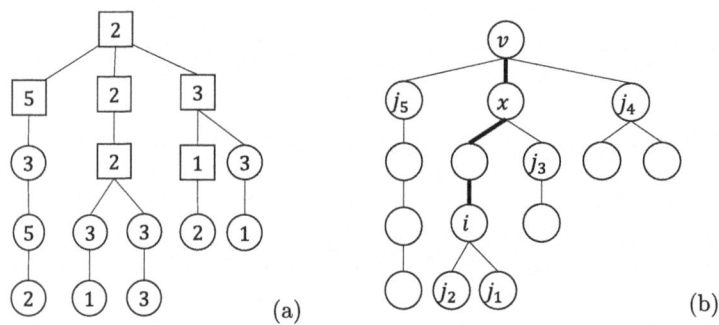

Fig. 1. (a) Illustration for window and outer boundary, where $w_0 = 7$. The numbers in the vertices are their weights. Squares represent the vertices in the window. (b) Illustration for $\mathsf{sum}(v,i)$. Here, $\mathsf{sum}(v,i) = F[j_1] + F[j_2] + F[j_3] + F[j_4] + F[j_5]$.

Recall Definition 1. We rewrite Eq. (1) as:

$$F[v] = \min_{i\in\mathsf{win}_v} \mathsf{cost}(v,i). \tag{2}$$

Recall next in Definition 1. A trivial formula of $\mathsf{cost}(v,i)$ is as follows:

$$\mathsf{cost}(v,i) = \begin{cases} \mathsf{sum}(v,i) + s_i & \text{if } i \in \mathsf{win}_v^* \\ \mathsf{sum}(v,i) + s_{\mathsf{next}_v(i)} & \text{if } i \in \mathsf{win}_v^-, \end{cases} \tag{3}$$

where

$$\mathsf{sum}(v,i) := \sum_{j\text{'s parent}\in T[v,i],j\notin T[v,i]} F[j]. \tag{4}$$

See Fig. 1 (b) for an illustration of $\mathsf{sum}(v,i)$.

By the definitions above, we can derive the following lemma, which is crucial for the design and analysis in this paper.

Lemma 1. *Assume $i \in \mathsf{win}_v^-$ and $x = \mathsf{next}_v(i)$. Then,*

$$i \in \mathsf{win}_x^- \text{ and } x = \mathsf{next}_x(i), \tag{5}$$

and more importantly,

$$\mathsf{cost}(v,i) - \mathsf{cost}(x,i) = \mathsf{cost}(v,x) - \mathsf{cost}(x,x). \tag{6}$$

Proof. We may assume $v \neq x$ (otherwise it is trivial).

Since $i \in \mathsf{win}_v^-$ and $x = \mathsf{next}_v(i)$, we know that $s_x \geq s_p$ for $p \in T[v, i]$. In particular, $s_x \geq s_p$ for $p \in T[x, i]$. Consequently, in $T[x, i]$, only x is s-maximal in T_x and thus we obtain Eq. (5).

Further applying Eq. (3), we get

$$\begin{cases} \mathsf{cost}(v, i) = \mathsf{sum}(v, i) + s_x \\ \mathsf{cost}(x, i) = \mathsf{sum}(x, i) + s_x. \end{cases}$$

Let p_x denote the parent of x.

$$\begin{aligned} \mathsf{cost}(v, i) - \mathsf{cost}(x, i) &= (\mathsf{sum}(v, i) + s_x) - (\mathsf{sum}(x, i) + s_x) \\ &= \mathsf{sum}(v, i) - \mathsf{sum}(x, i) \qquad\qquad (7) \\ &= \mathsf{sum}(v, p_x) - F[x]. \end{aligned}$$

The last equation follows from the definition (4).

Similarly, we have

$$\begin{aligned} \mathsf{cost}(v, x) - \mathsf{cost}(x, x) &= (\mathsf{sum}(v, x) + s_x) - (\mathsf{sum}(x, x) + s_x) \\ &= \mathsf{sum}(v, x) - \mathsf{sum}(x, x) \qquad\qquad (8) \\ &= \mathsf{sum}(v, p_x) - F[x]. \end{aligned}$$

Equations (7) and (8) together imply Eq. (6). $\square$

We will use a few heaps to compute the minimum cost $F[v]$. For simplicity and time complexity, all heaps in this paper refer to *binomial heaps* [26]. (The *leftist tree* [20] is not appropriate for our purpose.)

Table 1. Operations for the heaps.

Operation	Definition	Time required
find-min(max)(H)	finds the minimal(maximal) element in heap H.	$O(1)$
insert(H,i,k)	inserts vertex i in heap H with key k.	$O(\log n)$
delete-vertex(H,i)	deletes vertex i in heap H (if exists).	$O(\log n)$
meld(H_1,H_2)	melds heap H_1 with H_2 into H_1.	$O(\log n)$
add-all(H,v)	adds value v to all key values in H.	$O(\log n)$
key-value(H,i)	returns key value of i in H (if exists).	$O(\log n)$

Operations shown in Table 1 are needed for the binomial heaps in the algorithm. Most of the operations are originally available for the binomial heaps [26]. Additionally, we implement a lazy tag technique to create an add-all() operation in Table 1, which is discussed in the following subsection.

2.1 Binomial Heap with Lazy Tags

For each node in the binomial heap, we add a lazy tag to it, representing adding or subtracting a value uniformly for all elements below. In this section, we briefly explain the differences brought by the lazy tag technique.

Traditionally, when using lazy tags, we pushdown a tag to all children when the node is accessed. However, this would lead to a non-$O(1)$ pushdown operation in binomial heaps since there can be $O(\log n)$ children for one node, which may cause an $O(\log^2 n)$ factor in time bound.

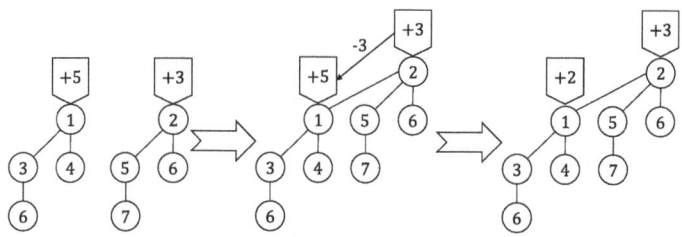

Fig. 2. Example of binomial heaps melding with lazy tag

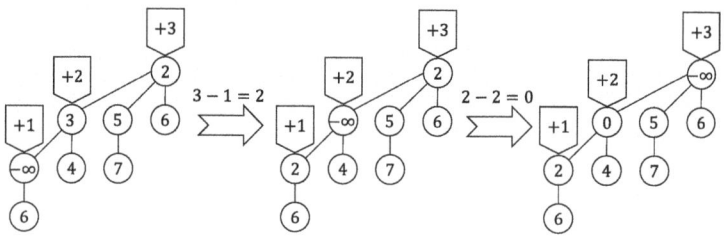

Fig. 3. Example of binomial heaps pushing up a node $-\infty$. In the first step, 3 is swapped down and becomes $3 - 1 = 2$ to counter the $+1$ tag's effect since it was not affected by $+1$ in the first place. Similarly, 2 is swapped down and becomes $2 - 2 = 0$ to counter the $+2$ tag.

Therefore, we do not pushdown a lazy tag unless it's a tag of a root vertex to be deleted. This change requires us to handle the tags in a slightly different way. First, we only exert a negative pushdown to the newly melded tree as shown in Fig. 2. Suppose tree B_a with a bigger root is linked to B_b, then the opposite number of the lazy tag of B_b will is added to the lazy tag of B_a during the linking operation. This trick allows us we to preserve the property of the heap while the negative pushdown is still an $O(1)$ operation.

Besides, when pushing up a node for a key value change. The key value of the node swapped down needs to be adjusted since the tag stays still. If a vertex is swapped down, it needs to counter the effect caused by the lower tag and thus add the opposite of the lower tag, see Fig. 3 for illustration.

Thanks to the $O(\log n)$ depth limit of binomial heap, even with lazy tags affecting the real key value of each node, we can still also acquire a node's key-value in $O(\log n)$ time by simple tree-climbing towards the root.

3 Algorithm for Finding the Optimal Chain Partition

Recall Eq. (2). This section shows how we compute $F[v]$ efficiently.

For computing $F[v]$, we use an interesting combined data structure, which we call *heap-over-heap*. Briefly, there are several disjoint heaps in the first layer (see $H_x^{(v)}$'s below), and the minimum elements of all heaps in the first layer are maintained in another heap (see $\mathbb{H}^{(v)}$ below) at the second layer.

Min-heap $H_x^{(v)}$ for $x \in \mathsf{win}_v^*$.
    $H_x^{(v)}$ stores i with key $\mathsf{cost}(x, i)$, if $(i = x)$ or $(i \in \mathsf{win}_v^-$ and $\mathsf{next}_v(i) = x)$.
Min-heap $\mathbb{H}^{(v)}$.
    $\mathbb{H}^{(v)}$ stores $i = \text{find-min}(H_x^{(v)})$ with key $\mathsf{cost}(v, i)$, for each $x \in \mathsf{win}_v^*$.

Each $i \in \mathsf{win}_v$ belongs to exactly one heap in $\{H_x^{(v)} : x \in \mathsf{win}_v^*\}$. See Fig. 4.

Remark 1. Be aware that i in $H_x^{(v)}$ is associated with a key value $\mathsf{cost}(x, i)$ instead of $\mathsf{cost}(v, i)$. This is crucial to the entire algorithm. If i is associated with $\mathsf{cost}(v, i)$ in the first layer, it is not easy to update the values when v is changed.

Lemma 2. $\arg\min_{i \in \mathsf{win}_v} \mathsf{cost}(v, i) = \text{find-min}(\mathbb{H}^{(v)})$. *In other words, computing $F[v] = \min_{i \in \mathsf{win}_v} \mathsf{cost}(v, i)$ reduces to computing the top element of $\mathbb{H}^{(v)}$.*

Proof. Take an arbitrary $i \in \mathsf{win}_v$. Assume i is in $H_x^{(v)}$.

Let $i' = \text{find-min}(H_x^{(v)})$. We have $\mathsf{cost}(x, i) \geq \mathsf{cost}(x, i')$.

Since i, i' are in $H_x^{(v)}$, respectively, applying Lemma 1,

$$\mathsf{cost}(v, i) - \mathsf{cost}(x, i) = \mathsf{cost}(v, x) - \mathsf{cost}(x, x)$$
$$\mathsf{cost}(v, i') - \mathsf{cost}(x, i') = \mathsf{cost}(v, x) - \mathsf{cost}(x, x).$$

Therefore, $\mathsf{cost}(v, i) - \mathsf{cost}(v, i') = \mathsf{cost}(x, i) - \mathsf{cost}(x, i') \geq 0$, namely, $\mathsf{cost}(v, i) \geq \mathsf{cost}(v, i')$. Since $a = \text{find-min}(\mathbb{H}^{(v)})$ and i' is also in $\mathbb{H}^{(v)}$, we have $\mathsf{cost}(v, i') \geq \mathsf{cost}(v, a)$. Thus for any $i \in \mathsf{win}_v$, $\mathsf{cost}(v, i) \geq \mathsf{cost}(v, a)$.

Further since $a \in \mathsf{win}_v$ (trivial), we have $a = \arg\min_{i \in \mathsf{win}_v} \mathsf{cost}(v, i)$. □

For maintaining the heap $\mathbb{H}^{(v)}$ over heaps $H_x^{(v)}$'s structure efficiently, we have to maintain two more types of heaps: (1) A heap (see $W^{(v)}$ below), which is used for detecting the vertices leaving the window win_v. (2) A bunch of heaps (see $S_x^{(v)}$'s below), which are used for organizing the s-maximal vertices.

Max-heap $W^{(v)}$.

$W^{(v)}$ stores i with key $\sum_{j \in T[v,i]} w_j$, for each $i \in \text{win}_v$.

Min-heap $S_x^{(v)}$ for $x \in \text{win}_v^*$.

$S_x^{(v)}$ stores i with key s_i, for each $i \in \text{win}_v^*$ with $\text{next}_v(i) = x$. See Fig. 4.

Remark 2. In our algorithm, for each x, we do not have to store all the $H_x^{(v)}$'s one by one for each v with $x \in \text{win}_v^*$. Instead, all such $H_x^{(v)}$'s can share one entity in the memory, denoted as H_x. The notation $H_x^{(v)}$ can be viewed as a historical version of H_x at the stage of computing v. The same logic holds for $S_x^{(v)}$.

3.1 Maintain the Heaps Efficiently

We compute $\{F[v]\}$ from bottom to up. Suppose $F[v]$ is to be computed in the algorithm. It means that we have already computed the $F[c]$ for all $c \in \text{Child}(v)$ and built all the aforementioned heaps related to T_c. Our current task is to obtain the heaps related to T_v from those heaps built for T_c's.

Consider an example as shown in Fig. 5. At the stage for computing $F[v]$,

1. vertices f, g, h are no longer in the window (namely, $f, g, h \in \text{win}_c \setminus \text{win}_v$), and hence should be eliminated from the heaps containing them;
2. vertices c, d are no longer s-maximal (as $s_v \geq s_c$ and $s_v \geq s_d$), and hence $\{S_x^{(v)} : x \in \text{win}_v^*\}$ changes accordingly; this leads to some merges of heaps in $\{H_x^{(v)} : x \in \text{win}_v^*\}$, and furthermore, $\mathbb{H}^{(v)}$ should be updated accordingly.

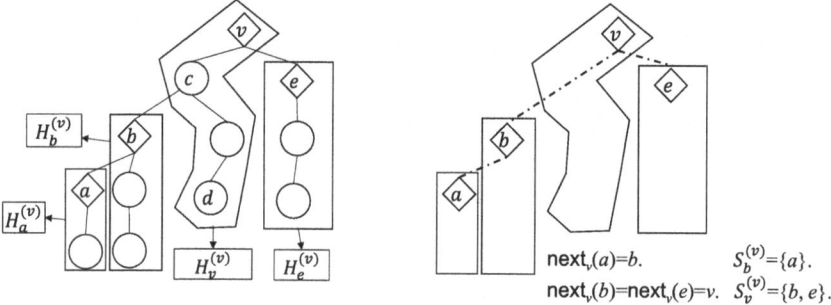

$\text{next}_v(a)=b.$ $S_b^{(v)}=\{a\}.$
$\text{next}_v(b)=\text{next}_v(e)=v.$ $S_v^{(v)}=\{b, e\}.$

Fig. 4. A partition of win_v into different parts, with each part organized by a heap. In this figure, rhombuses indicate s-maximal vertices and circles indicate other vertices.

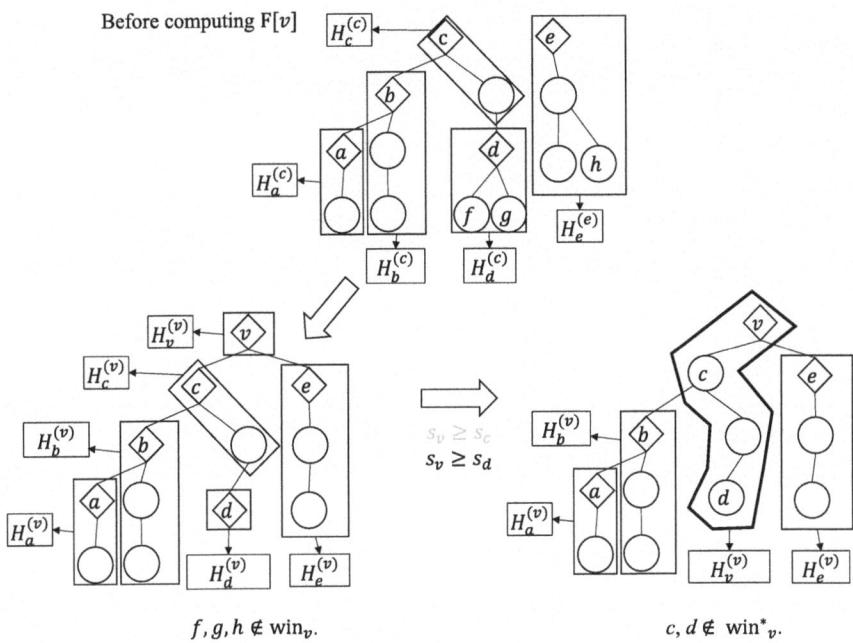

Fig. 5. Illustration of the change of the heaps at the stage for computing $F[v]$.

From the example, we see two major procedures are required at the moment:

(a) Find $i \in \mathsf{win}_c \setminus \mathsf{win}_v$. Then, eliminate them from corresponding heaps.

(b) Find $x \in \mathsf{win}_c^* \setminus \mathsf{win}_v^*$ (it loses s-maximality due to $s_x \leq s_v$), and find those i for which $\mathsf{next}_c(i) = x$ (due to $\mathsf{next}_v(i) = v \neq x$) (by utilizing $S_x^{(v)}$). Then, modify $\{S_x^{(v)} : x \in \mathsf{win}_v^*\}$ and the heap-over-heap structure accordingly.

We present the main process of our algorithm as Algorithm 1 below.

In the algorithm, we first add v into $H_v^{(v)}$ with key $s_v + \sum_{c \in \mathsf{Child}(v)} F[c]$ for initialization of v. Then, for every $c \in \mathsf{Child}(v)$, we do the following:

- First, add c to $S_v^{(v)}$ as part of the initialization of $S_v^{(v)}$, since c is s-maximal by default in T_c. We assume for the time being c is still s-maximal in T_v before we check s_i in the following process. Therefore, c first goes to $S_v^{(v)}$.

- We then add w_v to all keys in $W^{(v)}$, so that the vertices in $W^{(v)}$ with key greater than w_0 are vertices in $\mathsf{win}_c \setminus \mathsf{win}_v$ which need to be deleted (recall procedure (a)). We call a function Over-weight(c) to delete the corresponding vertices from all heaps; see appendix B.1 of the full paper [18] for details. After Over-weight(c), we meld $W^{(c)}$ into $W^{(v)}$.

- Next, we meld $\mathbb{H}^{(c)}$ into $\mathbb{H}^{(v)}$ as a part of the initialization of $\mathbb{H}^{(v)}$. By Lemma 1, since c is considered to be s-maximal in T_c by default, we add $(\sum_{c \in \mathsf{Child}(v)} F[c]) - F[i]$ to all vertices in $\mathbb{H}^{(c)}$ to update the $\mathsf{sum}(c, *)$ of every

```
for c ∈ Child(v) do                                                          1
    F[c] ← compute_F(c);                                                     2
a ← Σ_{c∈Child(v)} F[c];                                                     3
H_v^{(v)}, ℍ^{(v)}, S_v^{(v)}, W^{(v)} ← ∅;                                  4
insert(H_v^{(v)}, v, s_v + a);                                               5
for c ∈ Child(v) do                                                          6
    insert(S_v^{(v)}, c, s_c);                                               7
    add-all(W^{(v)}, w_v);                                                   8
    Over-weight(c);                                                          9
    meld(W^{(v)}, W^{(v)});                                                 10
    b ← a − F[c];                                                           11
    add-all(ℍ^{(c)}, b);                                                    12
    meld(ℍ^{(v)}, ℍ^{(c)});                                                 13
Update-s-maximal(v);                                                        14
insert(W^{(v)}, v, w_v);                                                    15
insert(ℍ^{(v)}, find-min(H_v^{(v)}), key-value(H_v^{(v)}, find-min(H_v^{(v)})));  16
return key-value(ℍ^{(v)}, find-min(ℍ^{(v)}));                               17
```

Algorithm 1: compute_F(v)

key to sum(v, *) before melding. As for updating the s_i part of the costs, we leave this issue to function Update-s-maximal(v) later.

After the above process for all $c \in$ Child(v), heap $S_v^{(v)}$ contains all Child(v) $\cap$ win$_v$. $W^{(v)}$ and all H's only contain vertices in win$_v$, which are the only vertices that we need for the heaps of T_v.

We now focus on the whole T_v. Here, we consider the vertices that lose its s-maximal identity after adding v as the new root (as mentioned in procedure (b)). We process these vertices in Update-s-maximal(v) with consecutive updating and melding (For more detailed description of functions Over-weight() and Update-s-maximal(), see the full version of this paper [18].)

After Update-s-maximal(v), we add v into $W^{(v)}$ with key w_v for further recursive calls and insert the top vertex of new $H_v^{(v)}$ into $\mathbb{H}^{(v)}$. Then, we have all heaps meeting their definitions in T_v (see appendix C of the full version of the paper [18]), and all we need is to extract minimum choice from $\mathbb{H}^{(v)}$ to obtain $F[v]$.

The reader can refer to Algorithm 2 and 3 by the end of Subsect. 3.2 for the pseudocode of Over-weight() and Update-s-maximal().

3.2   Analysis

Lemma 3. Over-weight() *takes* $O(n \log n)$ *time in total.*

Lemma 4. Update-s-maximal() *takes* $O(n \log n)$ *time in total.*

Lemmas above are proved in the full version of this paper (appendix B) [18].

Theorem 1. *The sum-of-max chain partition problem can be solved in* $O(n \log n)$ *time with our algorithm.*

Proof. By Lemma 3 and Lemma 4, the functions Over-weight() and Update-s-maximal() require $O(n \log n)$ time in total.

In the main process of each v, we insert v into $H_v^{(v)}$ with insert() in $O(\log n)$ time. Note that the acquisitions of $a = \sum_{c \in \mathsf{Child}(v)} F[c]$ in total take time

$$O((\sum_{v \in T} |\mathsf{Child}(v)|) \log n) = O(n \log n).$$

Befrore Update-s-maximal(v), for each $c \in \mathsf{Child}(v)$, we insert c into $S_v^{(v)}$ with insert() in $O(\log n)$ time. We then update $W^{(v)}$ in $O(\log n)$ time by add-all(). After Over-weight(c), we also update and meld $\mathbb{H}^{(c)}$ into $\mathbb{H}^{(v)}$ with add-all() and meld() in $O(\log n)$ time. Here, we conduct $O(1)$ times of $O(\log n)$ operations (excluding Over-weight(c)) for every $c \in \mathsf{Child}(v)$ as described above. These operations in total also take time

$$O((\sum_{v \in T} |\mathsf{Child}(v)|) \log n) = O(n \log n).$$

After Update-s-maximal(v), we insert v into $W^{(v)}$ and insert the top vertex of the new $H_v^{(v)}$ into $\mathbb{H}^{(v)}$ with insert() in $O(\log n)$ time. The extraction of $F[v]$ is also $O(\log n)$ with find-min() and key-value(). Here, we conduct $O(1)$ times of $O(\log n)$ operations (excluding Update-s-maximal(v)) for v, which is also $O(n \log n)$ in total.

Therefore, during the whole recursive process, the algorithm takes $O(n \log n)$ time. $\qquad\qquad\square$

---

while key-value($W^{(v)}$,find-max($W^{(v)}$)) $> w_0$ **do**	1
$\quad x \leftarrow$ find-max($W^{(v)}$);	2
$\quad$ delete-vertex($S_{\text{next}_c(x)}^{(v)}, x$);	3
$\quad$ **if** find-min($\mathbb{H}^{(c)}$) $= x$ **then**	4
$\quad\quad g \leftarrow$ key-value($\mathbb{H}^{(c)}, x$) $-$ key-value($H_{\text{next}_c(x)}^{(v)}, x$);	5
$\quad\quad$ delete-vertex($H_{\text{next}_c(x)}^{(v)}, x$);	6
$\quad\quad k \leftarrow$ find-min($H_{\text{next}_c(x)}^{(v)}$);	7
$\quad\quad$ insert($\mathbb{H}^{(c)}, k$, key-value($H_{\text{next}_c(x)}^{(v)}, k$) $+ g$);	8
$\quad\quad$ delete-vertex($\mathbb{H}^{(c)}, x$);	9
$\quad$ **else**	10
$\quad\quad$ delete-vertex($H_{\text{next}_c(x)}^{(v)}, x$);	11
$\quad$ delete-vertex($W^{(v)}, x$);	12

Algorithm 2: Function Over-weight(c).

---

while key-value$(S_v^{(v)}$,find-min$(S_v^{(v)})) \leq s_v$ **do**   1

   $x \leftarrow$ find-min$(S_v^{(v)})$;   2

   $z \leftarrow$ find-min$(H_x^{(v)})$;   3

   $g \leftarrow$ key-value$(\mathbb{H}^{(v)}, z) -$ key-value$(H_x^{(v)}, z) + s_v - s_x$;   4

   add-all$(H_x^{(v)}, g)$;   5

   meld$(H_v^{(v)}, H_x^{(v)})$;   6

   delete-vertex$(\mathbb{H}^{(v)}, z)$;   7

   meld$(S_v^{(v)}, S_x^{(v)})$;   8

   delete-vertex$(S_v^{(v)}, x)$;   9

Algorithm 3: Update-s-maximal(v)

Remark 3. The maintenance and acquisition of next$_*$() are $O(n\alpha(n)) < O(n \log n)$ in total. Therefore, function next$_*$() does not affect the $O(n \log n)$ time bound.

The acquisition of next$_*$() can be done in $O(\alpha(n)) < O(\log n)$ time with the find() operation of disjoint-set data structures. To achieve that, all we need to do is to set every vertex next$_c$(c) of $c \in$ Child(v) as v when $F[v]$ starts to be computed and merge the sets of vertex $x \in T_v$ and v when x loses its s-maximal identity in Update-s-maximal(v).

As a result, the find() operation can be seen as next$_c$() before Update-s-maximal(v) and next$_v$() after Update-s-maximal(v) by the definition of next. Thus we have next$_c$() and next$_v$() ready for every T_c of $c \in$ Child(v) in the main process of computing $F[v]$.

4　A Generalized Chain Partition Problem

This section discusses a generalization of our algorithm.

Suppose every vertex a of the tree is associated with a distinct rank r_a. The *dominant vertex* of a chain refers to the one in the chain with the largest rank.

A chain partition problem. Given a rooted tree T of n nodes, where vertex i is associated with a weight w_i, a cost s_i, and a rank r_i. Partition the tree into several disjoint chains $C_1, \ldots, C_k$, where k is arbitrary, so that the total weight of each chain is no more than a given threshold w_0 and the sum of the costs of the chains is minimized, where the cost of a chain is given by s_i – where i denotes the dominant vertex of the chain.

This problem contains the previous problem as a special case; by setting $r_i = s_i$ we obtain the original sum-of-max problem.

To tackle this problem, we replace the concept s-maximal with r-maximal, and this also leads to a change of the definition of $\mathsf{next}_v()$ and win^*:

Definition 2. *A vertex u in T_v is r-maximal if each vertex p in $T[v, u)$ admits that $r_p < r_u$. For $u \in T_v$ and $u \neq v$, let $\mathsf{next}_v(u)$ be the closest r-maximal proper ancestor of u in T_v. Let win_v^* denote the r-maximal vertices in win_v and win_v^- denote the vertices in $\mathsf{win}_v \setminus \mathsf{win}_v^*$.*

We now slightly modify our algorithm to solve the generalized problem:

One thing we have to change lies in the definition of heaps: $S_x^{(v)}$'s keys should be the r_i's of vertices instead of s_i's. Another change we make lies in the function Update-s-maximal() in appendix B.2 of the full version of the paper [18]. All we need to do is to change the criteria from s-maximal to r-maximal (by changing the condition $\leq s_v$ to $\leq r_v$).

The definitions above still fit Lemma 2, Eq. (2) and (3) and all relevant analysis in this paper. The heaps can be maintained and implemented to compute $F[v]$'s using the algorithm described above. As a summary, we conclude that

Theorem 2. *The chain partition problem above can be solved in $O(n \log n)$ time.*

5   Conclusions

In this paper, we present an $O(n \log n)$ algorithm for the sum-of-max chain partition of a tree under a knapsack constraint. The algorithm is based on non-trivial observations (such as Lemmas 1 and 2). The key ingredient of our algorithm is a delicately designed data structure that consists of two layers of heaps, where the elements of heaps in the second layer $\mathbb{H}^{(v)}$ come from the minimum elements of the heaps in the first layer $H^{(v)}$'s. We also generalize the algorithm to handle a more complex scenario (Sect. 4), and the time complexity remains $O(n \log n)$.

One question worth exploring in the future is how to find the sum-of-max path partition of a tree. Moreover, can we find the optimal chain partition of a tree when the cost of a chain C is defined as $\max_{i \in C} s_i - \min_{i \in C} s_i$?

References

1. Alstrup, S., Lauridsen, P.W., Sommerlund, P., Thorup, M.: Finding cores of limited length. In: Dehne, F., Rau-Chaplin, A., Sack, J.-R., Tamassia, R. (eds.) WADS 1997. LNCS, vol. 1272, pp. 45–54. Springer, Heidelberg (1997). https://doi.org/10.1007/3-540-63307-3_47
2. Bagheri, A., Keshavarz-Kohjerdi, F., Razzazi, M.: A linear-time approximation algorithm for the minimum-length geometric embedding of trees. Optimization 1–16 (2024)

3. Bagheri, A., Razzazi, M.: Minimum height path partitioning of trees. Sci. Iranica **17**(2) (2010)
4. Bender, M., Farach-Colton, M.: The level ancestor problem simplified. Theoret. Comput. Sci. **321**(1), 5–12 (2004)
5. Boesch, F., Chen, S., McHugh, J.: On covering the points of a graph with point disjoint paths. In: Graphs and Combinatorics: Proceedings of the Capital Conference on Graph Theory and Combinatorics at the George Washington University June 18–22, 1973, pp. 201–212. Springer (1974)
6. Brause, C., Krivoš-Belluš, R.: On a relation between k-path partition and k-path vertex cover. Discret. Appl. Math. **223**, 28–38 (2017)
7. Buchsbaum, A., Westbrook, J.: Maintaining hierarchical graph views. In: Proceedings of the Eleventh Annual ACM-SIAM Symposium on Discrete Algorithms, pp. 566–575 (2000)
8. Cai, Y., Zhang, X., Qian, J., Sun, Y.: An $O(n)$ Algorithm for ω-Path Partition of Edge-Weighted Forests. Ph.D. thesis (2003)
9. Chen, Y., Goebel, R., Lin, G., Su, B., Xu, Y., Zhang, A.: An improved approximation algorithm for the minimum 3-path partition problem. J. Comb. Optim. **38**(1), 150–164 (2019). https://doi.org/10.1007/s10878-018-00372-z
10. Goodman, S., Hedetniemi, S.: On the Hamiltonian completion problem. In: Graphs and Combinatorics: Proceedings of the Capital Conference on Graph Theory and Combinatorics at the George Washington University June 18–22, 1973, pp. 262–272. Springer (1974)
11. Harel, D., Tarjan, R.: Fast algorithms for finding nearest common ancestors. SIAM J. Comput. **13**(2), 338–355 (1984)
12. Hartmanis, J.: Computers and intractability: a guide to the theory of NP-completeness (Michael R. Garey and David S. Johnson). SIAM Rev. **24**(1), 90 (1982)
13. Jin, K., Zhang, D., Zhang, C.: Sum-of-max partition under a knapsack constraint. Comput. Electr. Eng. **105**, 108521 (2023)
14. Jin, Z., Li, X.: Vertex partitions of r-edge-colored graphs. Appl. Math.-A J. Chin. Univ. **23**(1), 120–126 (2008)
15. Klein, P.: Computing the edit-distance between unrooted ordered trees. In: European Symposium on Algorithms, pp. 91–102. Springer (1998)
16. Li, S., Yu, W., Liu, Z.: Improved approximation algorithms for the k-path partition problem. J. Global Optim. pp. 1–24 (2024)
17. Lu, C., Zhou, Q.: Path covering number and $l(2,1)$-labeling number of graphs. Discret. Appl. Math. **161**(13–14), 2062–2074 (2013)
18. Luo, R., Zhu, T., Jin, K.: Sum-of-max chain partition of a tree (2025). https://arxiv.org/abs/2503.11526
19. Manuel, P.: Revisiting path-type covering and partitioning problems. arXiv preprint arXiv:1807.10613 (2018)
20. Mehta, D., Sahni, S.: Handbook of Data Structures and Applications. Chapman and Hall/CRC, New York (2004)
21. Misra, J., Tarjan, R.: Optimal chain partitions of trees. Inf. Process. Lett. **4**(1), 24–26 (1975)
22. Monnot, J., Toulouse, S.: The path partition problem and related problems in bipartite graphs. Oper. Res. Lett. **35**(5), 677–684 (2007)
23. Pınar, A., Aykanat, C.: Fast optimal load balancing algorithms for 1D partitioning. J. Parallel Distrib. Comput. **64**(8), 974–996 (2004)
24. Slater, P.: Path coverings of the vertices of a tree. Discret. Math. **25**(1), 65–74 (1979)

25. Sleator, D., Tarjan, R.: A data structure for dynamic trees. In: Proceedings of the Thirteenth Annual ACM Symposium on Theory of Computing, pp. 114–122 (1981)
26. Vuillemin, J.: A data structure for manipulating priority queues. Commun. ACM **21**(4), 309–315 (1978)
27. Yan, J., Chang, G., Hedetniemi, S., Hedetniemi, S.: k-path partitions in trees. Discret. Appl. Math. **78**(1–3), 227–233 (1997)

Reconfiguring Multiple Connected Components with Size Multiset Constraints

Yu Nakahata[✉][iD]

Nara Institute of Science and Technology, 8916-5 Takayama-cho,
Ikoma, Nara 630-0192, Japan
yu.nakahata@is.naist.jp

Abstract. We propose a novel generalization of INDEPENDENT SET RECONFIGURATION (ISR): CONNECTED COMPONENTS RECONFIGURATION (CCR). In CCR, we are given a graph G, two vertex subsets A and B, and a multiset $\mathcal{M}$ of positive integers. The question is whether A and B are reconfigurable under a certain rule, while ensuring that each vertex subset induces connected components whose sizes match the multiset $\mathcal{M}$. ISR is a special case of CCR where $\mathcal{M}$ only contains 1. We also propose new reconfiguration rules: *component jumping* (CJ) and *component sliding* (CS), which regard *connected components as tokens*. Since CCR generalizes ISR, the problem is PSPACE-complete. In contrast, we show three positive results: First, CCR-CS and CCR-CJ are solvable in linear and quadratic time, respectively, when G is a path. Second, we show that CCR-CS is solvable in linear time for cographs. Third, when $\mathcal{M}$ contains only the same elements (i.e., all connected components have the same size), we show that CCR-CJ is solvable in linear time if G is chordal. The second and third results generalize known results for ISR and exhibit an interesting difference between the reconfiguration rules.

Keywords: Combinatorial reconfiguration · Graph algorithm · Connected component · Cograph · Chordal graph

1 Introduction

Imagine robots operating in a disaster area. Now that the work is done, the robots are set to be reassigned to their new positions. For safety reasons, we need to ensure that robots are not in the same area or adjacent to each other. How should we move the robots to change their positions from the current arrangement to the target arrangement? We assume the robots to be indistinguishable.

This situation can be modeled using *combinatorial reconfiguration* [10]. The aforementioned problem is INDEPENDENT SET RECONFIGURATION (ISR) [9,10, 13], the most well-studied problem in the field. In ISR, we are given a graph G and two vertex subsets, A and B. Imagine that a *token* is placed on each vertex in A. Can we move the tokens one by one to eventually lead to B while keeping

F. V. Fomin and M. Xiao (Eds.): COCOON 2025, LNCS 15984, pp. 68–80, 2026.
https://doi.org/10.1007/978-981-95-0218-9_6

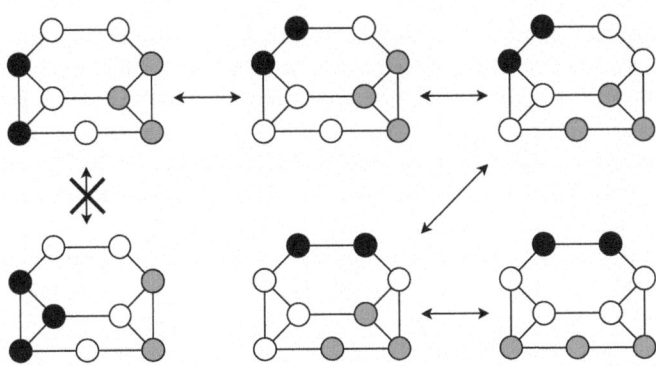

Fig. 1. A reconfiguration sequence in CCR-CJ. The start and target configurations A and B are shown in the upper left and lower right, respectively. $\mathcal{M}$ consists of one 2 and one 3. Black and gray vertices are in vertex subsets. Note that the upper left configuration and the lower left configuration are not adjacent under CJ because they exchange vertices between different connected components, which is allowed in TJ and TS. The reconfiguration sequence is also valid under CS.

the tokens not in the same vertex or adjacent vertices? In the above example, a robot corresponds to a token.

Let us generalize the above example to the following situation: Each robot has *size* k and occupies exactly k vertices. They can flexibly change their shapes like amoebas. However, the k vertices must be connected. Additionally, robots should not occupy the same vertex or adjacent vertices for safety reasons. How can we move the robots from the initial configuration to the target configuration? We assume that robots of the same size are indistinguishable from each other.

To model the above problem, we define a new reconfiguration problem, CON-NECTED COMPONENTS RECONFIGURATION (CCR). In CCR, we are given a graph G, two vertex subsets A and B, and a multiset $\mathcal{M}$ of positive integers. The question is whether A and B are reconfigurable under some rule while keeping each vertex subset induces connected components whose sizes are equal to $\mathcal{M}$. In the above example, A and B are the vertices occupied by robots in the initial and target configuration, respectively. $\mathcal{M}$ describes the sizes of robots. ISR is a special case of CCR where $\mathcal{M}$ only contains 1.

Reconfiguration rules define the valid movement of tokens in reconfiguration problems. The most well-studied reconfiguration rules are *token jumping* (TJ) [13] and *token sliding* (TS) [9]. TS allows us to move a token to an adjacent vertex where no other token exists. Additionally, TJ enables us to move a token to a non-adjacent vertex if no other token exists at that vertex. In the first example, moving a robot on the ground along an edge corresponds to TS, while TJ indicates that a robot can move to other distant vertices like a drone.

However, applying TJ and TS to CCR causes an issue. In TJ and TS, a token is a vertex. Therefore, tokens are allowed to move between different robots, as illustrated on the left of Fig. 1. We have assumed that the robots are flexible enough to change their shapes but not so flexible that they can self-repair. Therefore, we need rules that make robots move while maintaining their sizes. To this end, we propose two new reconfiguration rules: *component jumping* (CJ) and *component sliding* (CS), where *a token is a connected component*. When all the connected components have size 1, CCR-CJ and CCR-CS coincide with ISR-TJ and ISR-TS, respectively.[1] An example of a reconfiguration sequence under CJ and CS is shown in Fig. 1.

Since CCR-CJ and CCR-CS generalize ISR-TJ and ISR-TS, respectively, they are PSPACE-complete in general. In contrast, we show three positive results: First, CCR-CS and CCR-CJ are solvable in linear and quadratic time, respectively, when G is a path. The former is easy, but the algorithm for the latter is non-trivial and has a connection to the sorting problem with a buffer. Second, we show that CCR-CS is solvable in linear time for cographs. Third, when $\mathcal{M}$ contains only the same elements (i.e., all connected components have the same size), we show that CCR-CJ is solvable in linear time if G is chordal. The second and third results generalize known results for ISR and exhibit an interesting difference between the reconfiguration rules.

Due to the space limitation, proofs of the theorems and lemmas marked with an asterisk (*) are shown in the full version of this paper [16].

Related work. Combinatorial reconfiguration [10] is an emerging area in theoretical computer science. It is important from both practical and theoretical perspectives because many real-world systems are dynamic; additionally, it reveals the structure of the solution space. Combinatorial reconfiguration is studied for several problems, including independent sets [9,10,13], dominating sets [7,19], and cliques [12]. See the survey for more details [17].

Since ISR is the most well-studied problem in combinatorial reconfiguration, problems that generalize ISR have also been studied, such as REGULAR INDUCED SUBGRAPH RECONFIGURATION [6] and INDUCED ISOMORPHIC SUBGRAPH RECONFIGURATION [18]. Some studies generalize reconfiguration rules. As generalizations of TJ and TS, Hatano et al. [8] proposed d-Jump, where one token can be moved to within a distance d, Suga et al. [18] proposed k-TJ and k-TS, where at most k tokens can move simultaneously, and, Křišťan and Svoboda [14] proposed (k,d)-TJ, where at most k tokens can each move within a distance d. Note that, however, a token is a vertex in these papers. In this paper, we extend not only the problem (ISR to CCR) but also the token (a vertex to a connected component) and propose new reconfiguration rules CJ and CS.

2   Preliminaries

In this paper, we consider a simple undirected graph $G = (V, E)$. Throughout this paper, we denote the number of vertices by n. For $U \subseteq V$, we define $N[U] = U \cup \{v \in V : \exists u \in U, \{u, v\} \in E\}$, and $N(U) = N[U] \setminus U$. For $U, U' \subseteq V$, U *touches* U' if $U \cup U'$ is connected. Note that U touches U' if and only if U'

---

[1] Here, PROB-R indicates the problem PROB under the reconfiguration rule R.

touches U. For $u \in V$ and $U \subseteq V$, u *touches* U if $\{u\} \cup U$ is connected. Let $G[U]$ be an induced subgraph by U. When $G[U]$ has no edges, U is an *independent set*. We say that U is *connected* if $G[U]$ is connected. A subset of U that is connected and maximal is a *connected component* of U. The *size* of a connected component is the number of vertices in it. Let $\mathcal{C}(U)$ be the set of connected components of U; that is, $\mathcal{C}(U)$ is a partition of U and contains no empty set. Let $m(U)$ be a multiset of sizes of connected components of U. We refer to $m(U)$ as the *CC-multiset* of U in G.

For graphs G and H, if H is an induced subgraph of G, we say that G *contains* H as an induced subgraph. If G does not contain H as an induced subgraph, G is *H-free*. For instance, *cographs* are P_4-free graphs, where P_4 denotes a path with four vertices. A graph G is *chordal* if G is C_ℓ-free for every $\ell \geq 4$, where C_ℓ denotes a cycle with ℓ vertices. G is *even-hole-free* if G is C_ℓ-free for every even $\ell \geq 4$. By definition, chordal graphs are even-hole-free.

3 Our Problem and Reconfiguration Rules

In this section, we introduce CONNECTED COMPONENTS RECONFIGURATION (CCR) and propose two new reconfiguration rules: *component jumping* (CJ) and *component sliding* (CS).

Definition 1 (CCR). CONNECTED COMPONENTS RECONFIGURATION (*CCR*) *is defined as follows:*

> **Input** *A graph G, vertex subsets $A, B \subseteq V$, a multiset $\mathcal{M}$ consisting of positive integers, and a reconfiguration rule R.*
> **Output** *Is there a sequence of vertex subsets from A to B where (1) every vertex subset has a CC-multiset equal to $\mathcal{M}$ and (2) every two consecutive vertex subsets are adjacent in R?*

If the answer is YES, we say that A *and* B *are reconfigurable* and call the sequence satisfying (1) and (2) a *reconfiguration sequence from A to B*. The *length* of a reconfiguration sequence $\langle U_0 = A, U_1, \ldots, U_\ell = B \rangle$ is ℓ.

Using the multiset $\mathcal{M}$, we can express the solution spaces of several problems: If $\mathcal{M}$ only contains 1, the solution space is the independent sets. If $\mathcal{M}$ only contains 2, the solution space is the induced matchings. In this way, CCR generalizes the existing reconfiguration problems.

As for the reconfiguration rule R, TJ and TS are the most well-studied ones in the literature [9,10,13]. A reconfiguration rule defines the adjacency relation between solutions. The adjacency of $U, U' \subseteq V$ under a reconfiguration rule R is written as $U \xleftrightarrow{R} U'$. Then, TJ and TS are defined as follows.

TJ $U \xleftrightarrow{TJ} U' \Leftrightarrow |U \setminus U'| = |U' \setminus U| = 1$
TS $U \xleftrightarrow{TS} U' \Leftrightarrow |U \setminus U'| = |U' \setminus U| = 1$ and $\{u, u'\} \in E(G)$, where $U \setminus U' = \{u\}, U' \setminus U = \{u'\}$

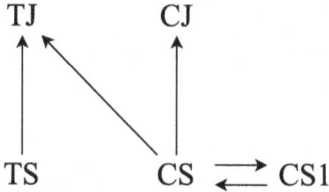

Fig. 2. Relation between the reconfiguration rules. A solid arrow R $\longrightarrow$ R$'$ means that, if the answer of CCR-R is YES, the answer of CCR-R$'$ is also YES.

TJ and TS are often explained using the notion of *token*. Imagine that a token is placed on each vertex in A. In each step, we can move one token to an unoccupied vertex; this is TJ, and in TS, we must move a token to its neighbor. The reconfiguration problem asks: Can we move tokens from A to B under a reconfiguration rule while keeping the token configuration *valid* (e.g., in ISR, tokens are not adjacent)?

In addition to TJ and TS, we define two new reconfiguration rules: *component jumping* (CJ) and *component sliding* (CS). We also define a special case of CS, CS1, which only allows component sliding by one vertex. In TJ and TS, a token is a vertex. In contrast, in CJ and CS, a *"token" is a connected component*. In the following, if $|\mathcal{C}(U) \setminus \mathcal{C}(U')| = |\mathcal{C}(U') \setminus \mathcal{C}(U)| = 1$, we denote the only connected components in $\mathcal{C}(U) \setminus \mathcal{C}(U')$ and $\mathcal{C}(U') \setminus \mathcal{C}(U)$ by C and C', respectively.

Definition 2 (CJ, CS, CS1). *For $U, U' \subseteq V(G)$ with $m(U) = m(U')$, the reconfiguration rules CJ, CS, and CS1 are defined as follows:*

CJ $U \xleftrightarrow{CJ} U' \Leftrightarrow |\mathcal{C}(U) \setminus \mathcal{C}(U')| = |\mathcal{C}(U') \setminus \mathcal{C}(U)| = 1$

CS $U \xleftrightarrow{CS} U' \Leftrightarrow U \xleftrightarrow{CJ} U'$ *and $C \cup C'$ is connected*

CS1 $U \xleftrightarrow{CS1} U' \Leftrightarrow U \xleftrightarrow{CS} U'$ *and $|C \setminus C'| = |C' \setminus C| = 1$*

CJ and CS are defined so that they generalize TS and TJ in ISR, respectively. Here, a token is individual but can change its body like an amoeba. Each token has its *size*, and a token with size k occupies exactly k vertices. In addition, vertices occupied by a token must be connected. In each step, CJ allows us to move one token (connected component) with size k to any vertex subset with k vertices and does not touch the other connected components. In CS, in addition, the union of vertices occupied by the connected component before and after the movement must be connected.

Let us consider the relation between reconfiguration rules in CCR. We write R $\longrightarrow$ R$'$ when: if the answer of CCR-R is YES, the answer of CCR-R$'$ is also YES. The following theorem shows the relation between reconfiguration rules. The summary is shown in Fig. 2.

Theorem 1 (*). *For a graph G and two vertex subsets $U, U' \subseteq V$ with $m(U) = m(U')$, the following holds:*

(1) CS1 $\longrightarrow$ CS and CS $\longrightarrow$ CS1,
(2) TS $\longrightarrow$ TJ, CS $\longrightarrow$ CJ, and CS $\longrightarrow$ TJ.

4   CCR-CJ and CCR-CS for Path Graphs

In this section, we show that for path graphs, CCR-CS and CCR-CJ are solvable in linear and quadratic time, respectively. Let $v_1, \ldots, v_n$ be the vertices of a path graph G aligned from left to right. For $U \subseteq V$, let h_U be the sequence of sizes of connected components along G from left to right. Using h_U, we can determine the reconfigurability under CS.

Lemma 1. *A and B are reconfigurable under* CS *if and only if* $h_A = h_B$.

Proof. $\Leftarrow$) If $h_A = h_B$, both A and B can be reconfigurable into the *left-most subset* L, where the vertices are kept to the left as much as possible. Since the reconfigurability is symmetric and transitive, it is possible to reconfigure from A to B by reconfiguring from A to L and from L to B.

$\Rightarrow$) We show the contraposition. When $h_A \neq h_B$, from $m(A) = m(B)$, there exists a size pair x and y ($x \neq y$) whose positions are reversed at h_A and h_B. Under CS, these pairs cannot be swapped. This means that A are B not reconfigurable. □

In the following, we show that not only the reconfigurability but also a reconfiguration sequence (if exists) can be computed in polynomial time. To output a reconfiguration sequence efficiently, we use a *compressed reconfiguration sequence*. In the sequence, we identify the position of a connected component (path) by its leftmost vertex. By outputting a pair of left-most positions of the connected component to be moved before and after each step, we can output a reconfiguration sequence efficiently.

Theorem 2. *If G is a path graph,* CCR-CS *is solvable in time $\mathcal{O}(n)$. If the answer is YES, we can find a compressed reconfiguration sequence in length $\mathcal{O}(n^2)$ in time $\mathcal{O}(n^2)$.*

Proof. From Lemma 1, output YES if $h_A = h_B$, NO otherwise. Since h_A and h_B can be computed in linear time, the algorithm runs in time $\mathcal{O}(n)$. If the answer is YES, by reconfiguring A to L and then L to B, we obtain a reconfiguration sequence. Since each connected component moves $\mathcal{O}(n)$ times, the length of the obtained reconfiguration sequence is $\mathcal{O}(n^2)$. □

The length bound of a reconfiguration sequence in Theorem 2 is tight because there are instances for which any reconfiguration sequence requires length $\Omega(n^2)$ in ISR-TS [5].

Next, we consider CJ. In contrast to CS, we can swap the positions of connected components when there is enough space. For $U \subseteq V$ having k connected components, we define the *buffer* of U in G as $b(U, k) = n - |U| - k$. Here, $|U| + k$ is the number of vertices occupied by k connected components of U when they are kept left. The buffer means the number of the right vertices we can freely use when the vertices in U are kept left.

Our algorithm for CCR-CJ imitates the *bubble sort*. However, our sequences may have duplicate elements. Thus, we need some notations before stating the algorithm. Let h'_A be a sequence of x_i's where x is an element in h_A, and i is the number of occurrences of x in the prefix of h_A until itself. For instance, if $h_A = \langle 2, 2, 3, 3 \rangle$, then $h'_A = \langle 2_1, 2_2, 3_1, 3_2 \rangle$. We refer to x as an *element* and x_i as a *subscripted element*. In addition, we define σ_A as a function from subscripted elements to integers. Here, $\sigma_A(x_i)$ is the *rank* of x_i in A, that is, its index in h_A. In the above example, $\sigma_A(3_1) = 3$ holds. We similarly define h'_B and σ_B.

The algorithm scans h'_A from left to right, and if we find adjacent subscripted elements x_i and y_j such that their ranks are reversed in h'_B, then the algorithm tries to *swap* the positions of x_i and y_j. However, since we can jump only one connected component in each step, we need a buffer with at least $\min\{x, y\}$. Conversely, if this amount of buffer exists, we can swap x_i and y_j as follows: Without loss of generality, we assume that $x < y$. (Note that $x \neq y$ because, if so, $i < j$ holds and $\sigma_A(x_i) < \sigma_A(x_j)$ and $\sigma_B(x_i) < \sigma_B(x_j)$.) First, we jump the connected component of size x to the right buffer. Second, we slide (actually jump) the connected component of size y to the left. Finally, we jump back the connected component of size x from the right buffer to the left vacant space occupied by the connected component of size y. We call this procedure the *swap* of x and y.

To establish the necessary and sufficient conditions for reconfigurability, we define an *inversion* between h'_A and h'_B as a pair of subscripted elements (x_i, y_j) such that the rank of x_i is smaller than y_j in h'_A, but the reverse holds for h'_B. We denote the set of inversions by $\mathrm{inv}(h'_A, h'_B)$, that is, $\mathrm{inv}(h'_A, h'_B) = \{(x_i, y_j) : \sigma_A(x_i) < \sigma_A(y_i), \sigma_B(x_i) > \sigma_B(y_i)\}$. The next lemma provides a necessary and sufficient condition for reconfigurability under CJ with respect to inversions.

Lemma 2. *A and B are reconfigurable under* CJ *if and only if*

$$\max_{(x_i, y_j) \in \mathrm{inv}(h'_A, h'_B)} \min\{x, y\} \leq b(A, k). \tag{1}$$

Proof. $\Rightarrow$) For each inversion (x_i, y_j), to swap their positions, we need a buffer with $\min\{x, y\}$. Thus, the only-if direction holds.

$\Leftarrow$) If Inequality (1) holds, we can *sort* h'_A to h'_B using the bubble sort. While there is an inversion, we repeat the following procedure: We scan h'_A from left to right, and if we find adjacent subscripted elements (x_i, y_j) such that they are an inversion, we swap them by using the right buffer. The correctness follows from the proof of bubble sort. Once h'_A equals h'_B, by Lemma 1, A and B are reconfigurable under CS and thus under CJ.   □

Theorem 3. *If G is a path graph,* CCR-CJ *is solvable in time $\mathcal{O}(n^2)$. If the answer is YES, there is a compressed reconfiguration sequence of length $\mathcal{O}(n^2)$ and we can output the sequence in time $\mathcal{O}(n^2)$.*

Proof. By Lemma 2, the answer is YES if Inequality (1) holds and NO otherwise. The algorithm first computes h'_A and h'_B in time $\mathcal{O}(n)$. Second, we compute inversions in time $\mathcal{O}(n^2)$. Then, Inequality (1) can be checked in the same time bound. To obtain a reconfiguration sequence, we first make A and B leftmost in $\mathcal{O}(n)$ steps. Then, we execute the bubble sort from h'_A to h'_B as shown in the proof of Lemma 2. Each swap consists of three steps; hence, the total number of steps is $\mathcal{O}(n^2)$. We obtain a reconfiguration sequence from A to B by concatenating the above sequences. We achieve the claimed time bound by using the compressed reconfiguration sequence described right before Theorem 2.   □

5   CCR-CS for Cographs

In this section, we show a linear-time algorithm for CCR-CS when G is a cograph. If the answer is YES, we can also find a *shortest* reconfiguration sequence of length $\mathcal{O}(|V|)$ under CS or CS1. Note that reconfigurability is equivalent for CS and CS1, but the lengths of shortest reconfiguration sequences may differ. Our algorithm is based on that for ISR-TS on a cograph by Kamiński et al. [13]. Note that ISR-TS is equivalent to CCR-CS when $\mathcal{M}$ only contains 1.

We prepare some notations. The *complement graph* $\overline{G}$ of a graph $G = (V, E)$ is defined as $\overline{G} = (V, \{\{u, v\} : u, v \in V, \{u, v\} \notin E\})$. A graph G is a cograph if and only if, for every induced subgraph F of G with at least two vertices, either F or $\overline{F}$ is disconnected [3]. A *co-component* of a graph G is the subgraph induced by the vertex set of a connected component in $\overline{G}$. Note that, if G is a connected cograph, $V(G)$ is partitioned into vertex sets each of which induces a co-component of G.

The algorithm solves the problem in divide-and-conquer manner using a *cotree* [4], a decomposition tree of a cograph with respect to taking components and co-components. As one of the base cases, we use the next two lemmas. In the following, for $X, Y \subseteq V$ and a reconfiguration rule R, we use $d_R(X, Y)$ as the shortest distance (length of a shortest reconfiguration sequence) between X and Y under R.

Lemma 3. *Let G be a connected cograph and X, Y be connected vertex subsets with $|X| = |Y|$. Then, the following holds.*

$$d_{CS}(X, Y) = \begin{cases} 0 & (X = Y) \\ 1 & (X \neq Y \text{ and } X \text{ touches } Y) \\ 2 & (X \text{ does not touch } Y) \end{cases} \tag{2}$$

Proof. The first case is obvious. If $X \neq Y$ and X touches Y, then $d_{CS}(X, Y) \geq 1$. Since $X \cup Y$ is connected, $X \xleftrightarrow{CS} Y$ and hence $d_{CS}(X, Y) = 1$. Otherwise, X does not touch Y, so $d_{CS}(X, Y) \geq 2$. As G is P_4-free, there exists a vertex z adjacent to both X and Y. Let Z be a connected vertex subset of size $|X|$ containing z. Then $X \xleftrightarrow{CS} Z$ and $Z \xleftrightarrow{CS} Y$, so $d_{CS}(X, Y) = 2$.   □

Lemma 4 (∗). *Let G be a connected cograph and X, Y be connected vertex subsets with $|X| = |Y|$. If X touches Y, $d_{CS1}(X, Y) = |X \setminus Y|$. Otherwise, $d_{CS1}(X, Y) = |X \setminus Y| + 1$.*

Theorem 4. CCR-*CS* *is solvable in time* $\mathcal{O}(|V| + |E|)$ *if the input graph* $G = (V, E)$ *is a cograph. If the answer is YES, we can compute a shortest reconfiguration sequence under* *CS* *or* *CS1* *in time* $\mathcal{O}(|V| + |E|)$ *and its length is* $\mathcal{O}(|V|)$.

Proof. Our algorithm is based on Kamiński et al. [13]. The differences are:

- The first pruning condition $|A| \neq |B|$ is replaced by $m(A) \neq m(B)$.
- The base case $|A| = |B| = 1$ is replaced by $|\mathcal{C}(A)| = |\mathcal{C}(B)| = 1$. In addition, the procedure for this base case is substituted by Lemmas 3 or 4.

Intuitively, these differences correspond to the notion of a token: a token changed from a vertex to a connected component.

If $m(A) \neq m(B)$, the answer is NO. If $|V(G)| = 1$, the problem is trivial. We assume that G has at least two vertices in the following. Now, either G or $\overline{G}$ is disconnected from the above characterization of cographs. If G is disconnected, let $C_1, \ldots, C_k$ be the connected components of G. We can solve the problem recursively for the connected components $C_1, \ldots, C_k$ with respective vertex subsets $(A \cap C_1, B \cap C_1), \ldots, (A \cap C_k, B \cap C_k)$. If one of the outputs is NO, then the answer is NO. Otherwise, the answer is YESand we merge subsequences to obtain a shortest reconfiguration sequence from A to B.

If $\overline{G}$ is disconnected, $G = (V, E)$ is connected. If in addition $|\mathcal{C}(A)| = |\mathcal{C}(B)| = 1$, by Lemmas 3 and 4, the answer is YES, and there exists a shortest reconfiguration sequence of $\mathcal{O}(|V|)$ from A to B. Otherwise, if A and B are contained in the same co-component of G, we solve the problem for A and B recursively on that co-component. Otherwise, the answer is NO because we cannot move any vertex in A to the other co-component. Since there is an edge for every two vertices of different co-components, if we move a vertex outside the co-component, some connected components in A will be merged. The correctness of the above procedure follows from Lemmas 3 and 4 and the characterizations of cographs.

Given a cograph G, it is known that a *cotree*, a decomposition tree with respect to taking components and co-components, can be computed in time $\mathcal{O}(|V| + |E|)$ [4]. Using this, the above algorithm can be implemented in time $\mathcal{O}(|V| + |E|)$. The total length of the reconfiguration sequence is $\mathcal{O}(|V|)$ with respect to the input graph because we divide the problem into disjoint ones. □

6   CCR-CJ for Connected Components of Equal Size

In this section, we consider CCR-CJ where A and B contain only connected components of equal size. ISR-TJ is its special case where the size of each connected component is 1. Our main tool is the *CC-Piran graph*, a generalization of the Piran graph used by Kamiński et al. [13]. They have shown that, given a graph G and two independent sets A and B, ISR-TJ is solvable in linear time if the Piran graph defined by G, A, B is even-hole-free. Analogously to their results, we show that given a graph G and two vertex subsets A, B whose all connected components have the same size, if the CC-Piran graph defined by G, A, B is even-hole-free, CCR-CJ is solvable in linear time.

For two independent sets A and B, the Piran graph $\Pi(A, B)$ of A and B is the subgraph of G induced by the vertex set $(A \setminus B) \cup (B \setminus A)$. We extend this definition to the CC-Piran graph.

Definition 3 (CC-Piran graph). *For a graph G and $A, B \subseteq V(G)$, the CC-Piran graph $\Pi^{cc}(A, B) = (V^{cc}, E^{cc})$ is defined as follows:*

- $V^{cc} = (\mathcal{C}(A) \setminus \mathcal{C}(B)) \cup (\mathcal{C}(B) \setminus \mathcal{C}(A))$
- $E^{cc} = \{\{C_A, C_B\} : C_A \in \mathcal{C}(A) \setminus \mathcal{C}(B), C_B \in \mathcal{C}(B) \setminus \mathcal{C}(A), C_A \text{ touches } C_B\}$

Note that the CC-Piran graph is equivalent to the Piran graph if A and B are independent sets. In the following, to avoid confusion, we call a vertex in the CC-Piran graph a *component*.

Theorem 5. *Let G be a graph and A and B be vertex subsets that only contain connected components of the same size. If the CC-Piran graph $\Pi^{cc}(A, B)$ is even-hole-free, then A and B are reconfigurable under CJ. Moreover, there exists an algorithm running in time $\mathcal{O}(|V| + |E|)$ (if the CC-Piran graph is even-hole-free) that finds a shortest reconfiguration sequence from A to B.*

Proof. We briefly review the idea of Kamiński et al. [13]. The Piran graph $\Pi(A, B)$ is bipartite, and as such it does not contain odd cycles. If, in addition, $\Pi(A, B)$ is also even-hole-free, the graph must be a forest. Since $|A \setminus B| = |B \setminus A|$, by analyzing the number of edges, there exists a vertex in $B \setminus A$ with at most one neighbor in $A \setminus B$. Using this property, a simple greedy algorithm works: Find a vertex v from $B \setminus A$ with at most one neighbor in $A \setminus B$. If v has a neighbor in $A \setminus B$, say w, jump w to v. Otherwise, jump an arbitrary token w from $A \setminus B$ to v. Replace A and B with $A \setminus \{w\}$ and $B \setminus \{v\}$, respectively. While $|A| \geq 1$, repeat the procedure. Since any reconfiguration sequence under TJ requires $|A \setminus B|$ steps, the obtained sequence is shortest.

We show that the same algorithm works for our case. Since $\Pi^{cc}(A, B)$ is bipartite and even-hole-free, the graph must be a forest. Since $|\mathcal{C}(A) \setminus \mathcal{C}(B)| = |\mathcal{C}(B) \setminus \mathcal{C}(A)|$, there exists a component in $\mathcal{C}(B) \setminus \mathcal{C}(A)$ with at most one neighbor in $\mathcal{C}(A) \setminus \mathcal{C}(B)$. Thus, the above algorithm also works for $\Pi^{cc}(A, B)$. The correctness is preserved because each token has the same size and any token of $\mathcal{C}(A)$ can be matched with any other of $\mathcal{C}(B)$. Note that a "token" is a component in our case while it is a vertex in the previous case.

As for the time complexity, we show that given a graph G and vertex subsets A and B, the CC-Piran graph $\Pi^{cc}(A, B)$ can be constructed in linear time. Assuming that the graph is given with adjacency lists, we construct $\Pi^{cc}(A, B)$ in two scans of the lists. First, we compute the connected components of A and B. Second, we compute the adjacency relation between components. □

Note that Theorem 5 does not say anything about the input graph class. The result of Kamiński et al. [13] indicates that ISR-TJ is in P if G is even-hole-free because even-hole-freeness is closed under taking induced subgraphs: when G is even-hole-free, every induced subgraph of G (and thus the Piran graph) is

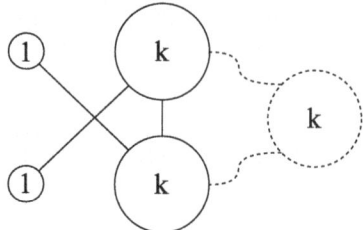

Fig. 3. The CC-Piran graph such that the greedy algorithm fails. The solid upper and lower circles are components in $\mathcal{C}(A) \setminus \mathcal{C}(B)$ and $\mathcal{C}(B) \setminus \mathcal{C}(A)$, respectively. The dotted circle indicates a connected component with k vertices in G outside the CC-Piran graph. In this instance, we can reconfigure A into B by first moving the token of size k to the right space, next moving the token of size 1, and then moving the token of size k to the lower right space.

even-hole-free. In contrast, the CC-Piran graph allows us to take a *minor* of G, which leads to the contraction of edges. Although even-hole-freeness is closed under taking induced subgraphs, it is not closed under taking minors because contracting an edge in an odd hole may create an even hole.

In addition, as demonstrated in the proof of Theorem 5, the assumption that all connected components have the same size is important for the greedy algorithm works. If there exist connected components with different sizes, there is a case we need a "detour", as illustrated in Fig. 3.

Then, our next question is: For what graph G, the CC-Piran graph is even-hole-free? The next lemma answers the question.

Lemma 5 (*). *If G is chordal, the CC-Piran graph is even-hole-free.*

We obtain the following corollary. Claimed time complexity is achieved by outputting a reconfiguration sequence using TJ on the CC-Piran graph.

Corollary 1. *If G is chordal and $A, B \subseteq V$ consists of only connected components of the same size, CCR-CJ is solvable in time $\mathcal{O}(|V| + |E|)$. In addition, we can find a shortest reconfiguration sequence in the same time bound.*

Corollary 1 exhibits an interesting contrast between CJ and CS: When $\mathcal{M}$ contains only 2, which corresponds to INDUCED MATCHING RECONFIGURATION (IMR), is known to be PSPACE-complete under TJ for chordal graphs [6]. Corollary 1 shows that, however, IMR is solvable in linear time for chordal graphs under CJ. This indicates that what we regard as a token largely affects computational complexity.

7   Concluding Remarks

Future work includes further analysis of the computational complexity of CCR. Some open problems are listed below:

- Are CCR-CJ and CCR-CS tractable for trees?
- Is CCR-CJ on cographs in P? ISR-TJ is in P for cographs [1,2].
- Is CCR-CJ tractable for even-hole-free graphs with general token sizes?
- Are CCR-CJ and CCR-CS in XP (i.e., solvable in time $n^{f(k)}$ for some computable function f) when parameterized by the number k of connected components? Since ISR-TJ and ISR-TS are W[1]-hard when parameterized by k [11,15], there unlikely exist FPT time (i.e., $f(k)n^{\mathcal{O}(1)}$) algorithms.

Acknowledgment. We thank anonymous reviewers for valuable comments. This work is partially supported by JSPS KAKENHI JP22K17851, JP23K24806, and JP24K02931.

References

1. Bonamy, M., Bousquet, N.: Reconfiguring independent sets in cographs. arXiv preprint arXiv:1406.1433 (2014)
2. Bonsma, P.: Independent set reconfiguration in cographs and their generalizations. J. Graph Theor. **83**(2), 164–195 (2016)
3. Corneil, D.G., Lerchs, H., Burlingham, L.S.: Complement reducible graphs. Discret. Appl. Math. **3**(3), 163–174 (1981)
4. Corneil, D.G., Perl, Y., Stewart, L.K.: A linear recognition algorithm for cographs. SIAM J. Comput. **14**(4), 926–934 (1985)
5. Demaine, E.D., et al.: Linear-time algorithm for sliding tokens on trees. Theoret. Comput. Sci. **600**, 132–142 (2015)
6. Eto, H., Ito, T., Kobayashi, Y., Otachi, Y., Wasa, K.: Reconfiguration of regular induced subgraphs. In: International Conference and Workshops on Algorithms and Computation, pp. 35–46. Springer (2022)
7. Haddadan, A., et al.: The complexity of dominating set reconfiguration. Theoret. Comput. Sci. **651**, 37–49 (2016)
8. Hatano, H., Kitamura, N., Izumi, T., Ito, T., Masuzawa, T.: Independent set reconfiguration under bounded-hop token jumping. In: International Conference and Workshops on Algorithms and Computation, pp. 215–228. Springer (2025)
9. Hearn, R.A., Demaine, E.D.: PSPACE-completeness of sliding-block puzzles and other problems through the nondeterministic constraint logic model of computation. Theoret. Comput. Sci. **343**(1–2), 72–96 (2005)
10. Ito, T., et al.: On the complexity of reconfiguration problems. Theoret. Comput. Sci. **412**(12–14), 1054–1065 (2011)
11. Ito, T., Kamiński, M., Ono, H., Suzuki, A., Uehara, R., Yamanaka, K.: On the parameterized complexity for token jumping on graphs. In: International Conference on Theory and Applications of Models of Computation, pp. 341–351. Springer (2014)
12. Ito, T., Ono, H., Otachi, Y.: Reconfiguration of cliques in a graph. Discret. Appl. Math. **333**, 43–58 (2023)
13. Kamiński, M., Medvedev, P., Milanič, M.: Complexity of independent set reconfigurability problems. Theoret. Comput. Sci. **439**, 9–15 (2012)
14. Křišťan, J.M., Svoboda, J.: Reconfiguration using generalized token jumping. In: International Conference and Workshops on Algorithms and Computation, pp. 244–265. Springer (2025)

15. Lokshtanov, D., Mouawad, A.E., Panolan, F., Ramanujan, M., Saurabh, S.: Reconfiguration on sparse graphs. J. Comput. Syst. Sci. **95**, 122–131 (2018)
16. Nakahata, Y.: Reconfiguring multiple connected components with size multiset constraints (2025). https://arxiv.org/abs/2505.07268
17. Nishimura, N.: Introduction Reconfiguration Algorithms **11**(4), 52 (2018)
18. Suga, T., Suzuki, A., Tamura, Y., Zhou, X.: Changing induced subgraph isomorphisms under extended reconfiguration rules. In: International Conference and Workshops on Algorithms and Computation, pp. 346–360. Springer (2025)
19. Suzuki, A., Mouawad, A.E., Nishimura, N.: Reconfiguration of dominating sets. J. Comb. Optim. **32**, 1182–1195 (2016)

Fault Diagnosability Evaluation of BCCC Data Center Networks

Baohua Niu[1] , Yan Wang[1] , Baolei Cheng[1] , Hai Liu[2] , Bai Yin[1],
Jianxi Fan[1(✉)] , and Xinyang Cai[3]

[1] School of Computer Science and Technology, Soochow University, Suzhou, China
jxfan@suda.edu.cn
[2] Department of Computer Science, The Hang Seng University of Hong Kong,
Hong Kong, China
[3] School of Mathematics and Physics, Xi'an Jiaotong-Liverpool University,
Suzhou, China

Abstract. The rapid development of cloud computing, artificial intelligence and Internet of Things has not only led to high-speed growth of data centers, but also put forward a higher level of demand for data centers. Data center networks (DCNs), as crucial components of data centers, play significant roles in the overall operation and function realization of data centers. As a server-centric DCN, BCube connected crossbars (BCCC) offer excellent network performance in terms of great scalability, low communication latency and high robustness to component failures. As the scale of BCCC increases, so does the likelihood of server failures, it is critical to identify and replace faulty servers promptly for network reliability. In this work, we first investigate the intermittent fault diagnosability and local diagnosability of $BCCC(n, k)$. On this basis, we also determine the traditional diagnosability of $BCCC(n, k)$ and the fact that $BCCC(n, k)$ has strong local diagnosability under the PMC model and MM^* model. Subsequently, we further obtain that the faulty $BCCC(n, k)$ with $\min\{n-1, k-1\}$ missing edges can still maintain strong local diagnosability under the PMC model. In addition, we present a corresponding local diagnosis algorithm to identify the vertex state and evaluate its performance by simulation experiments, which show that the algorithm keeps good diagnosis correctness even though the number of faulty vertices in $BCCC(n, k)$ reaches 30%.

Keywords: intermittent fault diagnosis · local diagnosability · PMC model · MM^* model · BCCC data center networks

1 Introduction

Based on the rapid development of scientific information technology, the big data ecosystem is increasingly improved. Cloud computing, as an emerging computing model, has been widely used in various fields. As the core of cloud computing, data centers undertake significant tasks of computing, storing and transmitting

© The Author(s), under exclusive license to Springer Nature Singapore Pte Ltd. 2026
F. V. Fomin and M. Xiao (Eds.): COCOON 2025, LNCS 15984, pp. 81–94, 2026.
https://doi.org/10.1007/978-981-95-0218-9_7

data with their advantages of low cost, communication scalability and high fault tolerance, which provide infrastructure services for GFS [1] and MapReduce [2]. Data center networks (DCNs), as a core component of data centers, have been the focus of academic and industrial attention as they directly or indirectly affect the quality of service for mobile communications and computing.

A number of DCNs have been proposed based on different application requirements and transmission mechanisms, which are mainly classified into two categories: switch-centric and server-centric. In the switch-centric DCNs, switches are primarily responsible for interconnect intelligence, while servers are only responsible for sending and receiving packets. In the server-centric DCNs, servers perform not only as end hosts for sending and receiving packets but also as relay vertices, while all switches are used only for forwarding packets. Recently, Li et al. [3] presented a new server-centric DCN topology named BCube connected crossbars (BCCC), by comparing it with other popular server-centric DCNs, BCCC provides significant advantages with respect to expandability, server port utilization, transmission latency, diameter and robustness against component failure. Subsequently, Li et al. [4] devised efficient algorithms to construct completely independent spanning trees in BCCC.

Technological development and massive data lead to the expansion of DCNs and a sharp increase in the number of servers. After long time operation, the server failure is inevitable, so identifying and replacing the faulty server is critical to improve the reliability of DCNs. The server diagnosis for DCNs follows system-level fault diagnosis, which refers to identifying all faulty processors in the system. The diagnosability of a system is the maximum number of faulty processors that can be correctly identified. There are some system-level fault diagnosis models previously suggested. Preparata et al. [5] introduced a test-based diagnosis model, denoted as the PMC model, which has attracted much attention from researchers. The other classical diagnosis model based on comparison diagnosis called the MM model was first presented by Maeng and Malek [6]. Subsequently, Sengupta and Dahbura [7] made further modifications to the MM model yielding the MM* model.

Faulty processors can be categorized into two types: permanent faults and intermittent faults. Intermittent faults are an important type of faults in multiprocessor systems with randomness, intermittency and strong concealment, which makes diagnosis more difficult. The research on system diagnosability from the perspective of intermittent faults is relatively scarce. Mallela et al. [8] studied the intermittent fault diagnosis capability of diagnosable systems. Subsequently, the intermittent fault diagnosability of crisp three-cycle networks [9], k-regular k-connected graphs [10], split-star networks [11] and product networks [12] under the PMC model were investigated. Lin et al. [13] determined the intermittent fault diagnosability of r-regular network under the comparison model.

In the classical measure of system-level diagnosis, a system is t-diagnosable if all faulty processors can be accurately identified and the number of faulty processors is at most t. However, when the number of faulty processors exceeds t, all faulty processors in the t-diagnosable system may also be correctly identified. Hence, Hsu et al. [14] suggested a measure of local diagnosability under the PMC model to determine the diagnosability of the system. Then, Chiang et al. [15] gave a sufficient condition for determining the vertex diagnosability of a given processor under the comparison model. Lin et al. established some conditions so that a t-regular system can be conditionally $(2t-1)$-diagnosable, provided every fault-free processor has at least one fault-free neighbor under the PMC model [16] and the comparison model [17]. Subsequently, the local diagnosability and strong local diagnosability of a class of Cayley graphs [18] and (n, k)-stars [19] have been investigated.

In this work, we focus on analyzing the intermittent fault and local diagnosability of BCCC(n, k) and prove that it achieves strong local diagnosability under both the PMC and MM* models. We further show that even with up to $\min\{n-1, k-1\}$ missing edges, BCCC(n, k) retains this property under the PMC model and provides a corresponding local diagnosis algorithm. Simulation experiments are conducted to evaluate the performance of the algorithm.

2 Preliminaries

2.1 Terminologies and Notations

Let $G = (V, E)$ be a simple undirected graph. For any vertex $v \in V(G)$, $N_G(v) = \{w \in V(G) \mid (v, w) \in E(G)\}$ and $d_G(v) = |N_G(v)|$ denote the open neighborhood and the degree of v, respectively. The minimum degree of G is denoted by $\delta(G) = \min\{d_G(v) \mid v \in V(G)\}$. For a subset $S \subseteq V(G)$, $N_G(S) = \cup_{u \in S} N_G(u) - S$ is the open neighbourhood of S and $N_G[S] = N_G(S) \cup S$ the closed neighbourhood. The distance between two vertices u and v in G, denoted by $dis_G(u, v)$, is the length of the shortest path between them. We denote by $d(G) = \max\{dis_G(u, v) \mid u, v \in V(G)\}$ the diameter of graph G. We often omit the subscript G when the context is clear. The connectivity $\kappa(G)$ of graph G is the minimum size of vertex set $S \subseteq V(G)$ such that $G - S$ is disconnected or trivial. Let F be the set of faulty vertices whose cardinality is denoted by $|F|$. We denote by K_n the complete graph of n vertices. For further unmentioned terminologies and concepts, the reader is referred to [20].

2.2 Two Diagnosis Models

In order to locate the faulty processor it is necessary to perform some tests. Different diagnosis models can be obtained based on different test definitions and assumptions about test results, two of the best known models are PMC model [5] and MM* model [7]. In the PMC model, all tests are conducted between adjacent processors and only fault-free processor can ensure reliable results. Obviously, the test set can be expressed as a directed graph. Denote by $\sigma(u, v)$ the test result of tester u on testee v. If $u, v \notin F$, then $\sigma(u, v) = 0$. If $u \notin F$ and $v \in F$, then $\sigma(u, v) = 1$. If $u \in F$, then $\sigma(u, v)$ may be 1 or 0.

MM* model supposes that each processor must test any two of its neighbors and then compare their responses. Let vertices u and v be adjacent to vertex w, then the comparison result is denoted as $\sigma_w(u, v)$. If $u, v, w \notin F$, then $\sigma_w(u, v) = 0$. If $w \notin F$ and $\{u, v\} \cap F \neq \emptyset$, then $\sigma_w(u, v) = 1$. If $w \in F$, then $\sigma_w(u, v)$ may be 1 or 0.

The collection of all outputs is called the syndrome σ. For a given syndrome, the vertex subset $F \subset V(G)$, is said to be consistent with σ if the syndrome σ can be produced from the faulty set F.

2.3   BCCC and Its Logic Topology

BCCC is built with two types of switches and a large number of dual-port servers. We refer to n servers connected to a single n-port switch as an element. We denote BCCC with order k as $BCCC(n, k)$, where n is the number of servers connected by each switch in each element.

Definition 1 [3]. *A $BCCC(n, 0)$ is simply constructed by one element and n type-B switches, in which each server in the element connects to one of the n type-B switches using its second port. For $k \geq 1$, $BCCC(n, k)$ is constructed by n $BCCC(n, k-1)s$ connected with n^k elements according to the following rules:*

(1) The addresses of dual-port servers are denoted by $a_{k+1}a_k \ldots a_0$, where $a_0 \in [0, k]$ and $a_i \in [0, n-1]$ for $1 \leq i \leq k+1$. In particular, if servers are in $BCCC(n, k-1)$, then $a_0 \in [0, k-1]$; otherwise, $a_0 = k$.

(2) The addresses of switches are represented as $s_k s_{k-1} \ldots s_0$, where $s_0 \in [0, n+k]$ and $s_i \in [0, n-1]$ for $1 \leq i \leq k$. In particular, for the less significant digit in the address, type-B switch has $s_0 \in [0, n-1]$ and type-A switch has $s_0 \in [n, n+k]$.

(3) Two servers $a_{k+1}a_k \ldots a_0$ and $a'_{k+1}a'_k \ldots a'_0$ are connected by a type-A switch if $a_i \neq a'_i$ for $i = a_0 + 1$ and $a_j = a'_j$ for all $j \in [0, k+1] \setminus \{i\}$. In this case, the type-A switch has address $s_k s_{k-1} \ldots s_0 = a_{k+1}a_k \ldots a_{i+1}a_{i-1} \ldots a_1(i-1+n)$.

(4) Two servers $a_{k+1}a_k \ldots a_0$ and $a'_{k+1}a'_k \ldots a'_0$ are connected by a type-B switch if $a_0 \neq a'_0$ and $a_j = a'_j$ for $1 \leq j \leq k+1$. In this case, the type-B switch has address $s_k s_{k-1} \ldots s_0 = a_{k+1}a_k \ldots a_1$.

Figure 1(a) illustrates an example of BCCC(4,1), which is composed of 4 BCCC(4,0) and 4 elements. All switches are cross-bars and can be considered transparent in the network, so in this case the $BCCC(n, 0)$ can be seen as a K_n. Based on the adjacency of servers connected by switches in Definition 1, the definition for the logic graph L-BCCC(n, k) of BCCC(n, k) is given below. An example of L-BCCC(5, 1) is shown in Fig. 1b).

Definition 2 [4]. *The logic graph of $BCCC(n, k)$, denoted by $L\text{-}BCCC(n, k)$, is composed of $(k+1)n^k$ K_n and it has the vertex set $V(L\text{-}BCCC(n, k)) = \{v_{k+1}v_k \ldots v_0 \mid v_0 \in [0, k] \text{ and } v_i \in [0, n-1] \text{ for } 1 \leq i \leq k+1\}$. Two vertices $v = v_{k+1}v_k \ldots v_0$ and $v' = v'_{k+1}v'_k \ldots v'_0$ are adjacent if and only if there is* *exactly one* $i \in [0, k+1]$ *with* $i = v_0 + 1$ *or* $i = 0$ *such that* $v_i \neq v'_i$ *and* $v_j = v'_j$ *for all* $j \in [0, k+1] \setminus \{i\}$. *Furthermore, we call the edge* (v, v') *an intra-edge if* $i \neq 0$ *and an inter-edge provided* $i = 0$.

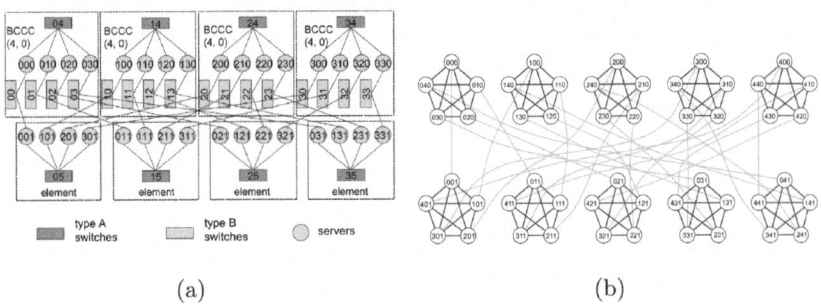

(a)   (b)

Fig. 1. BCCC(4,1) and L-BCCC(5, 1)

3   Intermittent Fault Diagnosis

In this section, we investigate the intermittent fault diagnosability $t_i(L\text{-}BCCC(n, k))$ of L-BCCC(n, k) under the PMC model.

Definition 3 [8]. *A system S is t_i-fault diagnosable if a fault-free vertex will never be diagnosed as faulty and the diagnosis at any time is at worst incomplete but never incorrect provided no more than t_i vertices are intermittently faulty.*

Lemma 1 [8]. *Under the PMC model, a system S is t_i-fault diagnosable without repair if and only if given any two distinct vertex subsets S_1 and S_2, $|S_1| \leq t_i$ and $|S_2| \leq t_i$, the set of the remaining vertices R is such that both S_1 and S_2 receive at least one testing link from R.*

The following lemma can be directly derived from the definition of L-BCCC(n, k).

Lemma 2. *Let x and y be any two different vertices in $L\text{-}BCCC(n, k)$, then*

$$|N(x) \cap N(y)| = \begin{cases} n - 2, & \text{if } (x, y) \text{ is an intra-edge;} \\ k - 1, & \text{if } (x, y) \text{ is an inter-edge;} \\ 1, & \text{if } dis(x, y) = 2; \\ 0, & \text{if } dis(x, y) > 2. \end{cases}$$

Lemma 3. *Let X be any vertex subset in $L\text{-}BCCC(n, k)$ with $|X| = 2$, if $k \geq n - 1$ then $|N(X)| \geq 2n + k - 3$, otherwise $|N(X)| \geq 2k + n - 2$.*

Proof. Without loss of generality, let $X = \{x, y\}$, we divide the discussion into two cases.

Case 1: $(x, y) \in E(L\text{-}BCCC(n, k))$.

If (x, y) is an intra-edge, by Lemma 2, then $|N(x) \cap N(y)| = n - 2$, so we have $|N(X)| = |N(x)| - 1 + |N(y)| - 1 - |N(x) \cap N(y)| = 2k + n - 2$. If (x, y) is an inter-edge, by Lemma 2, then $|N(x) \cap N(y)| = k - 1$, so we have $|N(X)| = |N(x)| - 1 + |N(y)| - 1 - |N(x) \cap N(y)| = 2n + k - 3$.

Case 2: $(x, y) \notin E(L\text{-}BCCC(n, k))$.

If $dis(x, y) = 2$, by Lemma 2, then $|N(x) \cap N(y)| = 1$, so we have $|N(X)| = |N(x)| + |N(y)| - |N(x) \cap N(y)| = 2k + 2n - 3$. If $dis(x, y) > 2$, by Lemma 2, then $|N(x) \cap N(y)| = 0$, so we have $|N(X)| = |N(x)| + |N(y)| - |N(x) \cap N(y)| = 2n + 2k - 2$.

Lemma 4. *Under the PMC model, the intermittent fault diagnosability of L-$BCCC(n, k)$ is $t_i(L\text{-}BCCC(n, k)) \leq n + k - 2$ for $k \geq 1$ and $n > 2$.*

Proof. Let L-BCCC(n, k) be G, v be any vertex in G, $S_1 = \{v\}$ and $S_2 = \{N_G(v)\}$. Then, we have $|S_1| \leq n + k - 1$, $|S_2| \leq n + k - 1$ and $S_1 \cap S_2 = \emptyset$. It is clear that there is no edge between $V(G) - S_1 - S_2$ and S_1. Hence, by Lemma 1, L-BCCC(n, k) is not $(n + k - 1)$-intermittent fault diagnosable without repair under the PMC model, i.e., $t_i(L\text{-}BCCC(n, k)) \leq n + k - 2$ for $k \geq 1$ and $n > 2$.

Lemma 5. *Let X be the vertex subset in L-$BCCC(n, k)$ with $1 \leq |X| \leq n+k-2$, then $|N(X)| \geq n + k - 1$ for $k \geq 1$ and $n > 2$.*

Proof. Denote L-BCCC(n, k) as G, based on the size of X, it suffices to distinguish three cases as follows.

Case 1: $|X| = 1$.

Without loss of generality, let $X = \{y\}$. Since G is $(n + k - 1)$-regular, it follows that $|N(y)| = n + k - 1$. Thus, the lemma holds.

Case 2: $|X| = 2$.

By Lemma 3, we have $|N(X)| \geq \min\{2n + k - 3, 2k + n - 2\} > n + k - 1$. Hence, the lemma holds.

Case 3: $3 \leq |X| \leq n + k - 2$.

Subcase 3.1: $(x, y) \notin E(G)$ for any two different vertices x and y in X.

By Lemmas 2 and 3, we have $|N(x) \cap N(y)| \leq 1$ and $|N(\{x, y\})| \geq 2k + 2n - 3$. Thus, $|N(X)| > |N(\{x, y\})| \geq 2k + 2n - 3 > n + k - 1$.

Subcase 3.2: There are at least two different vertices x and y in X such that $(x, y) \in E(G)$.

Let $D = (N(x) \cup N(y)) \cap X$, so we have $|D| \leq |X| - 2$. If (x, y) is an intra-edge, by Lemma 3, then we have $|N(\{x, y\})| = 2k + n - 2$. When $|D| < |X| - 2$, there must exist a vertex u such that u is not adjacent to x and y. By Lemma 2, there is at most one common neighbor between u and x, y. Hence, we have $|N(X)| \geq |N(\{x, y\})| - |D| + n + k - 3 > n + k - 1$. When $|D| = |X| - 2$, if all vertices in D are connected to x or y by the intra edge, then we have $|N(X)| \geq |N(\{x, y\})| - |D| + |D|k > n + k - 1$. Otherwise, there exists at least a vertex v such that v is connected to x or y by the inter edge. Then we have $|N(X)| \geq |N(\{x,y\})| - |D| + n - 1 > n + k - 1$.

The case where (x,y) is an inter-edge can be considered similarly and we omit it. To sum up, the lemma holds.

Theorem 1. *Under the PMC model, the intermittent fault diagnosability of L-BCCC(n,k) is $t_i(L\text{-}BCCC(n,k)) = n + k - 2$ for $k \geq 1$ and $n > 2$.*

Proof. Let L-BCCC(n,k) be G, S_1 and S_2 be any two different vertex subset of G such that $|S_1| \leq n+k-2$, $|S_2| \leq n+k-2$ and $S_1 \cap S_2 = \emptyset$. By Lemma 5, we have $|N(S_1)| \geq n+k-1$, which suggests that there is at least one edge between S_1 and $V(G) - S_1 - S_2$. Similarly, we have $|N(S_2)| \geq n + k - 1$ by Lemma 5, which suggests that there is at least one edge between S_2 and $V(G) - S_1 - S_2$. Then, by Lemma 1, L-BCCC(n,k) is $(n+k-2)$-intermittent fault diagnosable without repair under the PMC model, i.e., $t_i(L\text{-}BCCC(n,k)) \geq n + k - 2$. By virtue of Lemma 4, the theorem holds.

4   Diagnosability and Strong Local Diagnosability

In this section, we determine the diagnosability and strong local diagnosability of L-BCCC(n,k). Denote the neighbor connected to v by the inter (resp. intra) edge as the inter (resp. intra) neighbor of v.

Definition 4 [14]. *For $u \in V(G)$, G is locally t-diagnosable at u if, given a syndrome σ_F produced by a set of faulty vertices $F \subseteq V(G)$ containing u with $|F| \leq t$, every set of faulty vertices F' compatible with σ_F and $|F'| \leq t$ must also contain u. The local diagnosability of u is denoted by $t^l(u)$, which is defined as the maximum t such that G is locally t-diagnosable at u.*

Lemma 6 [14]. *The diagnosability $t(G)$ of graph G equals the minimum of the local diagnosability of each vertex in G, i.e., $t(G) = \min\{t^l(u) \mid u \in V(G)\}$.*

Definition 5 [14]. *For $u \in V(G)$, if $t^l(u) = d_G(u)$, then u has the strong local diagnosability. G has the strong local diagnosability if each vertex in G has the strong local diagnosability.*

Lemma 7 [14]. *For $u \in V(G)$ with $d_G(u) = k$, a Type I structure $T(u;k)$ of order k at u is defined as the graph with vertex set $V(T(u;k)) = \{u\}\cup\{x_i, y_i \mid 1 \leq i \leq k\}$ and edge set $E(T(u;k)) = \{(u, x_i), (x_i, y_i) \mid 1 \leq i \leq k\}$. Under the PMC model, if there exists a Type I structure $T(u;k)$ of order k at u in G, then the local diagnosability of u is k.*

Theorem 2. *Under the PMC model, the local diagnosability of vertex u in L-BCCC(n,k) is $t^l(u) = n + k - 1$ for $n \geq 2$ and $k \geq 1$.*

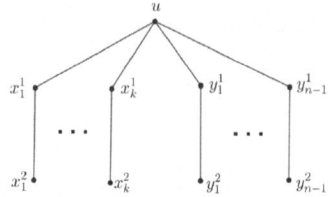

Fig. 2. $T(u; n + k - 1)$ of L-BCCC(n, k)

Proof. Let L-BCCC(n, k) be G and u be any vertex in G. Denote x_j^1 and y_i^1 as the inter and intra neighbor of u, respectively, where $1 \leq j \leq k$ and $1 \leq i \leq n-1$. Let x_j^2 be an intra neighbor of x_j^1 and y_i^2 be an inter neighbor of y_i^1. Hence, we construct the Type I structure $T(u; n + k - 1)$ of order $n + k - 1$ at u in G, where $V(T(u; n + k - 1)) = \{u\} \cup \{x_j^s \mid 1 \leq s \leq 2,\ 1 \leq j \leq k\} \cup \{y_i^t \mid 1 \leq t \leq 2,\ 1 \leq i \leq n - 1\}$ and $E(T(u; n + k - 1)) = \{(u, x_j^1),\ (u, y_i^1),\ (x_j^1,\ x_j^2),\ (y_i^1,\ y_i^2) \mid 1 \leq j \leq k,\ 1 \leq i \leq n - 1\}$ (see Fig. 2). By virtue of Lemma 7, the theorem holds.

The following corollary can be derived directly by means of Theorem 2, Lemma 6 and Definition 5.

Corollary 1. *Under the PMC model, the diagnosability of L-BCCC(n, k) is $t(L\text{-}BCCC(n, k)) = n + k - 1$ for $n \geq 2$ and $k \geq 1$, and L-BCCC(n, k) has strong local diagnosability.*

Theorem 3. *Let L-BCCC(n, k) be G and F be any set of missing edges in G with $|F| \leq \min\{n-1, k-1\}$. For any vertex $u \in V(G - F)$, the local diagnosability of u is equal to $d_{G-F}(u)$ and $G - F$ has the strong local diagnosability.*

Proof. We only consider the case of $n > k$, the case of $n \leq k$ is similar and thus omitted. We prove the result by induction. For $k = 2$, we have $|F| \leq 1$. When $|F| = 0$, it is easy to verify that the theorem holds according to Theorem 2. When $|F| = 1$, denote the missing edge as (x, y), then we have $d_{G-F}(x) = d_{G-F}(y) = n + k - 2$. Clearly, based on the Type I structure of order $n + k - 1$ at x in G without the missing edge, we can construct $T(x; n + k - 2)$ of order $n + k - 2$ at x in $G - F$. Similarly, $T(y; n + k - 2)$ of order $n + k - 2$ at y in $G - F$ can be constructed. For any vertex $z \in V(G) - \{x, y\}$, we have $d_{G-F}(z) = n + k - 1$. The inter neighbor of z has $n - 1$ intra neighbors and the intra neighbor of z has k inter neighbors. When $|F| = 1$, both the inter and intra neighbor of z have at least one neighbor due to $n > k = 2$, which constructs $T(z; n + k - 1)$ of order $n + k - 1$ at z in $G - F$. We assume that the theorem holds for $|F| = m < k - 1$, i.e., for any vertex $u \in V(G - F)$, there exists $T(u; d_{G-F}(u))$ of order $d_{G-F}(u)$ at u in $G - F$. If one edge connected to u in $T(u; d_{G-F}(u))$ is missing, the Type I structure of order $d_{G-F}(u) - 1$ at u is constructed. If one edge not connected to u in $T(u; d_{G-F}(u))$ is missing, then the inter neighbor of u has at least $n - 2 - m$ intra neighbors and the intra neighbor of u has at least $k - m - 1$ inter neighbors. Since $n - 2 - m > 0$ and $k - m - 1 > 0$, the Type I structure of order $d_{G-F}(u)$

at u can be constructed. Therefore, when $|F| = m + 1$, $G - F$ has the Type I structure $T(u; d_{G-F}(u))$ of order $d_{G-F}(u)$ at u. By Definition 5 and Lemma 7, the theorem holds.

Lin et al. [16] identified the state of the rooted vertex under the PMC model in terms of the Type I structure. Based on the $T(u; n+k-1)$ of order $n+k-1$ at u for L-BCCC(n,k) constructed in Theorem 2, we present an $O(n+k)$ algorithm LDT to ascertain the fault state of u under the PMC model.

Algorithm 1. LDT$(u, n + k - 1)$

Input: Type I structure $T(u; n+k-1)$ of order $n+k-1$ at vertex u in L-BCCC(n,k).
Output: 0 or 1 (The value is 0 if u is fault-free and 1 otherwise).
1: $\alpha_{0,0} \leftarrow |\{1 \leq j \leq k \mid (\sigma(x_j^1, u), \sigma(x_j^2, x_j^1)) = (0,0)\} \cup \{1 \leq i \leq n - 1 \mid (\sigma(y_i^1, u), \sigma(y_i^2, y_i^1)) = (0,0)\}|$;
2: $\alpha_{1,0} \leftarrow |\{1 \leq j \leq k \mid (\sigma(x_j^1, u), \sigma(x_j^2, x_j^1)) = (1,0)\} \cup \{1 \leq i \leq n - 1 \mid (\sigma(y_i^1, u), \sigma(y_i^2, y_i^1)) = (1,0)\}|$;
3: **if** $\alpha_{0,0} \geq \alpha_{1,0}$ **then**
4:     return 0;
5: **else**
6:     return 1;
7: **end if**

Lemma 8 [15]. *For $u \in V(G)$ with $d_G(u) = t$, an extended star $ES(u; t)$ of order t at u is defined as the graph with vertex set $V(ES(u; t)) = \{u\} \cup \{v_{ij} \mid 1 \leq i \leq n, 1 \leq j \leq 4\}$ and edge set $E(ES(u; t)) = \{(u, v_{k1}), (v_{k1}, v_{k2}), (v_{k2}, v_{k3}), (v_{k3}, v_{k4}) \mid 1 \leq k \leq n\}$. Under the MM* model, if there exists an extended star $ES(u; t)$ of order t at u in G, then the local diagnosability of u is t.*

Theorem 4. *Under the MM* model, the local diagnosability of vertex u in L-BCCC(n,k) is $t^l(u) = n + k - 1$ for $n \geq 2$ and $k \geq 1$.*

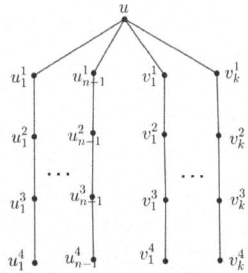

Fig. 3. $ES(u; n + k - 1)$ of L-BCCC(n,k)

Proof. Let L-BCCC(n,k) be G and u be any vertex in G. Denote v_j^1 and u_i^1 as the inter and intra neighbor of u, respectively, where $1 \le j \le k$ and $1 \le i \le n-1$. Let v_j^2 be an intra neighbor of v_j^1 and u_i^2 be an inter neighbor of u_i^1. Then, let v_j^3 be an inter neighbor of v_j^2 and u_i^3 be an intra neighbor of u_i^2. Moreover, denote that v_j^4 is an intra neighbor of v_j^3 and u_i^4 is an inter neighbor of u_i^3. Hence, we construct the extended star $ES(u; n+k-1)$ of order $n+k-1$ at u in G, where $V(ES(u; n+k-1)) = \{u\} \cup \{v_j^s \mid 1 \le s \le 4,\ 1 \le j \le k\} \cup \{u_i^t \mid 1 \le t \le 4,\ 1 \le i \le n-1\}$ and $E(ES(u; n+k-1)) = \{(u,v_j^1),\ (u,u_i^1),\ (v_j^1,\ v_j^2),\ (u_i^1,\ u_i^2),\ (v_j^2,\ v_j^3),\ (u_i^2,\ u_i^3),\ (v_j^3,\ v_j^4),\ (u_i^3,\ u_i^4) \mid 1 \le j \le k,\ 1 \le i \le n-1\}$ (see Fig. 3). By virtue of Lemma 8, the theorem holds.

The following corollary can be derived directly by means of Theorem 4, Lemma 6 and Definition 5.

Corollary 2. *Under the MM* model, the diagnosability of L-BCCC(n,k) is $t(L\text{-}BCCC(n,k)) = n+k-1$ for $n \ge 2$ and $k \ge 1$, and L-BCCC(n,k) has strong local diagnosability.*

Lin et al. [17] proposed the wind-bell-tree to determine the state of rooted vertex under the comparison model. Next, we construct the wind-bell-tree of L-BCCC(n,k). Let u be any vertex in L-BCCC(n,k). Denote x_j^1 and y_i^1 as the inter and intra neighbor of u, respectively, where $1 \le j \le k$ and $1 \le i \le n-1$. Let y_i^2 and w_i be the inter neighbor of y_i^1, and x_j^2 and z_j be the intra neighbor of x_j^1. In addition, denote that y_i^3 is an intra neighbor of y_i^2 and x_j^3 is an inter neighbor of x_j^2. Hence, the wind-bell-tree of order $n+k-1$ at u in L-BCCC(n,k) is defined as $WT(u; n+k-1)$, where $V(WT(u; n+k-1)) = \{u\} \cup \{x_j^s,\ z_j \mid 1 \le s \le 3,\ 1 \le j \le k\} \cup \{y_i^t,\ w_i \mid 1 \le t \le 3,\ 1 \le i \le n-1\}$ and $E(WT(u; n+k-1)) = \{(u,x_j^1),\ (u,y_i^1),\ (x_j^1,\ x_j^2),\ (x_j^1,\ z_j),\ (y_i^1,\ y_i^2),\ (y_i^1,\ w_i),\ (x_j^2,\ x_j^3),\ (y_i^2,\ y_i^3) \mid 1 \le j \le k,\ 1 \le i \le n-1\}$ (see Fig. 4). Based on the constructed $WT(u; n+k-1)$ in L-BCCC(n,k), we present an $O(n+k)$ algorithm LDWT to identify the state of u in L-BCCC(n,k) under the comparison model.

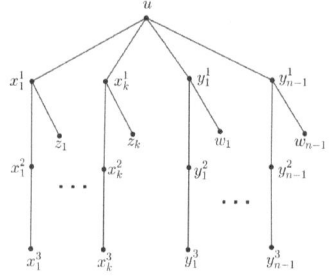

Fig. 4. $WT(u; n+k-1)$ of L-BCCC(n,k)

Algorithm 2. LDWT$(u, n + k - 1)$

Input: The wind-bell-tree $WT(u; n + k - 1)$ of order $n + k - 1$ at vertex u in L-BCCC(n, k).

Output: 0 or 1 (The value is 0 if u is fault-free and 1 otherwise).

1: $\beta_0 \leftarrow |\{1 \leq j \leq k \mid (\sigma_{x_j^1}(u, z_j), \sigma_{x_j^1}(u, x_j^2), \sigma_{x_j^2}(x_j^1, x_j^3)) = (0, 0, 0)\} \cup \{1 \leq i \leq n - 1 \mid (\sigma_{y_i^1}(u, w_i), \sigma_{y_i^1}(u, y_i^2), \sigma_{y_i^2}(y_i^1, y_i^3)) = (0, 0, 0)\}|;$

2: $\beta_1 \leftarrow |\{1 \leq j \leq k \mid (\sigma_{x_j^1}(u, z_j), \sigma_{x_j^1}(u, x_j^2), \sigma_{x_j^2}(x_j^1, x_j^3)) = (1, 1, 0)\} \cup \{1 \leq i \leq n - 1 \mid (\sigma_{y_i^1}(u, w_i), \sigma_{y_i^1}(u, y_i^2), \sigma_{y_i^2}(y_i^1, y_i^3)) = (1, 1, 0)\}|;$

3: **if** $\beta_0 \geq \beta_1$ **then**

4:     return 0;

5: **else**

6:     return 1;

7: **end if**

5   Performance Evaluation

In this section, we evaluate the performance of algorithms LDT and LDWT of L-BCCC(n, k) by simulation experiments. The required algorithms are implemented in python programming language and simulations are performed on 2.90 GHz Intel(R) Core(TM) i7-10700 CPU and 16 GB RAM under the Windows 10 operating system.

Let TP (resp. TN) be the number of vertices that are diagnosed by the algorithm as fault-free (resp. faulty) and actually are also fault-free (resp. faulty) and FP (resp. FN) be the number of vertices that are diagnosed by the algorithm as fault-free (resp. faulty) and actually are faulty (resp. fault-free). We measure the performance of the algorithm by the following four metrics.

Accuracy: the proportion of the number of correctly diagnosed vertices to the total number of vertices, denoted as ACC$= \frac{TP+TN}{TP+FP+FN+TN}$.

Precision: the proportion of the number of vertices diagnosed as fault-free and actually fault-free to the number of vertices diagnosed as fault-free, denoted as PRE$= \frac{TP}{TP+FP}$.

TPR (Recall): the proportion of the number of correctly diagnosed fault-free vertices to the total number of fault-free vertices, denoted as TPR$= \frac{TP}{TP+FN}$.

FPR: the proportion of the number of faulty vertices diagnosed as fault-free to the total number of faulty vertices, denoted as FPR$= \frac{FP}{FP+TN}$.

Figures 5 and 6 depict the experimental results of four evaluation metrics, ACC, PRE, FPR and TPR, for algorithms LDT and LDWT under different sizes of L-BCCC(n, k), respectively. The percentage of faulty vertices in L-BCCC(n, k) is set from 5% to 30%. From Figs. 5(a) and 5(b), it is clearly observed that ACC and PRE of algorithm LDT are approximated to be 100% when the number of faulty vertices does not exceed 11%. Then, ACC and PRE gradually decline with increasing number of faulty vertices in L-BCCC(n, k), but even if the number of faulty vertices reaches 30%, the values of ACC and PRE are still more than 99.6%. In Fig. 5(c), the FPR rises with the number of faulty vertices in L-BCCC(n, k), but in the case of no more than 11% of faulty vertices, the FPR

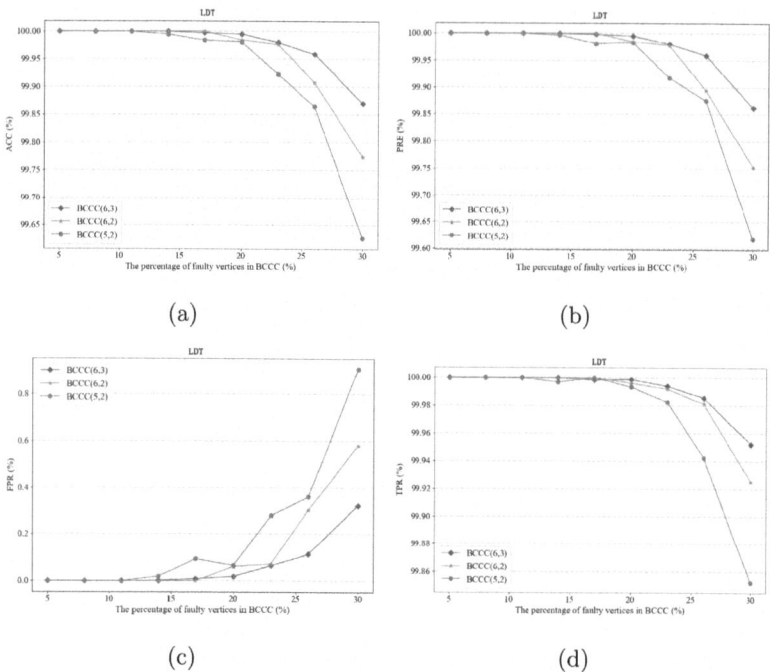

Fig. 5. Experimental results for algorithm LDT in L-BCCC(n, k)

is almost 0. When the number of faulty vertices reaches 30%, the diagnosis error rate of faulty vertices does not exceed 1%. Figure 5(d) depicts that when the number of faulty vertices in L-BCCC(n, k) does not exceed 11%, algorithm LDT is able to recognize almost all the fault-free vertices, but with the increasing number of faulty vertices, the TPR gradually declines. When the number of faulty vertices reaches 30%, the value of TPR still exceeds 99.8%.

From the four subgraphs of Fig. 6, it can be seen that the value of ACC, PRE and TPR declines with the increasing number of faulty vertices in L-BCCC(n, k) whereas the value of FPR rises with the increasing number of faulty vertices in L-BCCC(n, k). Figure 6(a) depicts that the ACC approximates 100% when the number of faulty vertices does not exceed 8%, and even when the number of faulty vertices reaches 30%, the percentage of correct diagnosis still reaches 84%. Figure 6(b) illustrates that when the number of faulty vertices in L-BCCC(n, k) does not exceed 14%, the PRE approximates 100%, and when the number of faulty vertices grows to 30%, the accuracy of vertices diagnosed as fault-free is more than 99.8%. In Fig. 6(c), when the number of faulty vertices in L-BCCC(n, k) does not exceed 14%, the FPR is almost 0. As the number of faulty vertices increases to 30%, the percentage of faulty vertices diagnosed as fault-free does not exceed 0.16%. From Fig. 6(d), when the number of faulty vertices in L-BCCC(n, k) does not exceed 8%, the TPR is approximated to be 100%,

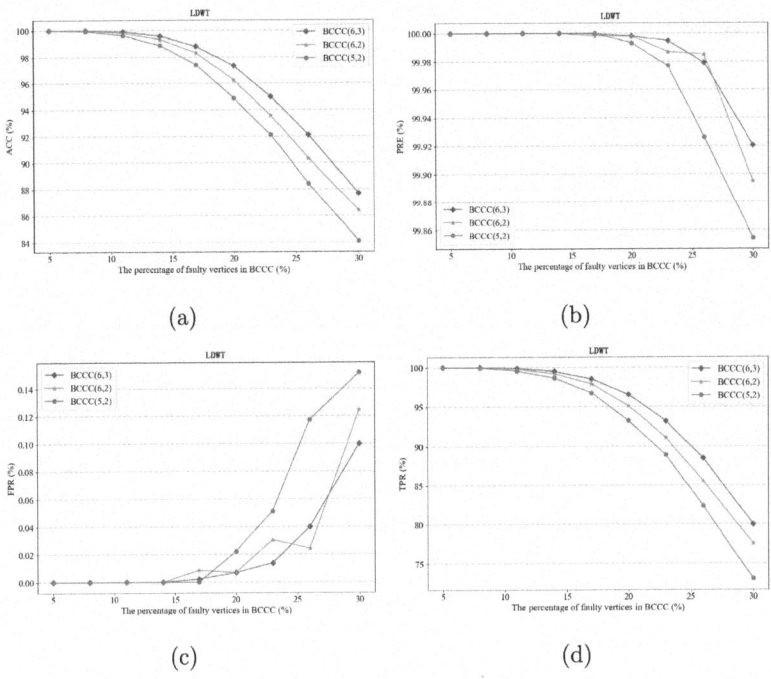

Fig. 6. Experimental results for algorithm LDWT in L-BCCC(n, k)

and as the number of faulty vertices reaches 30%, the proportion of fault-free vertices being correctly diagnosed exceeds 72%.

6   Conclusion

In this work, we focus on analyzing the intermittent fault and local diagnosability of BCCC(n, k) and prove that it achieves strong local diagnosability under both the PMC and MM* models. We further show that even with up to $\min\{n-1, k-1\}$ missing edges, BCCC(n, k) retains this property under the PMC model and provides a corresponding local diagnosis algorithm. Simulation experiments are conducted to evaluate the performance of the algorithm.

Acknowledgments. This work was partly supported by the National Natural Science Foundation of China (Nos. 62172291 and 62272333).

Disclosure of Interests. The authors have no competing interests to declare that are relevant to the content of this article.

References

1. Ghemawat, S., Gobioff, H., Leung, S.-T.: The Google file system. ACM SIGOPS Oper. Syst. Rev. **37**(5), 29–43 (2003)
2. Dean, J., Ghemawat, S.: MapReduce: simplified data processing on large clusters. Commun. ACM **51**(1), 107–113 (2008)
3. Li, Z., Guo, Z., Yang, Y.: BCCC: an expandable network for data centers. IEEE/ACM Trans. Netw. **24**(6), 3740–3755 (2016)
4. Li, X.-Y., Lin, W., Liu, X., Lin, C.-K., Pai, K.-J., Chang, J.-M.: Completely independent spanning trees on BCCC data center networks with an application to fault-tolerant routing. IEEE Trans. Parallel Distrib. Syst. **33**(8), 1939–1952 (2022)
5. Preparata, F.P., Metze, G., Chien, R.T.: On the connection assignment problem of diagnosable systems. IEEE Trans. Electron. Comput. EC **16**(6), 848–854 (1967)
6. Maeng, J., Malek, M.: A comparison connection assignment for self-diagnosis of multiprocessors systems. In: Proceedings of the 11th International Symposium on Fault-Tolerant Computing, pp. 173–175 (1981)
7. Sengupta, A., Dahbura, A.: On self-diagnosable multiprocessor systems: diagnosis by the comparison approach. IEEE Trans. Comput. **41**, 1386–1396 (1992)
8. Mallela, S., Masson, G.M.: Diagnosable systems for intermittent faults. IEEE Trans. Comput. **27**(6), 560–566 (1978)
9. Liang, J., Feng, H., Du, X.: Intermittent fault diagnosability of interconnection networks. J. Comput. Sci. Technol. **32**(6), 1279–1287 (2017)
10. Sun, X., Zhou, S., Lv, M., Liu, J., Lian, G.: Intermittent fault diagnosability of some general regular networks. Comput. J. **63**(1), 16–24 (2020)
11. Song, J., Lin, L., Huang, Y., Hsieh, S.-Y.: Intermittent fault diagnosis of split-star networks and its applications. IEEE Trans. Parallel Distrib. Syst. **34**(4), 1253–1264 (2023)
12. Feng, H., Wu, J., Chen, L.: Intermittent fault diagnosis of product network based on PMC model. Comput. J. **67**(9), 2777–2786 (2024)
13. Lin, L., Zhou, S., Hsieh, S.-Y.: Neural network enabled intermittent fault diagnosis under comparison model. IEEE Trans. Reliab. **72**(3), 1206–1219 (2023)
14. Hsu, G.-H., Tan, J.: A local diagnosability measure for multiprocessor systems. IEEE Trans. Parallel Distrib. Syst. **18**(5), 598–607 (2007)
15. Chiang, C.-F., Tan, J.: Using node diagnosability to determine t-diagnosability under the comparison diagnosis model. IEEE Trans. Comput. **58**(1), 251–259 (2009)
16. Lin, C.-K., Kung, T.L., Tan, J.: An algorithmic approach to conditional-fault local diagnosis of regular multiprocessor interconnected systems under the PMC model. IEEE Trans. Comput. **62**(3), 439–451 (2013)
17. Lin, C.-K., Teng, Y.-H., Tan, J., Hsu, L.-H.: Local diagnosis algorithms for multiprocessor systems under the comparison diagnosis model. IEEE Trans. Rel. **62**(4), 800–810 (2013)
18. Ren, Y., Wang, S.: The local diagnosability of a class of Cayley graphs with conditional faulty edges under the PMC model. Comput. J. **66**(8), 1913–1921 (2023)
19. Cheng, E., Lipták, L., Steffy, D.E.: Strong local diagnosability of (n, k)-star graphs and cayley graphs generated by 2-trees with missing edges, Inf. Process. Lett. **113**, 452–456 (2013)
20. West, D.B.: Introduction to Graph Theory. Prentice Hall, Upper Saddle River (2001)

Testing Some First-Order Logic Properties on Sparse Graphs

Pan Peng[(✉)] and Kefan Yu[(✉)]

School of Computer Science and Technology, University of Science and Technology of China, Hefei, PR, China
ppeng@ustc.edu.cn, ykf97@mail.ustc.edu.cn

Abstract. In this paper, we study the problem of testing some first-order logic properties on sparse graphs under the adjacency list model, including k-dominating set property, k-vertex cover property and diameter $\leq k$ property.

(1) For the k-dominating set property, we give a tester with query complexity $O\left(\frac{(kC)^{k+1}}{\varepsilon^{k+2}}\log\left(\frac{kC}{\varepsilon}\right)\right)$ on n-vertex graphs with at most Cn edges. Furthermore, if the input graph is planar, we improve the query complexity to $O\left(\frac{k^6}{\varepsilon^3}\log(\frac{k^2}{\varepsilon})\right)$.

(2) For the k-vertex cover property, we give a tester whose query complexity is $O\left(\frac{\alpha^3 k^2 \log k}{\varepsilon^2}\right)$ on graphs with arboricity bounded by α.

Previously, these properties were known to be testable with constant query complexity on general graphs under a stronger model with random edge sampling queries. By leveraging edge sampling simulation techniques, one can achieve poly($\log n$) query complexity in the adjacency model for bounded arboricity graphs. In contrast, our algorithms achieve constant query complexity (for fixed ε and k) on a broader class of sparse graphs or the same class of bounded arboricity graphs.

(3) For the diameter $\leq k$ property, we can distinguish whether a sparse graph has a diameter at most k or is ε-far from any graph that has a diameter at most $k+2$ with query complexity $O\left(\frac{C^{\lfloor k/2\rfloor+1}}{\varepsilon^{\lfloor k/2\rfloor+2}}\right)$.

Previously, a tester was known for general graphs that distinguishes between having a diameter at most k and being ε-far from any graph with diameter at most $\beta(k)$, where $\beta(k)$ ranges from $k+4$ to $4k+2$. We improve the upper bound on $\beta(k)$ to at most $k+2$ for sparse graphs, though at the cost of slightly higher query complexity.

1 Introduction

Graph property testing is a framework for studying sampling and querying based algorithms that solve a relaxation of classical decision problems on graphs in sublinear time. Given a graph G and a property Π, the goal of a property testing algorithm is to determine whether the input graph satisfies Π or is *far* from

The work is supported in part by NSFC grant 62272431.

F. V. Fomin and M. Xiao (Eds.): COCOON 2025, LNCS 15984, pp. 95–109, 2026.
https://doi.org/10.1007/978-981-95-0218-9_8

satisfying it. The notion of *far* depends on the specific model, which defines the types of allowed queries and the criteria for being ε-far from satisfying the property Π. The parameter $\varepsilon \in (0, 1)$ quantifies the proximity to the property. In each model, a (randomized) algorithm is called an ε-*tester* for a property Π if it accepts graphs that satisfy Π and rejects graphs that are ε-far from Π with high probability (e.g., at least $\frac{2}{3}$). A property Π is said to be *testable* if, for every fixed $\varepsilon > 0$, there exists an ε-tester for Π whose total number of queries is independent of the size of the input graph, though it may depend on ε or other bounded parameters. The *query complexity* of a tester is the number of oracle queries it makes.

The characterization of testable properties under the dense graph model is relatively well understood [2]. In contrast, analogous characterizations for sparse graphs remain underdeveloped. The *sparse graph class* under consideration consists of graphs whose edge counts are bounded by a constant multiple of their vertex counts. This class encompasses many significant graph families, including bounded-degree graphs, minor-closed class and graphs with bounded arboricity. In bounded-degree graphs, a few number of properties are known to be testable, including *k-connectivity* [13], *minor-freeness* [4,17], *hyperfinite* properties [21] and *subdivision-freeness* [16], *additive and monotone* planar properties [17] etc. A graph property is called monotone if it is closed under removal of edges and vertices. Recently, Adler, Köhler and Peng [1] showed that every first-order property of type $\exists * \forall *$ is testable on bounded-degree graphs, while there exists a first-order property of type $\forall * \exists *$ which is not testable in bounded-degree graphs.

Our understanding of which properties are testable on general unbounded-degree graphs remains substantially constrained. Research has primarily focused on special classes of unbounded-degree graphs, such as minor-closed graph classes, and graphs with bounded arboricity. Czumaj, Monemizadeh, Onak, and Sohler [7] initiated the study of property testing on arbitrary planar graphs, proving that the property of *bipartiteness* can be tested in constant queries. Czumaj and Sohler [8] demonstrated that the property of being H-free is testable with one-sided error for every finite graph H, under the random neighbor oracle model. Levi and Shoshan [19] showed that planar *hamiltonicity* can be tested using $\mathrm{poly}(\frac{1}{\varepsilon})$ queries under the general sparse model. Esperet and Norin [11] further established that for any proper minor-closed class $\mathcal{G}$, every *monotone* property (e.g., k-colorability) is testable for graphs on $\mathcal{G}$. Eden, Levi and Ron [9] showed that for any fixed k, the query complexity of testing C_k-freeness is upper bounded by $O\left(n^{1-1/\lfloor k/2 \rfloor}\right)$ under the general graph model. Moreover, for every fixed $k \geq 6$, any one-sided error algorithm for testing C_k-freeness must perform $\Omega\left(n^{1/3}\right)$ queries. Additionally, for $k = 4, 5$, the query complexity of one-sided error testing of C_k-freeness on constant arboricity graphs over n vertices is $\Theta\left(n^{1/4}\right)$. Currently, only general results for sparse graphs apply to simpler classes such as forests [18] and outerplanar graphs [3]. It was known that every property can be tested with $\mathrm{poly}(\log n)$ queries on both graph classes.

In this paper, we focus on testing some properties that are definable in first-order logic under the sparse graph model, including the k-dominating set property, the k-vertex cover property, and the diameter $\leq k$ property. As discussed earlier, [11] provides a constant-query tester for monotone properties on minor-closed graph classes, while [1] establishes a constant-query tester for first-order properties of type $\exists * \forall *$ in bounded-degree graphs. However, the properties we consider, such as the k-dominating set property and the diameter $\leq k$ property, are non-monotone and definable by the first-order logic (of type $\exists * \forall *$); and our focus is on unbounded-degree graphs. Consequently, the techniques developed in these works cannot be directly applied to these properties. Although the result from [11] can be applied to test the k-vertex cover property on planar graphs due to its monotonicity (yielding a query complexity bounded by a tower function of $O(d)$, where d denotes the maximum tree-depth of the forbidden graphs induced by the k-vertex cover property), our algorithm significantly reduces query complexity and extends to a broader class of graphs, particularly those with bounded arboricity, a more extensive family than minor-closed graph classes.

1.1 Our Contributions

All the graphs considered in this paper are finite, undirected and simple (for unexplained terminology and more details, see for example [6]). We use $G = (V, E)$ to denote a graph with vertex set V and edge set E and assume by default that $V(G) = [n]$ unless otherwise stated. For any vertex $v \in V$, denote the neighbors of v by Γ_v. Let X be a set, k be a positive integer, we use $\binom{X}{k}$ to denote the set of all k-subsets of the set X.

We focus on *the sparse graph model*, where the input graph $G = (V, E)$ is represented by an adjacency list, and we are granted an oracle $\mathcal{O}_G$ access to it [15]. Specifically, we have three types of queries:

- Random vertex queries: The oracle $\mathcal{O}_G$ returns a random vertex in $V(G)$.
- Degree queries: Specifies a vertex v, and $\mathcal{O}_G$ returns the degree of v.
- Neighbor queries: Specifies a vertex v and an index i, and the oracle $\mathcal{O}_G$ returns the i-th neighbor of v if it exists, or a special symbol $\perp$ otherwise.

Note that $|V(G)|$ is also known to $\mathcal{O}_G$. In this model, given a sparse graph G and a property Π, we say that G is ε-far from satisfying Π if more than εn edge insertions and/or deletions are required to make G satisfy Π.

Now we are ready to state our main results. Our first result is on testing the property of having a k-dominating set. A dominating set for a graph G is a subset D of its vertices, such that any vertex of G is either in D or has a neighbor in D. A graph G satisfies the *k-dominating set property* if it has a dominating set D of size no more than k.

Theorem 1. *Given a positive integer k and a constant C, the k-dominating set property is testable within the class of sparse graphs in which all the graphs satisfy $|E| \leq C \cdot |V|$, and the query complexity is $O\left(\frac{(kC)^{k+1}}{\varepsilon^{k+2}} \log\left(\frac{kC}{\varepsilon}\right)\right)$.*

If the input graph is restricted to the class of planar graphs, the query complexity can be significantly improved.

Theorem 2. *Given a positive integer k, the k-dominating set property is testable on planar graphs, and the query complexity is* $O\left(\frac{k^6}{\varepsilon^3}\log(\frac{k^2}{\varepsilon})\right)$.

Our next result is on testing the property of having a k-vertex cover. A vertex cover of a graph G is a subset C of its vertices, such that it includes at least one endpoint of every edge of the graph. A graph G satisfies *k-vertex cover property* if it has a vertex cover C of size no more than k. We study the problem of testing this property in graphs with bounded arboricity, where the *arboricity* $\alpha(G)$ of an undirected graph G is the minimum number of forests into which its edges can be partitioned. We remark that a graph with arboricity at most α has $O(\alpha n)$ edges and planar graphs have arboricity at most three.

Theorem 3. *Given a positive integer k, the k-vertex cover property is testable on α-bounded arboricity graph class, and the query complexity is* $O\left(\frac{\alpha^3 k^2 \log k}{\varepsilon^2}\right)$.

Our final result is on testing the property of having a constant diameter. The diameter of a graph G is the maximum distance between two vertices of G, denoted by $diam(G)$. Here, the distance between vertices u and v in G, denoted by $d_G(u,v)$, is the length of the shortest (u,v)-path in G. A graph G satisfies *the diameter $\leq k$ property* if its diameter no more than k.

Theorem 4. *Let $\varepsilon \in (0,1)$ be any constant, $C > 0$ be any fixed constant, $k \geq 2$ be any integer and G be a sparse graph of order n satisfying $|E(G)| \leq Cn$. There exists a testing algorithm such that*

- *return Accept with probability $\geq \frac{2}{3}$ if $diam(G) \leq k$;*
- *return Reject with probability $\geq \frac{2}{3}$ if G is ε-far from having a diameter $k+2$ (when k is even) or $k+1$ (when k is odd).*

Moreover, the query complexity is $O\left(\frac{C^{\lfloor k/2 \rfloor+1}}{\varepsilon^{\lfloor k/2 \rfloor+2}}\right)$.

We remark that in [22], Parnas and Ron present a family of algorithms in the general graph model to determine whether the diameter of a graph is bounded by k, or is ε-far from any graph with diameter at most $\beta(k)$, where $\beta(k)$ ranges between $k+4$ and $4k+2$. We improve the upper bound of $\beta(k)$ to at most $k+2$ if the input graph is sparse (with slightly worse query complexity).

Remark Iwama and Yoshida [14] study property testing for NP optimization problems with parameter k under the *augmented general graph model* which additionally allows the algorithm to sample an edge uniformly at random. It was shown that the k-dominating set, the k-vertex cover, the k-feedback vertex set, the k-multicut and the k-path-free properties are constant-query testable if k is constant in this model. Note that since sampling an edge from a distribution that is pointwise ε-close to the uniform distribution over $E(G)$ can be simulated by using at most $\text{poly}(\log n, 1/\varepsilon)$ queries random vertex sampling, neighbor queries and degree queries on bounded arboricity graphs [10], the algorithms in [14] can be transformed to our models for testing the properties studied here on bounded arboricity graphs. However, the resulting query complexities will be $\text{poly}(\log n, 1/\varepsilon)$, while ours are constant. Furthermore, it is known that the query complexity of sampling an edge almost uniformly at random algorithms requires at least $\Omega\left(\frac{\log n}{\log \log n}\right)$ queries, even when the input graph is a forest [10]. This indicates that the dependence on $\log n$ from the approach of combining [10] and [14] is unavoidable.

1.2   Our Techniques

The general framework of our property testing algorithms involves finding an appropriate quantity that can be approximated in sublinear queries to distinguish between two different cases. These quantities are defined based on *high-degree* vertices, and the definition of high-degree vertices varies depending on the specific properties being tested.

For the k-dominating set property on planar graphs (see Sect. 2), leveraging the fact that planar graphs are $K_{3,3}$-minor free and applying the Bonferroni

inequalities, we show that the quantity $\max\limits_{D}\left\{\sum\limits_{i \in D} \deg(i) - \sum\limits_{(i,j) \in \binom{D}{2}} |\Gamma_i \cap \Gamma_j|\right\}$

can be used to distinguish if G satisfies the property or is ε-far from having the property, where D is a k-subset of $V(G)$, $|\Gamma_i \cap \Gamma_j|$ is the number of common neighbors of vertex i and vertex j. Consequently, our task reduces to identifying all high-degree vertices and well-connected pairs with high probability and estimating the number of common neighbors between two high-degree vertices. Then, we estimate the quantity and check if it is above a certain threshold to decide whether to accept or reject the input graph. This threshold is chosen to differentiate between graphs that satisfy the property and those that are ε-far from it. For general sparse graphs (see the full version of the paper), we show that one only needs to determine whether a k-subset of V is $\frac{\varepsilon}{4}$-close to being a k-dominating set or ε-far from being a k-dominating set. Consequently, it suffices to verify if there exists a k-subset of vertices with degrees of at least $\frac{\varepsilon n}{4k}$ that is $\frac{\varepsilon}{4}$-close to being a k-dominating set. Since the number of high-degree vertices is constant, we can test this property with constant number of queries.

For the k-vertex cover property on planar graphs (see Sect. 3), we first design a subroutine to determine whether the number of edges in a planar graph of order n is less than $\frac{\varepsilon n}{10}$ or it is more than εn. This can be achieved by sampling a small number of vertices and checking whether there are sufficiently many isolated vertices. Subsequently, the task of testing k-vertex cover property in a planar graph can be reduced to finding all high-degree vertices in the input graph G. The previous subroutine is then employed to determine whether $G[V \setminus C]$ contains zero edges or many edges, where C contains the top k vertices with the largest degree. This algorithm can be extended to graphs with bounded arboricity (see Sect. 3.2).

For the diameter $\leq k$ property on sparse graphs satisfying $|E| \leq C \cdot |V|$, given the parameter k, we say a vertex v is *bad* if there is no high-degree vertex (of degree at least $\frac{4C}{\varepsilon}$) within its BFS queue of depth $\lfloor \frac{k}{2} \rfloor$ (whereas the algorithm in [22] for general graphs checks its BFS queue of depth $\frac{k}{2} + \frac{k}{2^{i+1}-2}$ for feasible i). We demonstrate that the number of bad vertices can be used to distinguish if the diameter of the input graph is $\leq k$, or if the input graph is ε-far from hav-

ing diameter $\leq \begin{cases} k+1 & \text{if } k \text{ is odd} \\ k+2 & \text{if } k \text{ is even} \end{cases}$. Subsequently, based on the algorithm

proposed in [22], we improve the upper bound of $\beta(k)$ in their result (see Sect. 4 and the full version of the paper).

Due to the space constraints, most of our proofs are deferred to the full version of this paper.

2   Testing the k-Dominating Set Property

In this section, we study algorithms for testing k-dominating set property. The k-dominating set property is a first-order logic property of $\exists * \forall *$ type. Due to the space constraints, we only provide a detailed proof for planar graphs, while deferring the discussion on general sparse graphs to the full version of this paper.

We first describe a key subroutine designed to identify all vertices with degrees of at least εn in a sparse graph G, where $|E(G)| \leq Cn$ for some constant C. This subroutine achieves a success probability of at least 0.9. It only requires sampling $O(1)$ vertices with degrees up to $O(1)$ and traversing all neighbors of the sampled vertices.

Algorithm 1: Find all high-degree vertices on sparse graphs

Input: a sparse graph G of order n satisfies $|E(G)| \leq Cn$ for some
    constant C, proximity parameter ε
Output: a vertex set T containing all vertices with degrees of at least εn
1 Sample $\frac{16}{\varepsilon} \ln \left(\frac{20C}{\varepsilon} \right)$ vertices u.a.r., denote them by S;
2 **foreach** $v \in S$ **do**
3 $\quad$ Query $\deg(v)$;
4 $\quad$ **if** $\deg(v) \geq \frac{4C}{\varepsilon}$ **then**
5 $\quad\quad$ Discard it;
6 $\quad$ **else**
7 $\quad\quad$ **foreach** $i \in \Gamma_v$ **do**
8 $\quad\quad\quad$ Query $\deg(i)$;
9 $\quad\quad\quad$ **if** $\deg(i) \geq \varepsilon n$ **then**
10 $\quad\quad\quad\quad$ Place i into set T;
11 **return** T.

Lemma 1. *Given $\varepsilon \in (0,1)$ and a sparse graph G satisfying $|E(G)| \leq Cn$ for some constant C, Algorithm 1 will output a vertex set T encompassing all*

vertices in G with degrees of at least εn with probability at least 0.9, and the query complexity is $O\left(\frac{C}{\varepsilon^2}\log\left(\frac{C}{\varepsilon}\right)\right)$.

For testing the k-dominating set property on planar graphs, we introduce the following lemma, a.k.a. Bonferroni inequalities.

Lemma 2. *(cf. [12]) Let $A_1, A_2, \cdots, A_n$ be finite sets. For each $k = 1, 2, \cdots, n$,*

$$\left|\bigcup_{r=1}^{n} A_r\right| \left\{\begin{matrix} \geq \\ \leq \end{matrix}\right\} \sum_{r=1}^{k}(-1)^{r+1}w(r) \ \text{if}\ k \ \text{is} \ \begin{cases} even \\ odd \end{cases},$$

where $w(r) = \displaystyle\sum_{1\leq i_1 < i_2 < \cdots < i_r \leq n} |A_{i_1} \cap A_{i_2} \cap \cdots \cap A_{i_r}|.$

We present the following lemma to distinguish if the input planar graph has a k-dominating set or is ε-far from satisfying the k-dominating set property, where $k \geq 2$.

Lemma 3. *Let $\varepsilon \in (0,1)$ be any constant, $k \geq 2$ be an integer and G be a planar graph. If G has a k-dominating set, then there exists a vertex subset D of size k such that $\displaystyle\sum_{i\in D}\deg(i) - \sum_{(i,j)\in\binom{D}{2}}|\Gamma_i \cap \Gamma_j| \geq n - k - 2\binom{k}{3}$. Conversely, if G is ε-far from satisfying the k-dominating set property, then for any vertex subset D' of size k, it holds that $\displaystyle\sum_{i\in D'}\deg(i) - \sum_{(i,j)\in\binom{D'}{2}}|\Gamma_i \cap \Gamma_j| \leq (1-\varepsilon)n$.*

Given an integer k and the proximity parameter $\varepsilon \in (0,1)$, we say a vertex v is *high-degree* if $\deg(v) \geq \frac{\varepsilon n}{k^2}$. Furthermore, a pair of high-degree vertices (i,j) is considered *well-connected* if $|\Gamma_i \cap \Gamma_j| \geq \frac{\varepsilon n}{k^2}$.

Testing k-dominating set property on planar graphs. We present the following algorithm. When $k \geq 2$ (for $k = 1$, we only need to test if there exists a vertex of degree $n - 1$ in G), the algorithm is executed in three phases: first, we invoke Algorithm 1 to find a set T that contains all high-degree vertices; next, we find all those well-connected pairs in G with high probability and estimate their common neighbors with adequate precision; finally, we check whether there exists a k'-set D' with sufficient large $\displaystyle\sum_{i\in D'}\deg(i) - \sum_{(i,j)\in\binom{D'}{2}}|\Gamma_v \cap \Gamma_s|$, where $k' \leq k$.

Algorithm 2: Testing the k-dominating set property on planar graphs

Input: a planar graph G, a constant k, proximity parameter ε
Output: Accept or Reject
1 Invoke Algorithm 1 to output a vertex set T that includes all high-degree vertices in G, where the proximity parameter is set to $\frac{\varepsilon}{k^2}$ and C is set to 3;
2 Initialize an all-zero vector V of length $|\binom{T}{2}|$, where each entry corresponds to a 2-set of T, an integer $N = 0$, and an empty set A;

Now we present the proof of Theorem 2.

---

3 Sample $\frac{10k^4}{\varepsilon^2} \cdot \ln\left(\frac{3600k^2}{\varepsilon}\right)$ vertices u.a.r., denote them by S';

4 foreach $i \in S'$ **do**

5 | Query $\deg(i)$;

6 | **if** $\deg(i) < \frac{24k^2}{\varepsilon}$ **then**

7 | | $N \leftarrow N + 1$, and initialize an empty set Γ_i^h;

8 | | **foreach** $j \in \Gamma_i$ **do**

9 | | | **if** $\deg(j) \geq \frac{\varepsilon n}{k^2}$ **then**

10 | | | | Place j into set Γ_i^h;

11 | | | **foreach** 2-*set* (u, v) *of* Γ_i^h **do**

12 | | | | Increment the corresponding entry of (u, v) in V by 1;

13 foreach *non-zero entry* (u, v) *of* V **do**

14 | **if** $\frac{V[(u,v)]}{N} > \frac{\varepsilon}{4k^2}$ **then**

15 | | Add element $\left((u,v), V[(u,v)] \cdot \frac{n}{N}\right)$ to A, where $|\widehat{\Gamma_u \cap \Gamma_v}| = V[(u,v)] \cdot \frac{n}{N}$ is the estimation value of $|\Gamma_u \cap \Gamma_v|$;

16 For each k'-subset D' of T with degree-sum $> (1 - \frac{5\varepsilon}{8})n$, where $1 \leq k' \leq k$, compute $\sum_{i \in D'} \deg(i) - \sum_{(i,j) \in \binom{D'}{2}} |\widehat{\Gamma_v \cap \Gamma_s}|$;

17 if $\max_{D'} \left\{ \sum_{i \in D'} \deg(i) - \sum_{(i,j) \in \binom{D'}{2}} |\widehat{\Gamma_v \cap \Gamma_s}| \right\} > (1 - \frac{5\varepsilon}{8})n$ **then**

18 | **return** Accept.

19 else

20 | **return** Reject.

Proof (of Theorem 2). In the first phase, according to Lemma 1, with a probability of at least 0.9, we will identify all high-degree vertices by employing Algorithm 1. Subsequently, we need to find all pairs of vertices in G that are well-connected with high probability, while simultaneously estimating the cardinality of their common neighbors with sufficient precision.

Let K be a threshold number, and let t_K denote the number of vertices in G whose degrees are at least K. Consider such a graph H, where $V(H)$ comprises all high-degree vertices in G and two vertices in H are adjacent if they are well-connected in G. Consequently, $|E(H)|$ is the number of well-connected pairs in G. Observing that H must be planar (if H contains a $K_{3,3}$ or K_5 as a minor, then G does as well), it follows that $|E(H)| \leq 3t_{\frac{\varepsilon n}{k^2}} - 6 < \frac{18k^2}{\varepsilon}$.

Sample $\frac{10k^4}{\varepsilon^2} \cdot \ln\left(\frac{3600k^2}{\varepsilon}\right)$ vertices in G uniformly at random, let N denote the number of sampled vertices with degrees less than $\frac{24k^2}{\varepsilon}$. Note that the number of vertices in G with degrees less than $\frac{24k^2}{\varepsilon}$ is at least $n - t_{\frac{24k^2}{\varepsilon}} \geq (1 - \frac{\varepsilon}{4k^2})n$. Define χ_i as an indicator such that $\chi_i = 1$ if the degree of the i-th sampled vertex is less than $\frac{24k^2}{\varepsilon}$; otherwise, $\chi_i = 0$. Then we have $\Pr[\chi_i = 1] \geq 1 - \frac{\varepsilon}{4k^2}$, and thus

$E[N] \geq \frac{9k^4}{\varepsilon^2} \cdot \ln\left(\frac{3600k^2}{\varepsilon}\right)$. Applying the Chernoff bound, we obtain

$$\Pr\left[N < \frac{8k^4}{\varepsilon^2} \cdot \ln\left(\frac{3600k^2}{\varepsilon}\right)\right] \leq \Pr[N < (1 - \frac{1}{9})E[N]] < 0.01.$$

For any well-connected pair (v, s), the probability that a randomly sampled vertex $r \in G$ belongs to $\Gamma_v \cap \Gamma_s$ is $\frac{|\Gamma_v \cap \Gamma_s|}{n} \geq \frac{\varepsilon}{k^2}$. By the Law of Total Probability,

$$\Pr\left[r \in \Gamma_v \cap \Gamma_s\right] = \Pr\left[r \in \Gamma_v \cap \Gamma_s \middle| \deg(r) < \frac{24k^2}{\varepsilon}\right] \cdot \Pr\left[\deg(r) < \frac{24k^2}{\varepsilon}\right] +$$

$$\Pr\left[r \in \Gamma_v \cap \Gamma_s \middle| \deg(r) \geq \frac{24k^2}{\varepsilon}\right] \cdot \Pr\left[\deg(r) \geq \frac{24k^2}{\varepsilon}\right]$$

$$\implies \Pr\left[r \in \Gamma_v \cap \Gamma_s\right] \middle/ \left(1 - \frac{\varepsilon}{4k^2}\right) \geq \Pr\left[r \in \Gamma_v \cap \Gamma_s \middle| \deg(r) < \frac{24k^2}{\varepsilon}\right] \quad \text{and}$$

$$\Pr\left[r \in \Gamma_v \cap \Gamma_s\right] - \frac{\varepsilon}{4k^2} \leq \Pr\left[r \in \Gamma_v \cap \Gamma_s \middle| \deg(r) < \frac{24k^2}{\varepsilon}\right].$$

Denote $\Pr\left[r \in \Gamma_v \cap \Gamma_s \middle| \deg(r) < \frac{24k^2}{\varepsilon}\right]$ by p, by combining the above two inequalities, we obtain $\left|p - \Pr\left[r \in \Gamma_v \cap \Gamma_s\right]\right| < \frac{\varepsilon}{2k^2}$.

Let $I_j^{v,s}$ be an indicator such that $I_j^{v,s} = 1$ if the degree of j-th sampled vertex in G is less than $\frac{24k^2}{\varepsilon}$ and the vertex lies in $\Gamma_v \cap \Gamma_s$; otherwise, $I_j^{v,s} = 0$. Let $V_N[(v, s)]$ denote the number of instances where the common neighbors of v and s are identified through N randomly sampled vertices in G, each with a degree less than $\frac{24k^2}{\varepsilon}$. We have $\Pr[I_j^{v,s} = 1] = p$, and thus $E[V_N[(v, s)]] = \sum_{j=1}^{N} I_j^{v,s} = Np$.

When $N \geq \frac{8k^4}{\varepsilon^2} \cdot \ln\left(\frac{3600k^2}{\varepsilon}\right)$, by additive Chernoff bound,

$$\Pr\left[|V_N[(v, s)] - p \cdot N| \geq \frac{\varepsilon N}{4k^2}\right] \leq 2\exp\left(-\frac{\varepsilon^2 N}{8k^4}\right)$$

$$\implies \Pr\left[\left|\frac{V_N[(v, s)] \cdot n}{N} - pn\right| \geq \frac{\varepsilon n}{4k^2}\right] \leq \frac{\varepsilon}{1800k^2}.$$

Given that there are at most $\frac{18k^2}{\varepsilon}$ well-connected pairs in G, with probability at least $(1 - 0.01) \cdot \left(1 - \frac{18k^2}{\varepsilon} \cdot \frac{\varepsilon}{1800k^2}\right) > 0.98$, Algorithm 2 will successfully identify all well-connected pairs in G. For each well-connected pair (v, s), the algorithm outputs the estimate $\widehat{|\Gamma_v \cap \Gamma_s|} = \frac{V_N[(v,s)] \cdot n}{N}$, satisfying

$$\left|\widehat{|\Gamma_v \cap \Gamma_i|} - |\Gamma_v \cap \Gamma_i|\right| \leq \left|\widehat{|\Gamma_v \cap \Gamma_i|} - pn\right| + \left|pn - |\Gamma_v \cap \Gamma_i|\right| < \frac{\varepsilon n}{4k^2} + \frac{\varepsilon n}{2k^2} = \frac{3\varepsilon n}{4k^2}.$$

Suppose that G has a k-dominating set D. The number of vertices in D with degrees less than $\frac{\varepsilon n}{k^2}$ cannot exceed $k - 1$. Let k' denote the number of high-degree vertices in D. Consequently, the degree-sum that we fail to incorporate is less than $\frac{\varepsilon n}{4}$. Hence, by Lemma 3 and the preceding proof, with a probability

of at least $0.9 \times 0.98 > \frac{2}{3}$, $\max_{D'} \left\{ \sum_{i \in D'} \deg(i) - \sum_{(i,j) \in \binom{D'}{2}} |\widehat{\Gamma_i \cap \Gamma_j|} \right\} > n - k -$

$2\binom{k}{3} - \frac{\varepsilon n}{4} - \binom{k}{2} \cdot \frac{3\varepsilon n}{4k^2} > (1 - \frac{5\varepsilon}{8})n$, where T comprises all the high-degree vertices identified by Algorithm 1, $1 \leq k' \leq k$ and D' is a k'-subset of T. Note that for any k'-subset D' of T, $\sum_{i \in D'} \deg(i) - \sum_{(i,j) \in \binom{D'}{2}} |\Gamma_i \cap \Gamma_j| \leq \sum_{i \in D'} \deg(i)$.

Thus, it suffices to consider those k'-subsets of T with a degree-sum greater than $(1 - \frac{5\varepsilon}{8})n$.

If G is ε-far from having a k-dominating set, then it is also ε-far from having a k'-dominating set, where $1 \leq k' \leq k$. By Lemma 3, for any vertex subset D' of size k' in G, the value $\sum_{i \in D'} \deg(i) - \sum_{(i,j) \in \binom{D'}{2}} |\Gamma_i \cap \Gamma_j|$ is at

most $(1 - \varepsilon)n$, following the same reasoning, with a probability of at least $\frac{2}{3}$,

$\max_{D'} \left\{ \sum_{i \in D'} \deg(i) - \sum_{(i,j) \in \binom{D'}{2}} |\widehat{\Gamma_i \cap \Gamma_j|} \right\} < (1 - \varepsilon)n + \binom{k}{2} \cdot \frac{3\varepsilon n}{4k^2} < (1 - \frac{5\varepsilon}{8})n.$

According to Lemma 1, by setting the proximity parameter to $\frac{\varepsilon}{k^2}$ and C to 3, the query complexity of Algorithm 2 is bounded by $O\left(\frac{k^4}{\varepsilon^2} \cdot \log(\frac{k^2}{\varepsilon})\right) + \frac{10k^4}{\varepsilon^2} \ln\left(\frac{3600k^2}{\varepsilon}\right) \cdot \frac{48k^2}{\varepsilon} = O\left(\frac{k^6}{\varepsilon^3} \log(\frac{k^2}{\varepsilon})\right)$.

3   Testing the k-Vertex Cover Property

Now we consider the problem of testing the k-vertex cover property on bounded arboricity graphs. The k-vertex cover property is a first-order logic property of $\exists * \forall *$ type and a monotone property.

3.1   Testing the k-Vertex Cover Property on Planar Graphs

We introduce a subroutine to determine whether a planar graph is either $\frac{\varepsilon}{10}$-close to being empty or ε-far from it. This procedure operates via constant-sized vertex sampling in the input graph and examines whether an adequate proportion of sampled vertices exhibit null degrees.

Algorithm 3: Testing a planar graphs is $\frac{\varepsilon}{10}$-close to being empty or ε-far from being empty

---

Input: a planar graph G of order n, proximity parameter ε
Output: Accept or Reject
1 Sample $\frac{1500}{\varepsilon}$ vertices in G uniformly at random, denote them by S'. Initialize $N = 0$;
2 **foreach** $i \in S'$ **do**
3   Query $\deg(i)$;
4   **if** $\deg(i) \neq 0$ **then**
5        $N \leftarrow N + 1$;
6 **if** $N < 400$ **then**
7   **return** Accept.
8 **else**
9   **return** Reject.

Theorem 5. *Let $\varepsilon \in (0,1)$ be any constant and G be a planar graph. Then Algorithm 3 will*

- *return Accept with probability ≥ 0.9 if G is $\frac{\varepsilon}{10}$-close to being empty;*
- *return Reject with probability ≥ 0.9 if G is ε-far from being empty;*

and the query complexity is $O\left(\frac{1}{\varepsilon}\right)$.

The following lemma distinguishes between two cases: when a planar graph satisfies the k-vertex cover property, and when it is ε-far from this property.

Lemma 4. *Let $\varepsilon \in (0,1)$ be any constant, k be any positive integer and G be a planar graph. If a planar graph G has a k-vertex cover, then there exists a vertex subset C' of size k such that $G[V \backslash C']$ is an empty graph. Conversely, if a planar graph G is ε-far from satisfying k-vertex cover property, then for any vertex subset C' of size k, it holds that $|E(G[V \backslash C'])| \geq \varepsilon n$.*

In this section, given an integer k and $\varepsilon \in (0,1)$, we say a vertex v is *high-degree* if $\deg(v) \geq \frac{\varepsilon n}{10k}$. Now, we present the following algorithm Algorithm 4 to finish the testing work. The algorithm is executed in two phases: first, with high probability, we identify the top k high-degree vertices in G and place them into set C'; second, we invoke Algorithm 3 to test whether $G[V \backslash C']$ is $\frac{\varepsilon}{10}$-close to being empty or ε-far from being empty. However, since we cannot query the information about graph $G[V \backslash C']$ directly, minor modifications are required in step 4 of Algorithm 3 during the invocation process.

Algorithm 4: Testing k-vertex cover property on planar graphs

---

Input: a planar graph G of order n, proximity parameter ε
Output: Accept or Reject

1 Sample $\frac{160k}{\varepsilon} \cdot \ln(100k)$ vertices uniformly at random, denote them by S;
   foreach $v \in S$ **do**
2 | Query $\deg(v)$;
3 | **if** $\deg(v) < \frac{120k}{\varepsilon}$ **then**
4 | | **foreach** $i \in \Gamma_v$ **do**
5 | | | Query $\deg(i)$. If $\deg(i) \geq \frac{\varepsilon n}{10k}$, then place i into set C';
6 | **else**
7 | | Discard it;
8 **if** $|C'| > k$ **then**
9 | **return** Reject.
10 **else**
11 | Invoke Algorithm 3 on $G[V \setminus C']$, while in the step 4, we augment N by
      one if $\deg(i) > |C'|$ or $\deg(i) \leq |C'|$ but $\Gamma_i \not\subseteq C'$;
12 | **if** *Algorithm 3 return Accept* **then**
13 | | **return** Accept.
14 | **else**
15 | | **return** Reject.

We have the following theorem on testing k-vertex cover on planar graphs.

Theorem 6. *Let $\varepsilon \in (0,1)$ be any constant, k be any positive integer. Then Algorithm 4 is a ε-tester for the k-vertex cover property on planar graphs, and the query complexity is $O(\frac{k^2 \log k}{\varepsilon^2})$.*

3.2 Extending the Analysis to Bounded Arboricity Graphs

In this subsection, we consider testing the k-vertex cover property on bounded arboricity graphs. It is worth noting that Lemma 4 remains applicable to sparse graphs. The major task is to adapt Algorithm 3 to distinguish between two cases: when the input graph is $\frac{\varepsilon}{\lambda}$-close to being empty, and it is ε-far from being empty, where λ is a constant determined by the proof.

For general sparse graphs, a hard example consists of a clique on $\sqrt{n}$ vertices plus $n - \sqrt{n}$ isolated vertices. Determining whether such a graph has a 1-vertex cover may require $\Theta(\sqrt{n})$ queries, as all edges are concentrated within a small subgraph. However, for certain significant subclasses, e.g. bounded arboricity class, the emptiness of the graph can be tested with a constant number of queries. The following characterization of arboricity due to Nash-Williams.

Theorem 7. *(cf. [20]) Let $G = (V, E)$ be a graph. For a subgraph H of G, let n_H and m_H denote the number of vertices and edges, respectively, in H. Then*

$$\alpha(G) = \max_H \{\lceil m_H/(n_H - 1) \rceil\},$$

where the maximum is taken over all subgraphs H of G.

This theorem demonstrates that a graph G with bounded arboricity exhibits sparsity uniformly across its structure. For instance, any planar graph with n vertices contains at most $3n-6$ edges. Consequently, by Nash-Williams' formula, the arboricity of planar graphs is at most 3. When G is ε-far from being empty, the number of vertices that are incident to edges is at least $\frac{\varepsilon|V(G)|}{\alpha}+1$. Therefore, Algorithm 4 can be appropriately modified to test the k-vertex cover property on graphs with bounded arboricity α (in this case, the algorithm will identify the top k vertices with degree at least $\frac{\varepsilon n}{10\alpha k}$), achieving a query complexity of $O\left(\frac{\alpha^3 k^2 \log k}{\varepsilon^2}\right)$, thereby establishing Theorem 3.

4   Testing the Diameter $\leq k$ Property

Now we consider the problem of testing the diameter $\leq k$ property on sparse graphs satisfying $|E(G)| \leq C|V(G)|$ for some constant C. The diameter $\leq k$ property is a first-order logic property of $\forall * \exists *$ type.

The following lemma demonstrates that diameter can be used to upper bound the number of vertices of a bounded-degree graph.

Lemma 5. *(cf. [5]) Let $n_{d,k}$ be the maximum possible number of vertices for a graph with degree at most d and diameter k. Then $n_{d,k} \leq M_{d,k}$, where $M_{d,k}$ is the Moore bound:*

$$M_{d,k} = \begin{cases} 1 + d\frac{(d-1)^k - 1}{d-2} & \text{if } k > 2 , \\ 1 + 2k & \text{if } k = 2 . \end{cases}$$

In this section, given a proximity parameter $\varepsilon \in (0,1)$, we say a vertex v is *high-degree* if $\deg(v) \geq \frac{4C}{\varepsilon}$. Define *the r-ball of a vertex v* to be the set of all vertices at a distance no more than r from v (including v itself), and denote it by $B_r(v) = \{u \mid d_G(u,v) \leq r\}$. We say a vertex v is *bad* if it lies outside the $\lfloor \frac{k}{2} \rfloor$-balls of all high-degree vertices, i.e. $v \in V(G) \setminus \bigcup_{i \in V^h} B_{\lfloor \frac{k}{2} \rfloor}(i)$, where V^h is the set of all high-degree vertices in G. Note that this definition is equivalent to saying that there is no high-degree vertex within $B_{\lfloor \frac{k}{2} \rfloor}(v)$ if v is bad.

Now, we present the following theorem to distinguish between two cases: when the diameter of a sparse graph $\leq k$, and when it is ε-far from having a diameter $k+2$ (when k is even) or $k+1$ (when k is odd).

Theorem 8. *Let $\varepsilon \in (0,1)$ be any constant, $k \geq 2$ be an integer and G be a sparse graph of order n satisfying $|E(G)| \leq Cn$. If $diam(G) \leq k$, then the number of bad vertices in G is at most $M_{\frac{4C}{\varepsilon},2k}$, where $M_{\frac{4C}{\varepsilon},2k}$ is the Moore bound as defined in Lemma 5. Conversely, if G is ε-far from having a diameter*
$$\begin{cases} k+1 & \text{if } k \text{ is odd} \\ k+2 & \text{if } k \text{ is even} \end{cases}, \text{ then the number of bad vertices in } G \text{ is at least } \frac{\varepsilon^2 n}{8C}.$$

Once we have the above theorem, we can design an algorithm for testing diameter $\leq k$ by checking whether the fraction of bad vertices is sufficiently large. We defer the algorithm and the proof of Theorem 4 (i.e., the result on testing diameter $\leq k$) to the full version of the paper.

References

1. Adler, I., Köhler, N., Peng, P.: On testability of first-order properties in bounded-degree graphs and connections to proximity-oblivious testing. SIAM J. Comput. **53**(4), 825–883 (2024)

2. Alon, N., Fischer, E., Newman, I., Shapira, A.: A combinatorial characterization of the testable graph properties: it's all about regularity. In: Proceedings of the Thirty-Eighth Annual ACM Symposium on Theory of Computing, pp. 251–260. STOC '06 (2006)

3. Babu, J., Khoury, A., Newman, I.: Every property of outerplanar graphs is testable. In: APPROX/RANDOM 2016. LIPIcs, vol. 60, pp. 21:1–21:19 (2016)

4. Benjamini, I., Schramm, O., Shapira, A.: Every minor-closed property of sparse graphs is testable. Adv. Math. **223**(6), 2200–2218 (2010)

5. Bollobás, B.: Modern Graph Theory. Graduate Texts in Mathematics 184, Springer-Verlag New York, 1 edn. (1998)

6. Bondy, J.A., Murty, U.S.R.: Graph theory with applications. North Holland (1976)

7. Czumaj, A., Monemizadeh, M., Onak, K., Sohler, C.: Planar graphs: random walks and bipartiteness testing. In: FOCS 2011, p. 423–432. IEEE Computer Society (2011)

8. Czumaj, A., Sohler, C.: A characterization of graph properties testable for general planar graphs with one-sided error (it's all about forbidden subgraphs). FOCS, pp. 1525–1548 (2019)

9. Eden, T., Levi, R., Ron, D.: Testing C_k-freeness in bounded-arboricity graphs. In: ICALP 2024. LIPIcs (2024)

10. Eden, T., Ron, D., Rosenbaum, W.: The arboricity captures the complexity of sampling edges. In: ICALP 2019. LIPIcs, vol. 132, pp. 52:1–52:14 (2019)

11. Esperet, L., Norin, S.: Testability and local certification of monotone properties in minor-closed classes. In: ICALP 2022. LIPIcs, vol. 229, pp. 58:1–58:15. Schloss Dagstuhl – Leibniz-Zentrum für Informatik (2022)

12. Galambos, J., Xu, Y.: A new method for generating bonferroni-type inequalities by iteration. Math. Proc. Cambridge Philos. Soc. **107**(3), 601–607 (1990)

13. Goldreich, O., Ron, D.: Property testing in bounded degree graphs. In: Proceedings of the Twenty-Ninth Annual ACM Symposium on Theory of Computing, pp. 406–415. STOC '97, Association for Computing Machinery (1997)

14. Iwama, K., Yoshida, Y.: Parameterized testability. ACM Trans. Comput. Theory **9**(4) (2017)

15. Kaufman, T., Krivelevich, M., Ron, D.: Tight bounds for testing bipartiteness in general graphs. SIAM J. Comput. **33**(6), 1441–1483 (2004)

16. Kawarabayashi, K.i., Yoshida, Y.: Testing subdivision-freeness: property testing meets structural graph theory. In: Proceedings of the Forty-Fifth Annual ACM Symposium on Theory of Computing, pp. 437–446. STOC '13, Association for Computing Machinery (2013)

17. Kumar, A., Seshadhri, C., Stolman, A.: Random walks and forbidden minors iii: poly $(d\varepsilon^{-1})$-time partition oracles for minor-free graph classes. In: FOCS 2021, pp. 257–268. IEEE Computer Society (2022)

18. Kusumoto, M., Yoshida, Y.: Testing forest-isomorphism in the adjacency list model. In: Automata. Languages, and Programming, pp. 763–774. Springer, Berlin Heidelberg (2014)

19. Levi, R., Shoshan, N.: Testing hamiltonicity (and other problems) in minor-free graphs. In: APPROX/RANDOM 2021. LIPIcs, vol. 207, pp. 61:1–61:23 (2021)

20. Nash-Williams, C.S.A.: Edge-disjoint spanning trees of finite graphs. J. London Math. Soc. **s1-36**(1), 445–450 (01 1961)
21. Newman, I., Sohler, C.: Every property of hyperfinite graphs is testable. In: Proceedings of the Forty-Third Annual ACM Symposium on Theory of Computing, pp. 675–684. STOC '11, Association for Computing Machinery (2011)
22. Parnas, M., Ron, D.: Testing the diameter of graphs. Random Struct. Algorithms **20**(2), 165–183 (2002)

Massively Parallel Approximate Steiner Tree Algorithms

Chilei Wang, Qiang-Sheng Hua$^{(\boxtimes)}$, and Hai Jin

National Engineering Research Center for Big Data Technology and System/Services Computing Technology and System Lab/Cluster and Grid Computing Lab, School of Computer Science and Technology, Huazhong University of Science and Technology, Wuhan 430074, People's Republic of China
qshua@hust.edu.cn

Abstract. This work studies the approximate Steiner tree problem in the *Massively Parallel Computation* (MPC) model where each machine has $O(n^\sigma)$ memory and $\sigma \in (0,1)$. n is the number of nodes in a graph. We focus on the undirected connected weighted graphs with shortest path diameter D and a terminal set S. The shortest path diameter is the minimum number of edges required for the shortest path constituting a weighted graph's diameter. The straightforward approach takes $O(n)$ rounds and $O(n^{3-\sigma/2})$ total memory, which is inefficient. To simplify the straightforward approach and reduce the round complexity, we design a constant-round subroutine to compute the routing table and combine algebraic strategies with recursive methods to compute the Steiner tree efficiently. By these techniques, we give the first parallel $2(1 - 1/|S|)$-approximate Steiner tree algorithm that requires $O(\sigma^{-1} \log n + D)$ rounds with the same memory size and the same approximation ratio. Moreover, we extend the straightforward approach to the MPC model with $O(n)$ memory per machine, which takes $O(\log n)$ rounds and significantly outperforms the existing algorithm [21] when $D \gg O(\log n)$.

Keywords: The MPC model · Steiner Tree · Round Complexity · SPF

1 Introduction

Over the last several years, the study on graph algorithms in the *Massively Parallel Computation* (MPC) model has become the focus of growing interest, including graph connectivity [2], shortest paths [10,11], and *minimum spanning tree* (MST) [9]. Furthermore, the significant development gap between storage capability and CPU performance causes communication to be a pivotal determinant of the algorithm's running speed. Hence, in the MPC model, reducing the communication complexity of graph algorithms is of utmost importance.

The *Steiner Tree* (ST) problem, a typical combinatorial optimization problem, has been widely applied in network design and computational biology [4]. The ST problem is defined as finding a tree with the smallest edge weights,

This work was supported in part by National Science and Technology Major Project 2022ZD0115301.

F. V. Fomin and M. Xiao (Eds.): COCOON 2025, LNCS 15984, pp. 110–124, 2026.
https://doi.org/10.1007/978-981-95-0218-9_9

connecting all nodes in a specific set S of an undirected graph. Actually, the ST problem is inherently connected to the MST and shortest path problems. Concretely, it reduces to the MST problem when S contains all nodes and to the shortest path problem if S comprises only two nodes. However, unlike MST and shortest path problems that have polynomial-time solutions, the ST problem is NP-hard [12], which cannot be solved in polynomial time unless $P = NP$.

Thus, most researchers concentrate on the approximate solutions for the ST problem [22]. In the last four decades, there have been numerous studies on the *approximate Steiner tree* (AST) problem in the sequential model [17] and the parallel and distributed models [4], including the Congested Clique model [21] and the Congest model [6]. These highlight the necessity and importance of studying the ST problem.

Although the MPC algorithms designed for the MST or shortest path problems are fairly rich, as far as we know, no MPC algorithm has been proposed for the ST problem so far. In addition, the study of the ST problem can be extended to related optimization issues. For example, the traveling salesman problem, which is used extensively in logistics management and other fields [18]. Therefore, this paper focuses on the AST problem for undirected weighted dense graphs in the MPC model and aims to design an efficient MPC algorithm for it.

1.1 Our Contributions

In summary, our contributions are below:

- This paper proposes the first parallel AST algorithm in the MPC model (refer to Theorem 1 and Algorithm 4 in Sect. 5 for details).
- Compared with the straightforward approach (see Sect. 4.1), our parallel algorithm can greatly reduce the round complexity when the graph's shortest path diameter [21] is smaller than n, where n is the total number of nodes.
- Technically, we design a simpler subroutine to compute the routing table (see Definition 4 and Algorithm 2 in Sect. 5.1) and simplify the process of the straightforward approach by combining the algebraic methods and the recursive strategies, leading to a reduction in round complexity (see Sect. 5.2).

Next, we present the main results below, which can be readily proved by combining Lemma 3, Lemma 5, and the analysis in Sect. 6:

Theorem 1. *In the MPC model, each machine has $O(n^\sigma)$ memory ($\sigma \in (0, 1)$ is a constant), given a graph $G = (V, E)$ with n nodes and a set S of terminals, there is a parallel algorithm that takes $O(\sigma^{-1} \log n + D)$ rounds and $O(n^{3-\sigma/2})$ total memory to solve the $2(1 - 1/|S|)$-approximate ST. Moreover, in the MPC model with $O(n)$ memory for each machine, the AST can be solved in $O(\log n)$ rounds and $O(n^{2.5})$ total memory.*

Then, we compare our algorithms with existing parallel AST algorithms. The results are outlined in Table 1. In the MPC model with $O(n^\sigma)$ memory assigned to each machine ($\sigma \in (0, 1)$ is a constant), our parallel algorithm has a lower

Table 1. Comparison between our parallel algorithms and the existing parallel approximate Steiner tree algorithms

Algo. (MPC($O(n^\sigma)$)†)	Round complexity	Total memory	Types of Algo.
The straightforward algorithm	$O(n)$	$O(n^{3-\sigma/2})$	Deterministic
Our work	$O(D + \sigma^{-1}\log n)$	$O(n^{3-\sigma/2})$	Deterministic
Algo. (MPC($O(n)$)‡)	Round complexity	Total memory	Types of Algo.
Saikia and Karmakar [21]	$O(\log\log n + D)$	$O(n^2)$	Deterministic
Our work	$O(\log n)$	$O(n^{2.5})$	Deterministic

Note: (1)† means the MPC model with $O(n^\sigma)$ memory per machine and $\sigma \in (0, 1)$;
(2) ‡ represents the MPC model with $O(n)$ memory for each machine.

round complexity compared to the straightforward approach, particularly when $D \ll n$. Furthermore, when $D = O(\log n)$, the round complexity required by our algorithm is $O(\log n)$, which matches the best deterministic MST algorithm with $O(n^2)$ total memory in [9]. In the MPC model with $O(n)$ memory for each machine, when $D \gg O(\log n)$, our algorithm outperforms that of Saikia and Karmakar [21][1], with the cost of increasing the total memory. However, it still has a certain gap compared to the best MST algorithm in [19], which takes only constant rounds.

2 Related Work

Over the past thirty years, designing parallel and distributed AST algorithms [1] has been primarily based on the sequential algorithms from [17] and [22]. Among these, the algorithms designed for the Congested Clique model and the Congest model are most relevant to those in the MPC model.

In the Congest model, to solve the AST problem, Chen et al. [7] proposed the first distributed algorithm, which takes $O(n^2)$ rounds[2] and has an approximation ratio of $2(1 - 1/|S|)$. Then, Chatermosook et al. [6] presented a distributed 2-approximate ST algorithm with $O(n \log n)$ rounds. The two algorithms above are deterministic. Later, Khan et al. [16] designed a randomized ST algorithm with $O(\log n)$-approximation ratio, using $O(D \log^2 n)$ rounds. Next, Saikia and Karmakar [20] presented a deterministic $2(1 - 1/|S|)$-approximate ST algorithm that requires $O(D + \sqrt{n} \log^* n)$ rounds, combining the methods from [17] and [22]. Following this, for the ST problem with the same approximation ratio in the Congested Clique model, Saikia and Karmakar [21] designed the first deterministic algorithm with $O(D + \log\log n)$ rounds, using the steps in [20]. Recently,

---

[1] The algorithms designed for the Congested Clique model with $O(n)$ restricted memory per machine can be applied directly in the MPC model with $O(n)$ memory for each machine [3].

[2] We only show the rounds required for these distributed algorithms since the first goal in the MPC model is reducing the round complexity.

Kerger et al. [15] extended the $2(1-1/|S|)$-approximate ST problem to the quantum Congested Clique model and designed a randomized distributed algorithm using $O(n^{1/4})$ rounds.

Compared to the nearly absent research on the AST problem in the MPC model, the MST problem has been well studied. For instance, in the MPC model with $O(n^\sigma)$ memory for each machine, Coy and Czumaj [9] proposed a deterministic parallel MST algorithm taking $O(\log n)$ rounds and $O(n+m)$ total memory or $O(\log D)$ rounds and $O((n+m)^{1+\Omega(1)})$ memory, where m is the number of edges in a graph. Moreover, the latter matches $\Omega(\log D)$ lower bound [9].

3 Preliminaries

In this section, we will first introduce the MPC model and the problem definitions. Then, we will describe the parallel *all-pairs shortest path* (APSP) algorithm designed by Hajiaghayi et al. [14] in Sect. 3.3, which is an essential component of our parallel AST algorithm.

3.1 The MPC Model

As a theoretical computation model, the MPC model evolves from the MapReduce model [13]. It contains three key parameters: the number P of machines required, the memory M of each machine, and the size N of the input, such that $P = \tilde{O}(N/M)$ ($\tilde{O}(\cdot)$ hides a logarithmic factor). The MPC algorithms are executed in synchronous rounds. Initially, the input data are distributed across P machines. Within each round, machines perform local computations without inter-machine communication. At the end of each round, machines send or receive messages with each other to obtain the data for the computation in next round. The data size sent or received from other machines is no larger than M. Machines communicate with each other in a pairwise interconnected manner.

Generally, the MPC model contains three memory settings [11]: 1) the strongly superlinear memory ($M = O(n^{1+\sigma})$, where $\sigma > 0$ is a constant); 2) the near-linear memory ($M = O(n)$); 3) the strongly sublinear memory ($M = O(n^\sigma)$, where $\sigma \in (0,1)$ is a constant). This paper considers the latter two memory settings. In particular, we focus on designing parallel algorithms in the MPC model with strongly sublinear memory, which is more difficult and more scalable.

Complexity Measurements: In the MPC model, the first and most critical objective is to minimize an algorithm's round complexity. Reducing the total memory required by the algorithm is the secondary goal.

There are many useful and efficient subroutines in the MPC model with strongly sublinear memory, such as sorting an ordered set, broadcasting messages from one machine to other machines, and finding the maximum or minimum in a set, which all take only $O(\sigma^{-1})$ rounds [10,13].

3.2   The Problem Definitions

This paper mainly considers the undirected connected weighted dense graphs $G = (V, E, W)$. Specifically, V denotes the set of n nodes, E represents the set of m edges, and $W : E \to \mathbb{R}^+$ is the weight function. When $m = \Theta(n^2)$, it is a dense graph. For any two nodes $u_1, u_2 \in V$, $d(u_1, u_2)$ denotes the distance from u_1 to u_2. In addition, in the MPC model, the IDs of the nodes in V are marked as $1, 2, \cdots, n$, and the edges related to a node, ordered by the ID of another endpoint, are stored in a continuous set of machines. Namely, the node and its edges with smaller IDs will be stored in the machines with smaller IDs [10].

Definition 1 (The ST problem). *Considering the graphs above, given a node (or terminal) subset $S \subseteq V$, the ST problem is defined as finding a tree $T_0 = (V_0, E_0)$ that minimizes $\sum_{e \in E_0} w(e)$, where $S \subseteq V_0 \subseteq V$ and $E_0 \subseteq E$.*

It is worth mentioning that the nodes belonging to $V \setminus S$ are called the non-terminals and the nodes that belong to $V_0 \setminus S$ are known as Steiner nodes. Next, we introduce some important definitions for computing the AST.

Definition 2 (The *Shortest Path Forest* (SPF) [7,21]). *Given a graph $G = (V, E, W)$ above and a node subset S in Definition 1, the SPF for S, denoted as $G_{SPF} = (V, E_{SPF}, w)$, is a subgraph of G that comprises $|S|$ disjoint trees $T_j = (V_j, E_j, W)$, where $j \in \{1, 2, \cdots, |S|\}$. Moreover, these trees satisfy the following conditions: 1) Each V_j contains only one node s_j from S; 2) For any node $v \in V_j$, s_j is the unique source of v, where $s_j \in S$; 3) $\bigcup_{j=1}^{|S|} V_j = V$ and different node subsets are pairwise disjoint; 4) $\bigcup_{j=1}^{|S|} E_j = E_{SPF} \subseteq E$; 5) The shortest path from $v \in V_j$ to $s_j \in V_j$ in T_j is the shortest path from v to s_j in G.*

Definition 3 (The *Complete Distance Graph* (CDG) [17,22]). *Considering the graphs above and the subset S in Definition 1, the CDG, denoted by $K_S = (V, E')$, consists of all nodes in G and is connected, undirected, and weighted. The weight of each edge $(u_1, u_2) \in E'$ is the corresponding shortest distance between u_1 and u_2 in the original graph G.*

Definition 4 (The *Routing Table* (RT) [5]). *Given a graph $G = (V, E)$ above and its adjacency matrix A_0 of size $n \times n$, the RT R is an $n \times n$ matrix. Each element $R[u_1, u_2] = z$ in R, where $u_1, u_2, z \in V$, satisfies that $(u_1, z) \in E$ and z is the first intermediate node on a shortest path from u_1 to u_2.*

3.3   A Known Parallel APSP Algorithm

Next, we introduce the parallel exact APSP algorithm from [14], which is Algorithm 1 below. The primary idea of Algorithm 1 is to iteratively compute the product $A \star A$ for $\lceil \log n \rceil$ times, where A is the $n \times n$ adjacency matrix of the graph $G = (V, E)$. In addition, $\star$ represents the multiplication over semiring $(\min, +)$. That is to say, for any nodes u_1, u_2 in V, $(A \star A)_{u_1, u_2} = \min_{z \in V} \{A_{u_1, z} + A_{z, u_2}\}$ (see lines 4 - 5 of Algorithm 1). After $\lceil \log n \rceil$ iterations (see line 1 of Algorithm 1), we obtain the distance matrix A^n (replaced by $Dist$ in line 7 of Algorithm 1). Lemma 1 below presents the rounds required by Algorithm 1.

Algorithm 1. The Parallel APSP algorithm [14]

Input: An $n \times n$ matrix A distributed among $O(n^{2-\sigma})$ machines numbered as $P_{i,j,1}$, each of which has a memory size $O(n^{\sigma})$, where $i, j \in \{1, 2, \cdots, n^{1-\sigma/2}\}$.
Output: The distance matrix $Dist$
1: **for** $h = 1, 2, \cdots, \lceil \log n \rceil$ **do**
2:     The machine $P_{i,j,1}$ broadcasts the sub-matrix $A_{i,j}^{2^{h-1}}$ with size $n^{\sigma/2} \times n^{\sigma/2}$ to other $O(n^{1-\sigma/2})$ machines numbered by $P_{i,j,k}$, where $k \in \{2, 3, \cdots, n^{1-\sigma/2}\}$. ▷ If $h = 1$, then $A_{i,j}^1 = A_{i,j}$
3:     The machine $P_{i,j,1}$ broadcasts the sub-matrix $A_{i,j}^{2^{h-1}}$ to other $O(n^{1-\sigma/2})$ machines numbered by $P_{k,i,j}$, where $k \in \{1, 2, \cdots, n^{1-\sigma/2}\}$.
4:     Each machine $P_{i,j,k}$ computes $A_{i,k}^{2^h} = A_{i,j}^{2^{h-1}} \star A_{j,k}^{2^{h-1}}$, where $i, j, k \in \{1, 2, \cdots, n^{1-\sigma/2}\}$. ▷ $A_{i,j}^{2^{h-1}} \star A_{j,k}^{2^{h-1}}$ represents the multiplication of two matrices $A_{i,j}^{2^{h-1}}$ and $A_{j,k}^{2^{h-1}}$ over semiring $(\min, +)$
5:     Execute the broadcast operation among machines numbered by $P_{i,j,k}$ to compute partial sums of sub-matrix $A_{i,k}^{2^h}$ to obtain $A_{i,k}^{2^h}$ $(i, j, k \in \{1, 2, \cdots, n^{1-\sigma/2}\})$.
6: **end for**
7: Return $Dist = A^n$

Lemma 1 (Algorithm *1* [14]). *In the MPC model with $O(n^{\sigma})$ memory for each machine ($\sigma \in (0, 1]$ is a constant), a graph $G = (V, E)$ that has n nodes and its adjacency matrix A, Algorithm 1 computes the distance matrix $Dist$ using $O(\sigma^{-1} \log n)$ rounds and $O(n^{3-\sigma/2})$ total memory.*

Proof. As we can see, in only one round, Algorithm 1 can compute all the partial sums of all pairs of nodes (refer to line 4 in Algorithm 1). Since there are only broadcast operations in each iteration (refer to lines 2, 3, and 5 in Algorithm 1), it takes only constant rounds. Therefore, we obtain the round complexity for Algorithm 1, as shown in Lemma 1. □

4 Techniques

4.1 A Straightforward Parallel AST Algorithm

The general steps of addressing the AST problem in the Congested Clique model [21] and the Congest model [20] are below: **1) Construct the SPF; 2) Modify the weights of edges in the graph; 3) Construct the MST on the modified graph; 4) Delete the non-terminal leaves from the MST.** When exact algorithms are used for each step [15,20,21], the approximation ratio of the ST is $2(1 - 1/|S|)$. Through these steps, we obtain a *straightforward AST* (SAST) algorithm in the MPC model with strongly sublinear memory below:

Step 1: Initially, we compute the CDG (see Definition 3 in Sect. 3.2) using Algorithm 1. Then, we solve the RT (refer to Definition 4 in Sect. 3.2). Specifically, we denote the RT for $A^{2^j} = A^{2^{j-1}} \star A^{2^{j-1}}$ as R_j ($j \in \{1, 2, \cdots, \lceil \log n \rceil\}$). According to [5], if R_j is known, then for $A^{2^{j+1}} = A^{2^j} \star A^{2^j}$, $R_{j+1}[u_1, u_2] = $

$R_{j+1}[u_1, R_j[u_1, u_2]]$, where $u_1, u_2 \in V$. The resulting RT is $R = R_{\lceil \log n \rceil}$. This is the same as computing the distance matrix in Algorithm 1. Thus, the rounds and total memory needed in this step are $O(\sigma^{-1} \log n)$ and $O(n^{3-\sigma/2})$, respectively.

Finally, we construct the SPF (the $|S|$ disjoint trees $T_j = (V_j, E_j)$, where $j \in \{1, 2, \cdots, |S|\}$, as defined in Definition 2). Specifically, we determine the source of every node $v \in V$ by comparing all distances from v to the terminals in S, using the distance matrix $Dist$. It requires $O(\sigma^{-1})$ rounds and $O(|S|n)$ total memory. Next, we can determine the edges in the tree T_j by the RT R in only one round (for each node v in T_j, we find and mark its connecting edge (u, v), where u is the direct predecessor.). This is implied by the fact that the subpath of the shortest path is also the shortest path [8], so u has the same source as v.

Step 2: We modify the edge weights of the original graph G as follows: 1) Set all tree edges with a weight of 0; 2) Set the weights of all non-tree edges whose endpoints have the same source to ∞; 3) For any edge (u_1, u_2) whose endpoints have different sources, we set its weight to $w'(u_1, u_2) = d(s_i, u_1) + w(u_1, u_2) + d(u_2, s_j)$, where s_i and s_j are the sources of u_1 and u_2 ($i, j \in \{1, 2, \cdots, |S|\}$ and $i \neq j$), respectively. Next, we analyze the round complexity required in this step. Since each node must modify its edge weights using information from all other nodes, which is at most n. As the memory of each machine is $O(n^\sigma)$ ($\sigma \in (0, 1)$), it takes $O(n^{1-\sigma}/\sigma)$ rounds and $O(n^2)$ total memory using the broadcast operation.

Step 3: We solve the MST on the modified graph with an existing deterministic MPC algorithm [9], which takes $O(\log n)$ rounds and $O(n^2)$ total memory.

Step 4: We prune the MST computed in Step 3. This results in at most n updates since deleting a non-terminal leaf will cause another new non-terminal leaf [1]. Thus, it takes up to $O(n)$ rounds and $O(n)$ total memory.

In summary, this algorithm takes $O(n)$ rounds and $O(n^{3-\sigma/2})$ total memory. In addition, the SAST algorithm computes the exact result for the $2(1 - 1/|S|)$-approximate Steiner tree without incurring any additional approximation ratio.

4.2   Challenges and Solutions

The challenges in reducing the rounds of the SAST algorithm are below:

Firstly, Step 2 is too complex and causes a high round complexity, which is inefficient. Simplifying this step and reducing the required rounds is important.

Secondly, the round complexity of deleting the non-terminal leaves in Step 4 is too high. It is a key challenge to reduce the round complexity.

The solutions to the challenges above are as follows:

To tackle the first challenge, we combine steps 2 and 3 in Sect. 4.1 into one step. We combine the recursive strategy with algebraic methods to connect these $|S|$ disjoint trees and then obtain the required MST. This simplifies both Step 2 and Step 3 and reduces the round complexity for computing the MST.

To address the second challenge, we mark the endpoints of these connecting edges that connect these $|S|$ trees. Then, using these endpoints and the routing table, we find the Steiner nodes, which reduces the round complexity when a graph's shortest path diameter is smaller than n.

Additionally, we design a simpler and novel algorithm to calculate the RT, which takes only constant rounds. This reduces the round complexity of the parallel algorithm that computes the RT in Sect. 4.1.

5   The Parallel Approximate Steiner Tree Algorithm

According to the strategies in Sect. 4.2, this section first computes the SPF in Sect. 5.1. Then, in Sect. 5.2, we address the AST problem.

5.1   The Construction of the SPF

The process of computing the SPF is similar to Step 1 in Sect. 4.1. The key difference is a new parallel algorithm designed for the RT.

Algorithm 2. The parallel RT algorithm

Input: The $n \times n$ distance matrix $Dist$, a matrix A with size $n \times n$, and the predecessor matrix $\prod^A$ of A, distributed among $O(n^{2-\sigma})$ machines numbered as $P_{i,j,1}$, where $i, j \in \{1, 2, \cdots, n^{1-\sigma/2}\}$, each of which has a memory size $O(n^\sigma)$.

Output: The routing table $R_{i,k}$ with size $n \times n$.

1: The machine $P_{i,j,1}$ broadcasts the sub-matrix $A_{i,j}$ with size $n^{\sigma/2} \times n^{\sigma/2}$ to other $O(n^{1-\sigma/2})$ machines numbered by $P_{i,j,k}$, where $k \in \{2, 3, \cdots, n^{1-\sigma/2}\}$.

2: The machine $P_{i,j,1}$ broadcasts the sub-matrix $Dist_{i,j}$ to other $O(n^{1-\sigma/2})$ machines numbered by $P_{k,i,j}$, where $k \in \{1, 2, \cdots, n^{1-\sigma/2}\}$

3: The machine $P_{i,j,1}$ broadcasts the $n^{\sigma/2} \times n^{\sigma/2}$ sub-matrices $\prod_{i,j}^A$ and $Dist_{i,j}$ to other $O(n^{1-\sigma/2})$ machines numbered by $P_{i,k,j}$, where $k \in \{2, 3, \cdots, n^{1-\sigma/2}\}$.

4: **for** all machines $P_{i,j,k}$ $(i, j, k \in \{1, 2, \cdots, n^{1-\sigma/2}\})$ execute locally in parallel **do**

5:     Set an empty matrix R with size $n^{\sigma/2} \times n^{\sigma/2}$

6:     **for** $i' = 1, 2, \cdots, n^{\sigma/2}$ **do**

7:         **for** $k' = 1, 2, \cdots, n^{\sigma/2}$ **do**

8:             **for** $j' = 1, 2, \cdots, n^{\sigma/2}$ **do**

9:                 **if** $i' = k'$ **then** Set $R_{i,k}[i', k'] = \prod_{i,k}^A[i', k']$ and stop

10:                **end if**

11:                $d(i', k') = A_{i,j}(i', j') + Dist_{j,k}(j', k')$

12:                **if** $d(i', k') = Dist_{i,k}(i', k')$ and $i' \neq j'$ **then** Set $R_{i,k}[i', k'] = j'$ and
stop   ▷ when $i' = j'$, this avoids the case of $Dist_{i,k}(i', k') = Dist_{j,k}(j', k')$, indicating equal distances between i'/j' and k'.

13:            **end if**

14:        **end for**

15:    **end for**

16:    **end for**

17: **end for**

18: Execute the broadcast operation among machines numbered by $P_{i,j,k}, j \in \{1, 2, \cdots, n^{1-\sigma/2}\}$ to obtain the final predecessor matrix $R_{i,k}$.

19: Return R ▷ R is made up of $n^{2-\sigma}$ submatrices $R_{i,k}$ of size $n^{\sigma/2} \times n^{\sigma/2}$.

A Simpler Algorithm for the RT: We compute the RT only once, using the distance matrix $Dist$ and the adjacency matrix A of G (Algorithm 2), instead of computing it $\lceil \log n \rceil$ times as in Step 1 in Sect. 4.1. First, we initialize the predecessor matrix $\prod^A$ of A such that for $u_1, u_2 \in V$, if $e(u_1, u_2) \in E$, then $\prod^A[u_1, u_2] = u_2$, otherwise, $\prod^A[u_1, u_2] = NIL$. Also, we set $\prod^A[u_1, u_1] = NIL$. Second, we compute $A \star Dist$ (line 12 in Algorithm 2), modifying the elements of $\prod^A$ accordingly. The resulting predecessor matrix $R = \prod^A$ is the RT. The principle behind this is simple: 1) $Dist = A \star Dist$ means that the shortest distances between nodes will not change anymore; 2) $(A \star Dist)_{u_1, u_2} = \min_{z \in V}\{w(u_1, z) + d(z, u_2)\}$ determines u_1's successor. The concrete process is shown in Algorithm 2. Moreover, we give a simple example in Fig. 1 to better understand Algorithm 2. Next, we analyze the rounds of Algorithm 2.

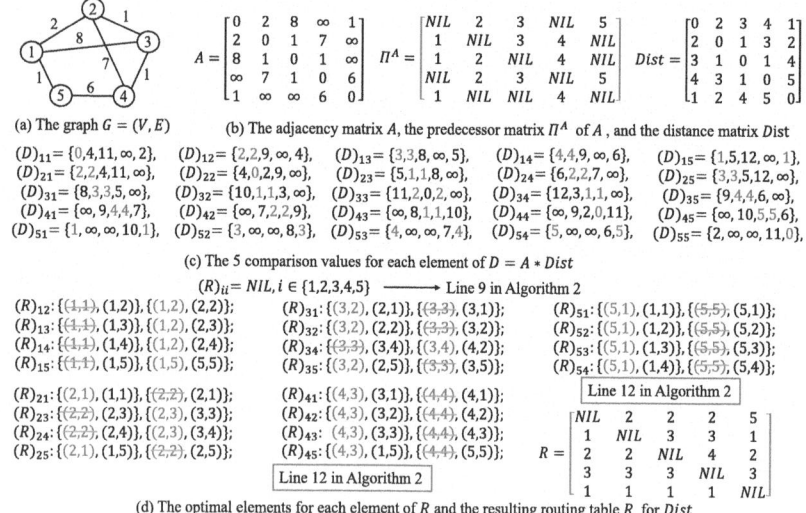

(a) The graph $G = (V, E)$

(b) The adjacency matrix A, the predecessor matrix Π^A of A, and the distance matrix $Dist$

$(D)_{11} = \{0,4,11,\infty,2\}$, $(D)_{12} = \{2,2,9,\infty,4\}$, $(D)_{13} = \{3,3,8,\infty,5\}$, $(D)_{14} = \{4,4,9,\infty,6\}$, $(D)_{15} = \{1,5,12,\infty,1\}$,
$(D)_{21} = \{2,2,4,11,\infty\}$, $(D)_{22} = \{4,0,2,9,\infty\}$, $(D)_{23} = \{5,1,1,8,\infty\}$, $(D)_{24} = \{6,2,2,7,\infty\}$, $(D)_{25} = \{3,3,5,12,\infty\}$,
$(D)_{31} = \{8,3,3,5,\infty\}$, $(D)_{32} = \{10,1,1,3,\infty\}$, $(D)_{33} = \{11,2,0,2,\infty\}$, $(D)_{34} = \{12,3,1,1,\infty\}$, $(D)_{35} = \{9,4,4,6,\infty\}$,
$(D)_{41} = \{\infty,9,4,4,7\}$, $(D)_{42} = \{\infty,7,2,2,9\}$, $(D)_{43} = \{\infty,8,1,1,10\}$, $(D)_{44} = \{\infty,9,2,0,11\}$, $(D)_{45} = \{\infty,10,5,5,6\}$,
$(D)_{51} = \{1,\infty,\infty,10,1\}$, $(D)_{52} = \{3,\infty,\infty,8,3\}$, $(D)_{53} = \{4,\infty,\infty,7,4\}$, $(D)_{54} = \{5,\infty,\infty,6,5\}$, $(D)_{55} = \{2,\infty,\infty,11,0\}$,

(c) The 5 comparison values for each element of $D = A \star Dist$

$(R)_{ii} = NIL, i \in \{1,2,3,4,5\}$ ⟶ Line 9 in Algorithm 2

$(R)_{12}$: $\{(1,1),(1,2)\},\{(1,2),(2,2)\}$; $(R)_{31}$: $\{(3,2),(2,1)\},\{(3,3),(3,1)\}$; $(R)_{51}$: $\{(5,1),(1,1)\},\{(5,5),(5,1)\}$;
$(R)_{13}$: $\{(1,1),(1,3)\},\{(1,2),(2,3)\}$; $(R)_{32}$: $\{(3,2),(2,2)\},\{(3,3),(3,2)\}$; $(R)_{52}$: $\{(5,1),(1,2)\},\{(5,5),(5,2)\}$;
$(R)_{14}$: $\{(1,1),(1,4)\},\{(1,2),(2,4)\}$; $(R)_{34}$: $\{(3,3),(3,4)\},\{(3,4),(4,2)\}$; $(R)_{53}$: $\{(5,1),(1,3)\},\{(5,5),(5,3)\}$;
$(R)_{15}$: $\{(1,1),(1,5)\},\{(1,5),(5,5)\}$; $(R)_{35}$: $\{(3,2),(2,5)\},\{(3,3),(3,5)\}$; $(R)_{54}$: $\{(5,1),(1,4)\},\{(5,5),(5,4)\}$;

$(R)_{21}$: $\{(2,1),(1,1)\},\{(2,2),(2,1)\}$; $(R)_{41}$: $\{(4,3),(3,1)\},\{(4,4),(4,1)\}$;
$(R)_{23}$: $\{(2,2),(2,3)\},\{(2,3),(3,3)\}$; $(R)_{42}$: $\{(4,3),(3,2)\},\{(4,4),(4,2)\}$;
$(R)_{24}$: $\{(2,2),(2,4)\},\{(2,3),(3,4)\}$; $(R)_{43}$: $\{(4,3),(3,3)\},\{(4,4),(4,3)\}$;
$(R)_{25}$: $\{(2,1),(1,5)\},\{(2,2),(2,5)\}$; $(R)_{45}$: $\{(4,3),(1,5)\},\{(4,4),(5,5)\}$;

Line 12 in Algorithm 2   Line 12 in Algorithm 2

(d) The optimal elements for each element of R and the resulting routing table R for $Dist$

Fig. 1. A concrete example of Algorithm 2

Lemma 2. (Algorithm 2). *In the MPC model with $O(n^\sigma)$ memory ($\sigma \in (0, 1]$) for each machine, for a graph $G = (V, E)$ that has n nodes, the adjacency matrix A, A's predecessor matrix $\prod^A$, and the distance matrix $Dist$, Algorithm 2 computes the RT R of $Dist$ using $O(\sigma^{-1})$ rounds and $O(n^{3-\sigma/2})$ total memory.*

Proof. As shown in Algorithm 2, there are only broadcast operations in lines 1 - 3 and line 20, which cost $O(\sigma^{-1})$ rounds. There is no communication between machines in lines 4 - 19 since each machine executes local computation. Similar to the computation of APSP in Algorithm 1, Algorithm 2 requires $O(n^{3-\sigma/2})$. □

The Parallel SPF Algorithm: Algorithm 3 computes the SPF in two steps. First, it computes the CDG (or $Dist$) in line 1 and the related RT R in line

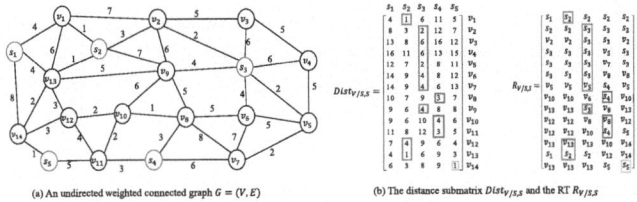

(a) An undirected weighted connected graph $G = (V, E)$ (b) The distance submatrix $Dist_{V/S,S}$ and the RT $R_{V/S,S}$

Fig. 2. An illustrative instance for Algorithm 3

Algorithm 3. The parallel SPF algorithm

Input: A given undirected connected graph $G = (V, E, W)$ and the terminal set S.
Output: The SPF $G_{SPF} = (V, E_{SPF}, w)$.
1: Execute Algorithm 1: compute the distance matrix $Dist$.
2: Execute Algorithm 2: compute the RT R of the distance matrix $Dist$.
3: The machines storing the distances of each $v \in V/S$ and nodes in S find v's source.
4: The machine that stores the tree edge of $v \in V/S$ whose source is $s \in S$, sends the
   tree edge and the distance $d(s, u)$ to the machine that stores $s \in S$.

---

2. Second, it finds the source of each node $v \in V/S$ by selecting the minimum distance from v to S in line 3. Using the RT, it constructs the tree T_j rooted at $s_j \in S$ ($j \in \{1, 2, \cdots, |S|\}$) in line 4. For clarity, we provide an illustrating example for Algorithm 3 in Fig. 2.

Lemma 3. *In the MPC model, each machine has $O(n^\sigma)$ memory and $\sigma \in (0, 1)$, for a graph G with n nodes and a terminal set S, Algorithm 3 computes the SPF $G_{SPF} = (V, E_{SPF})$ taking $O(\sigma^{-1} \log n)$ rounds and $O(n^{3-\sigma/2})$ total memory.*

Proof. By Lemmas 1 - 2, it requires $O(\sigma^{-1} \log n)$ rounds and $O(n^{3-\sigma/2})$ memory to run lines 1 - 2 in Algorithm 3. Sorting these distances and finding the minimum in line 3 takes $O(1/\sigma)$ rounds. Line 4 requires $O(1)$ rounds to send messages. $\square$

5.2   The Computation of the AST

This subsection computes the AST in the MPC model using the first two strategies proposed in Sect. 4.2, which will be detailed in Algorithm 4.

In Algorithm 4, we first label the states of all nodes in line 1, which are used to identify the Steiner nodes later. Then, we reorder the nodes whose IDs are the row and column indices of the adjacency matrix A such that querying the edges with endpoints in different trees (line 2) is more convenient. Next, we modify the weights of edges with endpoints in distinct trees in lines 3 - 5 (see Fig. 3(b)), the same as in Sect. 4.1. We select the edge with minimum weight connecting T_i ($i \in \{1, 2, \cdots, |S|\}$) to its closest tree (not T_i) in line 6 (see Fig. 3(c)). In lines 7 - 9 (refer to Fig. 4(a)), we delete the heaviest edge that causes a cycle or a duplicate edge among the $|S|$ trees. We find the necessary connecting edges (line 10 and Fig. 4(b)) and modify the states of non-terminals that are Steiner nodes (line 11). Since we utilize the RT, there is no resource contention between machines when multiple edges are selected for a tree T. Finally, we delete the non-terminal leaves in G_{SPF} in line 12 and obtain the AST (refer to Fig. 4(c)). Now, we analyze the correctness and round complexity for Algorithm 4.

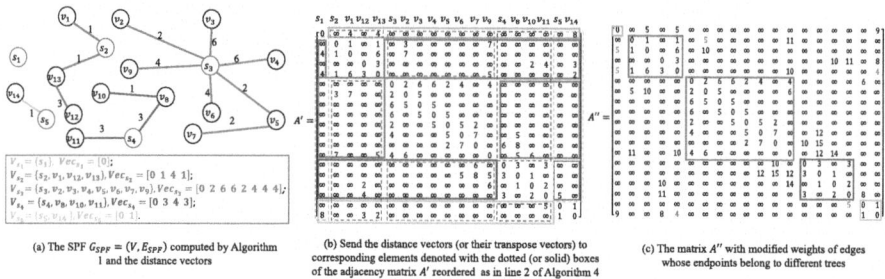

(a) The SPF $G_{SPF} = (V, E_{SPF})$ computed by Algorithm 1 and the distance vectors

(b) Send the distance vectors (or their transpose vectors) to corresponding elements denoted with the dotted (or solid) boxes of the adjacency matrix A' reordered as in line 2 of Algorithm 4

(c) The matrix A'' with modified weights of edges whose endpoints belong to different trees

Fig. 3. A straightforward example of the process of lines 2 - 6 in Algorithm 4

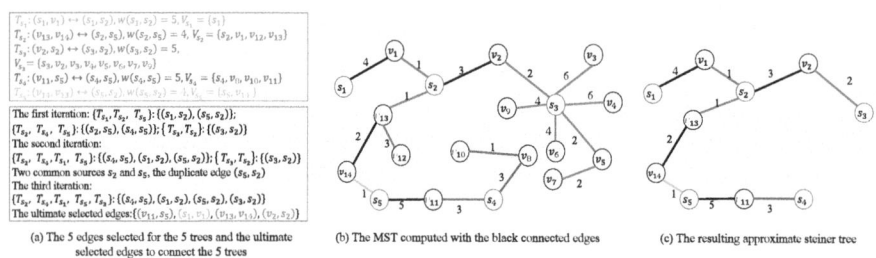

(a) The 5 edges selected for the 5 trees and the ultimate selected edges to connect the 5 trees

(b) The MST computed with the black connected edges

(c) The resulting approximate steiner tree

Fig. 4. A concrete instance of the process of lines 8 - 12 in Algorithm 4

Lemma 4. *Algorithm 4 can compute an MST correctly without line 10 in Algorithm 4. Additionally, the approximation ratio of the ST computed is $2(1-1/|S|)$.*

Proof. The core ideas behind the SAST algorithm and Algorithm 4 for computing the MST are the same. Firstly, the MSTs computed by the two algorithms both contain all the tree edges: the SAST algorithm sets the weights of all edges in G_{SPF} to 0 and Algorithm 4 retains the edges in G_{SPF}. Secondly, the two algorithms both modify the weights of edges whose endpoints are in different trees in the same way. The difference is as follows: each time, the SAST algorithm chooses the minimum weight edge. However, Algorithm 4 selects the minimum weight edge between each tree $T_i(i \in \{1, 2, \cdots, |S|\})$ and its closest tree. Consequently, we have $|S|$ selected edges, which may cause a cycle. However, Algorithm 4 either deletes the edge with maximum weight in the cycle or the duplicate edge in lines 7 - 9. This shows that Algorithm 4 computes the MST correctly. Thus, the approximation ratio of the ST solved by Algorithm 4 is $2(1 - 1/|S|)$, which is the same as that of the SAST algorithm.   □

Algorithm 4. The parallel AST algorithm

Input: The SPF $G_{SPF} = (V, E_{SPF}, w) = \bigcup_{j=1}^{|S|}(T_j = (V_j, E_j))$ obtained by Algorithm 3, the adjacency matrix A, the RT R, and the terminal set S.

Output: The AST T_0.

1: Label the states of the terminals in S and nodes in V/S as $true$ and $false$, respectively.

2: Reorder the row and column indices in A according to the ID values of terminals in S, each of which is followed by the ordered IDs of its child nodes. ▷ see Fig. 3(b)

3: The machines that store the $1 \times |V_{T_i}|$ distance vectors from $s_i \in S$ to nodes in the tree T_i ($i \in \{1, 2, \cdots, |S|\}$), broadcast them to the machines that store the $|V_{T_i}| \times |V_{T_j}|$ submatrix $A[V_{T_i}, V_{T_j}]$, where $j \neq i$ and $j \in \{1, 2, \cdots, |S|\}$.   ▷ $A[V_{T_i}, V_{T_j}]$ represents the submatrix of A with row indices in V_{T_i} and column indices in V_{T_j}.

4: The machines that store the $|V_{T_i}| \times 1$ distance vectors from nodes in the tree T_i ($i \in \{1, 2, \cdots, |S|\}$) to $s_i \in S$, broadcast them to the machines that store the $|V_{T_j}| \times |V_{T_i}|$ submatrix $A[V_{T_j}, V_{T_i}]$, where $j \neq i$ and $j \in \{1, 2, \cdots, |S|\}$.

5: The machines that store $A[V_{T_i}, V_{T_j}]$, add the elements of $A[V_{T_i}, V_{T_j}]$ with the corresponding elements in the distance vectors, where $i, j \in \{1, 2, \cdots, |S|\}$ and $i \neq j$. ▷ Modify the weights of edges whose endpoints belong to different trees.

6: The machines that store the modified submatrix $A[V_{T_i}, V/V_{T_i}]$ choose the minimum value of $A[V_{T_i}, V/V_{T_i}]$. ▷ Find the connecting edge for each tree to its closest tree.

7: Invoke other $O(|S|/n^{\sigma})$ machines (or one machine if $|S| \leq n^{\sigma}$) to receive the $|S|$ selected edges for the $|S|$ disjoint trees sent by the machines that store them.

8: The invoked machines modify the endpoints of each selected edge to their sources and then match these edges that have a common endpoint recursively, until there are two common sources. ▷ Find the duplicate edge or the cycle among the trees.

9: The invoked machines either delete the duplicate edge or compare all edge weights in the cycle and find the heaviest edge $e = (s_i, s_j)$ and delete the corresponding edge $e = (i, j)$.   ▷ Delete the heaviest edge in the cycle or the duplicate edge.

10: The machines that store the selected edges $e = (i, j)$, send these edges to the machines that store the tree T_i and its closest tree T_j, where $i, j \in \{1, 2, \cdots, |S|\}$ and $i \neq j$.   ▷ $e = (i, j)$ connects T_i and T_j, with $i \in T_i$ and $j \in T_j$, respectively.

11: The machines that store the tree T_i (T_j), find the nodes on the path from i (j) to the root of T_i (T_j, where $i, j \in \{1, 2, \cdots, |S|\}$ and $i \neq j$), using R. Meanwhile, these machines modify the states of these nodes to $true$. ▷ Find the Steiner nodes.

12: Delete all nodes in G_{SPF} whose states are $false$ and their corresponding edges. ▷ Delete the leaves in G_{SPF} that are non-terminals.

13: Return $T_0 = G_{SPF}$

Lemma 5. *In the MPC model, each machine has $O(n^{\sigma})$ ($\sigma \in (0, 1)$) memory, for the graph $G = (V, E)$ that has n nodes, a terminal set S, the adjacency matrix A, the RT R, and the SPF $G_{SPF} = (V, E_{SPF})$ computed by Algorithm 3, Algorithm 4 solves the AST in $O(\log |S| + D)$ rounds and $O(n^2)$ total memory.*

Proof. In Algorithm 4, executing line 1 takes only one round because each terminal in S is a root of a tree in $G_{SPF} = (V, E_{SPF})$. There is only a sorting operation in line 2 of Algorithm 4, which takes $O(\sigma^{-1})$ rounds. In lines 3 - 4 and 10, the broadcast operations require $O(\sigma^{-1})$ rounds. Finding the minimum and maximum weight edges in lines 5, 6, and 9 takes $O(\sigma^{-1})$ rounds. Dealing with $|S|$ edges recursively in line 8 requires $O(\log |S|)$ rounds. It takes only one round to send messages to the corresponding machines in line 10. The most expensive operation is to find the nodes on a path in line 11 of Algorithm 4, which requires $O(D)$ rounds since the graph's shortest path diameter is D. The total memory for this algorithm is $O(n^2)$, as the size of the adjacency matrix A is $O(n^2)$.   □

6   Extension to the Near-Linear Memory Setting

Now, we extend the straightforward method of computing AST to the MPC model with $O(n)$ memory for each machine. The four-step process remains unchanged: 1) In Step 1, we obtain the CDG and the RT R using Algorithm 1 and Algorithm 2, respectively. The subsequent process follows Sect. 4.1. We can store the edges of the $|S|$ disjoint trees in a single machine since the number of edges in these trees is less than n. By Lemmas 1 - 2, Step 1 takes $O(\log n)$ rounds and $O(n^{2.5})$ total memory. 2) For Step 2, since each node contains up to n edges, we can modify the edge weights of each node in a single machine, which costs only one round. As there are n nodes, the total memory needed is $O(n^2)$. 3) In Step 3, we utilize the existing fastest parallel MST algorithm proposed by Nowicki [19], requiring $O(1)$ rounds and $O(n^2)$ total memory. 4) Step 4 can be completed in one round, as a single machine is sufficient to store all edges of the MST.

Hence, we obtain a parallel AST algorithm in the MPC model with near-linear memory using $O(\log n)$ rounds and $O(n^{2.5})$ total memory. This greatly improves the result in [21], which takes $O(\log \log n + D)$ rounds when $D \gg \log n$.

7   Conclusion

This paper studies the approximate Steiner tree problem in the MPC model with $O(n^\sigma)$ memory for each machine and $\sigma \in (0,1)$. Specifically, to reduce the round complexity of the straightforward approximate Steiner tree algorithm, we use a simpler method to compute the routing table that takes only constant rounds and combine the recursive strategy with the algebraic methods to solve the approximate Steiner tree more efficiently. Then, we obtain a parallel $2(1 - 1/|S|)$-approximate Steiner tree algorithm that takes $O(D + \sigma^{-1} \log n)$ rounds and $O(n^{3-\sigma/2})$ total memory. Furthermore, we extend the straightforward approximate Steiner tree algorithm to the MPC model with $O(n)$ memory for each machine, requiring $O(\log n)$ rounds. This reduction in round complexity is highly beneficial when $D = \Omega(\log n)$.

References

1. Akbari, H., Iranmanesh, Z., Ghodsi, M.: Parallel minimum spanning tree heuristic for the Steiner problem in graphs. In: Proceedings of the ICPADS, pp. 1–8. IEEE (2007)
2. Andoni, A., Song, Z., Stein, C., Wang, Z., Zhong, P.: Parallel graph connectivity in log diameter rounds. In: Proceedings of the FOCS, pp. 674–685. IEEE (2018)
3. Behnezhad, S., Derakhshan, M., Hajiaghayi, M.: Brief announcement: semimapreduce meets congested clique. CoRR **abs/1802.10297** (2018)
4. Bezensek, M., Robic, B.: A survey of parallel and distributed algorithms for the steiner tree problem. Int. J. Parallel Program. **42**(2), 287–319 (2014)
5. Censor-Hillel, K., et al.: Algebraic methods in the congested clique. In: Proceedings of the PODC, pp. 143–152. ACM (2015)
6. Chalermsook, P., Fakcharoenphol, J.: Simple distributed algorithms for approximating minimum Steiner trees. In: Proceedings of the COCOON. pp. 380–389. Springer (2005)
7. Chen, G., Houle, M.E., Kuo, M.: The Steiner problem in distributed computing systems. Inf. Sci. **74**(1–2), 73–96 (1993)
8. Cormen, T.H., Leiserson, C.E., Rivest, R.L., Stein, C.: Introduction to Algorithms, 3rd Edition. MIT Press (2009)
9. Coy, S., Czumaj, A.: Deterministic massively parallel connectivity. In: Proceedings of the STOC, pp. 162–175. ACM (2022)
10. Dinitz, M., Nazari, Y.: Massively parallel approximate distance sketches. In: Proceedings of the OPODIS, pp. 35:1–35:17. Schloss Dagstuhl-Leibniz-Zentrum für Informatik (2019)
11. Dory, M., Matar, S.: Massively parallel algorithms for approximate shortest paths. In: Proceedings of the SPAA. pp. 415–426. ACM (2024)
12. Garey, M.R., Johnson, D.S.: Computers and Intractability: A Guide to the Theory of NP-Completeness. Freeman, W. H (1979)
13. Goodrich, M.T., Sitchinava, N., Zhang, Q.: Sorting, searching, and simulation in the mapreduce framework. In: Proceedings of the ISAAC. pp. 374–383 (2011)
14. Hajiaghayi, M., Lattanzi, S., Seddighin, S., Stein, C.: Mapreduce meets fine-grained complexity: mapreduce algorithms for apsp, matrix multiplication, 3-sum, and beyond. CoRR **abs/1905.01748** (2019)
15. Kerger, P.A., Neira, D.E.B., Izquierdo, Z.G., Rieffel, E.G.: Quantum distributed algorithms for approximate Steiner trees and directed minimum spanning trees. In: Proceedings of the QCE, pp. 1249–1259. IEEE (2023)
16. Khan, M., Kuhn, F., Malkhi, D., Pandurangan, G., Talwar, K.: Efficient distributed approximation algorithms via probabilistic tree embeddings. In: Proceedings of the PODC. pp. 263–272. ACM (2008)
17. Kou, L.T., Markowsky, G., Berman, L.: A fast algorithm for Steiner trees. Acta Informatica **15**, 141–145 (1981)
18. Lenzen, C., Patt-Shamir, B.: Improved distributed Steiner forest construction. In: Proceedings of the PODC. pp. 262–271. ACM (2014)
19. Nowicki, K.: A deterministic algorithm for the MST problem in constant rounds of congested clique. In: Proceedings of the STOC. pp. 1154–1165. ACM (2021)
20. Saikia, P., Karmakar, S.: A simple 2(1-1/l) factor distributed approximation algorithm for Steiner tree in the congest model. In: Proceedings of the ICDCN. pp. 41–50. ACM (2019)

21. Saikia, P., Karmakar, S.: Distributed approximation algorithms for Steiner tree in the congested clique. Int. J. Found. Comput. Sci. **31**(7), 941–968 (2020)
22. Wu, Y., Widmayer, P., Wong, C.K.: A faster approximation algorithm for the Steiner problem in graphs. Acta Informatica **23**(2), 223–229 (1986)

A Multi-start Variable Neighborhood Tabu Search Algorithm for the Cyclic Bandwidth Problem

Yuan Wang, Jianhang Sun, Zhipeng Lü, Zhouxing Su, Junwen Ding, and Qingyun Zhang[✉]

School of Computer Science and Technology, Huazhong University of Science and Technology, Wuhan, China
{wang_yuan,sunjianhang,zhipeng.lv,suzhouxing,junwending, qingyun_zhang}@hust.edu.cn

Abstract. The cyclic bandwidth problem (CBP) is a significant and challenging graph labeling problem with many real-world applications, including VLSI design, interconnection networks of parallel computers, and constraint satisfaction problems. Existing methods in the literature for solving the CBP still have room for improvement on large-scale instances. To address this issue and effectively solve the CBP, we present a novel multi-start variable neighborhood tabu search (MVNTS) algorithm with a greedy construction procedure and a reload strategy. Specifically, our algorithm employs tabu strategy and two types of neighborhoods—sampled and complete—to extensively explore the solution space. Moreover, the restart and reload strategies are used to ensure the trade-off between intensification and diversification of the search while increasing the scalability of the algorithm. Extensive experiments on 202 public benchmark instances demonstrate that MVNTS outperforms the state-of-the-art algorithms in the literature in terms of both solution quality and computational efficiency.

Keywords: Heuristics · Combinatorial optimization · Cyclic bandwidth minimization · Local search

1 Introduction

The cyclic bandwidth problem (CBP) is an important graph labeling problem that was first introduced by Leung et al. [1]. The original goal is to determine whether, given a graph G and a positive integer k, there exists a circular layout such that the distance in the cycle between each vertex pair connected by an edge is at most k. As a classical combinatorial optimization problem, the decision version of the CBP is proven to be NP-complete [2]. In addition, the CBP has wide-ranging practical applications in various fields. For example, it plays a crucial role in VLSI design, interconnection networks of parallel computers, and constraint satisfaction problems [3–7].

Supplementary Information The online version contains supplementary material available at https://doi.org/10.1007/978-981-95-0218-9_10.

Most of the early studies on the CBP are theoretical analyses, aiming to determine the exact lowest cyclic bandwidths for specific graphs or the lower and upper bounds for general graphs. For instance, for some specific graphs, such as 2D and 3D meshes, complete trees, and hypercubes, their optimal bandwidth and cyclic bandwidth are proven to be equal [8,9]. Hromkovič et al. [8] established a connection between the cyclic bandwidth $C_B(G)$ and the bandwidth $P_B(G)$ of a general graph G: $P_B(G) \geq C_B(G) \geq \frac{1}{2}P_B(G)$. In [9–11], researchers studied the optimal cyclic bandwidth of some special graphs, such as paths, cycles and trees, in extreme cases. A systematic method was proposed by Zhou [12] that obtained several lower bounds for $P_B(G)$ and $C_B(G)$ which are related to the parameters of the distance and degree. Chan et al. [13] obtained the upper bound of the cyclic bandwidth of graphs by adding a new edge and showed that the bound is sharp. Later, de Klerk et al. [14] obtained better lower bounds for $P_B(G)$ and $C_B(G)$ based on semidefinite programming (SDP) relaxations. Déprés et al. [15] provided three new improved lower bounds for the CBP. The length of the longest cycle within the cycle basis of G determines the first one, while the other two are determined by the density of neighboring vertices in G. Fertin et al. [16] improved the lower bounds on the cyclic bandwidth of many Harwell-Boeing benchmark graphs by combining existing knowledge and some solving techniques.

In addition to these theoretical studies, there are also studies of practical algorithms to solve the CBP. A branch and bound (B&B) based algorithm for solving small-scale instances of the CBP was proposed by Romero-Monsivais et al. [17]. Rodriguez-Tello et al. [18] proposed a tabu search algorithm (TS$_{\text{CB}}$) to solve the CBP, which obtained optimal solutions for 56 instances and found better upper bounds for all the remaining 57 instances. Later, an iterative three-phase search approach (ITPS) was introduced by Ren et al. [19], which contains three complementary search components and an enriched evaluation function to solve the CBP, and it improved 12 upper bounds. Then, Ren et al. [20] proposed a new iterated local search algorithm (NILS) to solve the CBP. NILS is an effective metaheuristic algorithm based on the iterated local search framework with the integration of dedicated search components and perturbation procedures.

Although the recent state-of-the-art algorithm NILS improved or matched the results of the previous methods on all benchmark instances, it showed limited performance on large-scale instances and inadequate robustness on challenging instances. To tackle large-scale instances more effectively, this paper introduces a novel multi-start variable neighborhood tabu search (MVNTS) algorithm. MVNTS is a combination of variable neighborhood search and tabu search strategies, aiming to strengthen the search capabilities with a restart strategy based on greedy construction and a reload strategy. The following are the main contributions of this paper:

- Our MVNTS is based on a new multi-start framework with a reload strategy that has not been used by any previous CBP method, which are used to ensure the trade-off between intensification and diversification of the search while increasing the scalability of the algorithm to different instance scales.

- MVNTS employs a tabu search strategy and a variable neighborhood search strategy based on a sampled neighborhood and a complete neighborhood to extensively explore the solution space. The sampled neighborhood offers higher diversity and is used for quick neighborhood search, while the complete neighborhood has stronger search intensity and is employed to find overall improvement moves.
- Tested on 202 public benchmark instances from 8 different families, MVNTS outperforms the state-of-the-art algorithms, since it finds better solutions on 20 instances, and matches the best solutions for all the remaining ones with significantly less computational time.

2   Problem Description

Formally, we consider a finite undirected graph $G(V, E)$, where V represents the vertex set with a cardinality of n, and E represents the edge set with a cardinality of m. We need to find a labeling solution φ for graph G to map the vertices from V to $\{1, ..., n\}$, and the cyclic bandwidth of graph G under the labeling φ is defined by Eqs. (1) – (2).

$$C_B(G, \varphi) = \max_{\{u,v\} \in E} C_B(\{u, v\}, \varphi) \tag{1}$$

$$C_B(\{u, v\}, \varphi) = |\varphi(u) - \varphi(v)|_n \tag{2}$$

where $\varphi(u)$ denotes the label assigned to vertex u, and $|x|_n = \min\{|x|, n - |x|\}$ represents the cyclic distance. In the case of $|x|_n = |x|$, it becomes the bandwidth problem [21].

The objective of the CBP is to find a labeling φ^* that minimizes the cyclic bandwidth of G, as illustrated in Eq. (3).

$$\varphi^* \leftarrow \arg\min_{\varphi \in \phi} \{C_B(G, \varphi)\} \tag{3}$$

where ϕ is the complete collection of all labeling schemes and φ^* is the optimal labeling scheme.

Figure 1 illustrates a small example of the CBP. Figure 1(a) gives a graph G ($n = 9$, $m = 9$) and a labeling φ, while Fig. 1(b) gives the corresponding chordal graph of graph G whose vertices are ordered from label 1 to label n circularly. In this case, the cyclic distances of all edges are listed as follows: $C_B(\{a, g\}, \varphi) = 4$, $C_B(\{b, e\}, \varphi) = 3$, $C_B(\{c, f\}, \varphi) = 3$, $C_B(\{d, g\}, \varphi) = 2$, $C_B(\{e, g\}, \varphi) = 1$, $C_B(\{e, h\}, \varphi) = 3$, $C_B(\{f, h\}, \varphi) = 2$, $C_B(\{g, i\}, \varphi) = 1$, $C_B(\{h, i\}, \varphi) = 4$. Thus, the cyclic bandwidth of G is the maximum value among these cyclic distances, i.e., $C_B(G, \varphi) = 4$.

It is noteworthy that the critical edge (say $\{x, y\}$) is an edge with the maximum cyclic distance in the graph G, i.e., $C_B(\{x, y\}, \varphi) = C_B(G, \varphi)$ and the critical vertices are the corresponding endpoints of the critical edges.

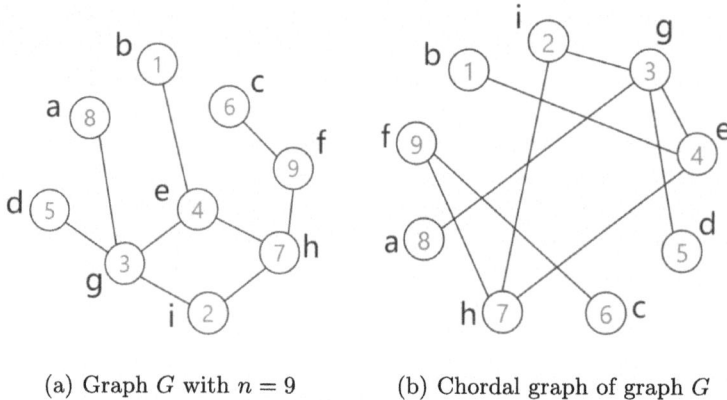

(a) Graph G with $n = 9$ (b) Chordal graph of graph G

Fig. 1. An illustrative example of the CBP.

3 Multi-start Variable Neighborhood Tabu Search Algorithm

In this section, we present the framework of our MVNTS algorithm and its main components. The pseudo-code of the MVNTS algorithm is shown in Algorithm 1. The outer *while* loop first invokes the greedy initialization algorithm to generate the initial solution φ based on the greedy strategy (line 2). At each iteration of the inner loop, the current solution φ is first improved by the variable neighborhood tabu search (line 5). Then, MVNTS compares the gap of the cyclic bandwidth between the best solution found so far $\varphi*$ and the current solution φ, and reloads the best found solution $\varphi*$ as the current solution φ when the gap between the current solution and the best found solution reaches a given threshold (line 6). Next, the perturbation procedure is invoked to perturb the current critical edges at random to escape from the local optimum (line 7). Finally, the best found solution $\varphi*$ during the search is returned.

3.1 Greedy Initialization

The MVNTS algorithm utilizes a greedy procedure based on breadth-first search to construct the initial solution. Initially, a starting vertex i is randomly chosen and assigned with label 1. Next, i is added to a queue called $OPEN$. The symbols max and min denote the largest and smallest unused labels, respectively. Now, max is equal to n, while min is equal to 2. We then dequeue vertices from $OPEN$ one by one (denoted by j), and assign the unlabeled neighbors of vertex j to the one between max and min that is closer to the label of j and add them to $OPEN$. If the distances for max and min are equal, ties are broken randomly. If position max (min) is occupied, we update $max \leftarrow max - 1$ ($min \leftarrow min + 1$). This process is repeated iteratively until all vertices have been assigned a label, which indicates that $OPEN$ is empty. The time complexity of the initialization procedure is $O(n + m)$.

Algorithm 1. The main framework of MVNTS

Input: Undirected graph $G(V, E)$, neighborhood controller K, local search intensity L_1, perturbation intensity L_2 and exploration limit L_3
Output: Best found solution $\varphi*$
1: **while** the stop condition is not met **do**
2:      $\varphi \leftarrow$ GreedyInitialization(G)   /* Section 3.1 */
3:      $step \leftarrow 0$
4:      **while** $step \leq L_3$ **do**
5:          $(\varphi, \varphi*) \leftarrow$ VNTS(φ, $\varphi*$, K, L_1) /* Local search optimization, Section 3.2 */
6:          $\varphi \leftarrow$ Reload(φ, $\varphi*$)   /* Section 3.2 */
7:          $(\varphi, \varphi*) \leftarrow$ Perturbation(φ, $\varphi*$, L_2)   /* Escape from the local optimum by perturbation, Section 3.2 */
8:          $step \leftarrow step + 1$
9: **return** $\varphi*$

3.2 Improvement Procedure

Tabu search [22] is an effective way to prevent the search from getting stuck in a loop by prohibiting revisiting recently explored solutions but may misjudge the tabu status of the solution making promising neighborhood moves incorrectly forbidden. Therefore, the settings of synthetic parameters such as the tabu tenure are sensitive. The variable neighborhood search [23] utilizes different neighborhoods to change the scope of the search thereby increasing the chances of finding better solutions. However, for variable neighborhood search, neighborhood selection is very critical. During the search process, some potential optimal solutions may be missed since only improved solutions are accepted. To address this, we aim to combine their advantages and balance the intensity and diversity of the search. As a result, MVNTS integrates both tabu search and variable neighborhood search strategies, which guide different search trajectories and help the search to jump out of the local optimum trap.

Variable Neighborhood Tabu Search. The variable neighborhood tabu search (VNTS) procedure (Algorithm 2) iteratively searches the neighborhood of the current solution φ and finds the best neighboring solution based on the evaluation function f_e to improve the current solution.

Neighborhood evaluation is an extremely important ingredient of the local search procedure. To evaluate neighborhood moves and obtain the best neighboring solution, MVNTS employs a bi-criteria evaluation function f_e, which is defined in Eq. (4).

$$f_e(\varphi) = C_B(G, \varphi) \times (m + 1) + M(G, \varphi) \tag{4}$$

where $M(G, \varphi)$ denotes the number of critical edges, i.e., $M(G, \varphi) = |\{\{u, v\} \in E : C_B(\{u, v\}, \varphi) = C_B(G, \varphi)\}|$. As the main objective that directly affects the quality of the solution, $C_B(G, \varphi)$ is the primary score. To distinguish multiple neighborhood moves with the same primary scoring value, we introduce an auxiliary score $M(G, \varphi)$, where the value of $M(G, \varphi)$ does not exceed the edge number m.

Let φ be the current solution, and $\varphi \oplus swap(u, v)$ be the neighboring solution obtained by swapping the labels of vertex u and vertex v. The reverse move of $swap(u, v)$ is $swap(v, u)$.

When considering a specific critical vertex u, we employ a straightforward yet classical method to identify candidate vertices to swap. Let $S(u)$ represent the set of vertices that are suitable for u to swap labels with, where 'suitable' means that $S(u)$ contains the vertices whose labels are closer to $mid(u)$ than $\varphi(u)$ as shown in Eq. (5).

$$S(u) = \{w \in V : |mid(u) - \varphi(w)|_n \leq |mid(u) - \varphi(u)|_n\} \tag{5}$$

where $mid(u)$ is the midpoint of the shortest path (here the distance depends on the circular order) in the chordal graph of G embodying all the neighboring vertices that are adjacent to u [18].

During the search process, we use two types of neighborhoods. One is called the sampled neighborhood (N_s) and the other is called the complete neighborhood (N_c). The difference between these two types of neighborhoods lies in the method of selecting a move. The sampled neighborhood chooses the best swap move for one critical vertex, while the complete neighborhood chooses the global best swap move for the entire critical vertex set. Both neighborhoods are based on the swap operator, i.e. swapping the labels of two vertices. The parameter K is used to set the evaluation frequency of the two neighborhoods, and the search alternates between $(K - 1)$ iterations using the sampled neighborhood and 1 iteration using the complete neighborhood.

The advantage of this approach is that it better balances the intensity and diversity of the search. The sampled neighborhood search emphasizes speed and diversity, allowing for rapid exploration of the solution space. In contrast, the complete neighborhood search prioritizes the quality and intensity of the search, focusing on finding superior solutions within a larger search area. For a specific critical vertex u, the neighborhood N_s comprises a collection of solutions obtained by exchanging the labels of u with a vertex selected from $S(u)$. Furthermore, when considering a given critical vertex set $C(\varphi)$, the neighborhood N_c consists of a set of solutions that can be derived by swapping the labels of a critical vertex $w \in C(\varphi)$ with a vertex from $S(w)$.

In Algorithm 2, during each iteration of the *while* loop, MVNTS initially calculates the critical vertex set based on the current solution (line 5). Then, it determines the neighborhood to be used for the current iteration (line 6). When using the sampled neighborhood (lines 7–11), for each critical vertex, it executes the best swap found by N_s to get the best neighborhood solution, and the number of swaps executed is equal to the number of critical vertices. Regarding using the complete neighborhood (lines 13–15), it moves to the best neighboring solution of the whole set of critical vertices found by evaluating N_c, and just makes one move. To prevent the search from revisiting previous solutions during recent iterations, each reversing move will be added to the tabu

---

Algorithm 2. VNTS

Input: Current solution φ, best found solution φ^*, and local search intensity L_1
Output: Current solution φ, best found solution $\varphi*$

1: $K_{counter}, unImp \leftarrow 0$
2: $\varphi', \varphi_b, \varphi_{ib} \leftarrow \varphi$
3: **while** $unImp < L_1$ **do**
4:     $K_{counter} \leftarrow K_{counter} + 1$
5:     $C(\varphi') \leftarrow \text{CriticalSet}(\varphi')$ /* Compute the critical vertex set */
6:     **if** $(K_{counter} \bmod K) \neq 0$ **then**
7:         **for** each u in $C(\varphi')$ **do**
8:             $\varphi \leftarrow \text{FindBestNeighbor}(N_s(\varphi, u))$
9:             Update tabu list
10:            $\varphi_{ib} \leftarrow \text{InnerUpdateBest}(\varphi_{ib}, \varphi)$
11:            $(\varphi_b, unImp) \leftarrow \text{OuterUpdateBest}(\varphi_{ib}, \varphi_b, unImp)$
12:    **else**
13:        $\varphi \leftarrow \text{FindBestNeighbor}(N_c(\varphi, C(\varphi')))$
14:        Update tabu list
15:        $(\varphi_b, unImp) \leftarrow \text{OuterUpdateBest}(\varphi, \varphi_b, unImp)$
16:    $\varphi' \leftarrow \varphi$
17:    $\varphi^* \leftarrow \text{UpdateBest}(\varphi^*, \varphi_b)$
18: **return** φ, φ^*

---

list [22]. Next, the number of consecutive non-improvement iterations $unImp$ is updated. The update rule is that if the local best solution φ_b is improved, counter $unImp$ is set to 0, otherwise, counter $unImp$ is increased by one. VNTS ends when the local best solution φ_b has not been improved for L_1 consecutive iterations and returns the current solution φ and the best solution φ^* found so far.

The FindBestNeighbor procedure (Algorithm 2, lines 8 and 13) finds the best swap pair that is not forbidden by the tabu list. In particular, the tabu status can only be ignored if the swap pair can improve the currently found best solution. For the swap pairs that can improve or match the current solution, the best one is selected. If multiple pairs have the same f_e value, ties are broken randomly. If no swap pair that can improve or match the current solution, it randomly chooses and executes a non-improving swap pair. After executing $swap(u, v)$, a new tabu record is appended to the tabu list. In the subsequent tl iterations, $swap(v, u)$ and the reverse operation $swap(u, v)$ are prohibited, with tl referred to as the tabu tenure. Here, to avoid looping among several solutions, we use a simple random tabu tenure method, where tl is a random value between 1 and W, and it is given by Eq. (6), where W is a positive integer.

$$tl \sim U(1, W) \tag{6}$$

In particular, during the FindBestNeighbor procedure, the perturbation procedure is immediately invoked to break ties when each critical vertex $w \in C(\varphi)$ is located in the extreme case of the position of $mid(w)$.

Algorithm 3. Perturbation procedure

Input: Current solution φ, best found solution $\varphi*$, and perturbation intensity L_2
Output: Perturbed solution φ, best found solution $\varphi*$

  1: $iter \leftarrow 0$
  2: **while** $iter < L_2$ **do**
  3:    $C_E(\varphi) \leftarrow$ CriticalEdgeSet(φ)   /* Compute the critical edge set */
  4:    **for** $e \in C_E(\varphi)$ **do**
  5:      $\varphi \leftarrow$ RandomizedShiftInsert(φ, e)
  6:      $iter \leftarrow iter + 1$
  7:      **if** $f_e(\varphi) < f_e(\varphi*)$ **then**
  8:        $\varphi* \leftarrow \varphi$
  9:      **if** $iter \geq L_2$ **then**
10:        break
11: **return** φ, $\varphi*$

Perturbation Procedure. The local search procedure stops when it cannot improve the best solution found for L_1 consecutive iterations. At this time, the current solution is considered to be trapped in a local optimum. Then, the perturbation procedure is employed to jump out of the trap. Our perturbation is inspired by the directed perturbation in [20] that is based on a randomized version of the ShiftInsert operator in [19]. The perturbation targets the critical edges, and more details about the RandomizedShiftInsert (Algorithm 3, line 5) can be found in [20].

The pseudo-code of our perturbation procedure is outlined in Algorithm 3. It applies the RandomizedShiftInsert operator (line 5) for a total of L_2 times for each critical edge e. Unlike the method in [20], we propose a delayed update scheme that focuses on a set of critical edges in the current solution (line 3), and the target edge set will not change until each edge in the set has been perturbed once, even if some of the edges are not critical edges anymore. Actually, in many cases, the reason for failing to improve a solution is the mutual influence of the critical edges. However, in [20], the critical edge to be perturbed is identified each time when the current solution has just been perturbed. This leads to the result that the perturbation process usually targets the newly generated critical edges, and it only has a limited effect on perturbing the structure of a solution. Therefore, it makes sense to adopt the delayed update mechanism.

Reload Strategy. To avoid wasting search time in the poor-quality solution space, we use the reload strategy to escape from the local solution space when the difference between the cyclic bandwidth of the current solution and that of the best solution found reaches a given threshold by restarting from the best solution. The function Reload in Algorithm 1 returns $\varphi*$ if the difference between $C_B(G, \varphi)$ and $C_B(G, \varphi*)$ is greater than the threshold value, and returns φ otherwise. The threshold value tv is dynamically set by Eq. (7).

$$tv = (C_B(G, \varphi) - C_B(G, \varphi*)) \times \alpha + \beta \tag{7}$$

where α is a decimal value between 0 and 1, and β is a positive integer. Here, the α part ensures the adaptability of the threshold for different scales of instances, while the β part makes it easier to activate the reload process on large-scale instances than on small-scale ones. Generally, for small-scale instances, local search and restart are effective ways to help the search jump out of the local optimal trap. Nevertheless, for large-scale instances, reload is more effective because diversifying the search by local search or restart takes a lot of time.

4   Experimental Results

This section reports the experimental outcomes of our MVNTS algorithm and compares its performance to other state-of-the-art algorithms. These comparisons serve to demonstrate the effectiveness and robustness of our algorithm.

4.1   Experimental Protocol

In order to evaluate the performance of the proposed MVNTS algorithm, we conduct comprehensive experiments on public benchmark instances of varying scales. There are 113 instances sourced from 7 families (paths, cycles, caterpillars, complete trees, 2D and 3D meshes, and hypercubes) and Harwell–Boeing family[1], which were previously used in [18–20]. Since these existing instances are relatively small and most can be easily solved to optimality, we introduce 89 new instances from Harwell–Boeing family. These new instances include larger and more challenging graphs, which will provide a more robust evaluation test bed for the algorithm's performance. Thus, there are a total of 202 CBP benchmark instances, which are categorized into three groups based on the number of vertices.

- **First group**: The first group includes 80 small-scale instances, with the number of vertices not exceeding 200.
- **Second group**: The second group includes 78 medium-scale instances, where the number of vertices ranges from 200 to 1,000.
- **Third group**: The third group includes 44 large-scale instances, with the number of vertices greater than 1,000.

Our proposed MVNTS is implemented in C++. For a fair comparison, we rerun the state-of-the-art algorithms ITPS [19] and NILS [20] (using the source codes of the reference algorithms). These algorithms are conducted 20 independent runs on each problem instance, with a time limit of 600 s per run. All experiments are carried out on Intel(R) Xeon(R) E5-2698v3 @ 2.30 GHz CPU.

---

[1] http://math.nist.gov/MatrixMarket/data/Harwell-Boeing.

Table 1. All the parameters, along with the range of tuning and their final values.

Parameter	Section	Tuning range	Final value
L_1	3.2.1	{1,2,5,10,15,20,50,100,300,500,1000,2500,5000}	100
L_2	3.2.2	{1,2,5,10,15,20,50,100,300,500,1000,2500,5000}	15
L_3	3	{1,2,5,10,15,20,50,100,300,500,1000,2500,5000}	2500
K	3.2.1	{2,3,4,5,6,7,8,9,10}	5
W	3.2.1	{1,3,5,10,20,30,50,100}	10
α	3.3	{0.1,0.15,0.2,0.25,0.3,0.35,0.4,0.45,0.5}	0.2
β	3.3	{1,3,5,10,30,50,100,300,500,1000}	100

4.2   Parameter Tuning

The proposed MVNTS algorithm has a total of 7 parameters (L_1, L_2, L_3, K, W, α, β). In detail, L_1 is the maximum unimproved move counter before stopping the local search, L_2 is the perturbation intensity, L_3 is the exploration limit, K is used to control the neighborhood, W is the tabu tenure controller, α and β are used to determine the reload threshold.

The parameters used in our algorithm are tuned using the automatic configuration tool irace [24]. Table 1 shows all the parameters and their tuning range. In particular, the increasing difference between neighboring candidate parameters is due to the fact that we believe the parameter values are more sensitive in smaller cases. In the tuning process, the number of times of running the algorithm is set to 2000, and for each run, the maximum runtime limit is set to 600 s. We also define a subset of classical and difficult instances for tuning. The subset includes some instances from the standard families (mesh2D20×50, mesh3D13, caterpillar44, hypercube11), and also includes some instances from the Harwell-Boeing family (dwt 503, 494 bus, can 634, bcsstk24, bcspwr10). Following the automatic tuning process, the parameter values were determined to be $L_1 = 100$, $L_2 = 15$, $L_3 = 2500$, $K = 5$, $W = 10$, $\alpha = 0.2$ and $\beta = 100$.

4.3   Comparison with Reference Algorithms

In this section, we provide the experimental comparison between our algorithm and the state-of-the-art algorithms ITPS and NILS[2].

Table 2 summarizes the comparison results in terms of the average objective function values on different instances. Column Instances shows the group of the benchmark instances, column $Num.$ presents the instance numbers of each group, and the following columns give the average values of the best cyclic bandwidth ($Avg.Cb_b$), the average values of all the results of the cyclic bandwidth ($Avg.Cb$), the average computation time ($Avg.T$). Columns + and = indicate

---

[2] The detailed comparison results are reported in https://github.com/HUST-Smart/MVNTS-CBP.

Table 2. Summary of comparative results organized by scale.

Instances	$Num.$	IPTS					NILS					MVNTS				
		$Avg.Cb_b$	$Avg.Cb$	$Avg.T$	+	=	$Avg.Cb_b$	$Avg.Cb$	$Avg.T$	+	=	$Avg.Cb_b$	$Avg.Cb$	$Avg.T$	+	=
First group	80	8.64	8.77	17.93	0	74	**8.56**	**8.56**	**0.36**	0	80	**8.56**	**8.56**	0.40	0	80
Second group	78	34.31	46.85	187.68	0	42	32.78	32.89	38.66	0	73	**32.72**	**32.72**	**18.55**	5	73
Third group	44	466.43	526.10	320.65	0	2	203.41	257.03	236.33	0	29	**153.11**	**167.12**	**164.72**	15	29

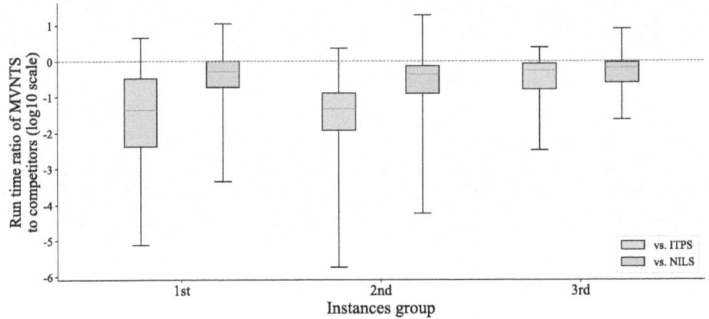

Fig. 2. The comparison of run time on three instance groups.

the number of instances on which the associated algorithm outperformed and obtained matched solutions compared to the best solution obtained by other algorithms, respectively.

From Table 2, we can observe that compared to ITPS, our algorithm shows better performance on all the three instance groups, and compared to NILS, it has advantage in terms of $Avg.Cb_b$ and $Avg.Cb$ on the medium and large-scale instances. In particular, compared to ITPS and NILS, our algorithm achieves better results on 20 out of the 202 instances, showing a significant advantage on the large-scale instances.

Moreover, when considering the average run time required for algorithms to reach their best solutions, it becomes evident that MVNTS outperforms ITPS and NILS in terms of computational efficiency on most instances as shown in Fig. 2. In fact, MVNTS is faster than NILS on about 75% of the instances and demonstrates an average speedup of 28.61 times over ITPS and 4.48 times over NILS.

Overall, our MVNTS algorithm shows comparable or even superior performance to NILS and ITPS in terms of solution quality, speed, and robustness on the all 202 benchmark instances.

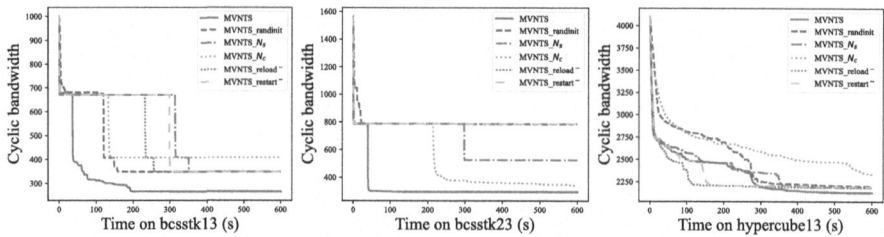

Fig. 3. Evolution of the cyclic bandwidth by MVNTS and its variants on three difficult instances.

4.4   Analysis of the Strategies

This section analyzes the effect of the greedy initialization strategy, the variable neighborhood strategy, the reload strategy, and the restart strategy. Our analysis is based on a comparison of the performance of MVNTS and that of its variants. Here we introduce the variants of MVNTS.

- **MVNTS_randinit**: The greedy initialization strategy is replaced by a completely random initialization strategy.
- **MVNTS_N_s**: The search procedure uses only the sampled neighborhood.
- **MVNTS_N_c**: The search procedure uses only the complete neighborhood.
- **MVNTS_reload⁻**: The reload strategy is removed.
- **MVNTS_restart⁻**: The restart strategy is removed.

The comparison experiment is conducted on three large and challenging instances. Figure 3 depicts the evolution of the cyclic bandwidth by MVNTS and its variants. Each point (x, y) on the curves represents that the best cyclic bandwidth y is found within x seconds. It can be observed that, compared to the variants, MVNTS achieved the best results on all three instances, with particularly significant advantages on bcsstk13 and bcsstk23. The experiments demonstrate the important role of the proposed strategies in enhancing the performance of MVNTS.

5   Conclusions

In this paper, we introduce a novel multi-start variable neighborhood tabu search algorithm with a greedy construction procedure and a reload strategy, which is named MVNTS, to tackle the CBP. MVNTS uses two types of neighborhoods that differ in the method of selecting a move. Experimental results demonstrate that the MVNTS algorithm outperforms the state-of-the-art ITPS and NILS on 84 (42%) and 20 (10%) of the 202 benchmark instances, respectively. Moreover, the time comparison box plots and detailed results further confirm that MVNTS performs well in terms of effectiveness, efficiency, and robustness.

Although many ideas in this study are proposed for solving the CBP, they are generic and might be applied to other combinatorial optimization problems.

We hope that this work will shed some light on further and better solutions to the CBP.

Acknowledgments. This work was supported in part by the National Natural Science Foundation of China (NSFC) under Grant 62402191, and 62202192.

References

1. Leung, J.Y.-T., Vornberger, O., Witthoff, J.D.: On Some Variants of the Bandwidth Minimization Problem. SIAM J. Comput. **13**(3), 650–667 (1984)
2. Lin, Y.X.: The cyclic bandwidth problem. Syst. Sci. Math. Sci. **7**(3), 282–288 (1994)
3. Rendl, F., Sotirov, R., Truden, C.: Lower bounds for the bandwidth problem. Comput. Oper. Res. **135**, 105422 (2021)
4. Lai, Y.-L., Williams, K.: A survey of solved problems and applications on bandwidth, edgesum, and profile of graphs. J. Graph Theor. **31**(2), 75–94 (1999)
5. Díaz, J., Petit, J., Serna, M.J.: A survey of graph layout problems. ACM Comput. Surv. **34**(3), 313–356 (2002)
6. Huang, R., Li, H., Zhang, Y.: Efficient bandwidth allocation and computation configuration in industrial IoT. ZTE Commun. **21**(1), 55–63 (2023)
7. Yang, H., Zhao, Y., Huang, S., Wang, D., Cao, X., Lin, X.: Multiple-constraint-aware RWA algorithms based on a comprehensive evaluation model: use in wavelength-switched optical networks. ZTE Commun. **10**(3), 55–61 (2012)
8. Hromkovič, J., Müller, V., Sýkora, O., Vrto, I.: On embedding interconnection networks into rings of processors. In: Etiemble, D., Syre, J.-C. (eds.) PARLE Conference 1992, LNCS, vol. 605, pp. 53–62. Springer (1992). https://doi.org/10.1007/3-540-55599-4_80
9. Lam, P., Shiu, W.C., Chan, W.H.: Characterization of graphs with equal bandwidth and cyclic bandwidth. Discret. Math. **242**(1), 283–289 (2002)
10. Lin, Y.: Minimum bandwidth problem for embedding graphs in cycles. Networks **29**(3), 135–140 (1997)
11. Lam, P.C.B., Shiu, W.C., Chan, W.H.: On bandwidth and cyclic bandwidth of graphs. Ars Combinatoria **47**, (1997)
12. Zhou, S.: Bounding the bandwidths for graphs. Theoret. Comput. Sci. **249**(2), 357–368 (2000)
13. Chan, W.H., Lam, P., Shiu, W.C.: Cyclic bandwidth with an edge added. Discret. Appl. Math. **156**(1), 131–137 (2008)
14. de Klerk, E., E.-Nagy, M., Sotirov, R.: On semidefinite programming bounds for graph bandwidth. Optim. Methods Softw. **28**(3), 485–500 (2013)
15. Déprés, H., Fertin, G., Monfroy, É.: Improved lower bounds for the cyclic bandwidth problem. In: Paszynski, M., Kranzlmüller, D., Krzhizhanovskaya, V.V., Dongarra, J.J., Sloot, P.M.A. (eds.) ICCS 2021, LNCS, vol. 12742, pp. 555–569. Springer (2021). 10.1007/978-3-030-77961-0_45
16. Fertin, G., Monfroy, É., Vasconcellos-Gaete, C.: Best of Both worlds: solving the cyclic bandwidth problem by combining pre-existing knowledge and constraint programming techniques. In: Franco, L., de Mulatier, C., Paszynski, M., Krzhizhanovskaya, V.V., Dongarra, J.J., Sloot, P.M.A. (eds.) ICCS 2024, LNCS, vol. 14836, pp. 197–211. Springer (2024). 10.1007/978-3-031-63775-9_14

17. Romero-Monsivais, H., Rodriguez-Tello, E., Ramírez-Torres, G.: A new branch and bound algorithm for the cyclic bandwidth problem. In: Batyrshin, I.Z., González-Mendoza, M. (eds.) MICAI 2012, LNCS, vol. 7630, pp. 139–150. Springer (2012). 10.1007/978-3-642-37798-3_13
18. Rodriguez-Tello, E., Romero-Monsivais, H., Ramirez-Torres, G., Lardeux, F.: Tabu search for the cyclic bandwidth problem. Comput. Oper. Res. **57**, 17–32 (2015)
19. Ren, J., Hao, J.-K., Rodriguez-Tello, E.: An iterated three-phase search approach for solving the cyclic bandwidth problem. IEEE Access **7**, 98436–98452 (2019)
20. Ren, J., Hao, J.-K., Rodriguez-Tello, E., Li, L., He, K.: A new iterated local search algorithm for the cyclic bandwidth problem. Knowl. Based Syst. **203**, 106136 (2020)
21. Chinn, P.Z., Chvatalova, J., Dewdney, A.K., Gibbs, N.E.: The bandwidth problem for graphs and matrices - a survey. J. Graph Theor. **6**(3), 223–254 (1982)
22. Glover, F., Laguna, M.: Tabu search. Springer (1998)
23. Mladenović, N., Hansen, P.: Variable neighborhood search. Comput. Oper. Res. **24**(11), 1097–1100 (1997)
24. López-Ibáñez, M., Dubois-Lacoste, J., Pérez Cáceres, L., Birattari, M., Stützle, T.: The irace package: Iterated racing for automatic algorithm configuration. Oper. Res. Perspect. **3**, 43–58 (2016)

Vertex-Critical (P_5, W_4)-Free Graphs

Wen Xia[1,2], Jorik Jooken[3], Jan Goedgebeur[3,4], Iain Beaton[5],
Ben Cameron[6] , and Shenwei Huang[7(✉)]

[1] College of Computer Science, Nankai University, Tianjin 300071, China
[2] Tianjin Key Laboratory of Network and Data Security Technology, Nankai
University, Tianjin 300071, China
[3] Department of Computer Science, KU Leuven Campus Kulak-Kortrijk, Kortrijk
8500, Belgium
[4] Department of Applied Mathematics, Computer Science and Statistics, Ghent
University, Ghent 9000, Belgium
[5] Department of Mathematics and Statistics, Acadia University,
Wolfville, Ns, Canada
[6] School of Mathematical and Computational Sciences, University of Prince Edward
Island, Charlottetown, PE, Canada
brcameron@upei.ca
[7] School of Mathematical Sciences and LPMC, Nankai University,
Tianjin 300071, China
shenweihuang@nankai.edu.cn

Abstract. A graph G is k-vertex-critical if $\chi(G) = k$ but $\chi(G - v) < k$ for all $v \in V(G)$. A graph is (H_1, H_2)-free if it contains no induced subgraph isomorphic to H_1 nor H_2. A W_4 is the graph consisting of a C_4 plus an additional vertex adjacent to all the vertices of the C_4. We show that there are finitely many k-vertex-critical (P_5, W_4)-free graphs for all $k \geq 1$ and we characterize all 5-vertex-critical (P_5, W_4)-free graphs. Our results imply the existence of a polynomial-time certifying algorithm to decide the k-colorability of (P_5, W_4)-free graphs for each $k \geq 1$ where the certificate is either a k-coloring or a $(k + 1)$-vertex-critical induced subgraph.

Keywords: Graph coloring · k-critical graphs · Strong perfect graph
theorem · Polynomial-time algorithms

1 Introduction

All graphs in this paper are finite, undirected and simple. Let P_t and C_t denote the path and the cycle on t vertices, respectively. Let K_n be the complete graph on n vertices. The complement of G is denoted by $\overline{G}$. For two graphs G and H, we use $G + H$ to denote the disjoint union of G and H. For a positive integer r, we use rG to denote the disjoint union of r copies of G. For $s, r \geq 1$, let $K_{r,s}$ be the complete bipartite graph with one part of size r and the other part of size s. The *clique number* of G, denoted by $\omega(G)$, is the size of the largest clique in G.

© The Author(s), under exclusive license to Springer Nature Singapore Pte Ltd. 2026
F. V. Fomin and M. Xiao (Eds.): COCOON 2025, LNCS 15984, pp. 139–152, 2026.
https://doi.org/10.1007/978-981-95-0218-9_11

A k-*coloring* of a graph G is an assignment of k colors to its vertices such that adjacent vertices are assigned different colors. A graph is k-*colorable* if it has a k-coloring. The *chromatic number* of G, denoted by $\chi(G)$, is the minimum number k for which G is k-colorable. A graph G is said to be k-*chromatic* if $\chi(G) = k$.

We say that G is *critical* if $\chi(H) < \chi(G)$ for every proper subgraph H of G. A k-*critical* graph is one that is k-chromatic and critical. For example, odd cycles are the only 3-critical graphs. Vertex-criticality is a weaker notion. A graph G is k-*vertex-critical* if $\chi(G) = k$ but $\chi(G - v) < k$ for any $v \in V(G)$.

For a fixed integer k, the k-*coloring problem* is to determine whether a given graph is k-colorable. It is known that this problem is NP-complete for all $k \geq 3$ [30] (see also [27]). But if we restrict the graph structure, then there may exist polynomial-time algorithms to solve the k-coloring problem for all k. One of the most popular structural restrictions is to forbid induced subgraphs. A graph G is $\mathcal{H}$-*free* if it does not contain any member in $\mathcal{H}$ as an induced subgraph. When $\mathcal{H}$ consists of a single graph H or two graphs H_1 and H_2, we write H-free and (H_1, H_2)-free instead of $\{H\}$-free and $\{H_1, H_2\}$-free, respectively. We say that G is k-vertex-critical $\mathcal{H}$-free if it is k-vertex-critical and $\mathcal{H}$-free. We say that G is k-critical $\mathcal{H}$-free if G is k-chromatic, $\mathcal{H}$-free and every proper $\mathcal{H}$-free subgraph of G is $(k - 1)$-colorable.

It is known that the k-coloring problem remains NP-complete on H-free graphs for all $k \geq 3$ when H contains an induced claw [28] or cycle [21,29]. This means that if the k-coloring problem can be solved in polynomial-time on H-free graphs for all $k \geq 3$, then H must be a linear forest (assuming P $\neq$ NP). In 2010, Hoàng et al. [19] proved that the k-coloring problem can be solved in polynomial-time on P_5-free graphs for all k. Later, Huang [22] showed that the problem remains NP-complete for P_6-free graphs for all $k \geq 5$ and for P_7-free graphs for all $k \geq 4$. So P_5 is the largest connected subgraph that can be forbidden for which k-coloring can be solved in polynomial-time for all k (again assuming P $\neq$ NP). Therefore, P_5-free graphs have attracted widespread attention from many scholars. The algorithms in [19] produce a k-coloring if the input graph is k-colorable; however, the algorithms do not produce an easily verifiable certificate when the input graph is not k-colorable.

An algorithm is *certifying* if it returns with each output a simple and easily verifiable certificate that the particular output is correct. For practical purposes, certifying algorithms are highly desirable as they allow for robust yet efficient testing of implementations of the algorithms [32]. For theoretical purposes, it is of interest to determine when certifying algorithms exist. For a k-colorability algorithm to be certifying, it should return a k-coloring if one exists and a $(k + 1)$-vertex-critical induced subgraph if the graph is not k-colorable. In fact, when the number of $(k + 1)$-vertex-critical graphs is finite in a given family, a polynomial-time k-colorability algorithm is readily implemented by searching the input graphs for induced $(k + 1)$-vertex-critical graphs (see [9] for details).

A linear-time certifying algorithm for 3-colorability of P_5-free graphs is provided by showing that the number of 4-vertex-critical P_5-free graphs is finite [7,31]. Given this result, one may consider whether there are only finitely many k-vertex-critical P_5-free graphs for $k \geq 5$. Unfortunately, for each $k \geq 5$, it was shown that there are infinitely many k-vertex-critical P_5-free graphs [20]. This led to significant interest in determining which subfamilies of P_5-free graphs admit polynomial-time k-colorability algorithms that are also certifying. In 2021, K. Cameron, Goedgebeur, Huang and Shi [12] obtained the dichotomy theorem that there are infinitely many k-vertex-critical (P_5, H)-free graphs for H of order 4 if and only if H is $2P_2$ or $P_1 + K_3$. In [12], the natural question was posed: which five-vertex graphs H lead to only finitely many k-vertex-critical (P_5, H)-free graphs?

Significant progress has already been made towards solving this question. The infinite family of k-vertex-critical P_5-free graphs was generalized to infinite families of k-vertex-critical (P_5, C_5)-free graphs for each $k \geq 6$ [10]. On the other hand, it is known that there are only finitely many 5-vertex critical (P_5, H)-free graphs when $H = C_5$ [20], bull [23] or chair [24]. For all $k \geq 1$, it has been shown that there are only finitely many k-vertex-critical (P_5, H)-free graphs when H = dart [35], banner [6], $P_2 + 3P_1$ [11], $P_3 + 2P_1$ [1], $\overline{P_3 + P_2}$ and gem [8], $K_{2,3}$ and $K_{1,4}$ [26], $K_{1,3} + P_1$ and $\overline{K_3 + 2P_1}$ [36] and $\overline{P_5}$ [15].

1.1   Our Contributions

In this paper, we continue to study the finiteness of vertex-critical (P_5, H)-free graphs when H has order 5. The 4-wheel W_4 is the graph consisting of a C_4 plus an additional vertex adjacent to all the vertices of the C_4 (see Fig. 1). Our main result is as follows.

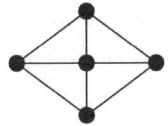

Fig. 1. The 4-wheel W_4.

Theorem 1. *For every fixed integer* $k \geq 1$, *there are only finitely many k-vertex-critical* (P_5, W_4)-*free graphs.*

Our result implies the existence of a polynomial-time certifying algorithm to decide the k-colorability of (P_5, W_4)-free graphs for $k \geq 1$. This algorithm works on an input (P_5, W_4)-free graph G by running the polynomial-time k-colorability algorithm for P_5-free graphs from [19] that also outputs a k-coloring of G as a certificate if the answer is yes. If the answer is no, we then we search for a $(k+1)$-vertex-critical (P_5, W_4)-free graph as induced subgraph of G. When one is found, we return the labeled induced subgraph as a certificate. Since by Theorem 1 there are only finitely many $(k + 1)$-vertex-critical (P_5, W_4)-free graphs for all

$k \geq 1$, and each can be checked as an induced subgraph of G in polynomial time, it follows that the entire certifying algorithm runs in polynomial time.

We note that W_4 is the densest graph of order 5 currently known that can be forbidden as an induced subgraph with P_5 and result in only finitely many k-vertex-critical graphs for all k. While the big picture of our proof technique is similar to previous work in this area, that is, to bound the structure around odd holes and odd antiholes and apply the Strong Perfect Graph Theorem, the density of W_4 made it especially difficult to induce around a C_5. This led us to develop novel arguments involving homogeneous components of sets that we expect to be of use for the structure of (P_5, H)-free graphs for $H \neq W_4$. We also characterize all 5-vertex-critical (P_5, W_4)-free graphs with the aid of exhaustive computer search based on a recursive graph generation algorithm, leading to the following theorem:

Theorem 2. *Let $\mathcal{S}_1$ and $\mathcal{S}_2 \subset \mathcal{S}_1$ be the set of 5-vertex-critical (P_5, W_4)-free graphs and 5-critical (P_5, W_4)-free graphs, respectively. We have $|\mathcal{S}_1| = 64$, $|\mathcal{S}_2| = 21$ and the largest graphs in $\mathcal{S}_1$ and $\mathcal{S}_2$ have 17 vertices.*

The remainder of the paper is organized as follows. We present some preliminaries in Sect. 2. We show that there are finitely many k-vertex-critical (P_5, W_4)-free graphs for all $k \geq 1$ in Sect. 3 assuming two lemmas. We will prove one of these lemmas in Sect. 4, while the proof of the other is omitted in the interest of space, but can be found in the extended preprint version of this paper [34]. We then characterize all such graphs for $k = 5$ in Sect. 5. Finally, we give a conclusion in Sect. 6.

2   Preliminaries

For general graph theory notation we follow [5]. For $k \geq 4$, an induced cycle of length k is called a *k-hole*. A k-hole is an *odd hole* (respectively *even hole*) if k is odd (respectively even). A *k-antihole* is the complement of a k-hole. Odd and even antiholes are defined analogously.

Let $G = (V, E)$ be a graph. If $uv \in E(G)$, we say that u and v are *neighbors* or *adjacent*, otherwise u and v are *nonneighbors* or *nonadjacent*. The *neighborhood* of a vertex v, denoted by $N_G(v)$, is the set of neighbors of v. For a set $X \subseteq V(G)$, let $N_G(X) = \cup_{v \in X} N_G(v) \setminus X$. We shall omit the subscript whenever the context is clear. For $x \in V(G)$ and $S \subseteq V(G)$, we denote by $N_S(x)$ the set of neighbors of x that are in S, i.e., $N_S(x) = N_G(x) \cap S$. For two sets $X, S \subseteq V(G)$, let $N_S(X) = \cup_{v \in X} N_S(v) \setminus X$.

For $X, Y \subseteq V(G)$, we say that X is *complete* (resp. *anticomplete*) to Y if every vertex in X is adjacent (resp. nonadjacent) to every vertex in Y. If $X = \{x\}$, we write "x is complete (resp. anticomplete) to Y" instead of "$\{x\}$ is complete (resp. anticomplete) to Y". If a vertex v is neither complete nor anticomplete to a set S, we say that v is *mixed* on S. For a vertex $v \in V$ and an edge $xy \in E$, if v is mixed on $\{x, y\}$, we say that v is *mixed* on xy. For a set $H \subseteq V(G)$, if no vertex in $V(G) \setminus H$ is mixed on H, we say that H is a *homogeneous set*, otherwise H is a *nonhomogeneous set*.

Lemma 1. *For disjoint sets $H, A \subseteq V(G)$. If $v \in A$ is not mixed on any edge in H, then v is not mixed on any connected component of H.*

Proof. Suppose that $v \in A$ is not mixed on any edge in H. Let H' be a connected component of H with edge xy. Let v be (anti)complete to xy. By the connectivity of H', for any vertex $u \in H'$ there exists a path $xy_1y_2 \cdots u$ from x to u. As v is (non)adjacent to x and v is not mixed on any edge of H' then v is (non)adjacent to y_1. Similarly, v must be (anti)complete to each edge in the path from x to u. Therefore v is (non)adjacent to u and hence (anti)complete to H'.

A vertex subset $S \subseteq V(G)$ is *stable* if no two vertices in S are adjacent. A *clique* is the complement of a stable set. Two nonadjacent vertices u and v are said to be *comparable* if $N(v) \subseteq N(u)$ or $N(u) \subseteq N(v)$. For an induced subgraph A of G, we write $G - A$ instead of $G - V(A)$. For $S \subseteq V$, the subgraph *induced* by S is denoted by $G[S]$. We say that a vertex w *distinguishes* two vertices u and v if w is adjacent to exactly one of u and v.

We proceed with a few useful results that will be used later. The first folklore property of vertex-critical graphs is that such graphs contain no comparable vertices. A generalization of this property was presented in [12].

Lemma 2 ([12]). *Let G be a k-vertex-critical graph. Then G has no two nonempty disjoint subsets X and Y of $V(G)$ that satisfy all the following conditions.*

- *X and Y are anticomplete to each other.*
- *$\chi(G[X]) \leq \chi(G[Y])$.*
- *Y is complete to $N(X)$.*

Another useful result is the following Lemma.

Lemma 3 ([35]). *Let G be a k-vertex-critical graph and S be a homogeneous set of $V(G)$. For each component A of $G[S]$, if $\chi(A) = m$ with $m < k$, then A is an m-vertex-critical graph.*

The following theorem tells us there are finitely many 4-vertex-critical P_5-free graphs.

Theorem 3 ([7,31]) *If $G = (V, E)$ is a 4-vertex-critical P_5-free graph, then $|V| \leq 13$.*

A graph G is *perfect* if $\chi(H) = \omega(H)$ for every induced subgraph H of G. We conclude this section with the famous Strong Perfect Graph Theorem (while we note that for the class of (P_5, W_4)-free graphs, the result was known earlier [3]).

Theorem 4 (The Strong Perfect Graph Theorem[13]). *A graph is perfect if and only if it contains no odd holes or odd antiholes.*

3   The Proof of Theorem 1

We prove Theorem 1 assuming the following two lemmas,the second of which will be proved in Sect. 4.

Lemma 4 (♠)[1] *If G is a k-vertex-critical (P_5, W_4)-free graph that contains an induced C_5, then G has finite order.*

Lemma 5. *If G is a k-vertex-critical (P_5, W_4)-free graph that contains an induced $\overline{C_7}$, then G has finite order.*

Proof (Proof of Theorem 1). Let G be a k-vertex-critical (P_5, W_4)-free graph. We show that $|G|$ is bounded. Let $\mathcal{L} = \{K_k, \overline{C_{2k-1}}\}$. If G has a subgraph isomorphic to a member $L \in \mathcal{L}$, then $|V(G)| = |V(L)|$ by the definition of vertex-critical and so we are done. So, we assume in the following G has no induced subgraph isomorphic to a member in $\mathcal{L}$. Then G is imperfect since the only perfect k-vertex-critical graph is K_k. Since G is W_4-free, G does not contain $\overline{C_{2t+1}}$ for $t \geq 4$. Moreover, since G is P_5-free, it does not contain C_{2t+1} for $t \geq 3$. It then follows from Theorem 4, G must contain C_5 or $\overline{C_7}$. The result now follows from Lemma 4 and Lemma 5.

4   Proof of Lemma 5: Structure Around an Induced 7-Antihole

We require some more notation before we can proceed with the proof.

Let $G = (V, E)$ be a graph and H be an induced subgraph of G. We partition $V \setminus V(H)$ into subsets with respect to H as follows: for any $X \subseteq V(H)$, we denote by $S(X)$ the set of vertices in $V \setminus V(H)$ that have X as their neighbors in $V(H)$, i.e.,

$$S(X) = \{v \in V \setminus V(H) : N_{V(H)}(v) = X\}.$$

We often drop the set braces of X for this notation and index with the number of neighbors on C (i.e., $S(\{v_1, v_2, v_6, v_7\})$ would be denoted $S_4(v_1, v_2, v_6, v_7)$). For $0 \leq m \leq |V(H)|$, we denote by S_m the set of vertices in $V \setminus V(H)$ that have exactly m neighbors in $V(H)$. Note that

$$S_m = \bigcup_{X \subseteq V(H):|X|=m} S(X).$$

Proof (Proof of Lemma 5). We prove the lemma by induction on k. If $1 \leq k \leq 4$, there are finitely many k-vertex-critical (P_5, W_4)-free graphs by Theorem 3. In the following, we assume that $k \geq 5$ and there are finitely many i-vertex-critical (P_5, W_4)-free graphs for $i \leq k - 1$.

Let G be a k-vertex-critical (P_5, W_4)-free graph such that $C = v_1, v_2, \ldots, v_7$ induces a $\overline{C_7}$ in G where $v_i v_j \in E(G)$ if and only if $1 < |i - j| < 6$ (all indices are modulo 7). We partition $V(G)$ with respect to C. Additionally, note if G has a C_5, then the proof is completed by Lemma 4. So in the following, we assume that G is (P_5, W_4, C_5)-free.

---

[1] Proofs of results marked with a ♠ have been omitted in the interest of space and can be found in the extended preprint version of this paper [34].

We begin by proving a few claims which will help show that only S_3 and S_5 are non-empty.

(I) For each $1 \leq i \leq 7$, no vertex $u \notin C$ is adjacent to v_i, v_{i+1}, v_{i-2}, and v_{i-3}. Then $\{u, v_i, v_{i+1}, v_{i-2}, v_{i-3}\}$ induces a W_4.

(II) For each $1 \leq i \leq 7$, no vertex $u \notin C$ is adjacent to v_i, v_{i+1} but not adjacent to v_{i-1}, v_{i+2}. Then $\{u, v_i, v_{i+2}, v_{i-1}, v_{i+1}\}$ induces a C_5.

(III) For each $1 \leq i \leq 7$, no vertex $u \notin C$ is adjacent to v_i, but not adjacent to v_{i-1}, v_{i+1}, v_{i+2}. Further, there is no adjacent vertex v_i, but not adjacent to v_{i+1}, v_{i-1}, v_{i-2}.
Then a P_5 is induced by the set $\{u, v_i, v_{i+2}, v_{i-1}, v_{i+1}\}$ in the first case and $\{u, v_i, v_{i-2}, v_{i+1}, v_{i-1}\}$ in the second case.

It follows from **(I)** that $S_7 = \emptyset$, $S_6 = \emptyset$, and additionally the only non-empty subsets of S_5 are $S_5(v_{i-2}, v_{i-1}, v_i, v_{i+1}, v_{i+3})$ for each $1 \leq i \leq 7$.

For subsets of S_4, we have that

$$S_4(v_{i-1}, v_i, v_{i+1}, v_{i+3}) \cup S_4(v_{i-2}, v_i, v_{i+1}, v_{i+2}) = \emptyset$$

from **(III)** and $S_4(v_{i-1}, v_i, v_{i+2}, v_{i+3}) = \emptyset$ from **(I)**. Thus, the only remaining non-empty sets of S_4 are $S_4(v_{i-2}, v_{i-1}, v_i, v_{i+1})$ and $S_4(v_{i-2}, v_i, v_{i+1}, v_{i+3})$. However, if $u \in S_4(v_{i-2}, v_{i-1}, v_i, v_{i+1})$ then $\{v_{i-2}, u, v_{i+1}, v_{i+3}, v_i\}$ induces a W_4 and if $u \in S_4(v_{i-2}, v_i, v_{i+1}, v_{i+3})$ then $\{u, v_{i+1}, v_{i-1}, v_{i+2}, v_i\}$ induces a C_5. Therefore $S_4 = \emptyset$.

From **(II)** it follows that the only non-empty subsets of S_3 are $S_3(v_{i-1}, v_i, v_{i+1})$ for each $1 \leq i \leq 7$. To show $S_2 = \emptyset$, let $u \in S_2(v_i, v_j)$. It follows from **(II)** that $|i - j| > 1$ modulo 7. Moreover, either $i - j > 2$ or $j - i > 2$ modulo 7 (since if $1 < i - j < 2$, then it must equal 2, and therefore $j - i = -2$ which is 5 modulo 7). It then follows from **(III)** that $S_2 = \emptyset$. Additionally it follows from **(III)** that $S_1 = \emptyset$. Finally, we will show that $S_0 = \emptyset$. Suppose not. By the connectivity of G, some vertex $u \in S_0$ has a neighbor v and since we just showed that $S_i = \emptyset$ for all $i \in \{1, 2, 4, 6, 7\}$, we must have $v \in S_3 \cup S_5$. It can be readily seen that there exists an index j such that v is adjacent to v_j, but not adjacent to v_{j+1} or v_{j+3}. Then $\{u, v, v_j, v_{j+3}, v_{j+1}\}$ is an induced P_5 (i.e., if $v \in S_5$, then from above, we know that $v \in S_5(v_{i-2}, v_{i-1}, v_i, v_{i+1}, v_{i+3})$, so the result follows with $j = i + 1$). This proves that $S_0 = \emptyset$.

Thus, $S_i = \emptyset$ for all $i \neq 3, 5$. We will now show that both S_5 and S_3 are bounded.

Claim. For each $1 \leq i \leq 7$, $S_5(v_{i-2}, v_{i-1}, v_i, v_{i+1}, v_{i+3})$ is bounded.

Proof. We will show that $S_5(v_{i-2}, v_{i-1}, v_i, v_{i+1}, v_{i+3})$ is a clique. Suppose not and let $u, v \in S_5(v_{i-2}, v_{i-1}, v_i, v_{i+1}, v_{i+3})$ such that $uv \notin E(G)$. Then $\{v_{i+3}, u, v_i, v, v_{i+1}\}$ induces a W_4. Thus $S_5(v_{i-2}, v_{i-1}, v_i, v_{i+1}, v_{i+3})$ is a clique. Moreover, $v_{i-1}, v_{i+1}, v_{i+3}$ induce a K_3 and are complete to $S_5(v_{i-2}, v_{i-1}, v_i, v_{i+1}, v_{i+3})$. Therefore $|S_5(v_{i-2}, v_{i-1}, v_i, v_{i+1}, v_{i+3})| \leq k - 4$.

Finally, we bound S_3. We will first prove some properties about S_3. All properties are proved for $i = 1$ due to symmetry.

(IV) For each $1 \leq i \leq 7$, $S_3(v_{i-1}, v_i, v_{i+1})$ is anticomplete to $S_3(v_i, v_{i+1}, v_{i+2}) \cup S_3(v_{i-2}, v_{i-1}, v_i)$.

Let $u \in S_3(v_7, v_1, v_2)$ and $v \in S_3(v_1, v_2, v_3)$. If $uv \in E(G)$, then $\{u, v, v_3, v_6, v_4\}$ is an induced P_5. By symmetry, $S_3(v_7, v_1, v_2)$ is anticomplete to $S_3(v_6, v_7, v_1)$.

(V) For each $1 \leq i \leq 7$, $S_3(v_{i-1}, v_i, v_{i+1})$ is anticomplete to $S_3(v_{i+1}, v_{i+2}, v_{i+3}) \cup S_3(v_{i-3}, v_{i-2}, v_{i-1})$.

Let $u \in S_3(v_7, v_1, v_2)$ and $v \in S_3(v_2, v_3, v_4)$. If $uv \in E(G)$, then $\{u, v_7, v_4, v, v_2\}$ is an induced W_4. By symmetry, $S_3(v_7, v_1, v_2)$ is anticomplete to $S_3(v_5, v_6, v_7)$.

(VI) For each $1 \leq i \leq 7$, $S_3(v_{i-1}, v_i, v_{i+1})$ is anticomplete to $S_3(v_{i+2}, v_{i+3}, v_{i+4}) \cup S_3(v_{i-4}, v_{i-3}, v_{i-2})$.

Let $u \in S_3(v_7, v_1, v_2)$ and $v \in S_3(v_3, v_4, v_5)$. If $uv \in E(G)$, then $\{v_6, v_2, u, v, v_3\}$ is a C_5. By symmetry, $S_3(v_7, v_1, v_2)$ is anticomplete to $S_3(v_4, v_5, v_6)$.

(VII) For each $1 \leq i \leq 7$, $S_5 \setminus S_5(v_{i+2}, v_{i+3}, v_{i+4}, v_{i+5}, v_i)$ is not mixed on any component of $S_3(v_{i-1}, v_i, v_{i+1})$.

By Lemma 1, it suffices to show that $S_5 \setminus S_5(v_{i+2}, v_{i+3}, v_{i+4}, v_{i+5}, v_i)$ is not mixed on any edge of $S_3(v_{i-1}, v_i, v_{i+1})$. Let $v \in S_5 \setminus S_5(v_3, v_4, v_5, v_6, v_1)$ and suppose that v is mixed on the edge uu' of $S_3(v_7, v_1, v_2)$. Without loss of generality let $u'v \in E(G)$ and $uv \notin E(G)$. Note that v is adjacent to two or three vertices in $\{v_3, v_4, v_5, v_6\}$. So there must exist two adjacent vertices $v_j, v_\ell \in \{v_3, v_4, v_5, v_6\}$ such that v is mixed on $\{v_j, v_\ell\}$. Without loss of generality let $vv_j \in E(G)$ and $vv_\ell \notin E(G)$. Then $\{u, u', v, v_j, v_\ell\}$ is an induced P_5.

(VIII) For each $1 \leq i \leq 7$, $S_3(v_{i_1}, v_i, v_{i+1})$ is complete to $S_5(v_{i+2}, v_{i+3}, v_{i+4}, v_{i+5}, v_i)$.

Let $u \in S_3(v_7, v_1, v_2)$ and $v \in S_5(v_3, v_4, v_5, v_6, v_1)$. If $uv \notin E(G)$, then $\{u, v_2, v_4, v, v_3\}$ is an induced P_5.

Claim. For each $1 \leq i \leq 7$, $S_3(v_{i-1}, v_i, v_{i+1})$ is bounded.

Proof. Let A be a component of $S_3(v_{i-1}, v_i, v_{i+1})$, then A is homogeneous by **(IV)**–**(VIII)**. By Lemma 3, each component of $S_3(v_{i-1}, v_i, v_{i+1})$ is an m-vertex-critical (P_5, W_4)-free graph where $1 \leq m \leq k - 3$. By the inductive hypothesis, it follows that there are finitely many m-vertex-critical (P_5, W_4)-free graphs for $1 \leq m \leq k - 3$. Now let H_1 be a component of $S_3(v_{i-1}, v_i, v_{i+1})$. Note that $N(H) = \{v_{i-1}, v_i, v_{i+1}\} \cup X$ for some $X \subseteq S_5$. Additionally, if two components of $S_3(v_{i-1}, v_i, v_{i+1})$ had the same neighborhood in S_5 then this would contradict Lemma 2. Therefore, each component of $S_3(v_{i-1}, v_i, v_{i+1})$ has a unique set of neighbors in S_5. By the Pigeonhole Principle, $|S_3(v_{i-1}, v_i, v_{i+1})| \leq 2^{|S_5|}$. Thus, $S_3(v_{i-1}, v_i, v_{i+1})$ is bounded.

This completes the proof of Lemma 5.

5   Characterizing All 5-Vertex-Critical (P_5, W_4)-Free Graphs

Hoàng et al. [19] proposed a recursive graph generation algorithm that can generate all k-vertex-critical $\mathcal{H}$-free graphs. This algorithm was further generalized, improved and its implementation was made more efficient and generic (by considering additional heuristics, pruning rules and parameters) in a series of papers [17, 18, 35, 36]. We used this algorithm to characterize all 5-vertex-critical (P_5, W_4)-free graphs.

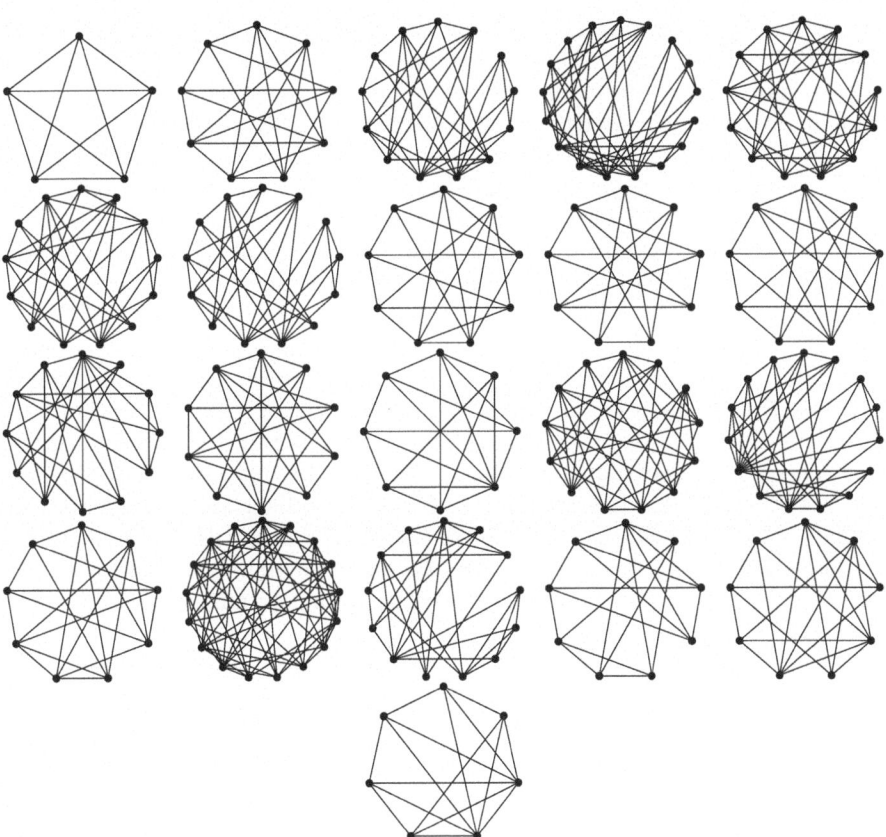

Fig. 2. All 5-critical (P_5, W_4)-free graphs.

To keep the current paper self-contained, we briefly sketch the main ideas of this algorithm. The algorithm receives as input a graph I and recursively generates all k-vertex-critical $\mathcal{H}$-free graphs for which I occurs as an induced subgraph. In order to do so, the algorithm adds a vertex u to I and adds edges between u and vertices in $V(I)$ in all possible ways and recurses for each of the obtained graphs. However, in order to make the algorithm more efficient and even terminate in some cases[2], several additional optimizations, heuristics and pruning rules are added while still ensuring that all k-vertex-critical $\mathcal{H}$-free graphs are enumerated, i.e. the algorithm is exhaustive. For example, it is clear that if the algorithm encounters a graph which is not $\mathcal{H}$-free, it does not need to recurse because such a graph cannot occur as a proper induced subgraph of any k-vertex-critical $\mathcal{H}$-free graph. Another example of a pruning rule is based on the following observation: if I occurs as an induced subgraph of a k-vertex-critical $\mathcal{H}$-free graph G and there are two vertices $x, y \in V(I)$ such that $N_I(x) \subseteq N_I(y)$, then there must exist a vertex $u \in V(G) \setminus V(I)$ such that u is adjacent to x, but not to y (this was in fact generalized to sets X and Y in Lemma 2). This observation restricts the set of edges that can be added between u and $V(I)$. For the sake of brevity, we refer the interested reader to [17] for a more complete overview of the different pruning rules, which are more complicated than the sketch in the current paper. When adding edges to a graph in multiple ways, often isomorphic copies will be obtained. The algorithm detects this by computing a canonical form of its input graph I using the *nauty* package [33] (two graphs are isomorphic if and only if they have the same canonical form) and prunes a graph if an isomorphic graph was already processed earlier. The pseudo code of the algorithm described in this paragraph is shown in Algorithm 1.

As mentioned in Sect. 3, each k-vertex-critical (P_5, W_4)-free graph is either isomorphic to K_k, isomorphic to $\overline{C_{2k-1}}$, contains C_5 as an induced subgraph or contains $\overline{C_7}$ as an induced subgraph. By additionally running[3] Algorithm 1 for $k = 5$, $\mathcal{H} = \{P_5, W_4\}$ and $I = C_5$ and $I = \overline{C_7}$ we obtain the characterization from Theorem 2. We summarize the counts of 5-vertex-critical (P_5, W_4)-free and 5-critical (P_5, W_4)-free graphs (see Fig. 2) in Table 1 and also make these graphs publicly available at the House of Graphs [14] at: https://houseofgraphs. org/meta-directory/critical-h-free. We also executed the algorithm for $k = 6$, $\mathcal{H} = \{P_5, W_4\}$ and $I = C_5$, but the algorithm did not terminate after one week of running.

---

[2] We remark that the algorithm will never terminate if there are infinitely many k-vertex-critical $\mathcal{H}$-free graphs, but if there are finitely many such graphs the algorithm sometimes terminates if the pruning rules are strong enough. In the latter case, it is guaranteed that there are only finitely many k-vertex-critical $\mathcal{H}$-free graphs and the algorithm enumerates all of them.

[3] In fact, we made two independent implementations of Algorithm 1 and obtained the same results for both implementations, which gives us great confidence that there are no bugs leading to incorrect results. These implementations are made publicly available at [16] and [25].

Algorithm 1. generateGraphs(Induced graph I, Integer k, Set of graphs $\mathcal{H}$)

1: // Generate all k-vertex-critical $\mathcal{H}$-free graphs that contain I as an induced subgraph
2: **if** I can be pruned by one of the pruning rules **then**
3:    return
4: **end if**
5: $C \leftarrow$ computeCanonicalForm(I)
6: // A graph isomorphic with I was already encountered before
7: **if** allCanonicalForms.contains(C) **then**
8:    return
9: **else**
10:    allCanonicalForms.add(C)
11:    **if** I is k-vertex-critical $\mathcal{H}$-free **then**
12:       Output I
13:    **end if**
14:    **for** every graph I' obtained by adding a vertex u to I and edges between u and
       vertices in $V(I)$ in all possible ways **do**
15:       generateGraphs(I',k,$\mathcal{H}$)
16:    **end for**
17: **end if**

---

6 Conclusion

In this paper, we proved that there are finitely many k-vertex-critical (P_5, W_4)-free graphs for all $k \geq 1$ and characterized all such graphs for $k = 5$. Our results gave an affirmative answer to the problem posed in [12] for $H = W_4$. In the future, it is natural to investigate the finiteness of the set of k-vertex-critical (P_5, H)-free graphs for other graphs H of order 5. From [10] and the recent work showing finiteness in the cases where $H = K_{1,3} + P_1$ and $H = \overline{K_3 + 2P_1}$ [36], the only remaining open cases are when H is any of the following eight graphs of order 5:

- co-gem
- chair

- $\overline{\text{diamond} + P_1}$

- $C_4 + P_1$

- bull
- $\overline{P_3 + 2P_1}$
- $K_5 - e$
- K_5

Of these, we expect the results in this paper will be most directly applicable to $\overline{P_3 + 2P_1}$, $K_5 - e$ and K_5 as these graphs are also dense, and so are more difficult to induce around a C_5. These are also the only remaining cases where the vertex-critical graphs are $\overline{C_{2\ell+1}}$-free for some ℓ (and in fact, we have either $\ell = 4$ or $\ell = 5$). We also note that cases when H is co-gem [4] and chair, $\overline{\text{diamond} + P_1}$, or bull [2] are known to be finite for all $k \geq 1$ when the graphs are further restricted to be $2P_2$-free.

Table 1. An overview of the number of pairwise non-isomorphic 5-vertex-critical (P_5, W_4)-free graphs and 5-critical (P_5, W_4)-free graphs.

n	# 5-vertex-critical (P_5, W_4)-free graphs	# 5-critical (P_5, W_4)-free graphs
5	1	1
6	0	0
7	1	1
8	1	1
9	44	7
10	4	1
11	0	0
12	1	1
13	8	6
14	0	0
15	2	1
16	0	0
17	2	2
Total	64	21

Acknowledgments. Shenwei Huang is supported by the Natural Science Foundation of China (NSFC) under Grant 12171256 and 12161141006. We also acknowledge the support of the joint NSFC-FWO scientific mobility project with grant number 12311530678 and the support of the joint FWO-NSFC scientific mobility project with grant number VS01224N. The research of Jan Goedgebeur was supported by Internal Funds of KU Leuven and an FWO grant with grant number G0AGX24N. Jorik Jooken is supported by a Postdoctoral Fellowship of the Research Foundation Flanders (FWO) with grant number 1222524N. The research of Ben Cameron was supported by the Natural Sciences and Engineering Research Council of Canada (NSERC), grants RGPIN-2022-03697 and DGECR-2022-00446. The research of Iain Beaton was also supported by NSERC grants RGPIN-2025-06012 and DGECR-2025-00001.

Disclosure of Interests.. The author(s) have no competing interests to declare that are relevant to the content of this article.

References

1. Abuadas, T., Cameron, B., Hoàng, C.T., Sawada, J.: Vertex-critical $P_3 + \ell P_1$-free and vertex-critical (gem, co-gem)-free graphs. Discrete Appl. Math. **344**, 179–187 (2024)
2. Adekanye, M., Bury, C., Cameron, B., Knodel, T.: On the Finiteness of k-Vertex-Critical $2P_2$-Free Graphs with Forbidden Induced Squids or Bulls. In Adele Anna Rescigno and Ugo Vaccaro editors, 35th International Workshop on Combinatorial Algorithms. IWOCA 2024 Ischia, Italy, July 1-3, 2024, Proceedings, volume 14764 of Lecture Notes in Computer Science, pages 301–313, 2024

3. Barré, V., Fouquet, J.L.: On minimal imperfect graphs without induced P_5. Discrete Appl. Math. **94**, 9–33 (1999)
4. Beaton, I., Cameron, B.: Vertex-critical graphs in co-gem-free graphs. Theoret. Comput. Sci. **1042**, 115234 (2025)
5. Bondy, J.A.., Murty, U.S.R..: Graph Theory. Springer, 2008
6. Brause, C., Geißer, M., Schiermeyer, I.: Homogeneous sets, clique-separators, critical graphs and optimal χ-binding functions. Discrete Appl. Math. **320**, 211–222 (2022)
7. Bruce, D., Hoàng, C.T., Sawada, J.: A certifying algorithm for 3-colorability of P_5-free graphs. In:Proceedings of 20th International Symposium on Algorithms and Computation, Lecture Notes in Computer Science 5878, pp. 594–604, 2009
8. Cai, Q., Goedgebeur, J., Huang, S.: Some results on k-critical P_5-free graphs. Discrete Appl. Math. **334**, 91–100 (2023)
9. Cai,Q., Huang, S., Li, T., Shi, Y.: Vertex-critical $(P_5,banner)$-free graph. In: Chen,Y., Deng, X., Mei Lu, (eds.) Frontiers in Algorithmics -13th International Workshop, FAW 2019, Sanya, China, April 29-May 3, 2019, Proceedings, volume 11458 of Lecture Notes in Computer Science, pp. 111–120, 2019
10. Cameron, B., Hoàng, C.T.: Infinite families of k-vertex-critical (P_5, C_5)-graphs. Graphs Combin. **40**, 30 (2024)
11. Cameron, B., Hoàng, C.T., Sawada, J.: Dichotomizing k-vertex-critical H-free graphs for H of order four. Discrete Appl. Math. **312**, 106–115 (2022)
12. Cameron, K., Goedgebeur, J., Huang, S., Shi, Y.: k-critical graphs in P_5-free graphs. Theoret. Comput. Sci. **864**, 80–91 (2021)
13. Chudnovsky, M., Robertson, N., Seymour, P., Thomas, R.: The strong perfect graph theorem. Ann. of Math. **164**, 51–229 (2006)
14. Coolsaet, K., D'hondt, S., Goedgebeur,J.: House of Graphs 2.0: A database of interesting graphs and more. Discrete Appl. Math. 325:97–107, 2023. Available at https://houseofgraphs.org
15. Dhaliwal, H.S., Hamel, A.M., Hoàng, C.T., Maffray, F., McConnell, T., Panait, S.A.: On color-critical $(P_5, co-P_5)$-free graphs. Discrete Appl. Math. **216**, 142–148 (2017)
16. Goedgebeur, J.: Homepage of generator for k-critical $\mathcal{H}$-free graphs. https://caagt.ugent.be/criticalpfree/
17. Goedgebeur, J., Schaudt, O.: Exhaustive generation of k-critical $\mathcal{H}$-free graphs. J. Graph Theory **87**, 188–207 (2018)
18. Goedgebeur,J., Jooken, J., Okrasa,K., Rzążewski, P., Schaudt, O.: Minimal Obstructions to C_5-Coloring in Hereditary Graph Classes In: Královič, R., Kučera, A. (eds.) 49th International Symposium on Mathematical Foundations of Computer Science (MFCS 2024), vol. 306 of LIPIcs, pp. 55:1–55:15. Schloss Dagstuhl – Leibniz-Zentrum für Informatik, 2024
19. Hoàng, C.T., Kamiński, M., Lozin, V.V., Sawada, J., Shu, X.: Deciding k-colorability of P_5-free graphs in polynomial time. Algorithmica **57**, 74–81 (2010)
20. Hoàng, C.T., Moore, B., Recoskie, D., Sawada, J., Vatshelle, M.: Constructions of k-critical P_5-free graphs. Discrete Appl. Math. **182**, 91–98 (2015)
21. Holyer, I.: The NP-completeness of edge-coloring. SIAM J. Comput. **10**, 718–720 (1981)
22. Huang, S.: Improved complexity results on k-coloring P_t-free graphs. European J. Combin. **51**, 336–346 (2016)
23. Huang, S., Li, J., Xia, W.: Critical $(P_5, bull)$-free graphs. Discrete Appl. Math. **334**, 15–25 (2023)

24. Huang, S., Li, Z.: Vertex-critical (P_5, *chair*)-free graphs. Discrete Appl. Math. **341**, 9–15 (2023)
25. Jooken, J.: GitHub page containing generator for k-vertex-critical $\mathcal{H}$-free graphs. https://github.com/JorikJooken/kVertexCriticalGraphs
26. Kamiński, M., Pstrucha, A.: Certifying coloring algorithms for graphs without long induced paths. Discrete Appl. Math. **261**, 258–267 (2019)
27. Karp, R.M.: Reducibility among combinatorial problems. In: Complexity of computer computations (Proc. Sympos., IBM Thomas J. Watson Res. Center, Yorktown Heigts, N.Y., 1972), pp. 85–103, 1972
28. Kamiński, M., Lazin, V.: Coloring edges and vertices of graphs without short or long cycles. Contrib. Discrete Math. **2**, 61–66 (2007)
29. Leven, D., Galil, Z.: NP completeness of finding the chromatic index of regular graphs. J. Algorithms **4**, 35–44 (1983)
30. Lovász,L.: Coverings and coloring of hypergraphs. In: Proceedings of the Fourth South-Eastern Conference on Combinatorics, Graph theory, and Computing. Congress. Numer., VIII, pp. 3–12, 1973
31. Maffray, F., Morel, G.: On 3-colorable P_5-free graphs. SIAM J. Discrete Math. **26**, 1682–1708 (2012)
32. McConnell, R., Mehlhorn, K., Näher, S., Schweitzer, P.: Certifying algorithms. Comput. Sci. Rev **5**(2), 119–161 (2011)
33. McKay, B.D., Piperno, A.: Practical graph isomorphism. II. J. Symbolic Comput. **60**, 94–112 (2014)
34. Xia, W., Jooken, J., Goedgebeur, J.,Beaton, I., Cameron, B., Huang, S.: Critical (P_5, W_4)-Free Graphs. arXiv:2501.04923 [math.CO], 2025
35. Xia, W., Jooken, J., Goedgebeur, J., Huang, S.: Critical (P_5, *dart*)-free graphs. In: 16th Annual International Conference on Combinatorial Optimization and Applications, COCOA 2023, pp. 390–402, 2024
36. Xia, W., Jooken, J., Goedgebeur, J., Huang, S.: Some Results on Critical (P_5, H)-free graphs. In: 30th International Conference on Computing and Combinatorics, COCOON 2024, pp. 433–444, 2025

Learning and Data-Related Theory

A Dynamic Working Set Method for Compressed Sensing

Siu-Wing Cheng[(⊠)] and Man Ting Wong

HKUST, Hong Kong, China
scheng@cse.ust.hk, mtwongaf@connect.ust.hk

Abstract. We propose a dynamic working set method (DWS) for the problem $\min_{x \in \mathbb{R}^n} \frac{1}{2}\|Ax - b\|^2 + \eta\|x\|_1$ that arises from compressed sensing. DWS manages the working set while iteratively calling a regression solver to generate progressively better solutions. Our experiments show that DWS is more efficient than other state-of-the-art software in the context of compressed sensing. Scale space such that $\|b\| = 1$. Let s be the number of non-zeros in the unknown signal. We prove that for any given $\varepsilon > 0$, DWS reaches a solution with an additive error ε/η^2 such that each call of the solver uses only $O(\frac{1}{\varepsilon} s \log s \log \frac{1}{\varepsilon})$ variables, and each intermediate solution has $O(\frac{1}{\varepsilon} s \log s \log \frac{1}{\varepsilon})$ non-zero coordinates.

Keywords: Compressed sensing · working set · linear regression

1 Introduction

Compressed sensing allows for the recovery of sparse signals using very few observations. Applications include multislice brain imaging [19], wavelet-based image/signal reconstruction and restoration [6], the single-pixel Camera [11], and hyperspectral imaging [18]. There are two components in compressed sensing. First, a matrix $A \in \mathbb{R}^{k \times n}$ is designed such that for any unknown signal $z \in \mathbb{R}^n$, a small number of k noisy observations are taken as $b = Az + n \in \mathbb{R}^k$, where n denotes Gaussian noise. Second, an algorithm is run on A and b to recover z.

Let s be the number of non-zeros in the unknown $z \in \mathbb{R}^n$. In many applications, s is no more than 8% of n (e.g. [11,18]), and it has been argued [9] that certain images with n pixels can be reconstructed with $O(\sqrt{n} \log^3 n)$ observations, i.e., $s = o(n)$. If A has the *restricted isometry properties* (RIP), it has been proved that z can be recovered with high probability by solving

$$\min_{x \in \mathbb{R}^n} F(x) = \min_{x \in \mathbb{R}^n} \frac{1}{2}\|Ax - b\|^2 + \eta\|x\|_1 \tag{1}$$

for an appropriate $\eta > 0$ with $k = Cs\ln(n/s)$ for some constant C [1,4,9]. It is popular to use a random matrix A to achieve RIP with high probability. For

Research supported by Research Grants Council, Hong Kong, China (project no. 16203718).

example, sample each matrix entry independently from the normal distribution $\mathcal{N}(0, 1)$ and then orthonormalize the rows [12]; all non-zero singular values of A are thus equal to 1. A detailed discussion of RIP can be found in [1,4,9].

In this paper, we are concerned with solving $\min_{x \in \mathbb{R}^n} F(x)$ when $s \ll n$, A is an arbitrary $k \times n$ matrix with $\|A\| \leq 1$, and $\eta = \alpha \|A^t b\|_\infty$ for some fixed $\alpha \in (0, 1)$.[1] We propose a dynamic working set method and show that it gives superior performance than several state-of-the-art solvers in compressed sensing experiments when A is generated randomly as described above. We also mathematically analyze the convergence and efficiency of our method.

Related Work. If A in (1) is an arbitrary matrix, the problem is generally known as Lasso [27], which is originally proposed for regularized regression and variable selection. The sparsity level for Lasso to yield the best fit is typically unknown, whereas the compressed sensing applications often give a specific sparsity range for the unknown signal. Problem (1) can be transformed to a convex quadratic programming problem (e.g. [12]) that can be solved in $O(n^3 L)$ time [21], where L is the total number of bits representing the instance. Tailormade algorithms have also been developed. The earlier ones include gradient projection for sparse reconstruction (GPSR) [12], iterated thresholding (IST) [8], L1_LS [17], the homotopy method [10], and L1-magic [5]. In compressed sensing experiments, L1_LS runs faster than L1-magic and the homotopy method [5], and GPSR runs faster than IST and L1_LS [12].

Recently, coordinate descent algorithms with theoretical guarantees have been effective in solving large convex optimization problems with sparse solution [23,29]. Two solvers in this category are glmnet [13] and scikit-learn [24]. To solve problems with even more variables, working set strategies have been combined with coordinate descent or other solvers. They iteratively call a solver to generate progressively better solutions, and a small set of free variables is maintained to reduce the execution time of each call. Algorithms that employ the working set methods include Picasso [15], Blitz [16], Fireworks [25], Celer [20], and Skglm [2]. The convergence of these methods has been proven. In Lasso experiments, Blitz runs faster than L1_LS and glmnet [16], Celer runs faster than Blitz and scikit-learn [20], and Skglm performs better than Celer, Blitz, and Picasso [2].

According to the literature, GPSR, Skglm, and Celer would be the major competing solvers for compressed sensing problems.

Our Contributions. We propose a dynamic working set (DWS) algorithm for solving problem (1) when $s \ll n$, A is an arbitrary $k \times n$ matrix with $\|A\| \leq 1$, and $\eta = \alpha \|A^t b\|_\infty$ for a fixed $\alpha \in (0, 1)$.

Define the *support set* of a solution to be the subset of non-zero variables in it. DWS checks how well the support set size matches the working set size in the

---

[1] Whenever $\eta \geq \|A^t b\|_\infty$, $x = 0$ is the optimal solution [14].

previous iteration. The result determines the number of free variables that will be added to the previous support set to form the next working set.

We ran compressed sensing experiments on DWS with GPSR as the solver. We set s to be 1%, 4%, and 8% of n which is similar to the ranges of s used in previous works [3,12,28]. DWS is 1.91× faster than Skglm, 3× faster than Celer, and 2.45× faster than running GPSR alone on average. Similar trends are observed for other values of s in the range of 1% to 8% of n.

Scale space such that $\|\mathbf{b}\| = 1$. Take any $\varepsilon \in (0,1)$. Let U be an upper bound on any working set size before DWS reaches a solution $\mathbf{x}_r$ such that $F(\mathbf{x}_r) \leq$ optimum $+ \varepsilon/\eta^2$. We prove that $U = O(\frac{1}{\varepsilon}s \log s \log \frac{\eta}{\varepsilon})$ if ε is given beforehand and $U = O(\frac{1}{\varepsilon}k \log k \log \frac{\eta}{\varepsilon})$ otherwise. There are two implications. First, DWS can converge to any positive error. Second, if ε is given beforehand or $k = \Theta(s \log(n/s))$ (which allows the recovery of the sparse signal), then DWS uses provably small working sets and produces provably sparse solutions until $\mathbf{x}_r$.

Notations. Matrices are represented by uppercase letters in typewriter font. Vectors are represented by lowercase letters in typewriter font or lowercase Greek symbols. The inner product of $\mathbf{x}$ and $\mathbf{y}$ is $\langle \mathbf{x}, \mathbf{y} \rangle$ or $\mathbf{x}^t \mathbf{y}$. We use $(\mathbf{x})_i$ to denote the i-th coordinate of a vector $\mathbf{x}$. Define the support set of $\mathbf{x}$ to be $\text{supp}(\mathbf{x}) = \{i : (\mathbf{x})_i \neq 0\}$. Given a matrix $\mathbf{M}$ and a vector $\mathbf{x}$, we use $\|\mathbf{M}\|$ and $\|\mathbf{x}\|$ to denote their L_2-norms, and we use $\|\mathbf{x}\|_1$ and $\|\mathbf{x}\|_\infty$ to denote the L_1-norm and L_∞-norm of $\mathbf{x}$, respectively. Let n be the total number input variables. Let s be the support set size of the optimal solution.

2   Algorithm DWS

Let $f(\mathbf{x}) = \frac{1}{2}\|\mathbf{Ax} - \mathbf{b}\|^2$. Let $g(\mathbf{x}) = \eta\|\mathbf{x}\|_1$. The objective function is $F(\mathbf{x}) = f(\mathbf{x}) + g(\mathbf{x})$. DWS calls a solver iteratively. In each iteration, some variables are free, forming the *working set*, and the others are fixed at zero. We use $\mathbf{x}_r$ to denote the solution returned by the solver in the r-th iteration.

Algorithm 1 gives the pseudocode of DWS. We define $\mathbf{x}_0 = 0$. For $r \geq 0$, we extract a subset of variables

$$E_r = \left\{j \in [n] : \left|\frac{\partial f(\mathbf{x}_r)}{\partial(\mathbf{x})_j}\right| > \eta\right\}.$$

We will prove that for all $j \in E_r$, if $\partial f(\mathbf{x}_r)/\partial(\mathbf{x})_j < 0$, the j-th positive axis is a descent direction from $\mathbf{x}_r$; otherwise, if $\partial f(\mathbf{x}_r)/\partial(\mathbf{x})_j > 0$, the j-th negative axis is a descent direction from $\mathbf{x}_r$. The *weight* of $j \in [n]$ is $|\partial f(\mathbf{x}_r/\partial(\mathbf{x})_j|$. An element j is *heavier* than another if its weight is larger. DWS uses a parameter p_0 to initialize the first working set W_1 to consist of the p_0 elements of $[n]$ with the p_0 largest $|\partial f(\mathbf{x}_0)/\partial(\mathbf{x})_j|$. When $p_0 \leq |E_0|$, W_1 consists of the p_0 heaviest elements of E_0, and the same initialization is done in Skglm. When $p_0 > |E_0|$, Skglm selects $p_0 - |E_0|$ variables outside E_0 in some order and inserts them into W_1, which is similar to what we do. Celer also starts with a working set of size p_0 by some selection criterion. The working set of DWS for the $(r + 1)$-th iteration for $r \geq 1$ is $W_{r+1} = \text{supp}(\mathbf{x}_r) \cup \{\text{the } \tau_{r+1} \text{ heaviest elements in } E_r\}$, where τ_{r+1} is defined in lines 10–13 of Algorithm 1. DWS uses a basic step size τ for increasing the working set size, and τ_{r+1} is equal to $\min\{h^{a_r}\tau, k, |E_r|\}$ for some appropriate integer a_r. By our assumption that $k > s$, we will not release more than k variables from E_r to W_{r+1}. The variables in W_r that are zero will be kicked out of W_{r+1}. This can significantly reduce the running time of the next iteration. The rationale behind the setting of a_r is:

-   If $|\text{supp}(\mathbf{x}_r)| \leq |\text{supp}(\mathbf{x}_{r-1})| + \tau/h$, lines 10–13 of Algorithm 1 set $a_r = 0$, i.e., $\tau_{r+1} = \min\{\tau, k, |E_r|\}$. The slow growth in the support set size suggests that the working set size may be close to the ideal. We should not increase the working set size so much to slow down the next iteration.
-   Otherwise, let m be the smallest non-negative integer such that $|\text{supp}(\mathbf{x}_r)| \leq |\text{supp}(\mathbf{x}_{r-1})| + h^m \tau$. We can release $h^{m+1}\tau$ or $h^{a_{r-1}+1}\tau$ variables from E_r, i.e., a factor h more. To avoid a large increase in the working set size, lines 10–13 set $a_r = \min\{m + 1, a_{r-1} + 1\}$.

Algorithm 1. DWS

1: $h \leftarrow$ any constant in $(1, 2]$ /* $h = 2$ in the experiments. */
2: $\tau \leftarrow$ any integer in $[k]$ /* $\tau = \lfloor 4\ln^2 n \rfloor$ in the experiments */
3: $\mathbf{x}_0 \leftarrow 0$
4: compute $\nabla f(\mathbf{x}_0) = -\mathbf{A}^t \mathbf{b}$ to generate E_0
5: $W_1 \leftarrow \left\{ \text{the } p_0 \text{ elements of } [n] \text{ with the } p_0 \text{ largest } \left| \frac{\partial f(\mathbf{x}_0)}{\partial (\mathbf{x})_j} \right| \right\}$ /* $p_0 = 10$ in the experiments */
6: $a_0 \leftarrow 0; r \leftarrow 1$
7: **while** $E_{r-1} \neq \emptyset$ **do**
8:      $\mathbf{A}_r \leftarrow$ submatrix of $\mathbf{A}$ with columns corresponding to W_r
9:      $\mathbf{x}_r \leftarrow$ optimal solution obtained by calling the solver with $\mathbf{A}_r$ and $\mathbf{b}$
10:     $m \leftarrow$ the smallest integer in $[-1, \infty)$ s.t. $|\text{supp}(\mathbf{x}_r)| \leq h^m \tau + |\text{supp}(\mathbf{x}_{r-1})|$
11:     $a_r \leftarrow \min\{m + 1, a_{r-1} + 1\}$
12:     compute $\nabla f(\mathbf{x}_r) = \mathbf{A}^t \mathbf{A} \mathbf{x}_r - \mathbf{A}^t \mathbf{b}$ to generate E_r
13:     $\tau_{r+1} \leftarrow \min\{h^{a_r}\tau, k, |E_r|\}$
14:     $W_{r+1} \leftarrow \text{supp}(\mathbf{x}_r) \cup \{\text{the } \tau_{r+1} \text{ heaviest elements in } E_r\}$
15:     $r \leftarrow r + 1$
16: **end while**
17: **return** $\mathbf{x}_r$

---

3   Experimental Results

In our experiments, we generate a random matrix as described in the introduction. All non-zero singular values of $\mathbf{A}$ are equal to 1. To generate a vector $\mathbf{b}$, we first generate a true signal $\mathbf{z} \in \mathbb{R}^n$ by sampling s coordinates uniformly at random, setting each to -1 or 1 with probability $1/2$, and setting the other $n - s$ coordinates to zero. Then, compute $\mathbf{b} = \mathbf{Az} + \mathbf{n}$, where each entry of $\mathbf{n}$ is drawn independently from $\mathcal{N}(0, 10^{-4})$.

We follow the experimental set up in GPSR [12] to set $\eta = 0.1 \cdot \|\mathbf{A}^t\mathbf{b}\|_\infty$. Note that if $\eta \geq \|\mathbf{A}^t\mathbf{b}\|_\infty$, then $\mathbf{x} = 0$ is the optimal solution [14]. We will report our experimental results with $n \in \{15000, 30000, 45000, 60000\}$, $s \in \{0.01n, 0.04n, 0.08n\}$, and $k = 2s \ln(n/s)$. A similar range of s has been used in previous works [3,12,28] and some compressed sensing applications such as Single Pixel Camera [11] and hyperspectral imaging [18]. We also tried random inputs with $k = Cs \ln(n/s)$ for $C \in \{1.6, 3, 4\}$ and other values of s in the range $[0.01n, 0.08n]$. Similar trends have been observed. All experiments were run on a 12th Gen Intel® Core™ i9-12900KF CPU (3.19 GHz and 64 GB RAM).

We use BenchOpt [22] to conduct experiments. It comes with Celer and Skglm. It allows the user to add new methods. It generates informative graphs, such as the support set size against iteration, the working set size against iteration, and the *suboptimality curve*, i.e., $F(\mathbf{x}_r) - F(\mathbf{x}_*)$ against running time.

BenchOpt does not simply run a working set method $\mathcal{A}$ to completion. It starts with a variable $i = 1$, runs the first i iterations of $\mathcal{A}$, produces a data point, increments i, and repeats the above on the same input. For example, a data point for the suboptimality curve is the tuple formed by the running time of the i iterations and $F(\mathbf{x}_i) - F(\mathbf{x}_*)$. As mentioned in [2], different runs of a solver on the same input may have different running times. So a plot for $\mathcal{A}$ may not be monotone with respect to the x-axis (e.g. the suboptimality curves for Skglm and Celer in Fig. 1). BenchOpt does not use the termination condition prescribed by $\mathcal{A}$; instead, it stops running $\mathcal{A}$ when the objective function value does not decrease for several consecutive iterations. The final error is thus clear for comparison. For clarity, we circle the data points in all graphs at which the corresponding methods should have terminated. BenchOpt uses the smallest objective function value V among all solvers tested and take $F(\mathbf{x}_*)$ to be $V - 10^{-10}$.

In implementing DWS, we use the GPSR-BB version of the GPSR package as the solver. For simplicity, we refer to the GPSR-BB version as GPSR. In the full version, we show that DWS is significantly faster than GPSR when s is 1% or 4% of n; DWS has a similar efficiency as GPSR when s is 8% of n; the average speedup achieved by DWS is roughly $2.45\times$.

Skglm and Celer update the working set using a doubling strategy [2,20] that sets the working set size for iteration $r + 1$ to be $2 \cdot |\mathrm{supp}(\mathbf{x}_r)|$. The variables in the working set for iteration r that are zero may be excluded from the working set for iteration $r + 1$. Celer also supports a non-pruning mode that sets the working set size for iteration $r + 1$ to be twice the working set size for iteration r, and all variables in the working set for iteration r are kept. In the full version, we show that Celer is not more efficient in the non-pruning mode as s increases. Therefore, we will ignore the non-pruning mode of Celer.[2]

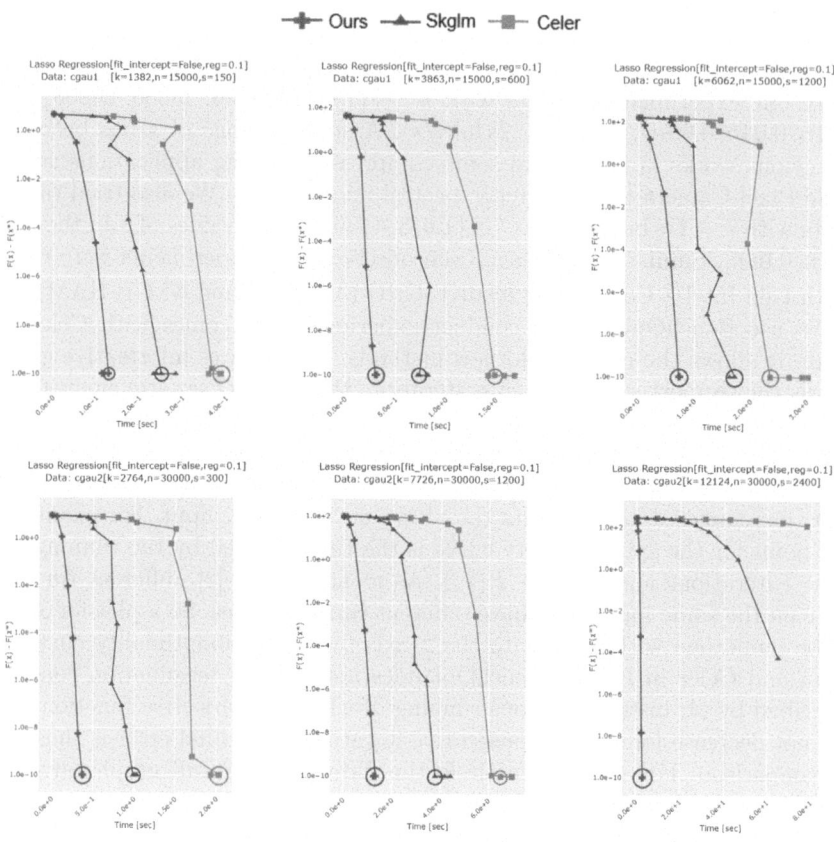

Fig. 1. Plots of $F(\mathbf{x}_r) - F(\mathbf{x}_*)$ against running time.

We assume no knowledge of s. As in Skglm and Celer [2,20], DWS starts with a working set of size $p_0 = 10$ ($|E_0|$ is typically larger than 10). Figure 1 shows the running times for some random inputs for $n \in \{15000, 30000\}$. Skglm and Celer timed out in some runs; in those cases, no data point of their plots is circled (which indicates termination). When Skglm and Celer did not time out, DWS is at least $1.91\times$ faster than Skglm and at least $3.0\times$ faster than Celer. The top two rows in Fig. 2 show the plots of the support set sizes. The three methods give the same final support set size which is about 38% larger than s on average.

---

[2] For Skglm, there is a discrepancy between the doubling strategies in the publicly available code and the paper. Our description follows the code. The convergence of Skglm is proved for the version in the paper that grows a working set monotonically. The convergence of Celer is proved for its non-pruning mode.

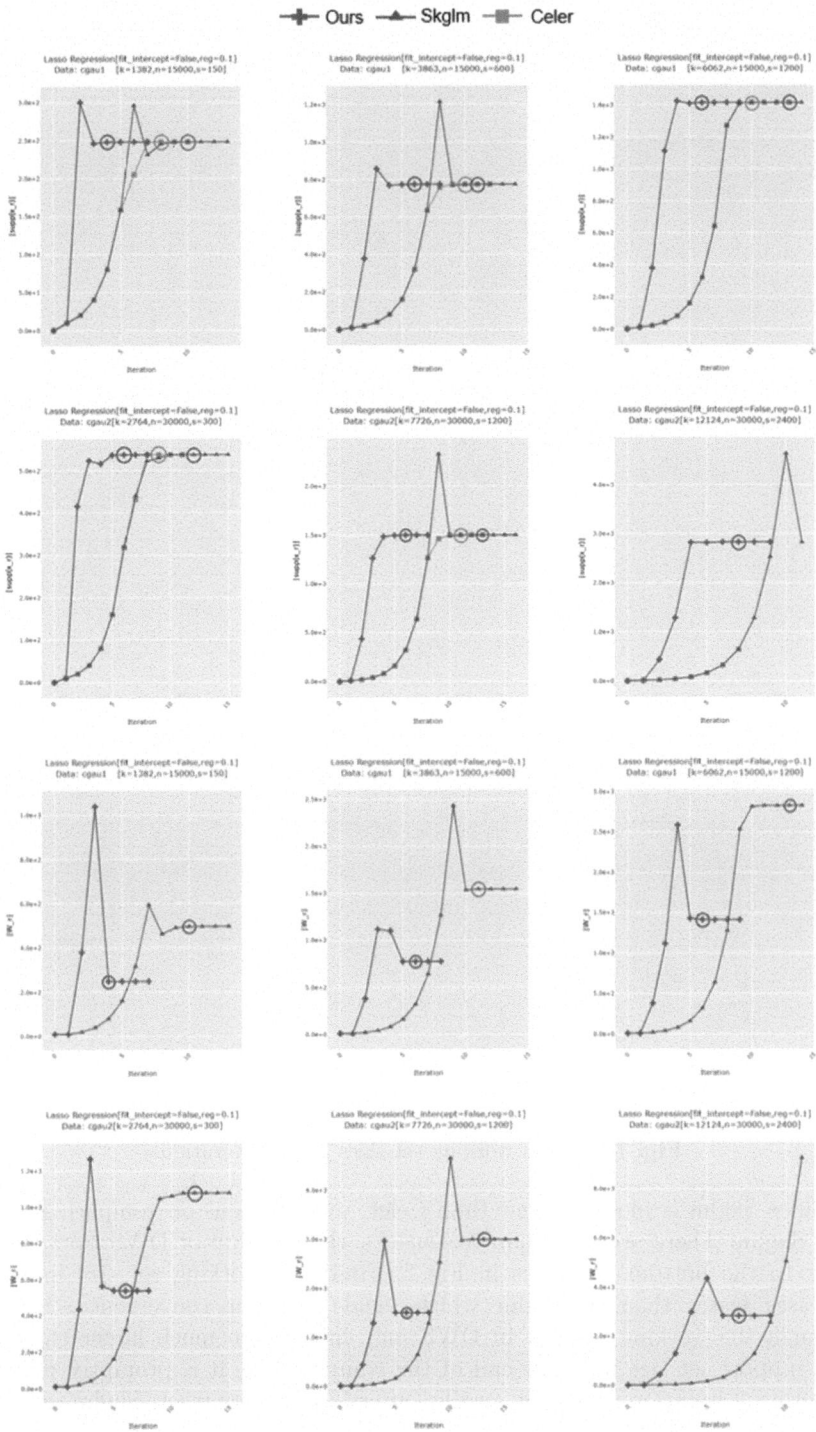

Fig. 2. The top two rows show the plots of support set sizes against iteration. The bottom two rows show the plots of working set sizes against iteration.

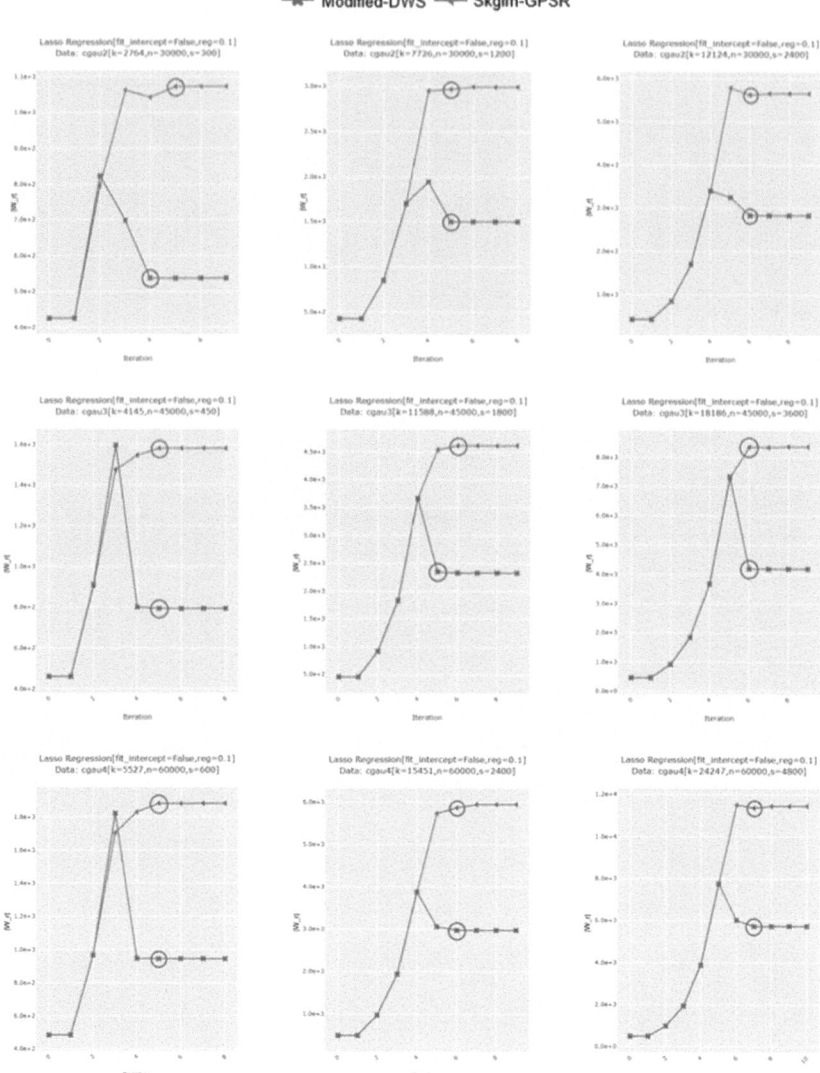

Fig. 3. Plots of working set sizes against iteration.

Since Skglm is more efficient than Celer, we will focus on comparing DWS with Skglm. There are two main reasons for the speedup of DWS over Skglm. Refer to the bottom two rows in Fig. 2. First, the working set size in DWS increases faster than in Skglm which yields a faster convergence. Second, although the working set size in DWS may increase to much larger than the final support set size near the end of the computation, it is promptly reduced in the next iteration and kept smaller afterward. In contrast, Skglm sustains a much larger working set (roughly twice as large) over multiple iterations

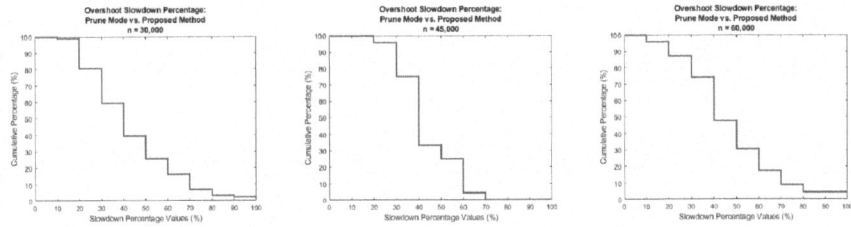

Fig. 4. The vertical axes show the cumulative percentages of test cases. The horizontal axes show the slowdown percentage defined as ((runtime of Skglm-GPSR − runtime of modified-method)/runtime of modified-method) × 100%.

near the end of the computation, which makes these iterations run significantly slower. In the full version, we show similar trends in the experimental results for $n \in \{45000, 60000\}$.

How important is the ability of DWS to scale back the working set size? We study this question as follows. First, we implemented the doubling method of Skglm with GPSR as the solver so that the comparison is on the same footing. We refer to the resulting variant as Skglm-GPSR. Second, we set $p_0 = \tau$ for both Skglm-GPSR and DWS, and we pretend that $|\text{supp}(x_0)| = \tau$ in DWS although x_0 is still the zero vector. We refer to the resulting variant as modified-DWS. Figure 3 shows that Skglm-GPSR and modified-DWS have a nearly common working set size (around $2^{r-1}\tau$ in the r-th iteration) until the computation is near the end. Therefore, there is no issue with the working set size increasing faster in modified-DWS or Skglm-GPSR. Near the end of the computation, the working set size is scaled back in modified-DWS, whereas the working set size in Skglm-GPSR is roughly twice as large. We tried 269 cases for $n = 30000$, 74 cases for $n = 45000$, and 74 cases for $n = 60000$. Refer to Fig. 4. Skglm-GPSR is slower by 20% or more in at least 80% of the cases, by 30% or more in at least 59% of the cases, and by 40% or more in at least 30% of the cases. The ability of DWS to scale back the working set size improves efficiency significantly.

4   Theoretical Analysis

Given any function $\varphi : \mathbb{R}^n \to \mathbb{R}$, a vector $\xi \in \mathbb{R}^n$ such that $\varphi(y) \geq \varphi(x) + \langle \xi, y-x \rangle$ for all $y \in \mathbb{R}^\nu$ is called a *subgradient* of φ at x [26]. For a smooth function, the subgradient at a point is unique and equal to the gradient, which is denoted by $\nabla \varphi$. There are multiple subgradients at a non-smooth point x; we use $\partial \varphi(x)$ to denote the set of all subgradients of φ at x.

We have $\partial F(x) = \{\nabla f(x) + \xi : \xi \in \partial g(x)\}$. A vector n is a *descent direction* from x if and only if $\sup\{\langle \gamma, n \rangle : \gamma \in \partial F(x)\} < 0$. The function F is minimized at x if and only if $\partial F(x)$ contains the zero vector [26].

Lemma 2 proves the termination condition of $E_r = \emptyset$. Theorem 1 analyzes the sizes of the working and support sets. Missing proofs can be found in the full version [7].

Lemma 1.

(i) *Take any* $\mathbf{x} \in \mathbb{R}^n$.
   (a) $\forall \xi \in \partial g(\mathbf{x})$, $\forall i \in [n]$, *if* $(\mathbf{x})_i = 0$, *then* $(\xi)_i \in [-\eta, \eta]$; *otherwise*, $(\xi)_i = \text{sign}((\mathbf{x})_i) \cdot \eta$.
   (b) *Every vector that satisfies the conditions in* (i)(a) *is a subgradient in* $\partial g(\mathbf{x})$.

(ii) $\forall i \in E_r$, $\forall \gamma \in \partial F(\mathbf{x}_r)$, $\text{sign}((\gamma)_i) = \text{sign}((\nabla f(\mathbf{x}_r))_i) \in \{-1, 1\}$.
(iii) $E_r = \{ i \in [n] : \forall \gamma \in \partial F(\mathbf{x}_r), (\gamma)_i \neq 0 \}$.

Lemma 2. *Let* $\mathbf{e}_i$ *be the unit vector in the direction of the positive* i-*th axis. Every unit conical combination of* $\{ -\text{sign}((\nabla f(\mathbf{x}_r))_i) \cdot \mathbf{e}_i : i \in E_r \}$ *is a descent direction from* $\mathbf{x}_r$. *If* $E_r = \emptyset$, *then* $\mathbf{x}_r$ *is the global minimum.*

Proof. Let $\rho = \min_{i \in E_r} |\nabla f(\mathbf{x}_r)_i| - \eta$ which is positive. Take any $i \in E_r$ and any $\gamma \in \partial F(\mathbf{x}_r)$. By Lemma 1(i), the i-th coordinate of any subgradient in $\partial g(\mathbf{x}_r)$ is in the range $[-\eta, \eta]$, which implies that $|(\gamma)_i| \geq \rho$. Let $\mathbf{s}_i = -\text{sign}((\nabla f(\mathbf{x}_r))_i) \cdot \mathbf{e}_i$. By Lemma 1(ii), $\langle \gamma, \mathbf{s}_i \rangle = -|(\gamma)_i| \leq -\rho < 0$. For every unit conical combination $\sum_{i \in E_r} \alpha_i \mathbf{s}_i$, some coefficient α_i is at least $1/\sqrt{n}$. Thus, $\sup_{\gamma \in \partial F(\mathbf{x}_r)} \langle \gamma, \sum_{i \in E_r} \alpha_i \mathbf{s}_i \rangle \leq -\rho/\sqrt{n} < 0$, proving that $\sum_{i \in E_r} \alpha_i \mathbf{s}_i$ is a descent direction. If E_r is empty, by Lemma 1(iii), for every $i \in [n]$, there exists $\xi \in \partial g(\mathbf{x}_r)$ such that $(\xi)_i = -(\nabla f(\mathbf{x}_r))_i$, that is, $-(\nabla f(\mathbf{x}_r))_i$ is a legitimate i-th coordinate of a subgradient in $\partial g(\mathbf{x}_r)$. It follows from Lemma 1(i)(b) that the zero vector belongs to $\partial F(\mathbf{x}_r)$, which implies that $\mathbf{x}_r$ is the global minimum. $\square$

Define a vector $\zeta_{\mathbf{x}_r} \in \mathbb{R}^n$ such that for all $i \in [n]$, if $i \notin E_r$, then $(\zeta_{\mathbf{x}_r})_i = -(\nabla f(\mathbf{x}_r))_i$, and if $i \in E_r$, then $(\zeta_{\mathbf{x}_r})_i = -\text{sign}((\nabla f(\mathbf{x}_r))_i) \cdot \eta$. Define $\gamma_{\mathbf{x}_r} = \nabla f(\mathbf{x}_r) + \zeta_{\mathbf{x}_r}$. Let $\mathbf{x}_*$ denote the optimal solution that minimizes F. Given a vector $\mathbf{v}$ and a subset $S \subseteq [n]$, $\mathbf{v} \downarrow S$ denotes the orthogonal projection of $\mathbf{v}$ in the subspace spanned by $\{ \mathbf{e}_i : i \in S \}$.

Lemma 3. $\zeta_{\mathbf{x}_r} \in \partial g(\mathbf{x}_r)$ *and* $\gamma_{\mathbf{x}_r} \in \partial F(\mathbf{x}_r)$.

Proof. By Lemma 1(ii), for all $i \in E_r$, $(\zeta_{\mathbf{x}_r})_i \in \{-\eta, \eta\}$. For all $i \in E_r$, $(\mathbf{x}_r)_i = 0$ as $E_r \cap W_r = \emptyset$. By Lemma 1(i)(b), for all $i \in E_r$, $(\zeta_{\mathbf{x}_r})_i$ is a legitimate i-th coordinate for a subgradient in $\partial g(\mathbf{x}_r)$. By Lemma 1(iii), for all $i \notin E_r$, there exists $\gamma \in \partial F(\mathbf{x}_r)$ such that $(\gamma)_i = 0$. It means that for all $i \notin E_r$, $(\zeta_{\mathbf{x}_r})_i = -(\nabla f(\mathbf{x}_r))_i$ must a legitimate i-th coordinate for a subgradient in $\partial g(\mathbf{x}_r)$ so that it cancels $(\nabla f(\mathbf{x}_r))_i$. Thus, $\zeta_{\mathbf{x}_r} \in \partial g(\mathbf{x}_r)$. Then, $\gamma_{\mathbf{x}_r} \in \partial F(\mathbf{x}_r)$ by definition. $\square$

Lemma 4. *Let* $\mathbf{n}_r$ *be any unit conical combination of* $\{ -\text{sign}((\nabla f(\mathbf{x}_r))_i) \cdot \mathbf{e}_i : i \in E_r \}$. *Let* $\mathbf{y}_r$ *be the point in direction* $\mathbf{n}_r$ *from* $\mathbf{x}_r$ *that minimizes* F.

(i) *There exists* $\xi \in \partial g(\mathbf{y}_r)$ *such that* $\langle \nabla f(\mathbf{y}_r) + \xi, \mathbf{y}_r - \mathbf{x}_r \rangle = 0$.
(ii) *For every* $\xi \in \partial g(\mathbf{y}_r)$, *both* $\langle \zeta_{\mathbf{x}_r} - \xi, \mathbf{x}_r \rangle$ *and* $\langle \zeta_{\mathbf{x}_r} - \xi, \mathbf{y}_r \rangle$ *are zero.*
(iii) *For every* $\mathbf{z} \in \mathbb{R}^n$, $F(\mathbf{x}_r) - F(\mathbf{z}) \leq -\langle \gamma_{\mathbf{x}_r}, \mathbf{z} - \mathbf{x}_r \rangle$.

(iv) $F(\mathbf{x}_r) - F(\mathbf{y}_r) = -\frac{1}{2}\langle\gamma_{\mathbf{x}_r}, \mathbf{y}_r - \mathbf{x}_r\rangle$.

Lemma 5. *Let $\mathbf{n}_r$ be any unit conical combination of $\{-\mathrm{sign}((\nabla f(\mathbf{x}_r))_i) \cdot \mathbf{e}_i :$ $i \in E_r\}$. Let $\mathbf{y}_r$ be the point in direction $\mathbf{n}_r$ from $\mathbf{x}_r$ that minimizes F. Let $\mathbf{n}_* = (\mathbf{x}_* - \mathbf{x}_r)/\|\mathbf{x}_* - \mathbf{x}_r\|$. If $F(\mathbf{x}_r) > F(\mathbf{x}_*) + \varepsilon\|\mathbf{x}_* - \mathbf{x}_r\|^2$ for some $\varepsilon \in (0,1)$, then $\langle\gamma_{\mathbf{x}_r}, \mathbf{y}_r - \mathbf{x}_r\rangle/\langle\gamma_{\mathbf{x}_r}, \mathbf{x}_* - \mathbf{x}_r\rangle \geq \varepsilon \cdot \langle\gamma_{\mathbf{x}_r}, \mathbf{n}_r\rangle^2/\langle\gamma_{\mathbf{x}_r}, \mathbf{n}_*\rangle^2$.*

Proof. By Lemma 4(i), there exists $\xi \in \partial g(\mathbf{y}_r)$ such that $\langle\nabla f(\mathbf{y}_r) + \xi, \mathbf{n}_r\rangle = 0$. We have $\langle\zeta_{\mathbf{x}_r}, \mathbf{y}_r - \mathbf{x}_r\rangle - \langle\xi, \mathbf{y}_r - \mathbf{x}_r\rangle = \langle\zeta_{\mathbf{x}_r} - \xi, \mathbf{y}_r\rangle - \langle\zeta_{\mathbf{x}_r} - \xi, \mathbf{x}_r\rangle$ which is zero by Lemma 4(ii). It implies that $\langle\zeta_{\mathbf{x}_r}, \mathbf{n}_r\rangle = \langle\xi, \mathbf{n}_r\rangle$, and hence $\langle\nabla f(\mathbf{y}_r) + \zeta_{\mathbf{x}_r}, \mathbf{n}_r\rangle = 0$. Substituting $\nabla f(\mathbf{y}_r)$ by $\mathbf{A}^t\mathbf{A}\mathbf{y}_r - \mathbf{A}^t\mathbf{b}$, we obtain $\langle\mathbf{A}^t\mathbf{A}\mathbf{y}_r - \mathbf{A}^t\mathbf{b} + \zeta_{\mathbf{x}_r}, \mathbf{n}_r\rangle = 0$. Rearranging terms gives $\langle\mathbf{A}^t\mathbf{A}(\mathbf{y}_r - \mathbf{x}_r), \mathbf{n}_r\rangle = \langle-\mathbf{A}^t\mathbf{A}\mathbf{x}_r + \mathbf{A}^t\mathbf{b} - \zeta_{\mathbf{x}_r}, \mathbf{n}_r\rangle = \langle-\nabla f(\mathbf{x}_r) - \zeta_{\mathbf{x}_r}, \mathbf{n}_r\rangle = \langle-\gamma_{\mathbf{x}_r}, \mathbf{n}_r\rangle$. Therefore, $\|\mathbf{y}_r - \mathbf{x}_r\| \cdot \|\mathbf{A}\mathbf{n}_r\|^2 = \langle\mathbf{A}^t\mathbf{A}(\mathbf{y}_r - \mathbf{x}_r), \mathbf{n}_r\rangle = \langle-\gamma_{\mathbf{x}_r}, \mathbf{n}_r\rangle$. Hence, $\|\mathbf{y}_r - \mathbf{x}_r\| \geq \langle-\gamma_{\mathbf{x}_r}, \mathbf{n}_r\rangle$ as $\|\mathbf{A}\mathbf{n}_r\|^2 \leq \|\mathbf{A}\|^2 \leq 1$.

By Lemma 4(iii), $F(\mathbf{x}_r) - F(\mathbf{x}_*) \leq \|\mathbf{x}_* - \mathbf{x}_r\| \cdot \langle-\gamma_{\mathbf{x}_r}, \mathbf{n}_*\rangle$. If $\langle-\gamma_{\mathbf{x}_r}, \mathbf{n}_*\rangle < \varepsilon\|\mathbf{x}_* - \mathbf{x}_r\|$, then $F(\mathbf{x}_r) - F(\mathbf{x}_*) \leq \varepsilon\|\mathbf{x}_* - \mathbf{x}_r\|^2$, contradicting the assumption of the lemma. Hence, $\langle-\gamma_{\mathbf{x}_r}, \mathbf{n}_*\rangle \geq \varepsilon\|\mathbf{x}_* - \mathbf{x}_r\|$. Combining this inequality with $\|\mathbf{y}_r - \mathbf{x}_r\| \geq \langle-\gamma_{\mathbf{x}_r}, \mathbf{n}_r\rangle$ gives $\frac{\langle\gamma_{\mathbf{x}_r}, \mathbf{y}_r - \mathbf{x}_r\rangle}{\langle\gamma_{\mathbf{x}_r}, \mathbf{x}_* - \mathbf{x}_r\rangle} = \frac{\|\mathbf{y}_r - \mathbf{x}_r\|}{\|\mathbf{x}_* - \mathbf{x}_r\|} \cdot \frac{\langle\gamma_{\mathbf{x}_r}, \mathbf{n}_r\rangle}{\langle\gamma_{\mathbf{x}_r}, \mathbf{n}_*\rangle} \geq \varepsilon \cdot \frac{\langle\gamma_{\mathbf{x}_r}, \mathbf{n}_r\rangle^2}{\langle\gamma_{\mathbf{x}_r}, \mathbf{n}_*\rangle^2}$. □

Lemma 6. *Let G_r be the set of the τ_{r+1} heaviest elements in E_r. Let $H_r = G_r \cup (\mathrm{supp}(\mathbf{x}_*) \cap E_r)$. Then, $\langle\gamma_{\mathbf{x}_r}, \mathbf{x}_* - \mathbf{x}_r\rangle = \langle\gamma_{\mathbf{x}_r} \downarrow H_r, \mathbf{x}_* - \mathbf{x}_r\rangle$.*

Proof. For any $i \in H_r$, $(\gamma_{\mathbf{x}_r})_i = (\gamma_{\mathbf{x}_r} \downarrow H_r)_i$ by definition. Therefore, $(\gamma_{\mathbf{x}_r})_i \cdot (\mathbf{x}_* - \mathbf{x}_r)_i = (\gamma_{\mathbf{x}_r} \downarrow H_r)_i \cdot (\mathbf{x}_* - \mathbf{x}_r)_i$.

Take any $i \notin E_r$. We have $(\gamma_{\mathbf{x}_r})_i = 0$ because $\gamma_{\mathbf{x}_r} = \zeta_{\mathbf{x}_r} + \nabla f(\mathbf{x}_r)$ and $(\zeta_{\mathbf{x}_r})_i = -(\nabla f(\mathbf{x}_r))_i$ by definition. Therefore, both $(\gamma_{\mathbf{x}_r})_i \cdot (\mathbf{x}_* - \mathbf{x}_r)_i$ and $(\gamma_{\mathbf{x}_r} \downarrow H_r)_i \cdot (\mathbf{x}_* - \mathbf{x}_r)_i$ are zero.

For any $i \in E_r \setminus H_r$, $i \notin \mathrm{supp}(\mathbf{x}_*)$ as $\mathrm{supp}(\mathbf{x}_*) \cap E_r \subseteq H_r$. So $(\mathbf{x}_*)_i = 0$. Also, $(\mathbf{x}_r)_i = 0$ as $i \in E_r$. So $(\gamma_{\mathbf{x}_r})_i \cdot (\mathbf{x}_* - \mathbf{x}_r)_i$ and $(\gamma_{\mathbf{x}_r} \downarrow H_r)_i \cdot (\mathbf{x}_* - \mathbf{x}_r)_i$ are zero. □

Lemma 7. *If $F(\mathbf{x}_r) > F(\mathbf{x}_*) + \varepsilon\|\mathbf{x}_* - \mathbf{x}_r\|^2$, then*

$$\frac{F(\mathbf{x}_{r+1}) - F(\mathbf{x}_*)}{F(\mathbf{x}_r) - F(\mathbf{x}_*)} \leq 1 - \frac{\varepsilon\tau_{r+1}}{8(s + \tau_{r+1})\ln\tau_{r+1}}.$$

Proof. Let G_r be the set of the τ_{r+1} heaviest elements in E_r. Let $H_r = G_r \cup (\mathrm{supp}(\mathbf{x}_*) \cap E_r)$. We prove in the full version [7] that there exists a unit descent direction $\mathbf{n}_r$ from $\mathbf{x}_r$ such that $\mathbf{n}_r$ is a conical combination of $\{-\mathrm{sign}((\nabla f(\mathbf{x}_r))_i) \cdot \mathbf{e}_i : i \in G_r\}$ and $\langle-\gamma_{\mathbf{x}_r}, \mathbf{n}_r\rangle \geq \|\gamma_{\mathbf{x}_r} \downarrow H_r\| \cdot \sqrt{\frac{\tau_{r+1}}{4(s + \tau_{r+1})\ln\tau_{r+1}}}$. Let $\mathbf{n}_* = (\mathbf{x}_* - \mathbf{x}_r)/\|\mathbf{x}_* - \mathbf{x}_r\|$. By Lemma 6, $\langle-\gamma_{\mathbf{x}_r}, \mathbf{n}_*\rangle = \langle-\gamma_{\mathbf{x}_r} \downarrow H_r, \mathbf{n}_*\rangle$. Then, $\langle-\gamma_{\mathbf{x}_r}, \mathbf{n}_*\rangle \leq \|\gamma_{\mathbf{x}_r} \downarrow H_r\|$. Hence, $\frac{\langle\gamma_{\mathbf{x}_r}, \mathbf{n}_r\rangle}{\langle\gamma_{\mathbf{x}_r}, \mathbf{n}_*\rangle} \geq \sqrt{\frac{\tau_{r+1}}{4(s + \tau_{r+1})\ln\tau_{r+1}}}$.

Let $\mathbf{y}_r$ be the point in the direction $\mathbf{n}_r$ from $\mathbf{x}_r$ that minimizes F. We have $\frac{F(\mathbf{x}_{r+1}) - F(\mathbf{x}_*)}{F(\mathbf{x}_r) - F(\mathbf{x}_*)} = 1 - \frac{F(\mathbf{x}_r) - F(\mathbf{x}_{r+1})}{F(\mathbf{x}_r) - F(\mathbf{x}_*)} \leq 1 - \frac{F(\mathbf{x}_r) - F(\mathbf{y}_r)}{F(\mathbf{x}_r) - F(\mathbf{x}_*)}$. By Lemma 4(iii) and (iv), $F(\mathbf{x}_r) - F(\mathbf{y}_r) = -\frac{1}{2}\langle\gamma_{\mathbf{x}_r}, \mathbf{y}_r - \mathbf{x}_r\rangle$ and $F(\mathbf{x}_r) - F(\mathbf{x}_*) \leq -\langle\gamma_{\mathbf{x}_r}, \mathbf{x}_* - \mathbf{x}_r\rangle$. Therefore, $\frac{F(\mathbf{x}_{r+1}) - F(\mathbf{x}_*)}{F(\mathbf{x}_r) - F(\mathbf{x}_*)} \leq 1 - \frac{1}{2} \cdot \frac{\langle\gamma_{\mathbf{x}_r}, \mathbf{y}_r - \mathbf{x}_r\rangle}{\langle\gamma_{\mathbf{x}_r}, \mathbf{x}_* - \mathbf{x}_r\rangle} \leq 1 - \frac{\varepsilon}{2} \cdot \frac{\langle\gamma_{\mathbf{x}_r}, \mathbf{n}_r\rangle^2}{\langle\gamma_{\mathbf{x}_r}, \mathbf{n}_*\rangle^2}$ by Lemma 5, which is at most $1 - \frac{\varepsilon\tau_{r+1}}{8(s + \tau_{r+1})\ln\tau_{r+1}}$ by the lower bound on $\frac{\langle\gamma_{\mathbf{x}_r}, \mathbf{n}_r\rangle}{\langle\gamma_{\mathbf{x}_r}, \mathbf{n}_*\rangle}$. □

Recall the parameter $h \in (1, 2]$ in Algorithm 1. The next result bounds the working set sizes up to the first solution with an additive error at most ε/η^2.

Theorem 1. *Suppose that* $\|\mathsf{A}\| \leq 1$ *and* $\eta = \alpha \|\mathsf{A}^t\mathsf{b}\|_\infty$ *for a fixed* $\alpha \in (0, 1)$. *Scale space such that* $\|\mathsf{b}\| = 1$. *Let* $\kappa + 1$ *be the minimum index such that* $F(\mathsf{x}_{\kappa+1}) - F(\mathsf{x}_*) \leq \varepsilon/\eta^2$.

  - *If* $h \leq 2^{O(\varepsilon/(\ln n \ln(\eta/\varepsilon)))}$, *then* $\sum_{i=1}^{\kappa} \tau_i = O\left(\frac{1}{\varepsilon}(s+\tau)\log(s+\tau)\log\frac{\eta}{\varepsilon}\right)$.
  - *Otherwise,* $\sum_{i=1}^{\kappa} \tau_i = O\left(\frac{1}{\varepsilon}k\log k\log\frac{\eta}{\varepsilon}\right)$.

Proof. Take any constant $c \geq 1$. Divide $[\kappa]$ into two disjoint subsets I and J such that for all $i \in I$, $\tau_i \leq cs$, and for all $i \in J$, $\tau_i > cs$.

For any x_r, $\eta\|\mathsf{x}_r\| \leq \eta\|\mathsf{x}_r\|_1 \leq F(\mathsf{x}_r) \leq F(\mathsf{x}_0) = \frac{1}{2}\|\mathsf{b}\|^2 = \frac{1}{2}$. The same argument works for x_*. It means that $\varepsilon\|\mathsf{x}_* - \mathsf{x}_r\|^2 \leq \varepsilon/\eta^2$. Therefore, for any $r \in [\kappa]$, $F(\mathsf{x}_r) - F(\mathsf{x}_*) > \varepsilon/\eta^2 \geq \varepsilon\|\mathsf{x}_* - \mathsf{x}_r\|^2$ by assumption, which makes Lemma 7 applicable for all $r \in [\kappa]$.

View I as a chronological sequence. Let i be the largest index in I. Note that $F(\mathsf{x}_0) - F(\mathsf{x}_*) \leq F(\mathsf{x}_0) \leq \frac{1}{2}\|\mathsf{b}\|^2 = \frac{1}{2}$. Then, by Lemma 7, $F(\mathsf{x}_i) - F(\mathsf{x}_*) \leq \frac{1}{2}\prod_{i \in I}\left(1 - \frac{\varepsilon\tau_i}{8(s+\tau_i)\ln(cs)}\right)$, which is at most $\frac{1}{2}\prod_{i \in I}\left(1 - \frac{\varepsilon\tau_i}{8(c+1)s\ln(cs)}\right)$. Let $\tau_{\mathrm{avg}} = \sum_{i \in I} \tau_i/|I|$. It is well known that the geometric mean is at most the arithmetic mean. Therefore, $\frac{1}{2}\prod_{i \in I}\left(1 - \frac{\varepsilon\tau_i}{8(c+1)s\ln(cs)}\right) \leq \frac{1}{2}\left(1 - \frac{\varepsilon\tau_{\mathrm{avg}}}{8(c+1)s\ln(cs)}\right)^{|I|} \leq \frac{1}{2}e^{-\varepsilon\tau_{\mathrm{avg}}|I|/(8(c+1)s\ln(cs))}$. This upper bound is at least ε/η^2 so that $\mathsf{x}_{\kappa+1}$ is the first solution that satisfies $F(\mathsf{x}_{\kappa+1}) - F(\mathsf{x}_*) \leq \varepsilon/\eta^2$. Hence, $\frac{\varepsilon\tau_{\mathrm{avg}}|I|}{8(c+1)s\ln(cs)} \leq \ln\frac{\eta^2}{2\varepsilon}$, which implies that $\sum_{i \in I} \tau_i = \tau_{\mathrm{avg}}|I| = O\left(\frac{1}{\varepsilon}s\log s\log\frac{\eta}{\varepsilon}\right)$.

View J as a chronological sequence. Let $\tau_{\max} = \max_{i \in J} \tau_i$. Take a contiguous subsequence of J of length $(16\ln\tau_{\max})/\varepsilon$. Let i and j be the minimum and maximum indices in this subsequence, respectively. By Lemma 7, $F(\mathsf{x}_j) - F(\mathsf{x}_*) \leq e^{-1} \cdot (F(\mathsf{x}_i) - F(\mathsf{x}_*))$. Since $F(\mathsf{x}_0) - F(\mathsf{x}_*) \leq F(\mathsf{x}_0) = \frac{1}{2}\|\mathsf{b}\|^2 = \frac{1}{2}$, we can divide J into no more than $\ln\frac{\eta^2}{2\varepsilon}$ contiguous subsequences of length $(16\ln\tau_{\max})/\varepsilon$. It follows that $|J| \leq \frac{16}{\varepsilon}\ln\tau_{\max} \cdot \ln\frac{\eta^2}{2\varepsilon}$. The algorithm ensures that $\tau_{i+1} \leq h\tau_i$. Extract the longest subsequence of J (not necessarily contiguous) in which τ_i strictly increases. Every consecutive τ_i's in this subsequence differ by a factor h. This subsequence starts with $\min_{i \in J} \tau_i \leq \max\{hcs, \tau\}$. If $h \leq 2^{O(\varepsilon/(\ln n \ln(\eta/\varepsilon)))}$, then $\tau_{\max} \leq h^{|J|} \cdot \max\{hcs, \tau\} \leq 2^{O(1)} \cdot (s + \tau)$. So $\sum_{i \in J} \tau_i \leq O(s + \tau) \cdot |J| = O\left(\frac{1}{\varepsilon}(s+\tau)\log(s+\tau)\log\frac{\eta}{\varepsilon}\right)$. If $h > 2^{O(\varepsilon/(\ln n \ln(\eta/\varepsilon)))}$, we still have $\tau_{\max} \leq k$ and hence $\sum_{i \in J} \tau_i \leq k|J| = O\left(\frac{1}{\varepsilon}k\log k\log\frac{\eta}{\varepsilon}\right)$. $\qquad \square$

Remark 1. Clearly $\kappa \leq \sum_{i=1}^{\kappa} \tau_i$. One can work out the exact upper bound for $\sum_{i=1}^{\kappa} \tau$ and hence κ. DWS can be stopped after κ iterations to obtain an error at most ε/η^2, although DWS has probably terminated earlier in practice.

Remark 2. Starting from p_0, we add at most $\sum_{i=1}^{\kappa} \tau_i$ free variables to any working set before reaching $\mathsf{x}_{\kappa+1}$. Suppose that we set p_0 and τ to be $O(1)$. If ε is given beforehand, we can ensure that every working set has $O(\frac{1}{\varepsilon}s\log s\log\frac{\eta}{\varepsilon})$ variables before reaching x_κ. So each call of the solver runs provably faster than using all n variables. When ε is not given, if $k = \Theta(s \log(n/s))$ (sufficient for the true signal to be recovered with high probability), every working set still has only $O(\frac{1}{\varepsilon}s \cdot \text{polylog}(n))$ variables. Clearly, $|\text{supp}(\mathbf{x}_r)| \le |W_r|$. It follows that $|\text{supp}(\mathbf{x}_r)| \le p_0 + \sum_{i=1}^{\kappa} \tau_i$. We conclude that all solutions $\mathbf{x}_r$, $r \in [\kappa + 1]$, are provably sparse if ε is given beforehand or $k = \Theta(s \log(n/s))$.

Remark 3. Figure 2 shows that the working set sizes are at most cs for some small constant c in the experiments. That is, $\max_{i \in [\kappa]} \tau_i = O(s)$. Under this assumption, the proof of Theorem 1 reveals that $\sum_{i=1}^{\kappa} \tau_i = O(\frac{1}{\varepsilon}s \log s \log \frac{\eta}{\varepsilon})$ even if ε is not given beforehand. Then, the working set sizes and support set sizes can be bounded by $O(\frac{1}{\varepsilon}s \log s \log \frac{\eta}{\varepsilon})$ even if ε is not given beforehand.

Remark 4. To prepare for the next iteration, we need to compute $\nabla f(\mathbf{x}_r) = \mathbf{A}^t \mathbf{A} \mathbf{x}_r - \mathbf{A}^t \mathbf{b}$. We precompute $\mathbf{A}^t \mathbf{b}$ in $O(kn)$ time. Let $w = |W_r| \le \sum_{i=1}^{\kappa} \tau_i$. Note that $|\text{supp}(\mathbf{x}_r)| \le w$. To obtain $\mathbf{A}^t \mathbf{A} \mathbf{x}_r$, we use $\mathbf{A}_r \in \mathbb{R}^{k \times w}$ and $\text{supp}(\mathbf{x}_r)$ to obtain $\mathbf{A} \mathbf{x}_r$ in $O(kw)$ time, and then we compute $\mathbf{A}^t(\mathbf{A} \mathbf{x}_r)$ in $O(kn)$ time. We extract $\mathbf{A}_{r+1}$ from $\mathbf{A}$ corresponding to W_{r+1} is $O(k|W_{r+1}|) = O(k \cdot \sum_{i=1}^{\kappa} \tau_i)$ time.

Remark 5. If $\|\mathbf{A}\mathbf{x}_*\| \le \|\mathbf{b}\|/4$, then $F(\mathbf{x}_*) \ge \frac{1}{2}\|\mathbf{A}\mathbf{x} - \mathbf{b}\|^2 \ge \frac{9}{32}\|\mathbf{b}\|^2$. If $\|\mathbf{A}\mathbf{x}_*\| > \|\mathbf{b}\|/4$, then $F(\mathbf{x}_*) \ge \eta\|\mathbf{x}_*\|_1 \ge \eta\|\mathbf{x}_*\| \ge \eta\|\mathbf{A}\mathbf{x}_*\| > \eta\|\mathbf{b}\|/4$. Recall that $\eta < \|\mathbf{A}^t\mathbf{b}\|_\infty \le \|\mathbf{b}\|$. We conclude that $F(\mathbf{x}_*) \ge \eta\|\mathbf{b}\|/4$ which is at least $\eta/4$ after $\|\mathbf{b}\|$ is scaled to 1. Therefore, the additive error of ε/η^2 in Theorem 1 is at most $\frac{4\varepsilon}{\eta^3}F(\mathbf{x}_*)$.

References

1. Baraniuk, R., Davenport, M., DeVore, R., Wakin, M.: A simple proof of the restricted isometry property for random matrices. Constr. Approx. **28**, 253–263 (2008)
2. Bertrand, Q., Klopfenstein, Q., Bannier, P.A., Gidel, G., Massias, M.: Beyond L1: faster and better sparse models with skglm. In: NeurIPS, pp. 38950–38965 (2022)
3. Bruckstein, A.M., Donoho, D.L., Elad, M.: From sparse solutions of systems of equations to sparse modeling of signals and images. SIAM Rev. **51**(1), 34–81 (2009)
4. Candes, E., Romberg, J., Tao, T.: Robust uncertainty principles: exact signal reconstruction from highly incomplete frequency information. IEEE Trans. Inf. Theory **52**(2), 489–509 (2006)
5. Candes, E., Romberg, J.: L1-magic: Recovery of sparse signals via convex programming (2005). https://candes.su.domains/software/l1magic/
6. Candes, E.J., Tao, T.: Near-optimal signal recovery from random projections: universal encoding strategies? IEEE Trans. Inf. Theory **52**(12), 5406–5425 (2006)
7. Cheng, S.W., Wong, M.: A dynamic working set method for compressed sensing. arXiv preprint arXiv:2505.09370 (2025)
8. Daubechies, I., Defrise, D., Mol, C.D.: An iterative thresholding algorithm for linear inverse problems with a sparsity constraint. Commun. Pure Appl. Math. **57**(11), 1413–1457 (2004)
9. Donoho, D.: Compressed sensing. IEEE Trans. Inf. Theory **52**, 1289–1306 (2006)
10. Donoho, D.L., Tsaig, Y.: Fast solution of ℓ_1-norm minimization problems when the solution may be sparse. IEEE Trans. Inf. Theory **54**(11), 4789–4812 (2008)

11. Duarte, M.F., Davenport, M.A., Takhar, D., Laska, J.N., Sun, T., Kelly, K.F., Baraniuk, R.G.: Single-pixel imaging via compressive sampling. IEEE Signal Process. Mag. **25**(2), 83–91 (2008)

12. Figueiredo, M., Nowak, R.D., Wright, S.J.: Gradient projection for sparse reconstruction: application to compressed sensing and other inverse problems. IEEE J. Sel. Topics Signal Process. **1**(4), 586–597 (2007)

13. Friedman, J.H., Hastie, T., Tibshirani, R.: Regularization paths for generalized linear models via coordinate descent. J. Stat. Softw. **33**(1), 1–22 (2010). https://doi.org/10.18637/jss.v033.i01

14. Fuchs, J.: More on sparse representations in arbitrary bases. IEEE Trans. Inf. Theory **50**, 1341–1344 (2004)

15. Ge, J., et al.: Picasso: a sparse learning library for high dimensional data analysis in R and Python. J. Mach. Learn. Res. **20**(44), 1–5 (2019)

16. Johnson, T., Guestrin, C.: Blitz: a principled meta-algorithm for scaling sparse optimization. In: Proceedings of ICML. Proceedings of Machine Learning Research, vol. 37, pp. 1171–1179. PMLR, Lille (2015). https://proceedings.mlr.press/v37/johnson15.html

17. Kim, S.J., Koh, K., Lustig, M., Boyd, S., Gorinevsky, D.: An interior-point method for large-scale ℓ_1-regularized least squares. IEEE J. Sel. Topics Signal Process. **1**(4), 606–617 (2007)

18. Lin, X., Liu, Y., Wu, J., Dai, Q.: Spatial-spectral encoded compressive hyperspectral imaging. ACM Trans. Graph. **33**(6), 233:1–233:11 (2014)

19. Lustig, M., Donoho, D., Pauly, J.M.: Sparse MRI: the application of compressed sensing for rapid MR imaging. Magn. Reson. Med. **58**(6), 1182—1195 (2007). https://doi.org/10.1002/mrm.21391. https://onlinelibrary.wiley.com/doi/pdfdirect/10.1002/mrm.21391

20. Massias, M., Vaiter, S., Gramfort, A., Salmon, J.: Dual extrapolation for sparse glms. J. Mach. Learn. Res. **21**(234), 1–33 (2020)

21. Monteiro, R.D.C., Adler, I.: Interior path following primal-dual algorithms. Part II: convex quadratic programming. Math. Program. **44**, 43–66 (1989)

22. Moreau, T., et al.: Benchopt: reproducible, efficient and collaborative optimization benchmarks. In: Proceedings of NeurIPS, pp. 25404–25421 (2022)

23. Nesterov, Y.: Efficiency of coordinate descent methods on huge-scale optimization problems. SIAM J. Optim. **22**(2), 341–362 (2012)

24. Pedregosa, F., et al.: Scikit-learn: machine learning in python. J. Mach. Lear. Res. **12**, 2825–2830 (2011)

25. Rakotomamonjy, A., Flamary, R., Salmon, J., Gasso, G.: Convergent working set algorithm for Lasso with non-convex sparse regularizers. In: Proceedings of AISTAT, pp. 5196–5211 (2022)

26. Ruszczynski, A.: Nonlinear Optimization. Princeton University Press, Princeton (2006)

27. Tibshirani, R.: Regression shrinkage and selection via the Lasso. J. R. Stat. Soc. Ser. B (Methodol.) **58**(1), 267–288 (1996)

28. Tropp, J.A., Gilbert, A.C.: Signal recovery from random measurements via orthogonal matching pursuit. IEEE Trans. Inf. Theory **53**(12), 4655–4666 (2007)

29. Tseng, P., Yun, S.: A coordinate gradient descent method for nonsmooth separable minimization. Math. Program. **117**, 387–423 (2009)

Data Debugging Is NP-Hard for Classifiers Trained with SGD

Zizheng Guo[1], Jun Wu[2], Pengyu Chen[1], Yanzhang Fu[1], and Dongjing Miao[1]($\boxtimes$)

[1] Harbin Institute of Technology, Harbin 150001, China
zguo.research@gmail.com,pchen.research@gmail.com
{fuyanzhang,miaodongjing}@stu.hit.edu.cn
[2] Exploration and Development Rearch Institute of Daqing Oilfield, Daqing, China
wujun1@petrochina.com.cn

Abstract. Data debugging is to find a subset of the training data such that the model obtained by retraining on the subset has a better accuracy.A bunch of heuristic approaches are proposed, however, none of them are guaranteed to solve this problem effectively.This leaves an open issue whether there exists an efficient algorithm to find the subset such that the model obtained by retraining on it has a better accuracy.To answer this open question and provide theoretical basis for further study on developing better algorithms for data debugging, we investigate the computational complexity of the problem named DEBUGGABLE.Given a machine learning model $\mathcal{M}$ obtained by training on dataset D and a test instance $(\mathbf{x}_{\text{test}}, y_{\text{test}})$ where $\mathcal{M}(\mathbf{x}_{\text{test}}) \neq y_{\text{test}}$, DEBUGGABLE is to determine whether there exists a subset D' of D such that the model $\mathcal{M}'$ obtained by retraining on D' satisfies $\mathcal{M}'(\mathbf{x}_{\text{test}}) = y_{\text{test}}$. To cover a wide range of commonly used models, we take SGD-trained linear classifier as the model and derive the following main results.(1) If the loss function and the dimension of the model are not fixed, DEBUGGABLE is NP-complete regardless of the training order in which all the training samples are processed during SGD.(2) For hinge-like loss functions, a comprehensive analysis on the computational complexity of DEBUGGABLE is provided;(3) If the loss function is a linear function, DEBUGGABLE can be solved in linear time. These results not only highlight the limitations of current approaches but also offer new insights into data debugging.

1 Introduction

Given a machine learning model, data debugging is to find a subset of the training data such that the model will have a better accuracy if retrained on that subset [9]. Data debugging serves as a popular method of both data cleaning and machine learning interpretation. In the context of data cleaning, data debugging (*a.k.a.* training data debugging [25] or data cleansing [9]) can be used to improve the quality of the training data by removing the flaws leading to mispredictions [14,18,21]. When it comes to ML interpretation, data debugging locates

© The Author(s), under exclusive license to Springer Nature Singapore Pte Ltd. 2026
F. V. Fomin and M. Xiao (Eds.): COCOON 2025, LNCS 15984, pp. 169–180, 2026.
https://doi.org/10.1007/978-981-95-0218-9_13

the part of the training data responsible for unexpected predictions of an ML model. Therefore it is also studied as a training data-based (*a.k.a.* instance-based [1]) interpretation, which is crucial for helping system developers and ML practitioners to debug ML system by reporting the harmful part of training data [22].

To solve the data debugging problem, existing researches adopt a two-phase score-based heuristic approach [25]. In the first phase, a score representing the estimated impact on the model accuracy is assigned to each training sample in the training data. It is hoped that the harmful part of training data gets a lower score than the other part. In the second phase, training samples with lower scores are removed and the model is retrained on the modified training data. The two phases are carried out iteratively until a well-trained model is obtained. Most of the related works focus on developing algorithms to estimate the scores efficiently in the first phase [2, 7, 8, 11–13, 15–17], but rarely study the effectiveness of the entire two-phase approach.

Since it is computationally intractable to estimate the score for all possible subsets of the training data, it is often assumed that the score representing the impact of a subset is approximately equal to the sum of the scores of each individual training samples from the subset. However, Koh et al. [16] showed this is not always the case. For a bunch of subsets sampled from the training data, they empirically studied the difference between the estimated impact and the actual impact of each subset by taking influence functions as the scoring method. The estimated impact is calculated by summing up the score by influence function of each training samples in the subset, and the actual impact is measured by the improvement of accuracy of the model retrained after removing the subset from training data. They found that the estimated impact tends to underestimate the actual impact. Removing a large number of training samples could result in a large deviation between estimated and actual impacts. Although an upper bound of the deviation under certain assumptions has been derived, it is still unknown whether the deviation can be reduced or eliminated efficiently.

The above deviation also poses challenges to the effectiveness of the entire approach. Suppose the influence function is adopted as the scoring method, the accuracy of the model is not guaranteed to improve due to the deviation reported in [16] if a large group of training samples are removed during each iteration. Moreover, there is no theoretical analysis for the effectiveness of the greedy approach in the second phase. Even if only one training sample is removed during each iteration of the two-phase approach, the accuracy of the model is still not guaranteed to be improved. The effectiveness of the entire two-phase approach is therefore not assured. This leaves the following open problem:

Problem 1. Is there an efficient algorithm to find the subset of the training data, such that the model obtained by retraining on it has a better accuracy?

The computational complexity results presented in this paper demonstrate that it is unlikely to solve the data debugging problem efficiently in polynomial time. To figure out its hardness, we study the problem DEBUGGABLE which is the decision version of data debugging when the test set consists of only one instance. Formally, DEBUGGABLE is defined as follows:

Problem 2. (DEBUGGABLE) Given a classifier $\mathcal{M}$, its training data T, a test instance $(\mathbf{x}, y)$. Is there a $T' \subseteq T$, such that $\mathcal{M}$ predicts y on $\mathbf{x}$ if retrained on T'?

Basically, we prove that DEBUGGABLE is NP-complete, which means data debugging is unlikely to be solved in polynomial time. This result answers the open question mentioned above directly, this is, the large deviation of estimated impacts [16] cannot be reduced or eliminated efficiently. This is because if the impact of a subset of the training data could be accurately estimated as the sum of the impact of each training sample in the subset, data debugging can be solved in polynomial time, which is impossible unless P=NP.

Although DEBUGGABLE is generally intractable, we still hope to develop efficient algorithms tailored to specific cases. Thus it is necessary to figure out the root cause of the hardness for DEBUGGABLE. Previous research are always conducted based on the belief that the complexity of data debugging is complicated due to the chosen model architecture. However, we show that at least for models trained by stochastic gradient descent (SGD), the hardness stems from the hyper-parameter configuration selected for the SGD training, which was not yet aware of by previous work. To cover a wide range of commonly used machine learning models, we take linear classifiers as the model and show that even for linear classifiers, DEBUGGABLE is NP-hard as long as they are trained by SGD. Moreover, we provided a comprehensive analysis on hyper-parameter configurations that affect the computational complexity of DEBUGGABLE, including the loss function, the model dimension and the training order. Training order, *a.k.a.* training data order [20] or order of training samples [4], refers to the order in which each training sample is considered during the SGD. Detailed complexity results are shown in Table 1.

Our contribution can be concluded as follows:

- We studied the computational complexity of data debugging and showed that data debugging is NP-hard for linear classifiers in the general setting for *all possible training orders*.
- We studied the complexity of DEBUGGABLE when the loss is fixed as the hinge-like function. For 2 or higher dimension, DEBUGGABLE is NP-complete when the training order is adversarially chosen; For one-dimensional cases, DEBUGGABLE can be NP-hard when the interception $\beta < 0$, and is solvable in linear time when $\beta \geq 0$.
- We proved that DEBUGGABLE is solvable in linear time when the loss function is linear.

Moreover, we have a discussion on the implications of these complexity results for machine learning interpretability and data quality, as well as limitations of score-based greedy methods. Our results suggest the further study as follows. (1) It is better to characterize the training sample and find the criterion which can be used to decide the existence of efficient algorithms; (2) Designing algorithms with CSP-solver is a potential way to solve data debugging more efficiently than the brute-force one; (3) Developing random algorithms is a potential way to solve data debugging successfully with high probability.

Table 1. Computational complexity of the data debugging problem

Loss Function	Dimension	Training Order	Complexity
Not Fixed	Not Fixed	-	NP-hard
Hinge-like	≥ 2	Adversarially Chosen	NP-hard
Hinge-like, $\beta < 0$	1	Adversarially Chosen	NP-hard
Hinge-like, $\beta \geq 0$	1	-	Linear Time
Linear	-	-	Linear Time

1.1   Related Works

The solution of data debugging has applications in database query results reliability enhancement [19,25], training data cleaning [9] and machine learning interpretation [3,15–17,23]. Existing works on data debugging mainly adopt a two-phase approach, which scores the training samples in the first phase and greedily deletes training samples with lower scores in the second phase. Most of the research focus on the first phase. There are mainly two ways of scoring adopted for data debugging in practice. Leave-one-out (LOO) retraining is a widely studied way, which evaluates the contribution of a training sample through the difference in the model's accuracy trained without that training sample. To avoid the cost of model retraining, Koh and Liang took influence functions as an approximation of LOO [17]. After that, various extensions and improvements of the influence function based method are proposed, such as Fisher kernel [15], influence function for group impacts [16], second-order approximations [2] and scalable influence functions [8]. Another way is Shapley-based scoring, where the impact of a training sample is measured by its average marginal contribution to all subsets of the training data [7]. Since Shapley-base scoring suffers from expensive computational cost [6], recent works focus on techniques that efficiently estimate the Shapley value, including Monte-Carlo sampling [7], group testing [11,13] and using proxy models such as k-NN [12,14]. However, those methods do not admit any theoretical guarantee on the effectiveness. This paper discusses the limitations of the above methods and suggests some future directions on data debugging.

2   Preliminaries and Problem Definition

Linear classifiers. Formally, a (binary) linear classifier is a function $\lambda_{\mathbf{w}} : \mathbb{R}^d \to \{-1, 1\}$, where d is called its *dimension* and $\mathbf{w} \in \mathbb{R}^d$ its parameter. Without loss of generality, the bias term of a linear classifier is set as zero in this paper. All vectors in this paper are assumed to be *column* vectors. For an input $\mathbf{x}$, the value of $\lambda_\mathbf{w}$ is defined as

$$\lambda_\mathbf{w}(\mathbf{x}) = \begin{cases} 1 & \text{if } \mathbf{w}^\top \mathbf{x} \geq 0 \\ -1 & \text{otherwise.} \end{cases}$$

We denote the class of linear models as Λ.

Training data. A *training sample* is a pair $(\mathbf{x}, y)$ in which $\mathbf{x} \in \mathbb{R}^d$ is the input and $y \in \{-1, 1\}$ is the label of $\mathbf{x}$. The *training data* is a multiset of training samples. We employ $\mathbf{w} \xrightarrow{T} \mathbf{w}'$ to denote that the parameter $\mathbf{w}'$ is obtained by training the parameter $\mathbf{w}$ on the training data T, and employ $\mathbf{w} \xrightarrow{(\mathbf{x},y)} \mathbf{w}'$ to denote that $\mathbf{w}'$ is obtained by training $\mathbf{w}$ on the training sample $(\mathbf{x}, y)$.

Loss functions and learning rates. Binary linear classifiers typically use unary functions on $y\mathbf{w}^\top\mathbf{x}$ as their loss functions [24]. Therefore we only consider loss functions of the form $\mathcal{L} : y\mathbf{w}^\top\mathbf{x} \mapsto \mathbb{R}$ for the rest of the paper.

The *linear* loss is in the form of

$$\mathcal{L}_{\text{lin}}(y\mathbf{w}^\top\mathbf{x}) = -\alpha(y\mathbf{w}^\top\mathbf{x} + \beta).$$

The *hinge-like* loss function is defined as the following form

$$\mathcal{L}_{\text{hinge}}(y\mathbf{w}^\top\mathbf{x}) = \begin{cases} -\alpha(y\mathbf{w}^\top\mathbf{x} - \beta), & y\mathbf{w}^\top\mathbf{x} < \beta \\ 0, & \text{otherwise.} \end{cases}$$

We call β as the *interception* of $\mathcal{L}_{\text{hinge}}$. We represent the learning rate of a model using a vector $\boldsymbol{\eta} = (\eta_1, \ldots, \eta_d)$, where $\eta_i \geq 0$ and each parameter w_i can be updated with the corresponding learning rate η_i.

Stochastic gradient descent. The stochastic gradient descent (SGD) method updates parameter $\mathbf{w}$ from its initial value $\mathbf{w}^{(0)}$ through several epochs. During each epoch, the SGD goes through the entire set of training samples in some training order through several iterations. The training order is defined as a sequence of training samples, in the form of $(\mathbf{x}_1, y_1) \ldots (\mathbf{x}_n, y_n)$. For $1 \leq i < j \leq n$, $(\mathbf{x}_i, y_i)$ is considered before $(\mathbf{x}_j, y_j)$ during the SGD. We use w_i to denote the i-th coordinate of $\mathbf{w}$. We also use $\mathbf{w}^{(e,k)}$ to denote the value of $\mathbf{w}$ at the end of k-th iteration of epoch e and use $\mathbf{w}^{(e)}$ to denote the value of $\mathbf{w}$ after the end of epoch e. Assuming $(\mathbf{x}, y)$ to be the training sample considered at iteration k, the stochastic gradient descent (SGD) method updates parameter w_i for each i by

$$w_i^{(e,k)} \leftarrow w_i^{(e,k-1)} - \eta_i \cdot \frac{\partial \mathcal{L}(y(\mathbf{w}^{(e,k-1)})^\top\mathbf{x})}{\partial w_i} \tag{1}$$

In other words, we have

$$\mathbf{w}^{(e,k)} \leftarrow \mathbf{w}^{(e,k-1)} - \boldsymbol{\eta} \otimes \nabla\mathcal{L}(y(\mathbf{w}^{(e,k-1)})^\top\mathbf{x})$$

where $\boldsymbol{\eta} \otimes \nabla \mathcal{L} = (\eta_1 \frac{\partial \mathcal{L}}{\partial w_1}, \ldots, \eta_d \frac{\partial \mathcal{L}}{\partial w_d})$ is the Hadamard product. We say a training sample $\mathbf{x}$ is *activated* at iteration k during epoch e if $\nabla \mathcal{L}(y(\mathbf{w}^{(e,k-1)})^\top \mathbf{x}) \neq 0$. The SGD terminates at the end of epoch e if $\|\mathbf{w}^{(e-1)} - \mathbf{w}^{(e)}\| < \varepsilon$ for threshold ε or e reached some predetermined value. We denote $\mathbf{w}^* = \mathbf{w}^{(e)}$. A linear classifier trained by SGD with the meta-parameters mentioned above is denoted as $\text{SGD}_\Lambda(\mathcal{L}, \boldsymbol{\eta}, \varepsilon, T) = \lambda_{\mathbf{w}^*}$. With a slight abuse of notation, we define $\text{SGD}_\Lambda(\mathcal{L}, \boldsymbol{\eta}, \varepsilon, T, \mathbf{x}) = \lambda_{\mathbf{w}^*}(\mathbf{x})$. We also use $\text{SGD}_\Lambda(T, \mathbf{x})$ to avoid cluttering when the context is clear.

Problem definition. With the above definitions, DEBUGGABLE for SGD-trained linear classifiers can be formalized as follows:

> DEBUGGABLE-LIN
> **Input:** Training data T, loss function $\mathcal{L}$, initial parameter $\mathbf{w}^{(0)}$, learning rate $\boldsymbol{\eta}$, threshold ε and instance $(\mathbf{x}_{\text{test}}, y_{\text{test}})$.
> **Output:** "Yes": if $\exists \Delta \subseteq T$ such that $\text{SGD}_\Lambda(\mathcal{L}, \boldsymbol{\eta}, \varepsilon, T \setminus \Delta, \mathbf{x}_{\text{test}}) = y_{\text{test}}$;
>   "No": otherwise.

We say $\text{SGD}_\Lambda(\mathcal{L}, \boldsymbol{\eta}, \varepsilon, T)$ is *debuggable* on $(\mathbf{x}_{\text{test}}, y_{\text{test}})$ if $(\mathcal{L}, \mathbf{w}^{(0)}, \boldsymbol{\eta}, \varepsilon, T, \mathbf{x}_{\text{test}}, y_{\text{test}})$ is a yes-instance of DEBUGGABLE-LIN, and not *debuggable* on $(\mathbf{x}_{\text{test}}, y_{\text{test}})$ otherwise.

3   Results for Unfixed Loss Functions

In this section, we prove the NP-hardness of DEBUGGABLE-LIN. Intuitively, DEBUGGABLE-LIN is to determine whether there exists a subset $T' \subseteq T$ where activated training samples within T' drive the parameter $\mathbf{w}$ toward the region defined by $y_{\text{test}} \mathbf{w}^\top \mathbf{x}_{\text{test}} > 0$. The activation of training samples depends on the complex interaction between the training data and the model.

Theorem 1. DEBUGGABLE-LIN *is NP-hard for all training orders.*

We only show the proof sketch and leave the details in the appendix.

Proof (Sketch). We build a reduction from an NP-hard problem MONOTONE 1-IN-3 SAT [5]:

> MONOTONE 1-IN-3 SAT
> **Input:** A 3-CNF formula φ with no negation signs.
> **Output:** "Yes": if φ has a 1-in-3 assignment, under which each clause contains exactly one true literal;
>   "No": otherwise.

For example, $\varphi_1 = (x_1 \vee x_2 \vee x_3) \wedge (x_2 \vee x_3 \vee x_4)$ is a yes-instance because $(x_1, x_2, x_3, x_4) = (\text{T,F,F,T})$ is an 1-in-3 assignment; $\varphi_2 = (x_1 \vee x_2 \vee x_3) \wedge (x_2 \vee x_3 \vee x_4) \wedge (x_1 \vee x_2 \vee x_4) \wedge (x_1 \vee x_3 \vee x_4)$ is a no-instance.

Given a 3-CNF formula φ, our goal is to construct a configuration of the training process, such that the resulting model outputs the correct answer if and only if its training data T' encodes an 1-in-3 assignment ν of φ. This can be done by carefully designing the encoding so that for each $x_i \in \varphi$, $\nu(x_i) = \text{TRUE}$ if and only if $\mathbf{t}_{x_i} \in T'$. Finally, we can construct some T with $T \supseteq T' \cup \{\mathbf{t}_{x_i} | x_i \in \varphi\}$, such that some classifier trained on T is a yes-instance of DEBUGGABLE-LIN if and only if φ is a yes-instance of MONOTONE 1-IN-3 SAT, thereby finishing our proof.

The reduction. Suppose φ has m clauses and n variables, let $N = n+2m+1$. We set the dimension of the linear classifier to N.

The input. Each coordinate of the input is named as

$$\mathbf{x} = (x_{c_1}, \ldots, x_{c_m}, x_{x_1}, \ldots, x_{x_n}, x_{b_1}, \ldots, x_{b_m}, x_{\text{dummy}})^\top$$

We also use x_i to denote the i-th coordinate of $\mathbf{x}$.

The parameters. Each coordinate of the parameter is named as

$$\mathbf{w} = (w_{c_1}, \ldots, w_{c_m}, w_{x_1}, \ldots, w_{x_n}, w_{b_1}, \ldots, w_{b_m}, w_{\text{dummy}})^\top$$

We also use w_i to denote the i-th coordinate of $\mathbf{w}$. Each w_{x_j} represents the truth value of variable x_j, where 1 represents TRUE and -1 represents FALSE.Similarly, each w_{c_j} represents the truth value of clause c_j based on the value of its variables. w_{b_j} and w_{dummy} are used for convenience of proof. The initial value of the parameter is set to

$$\mathbf{w}^{(0)} = (\overbrace{\frac{1}{2}, \ldots, \frac{1}{2}}^{m}, \overbrace{-1, \ldots, -1}^{n}, \overbrace{-1, \ldots, -1}^{m}, 1)^\top$$

Loss function. We denote $U(x_0, \delta) := \{x | x_0 - \delta < x < x_0 + \delta\}$ as the δ-neighborhood of x_0 and define $U(\pm x_0, \delta) = U(x_0, \delta) \cup U(-x_0, \delta)$. We define the *local ramp function* as

$$r_{x_0, \delta}(x) = \begin{cases} 0 & , x \leq x_0 - \delta; \\ x - x_0 + \delta & , x \in U(x_0, \delta); \\ 2\delta & , x \geq x_0 + \delta. \end{cases}$$

The loss function is defined as

$$\mathcal{L} = -\frac{12N}{5} r_{-5, 0.01}(y\mathbf{w}^\top\mathbf{x}) - r_{-\frac{1}{2}, 0.26}(y\mathbf{w}^\top\mathbf{x}) - \frac{1}{1000N} \sum_{x_0 \in \{\pm 1, \pm 3\}} r_{x_0, 0.01}(y\mathbf{w}^\top\mathbf{x}).$$

$\mathcal{L}$ is monotonically decreasing with derivatives

$$\frac{\partial \mathcal{L}}{\partial w_i} = \begin{cases} -\frac{12N}{5} \cdot yx_i & , y\mathbf{w}^\top\mathbf{x} \in U(-5, 0.01); \\ -yx_i & , y\mathbf{w}^\top\mathbf{x} \in U(-\frac{1}{2}, 0.26); \\ -\frac{1}{1000N} yx_i & , y\mathbf{w}^\top\mathbf{x} \in \bigcup_{x_0 \in \{\pm 1, \pm 3\}} U(x_0, 0.01); \\ 0 & , \text{otherwise.} \end{cases} \quad (2)$$

Learning rate. The learning rate for SGD is set to be

$$\boldsymbol{\eta} = (\overbrace{5,\ldots,5}^{m}, \overbrace{\frac{1}{6N},\ldots,\frac{1}{6N}}^{n}, \overbrace{2000N,\ldots,2000N}^{m}, 1)^{\top}.$$

Training data. We define two gadgets, var(i) and clause(i, i_1, i_2, i_3), as illustrated in Table 2 and 3. All the unspecified coordinates are set to zero. We use T_0 to denote the training data. var(i) is contained in T_0 if and only if $x_i \in \varphi$, and clause(i, i_1, i_2, i_3) is contained in T_0 if and only if $c_i = (x_{i_1} \lor x_{i_2} \lor x_{i_3}) \in \varphi$.

<table>
<tr><td colspan="2" align="center">**Table 2.** var(i)</td><td colspan="6" align="center">**Table 3.** clause(i, i_1, i_2, i_3)</td></tr>
<tr><td>x_{x_i}</td><td>y</td><td></td><td>x_{c_i}</td><td>$x_{x_{i_1}}$</td><td>$x_{x_{i_2}}$</td><td>$x_{x_{i_3}}$</td><td>x_{b_i}</td><td>y</td></tr>
<tr><td>5</td><td>1</td><td></td><td>1</td><td>1</td><td>1</td><td>1</td><td>1/2</td><td>1</td></tr>
</table>

Threshold and instance. The threshold ε can be any fixed value in $\mathbb{R}_+$. The instance is defined as $(\mathbf{x}_{\text{test}}, y_{\text{test}})$, where $y_{\text{test}} = 1$ and

$$\mathbf{x}_{\text{test}} = (\overbrace{1,\ldots,1}^{m}, \overbrace{0,\ldots,0}^{n+m}, \frac{-11\,m + 5}{2})^{\top}.$$

The following reduction works for all possible training orders. Intuitively, during the training process, each var(i) in the training data will set w_{x_i} to around 1 (that is, mark x_i as TRUE) in the first epoch, and each clause(i, i_1, i_2, i_3) will set w_{c_i} to near $\frac{11}{2}$ in the second epoch, if and only if exactly one of $w_{x_{i_1}}, w_{x_{i_2}}, w_{x_{i_3}}$ is near 1 and the others near -1 (that is, mark c_i as satisfied if exactly one of its literals is TRUE and the others FALSE). The training process terminates at the end of the second epoch.

4   Results for Fixed Loss Functions

We have proved the NP-hardness for DEBUGGABLE-LIN when the loss function is not fixed. In this section, we study the complexity when the loss function is fixed as linear and hinge-like functions. Assuming that SGD terminates after only one epoch with a fixed order, we will show that DEBUGGABLE-LIN is solvable in linear time for linear loss. For hinge-like loss functions, DEBUGGABLE-LIN can be solved in linear time only when the dimension $d = 1$ and the interception $\beta \geq 0$. For the rest cases, DEBUGGABLE-LIN becomes NP-hard.

4.1   The Easy Case

We start with the linear loss function $\mathcal{L} = -\alpha(y\mathbf{w}^{\top}\mathbf{x} + \beta)$, with which all the training data are activated and $\mathbf{w}^* = \mathbf{w}^*(T) = \mathbf{w}^{(0)} + \sum_{(\mathbf{x},y)\in T} \alpha y \boldsymbol{\eta} \otimes \mathbf{x}$. Since $y_{\text{test}} \in \{-1, 1\}$, DEBUGGABLE-LIN is equivalent to deciding whether

$$\max_{T' \subseteq T}\{y_{\text{test}}(\mathbf{w}^*(T'))^{\top}\mathbf{x}_{\text{test}}\} > 0.$$

A training sample $(\mathbf{x}, y)$ is "good" if $y_{\text{test}}(\alpha y \boldsymbol{\eta} \otimes \mathbf{x})^{\top} \mathbf{x}_{\text{test}} > 0$ and "bad" otherwise. The *good training-sample assessment* (GTA) algorithm, as shown in Algorithm 1, deals with this situation by greedily picking all "good" training samples.

Denoting T^* as the set of all good data in T, it follows that

$$y_{\text{test}}(\mathbf{w}^*(T^*))^{\top} \mathbf{x}_{\text{test}} = y_{\text{test}}(\mathbf{w}^{(0)})^{\top} \mathbf{x}_{\text{test}} + \sum_{(\mathbf{x},y) \in T^*} y_{\text{test}}(\alpha y \boldsymbol{\eta} \otimes \mathbf{x})^{\top} \mathbf{x}_{\text{test}}$$

$$\geq y_{\text{test}}(\mathbf{w}^{(0)})^{\top} \mathbf{x}_{\text{test}} + \sum_{(\mathbf{x},y) \in T'} y_{\text{test}}(\alpha y \boldsymbol{\eta} \otimes \mathbf{x})^{\top} \mathbf{x}_{\text{test}}$$

for all $T' \subseteq T$. Hence $\max_{T' \subseteq T} \{ y_{\text{test}}(\mathbf{w}^*(T'))^{\top} \mathbf{x}_{\text{test}} \} = y_{\text{test}}(\mathbf{w}^*(T^*))^{\top} \mathbf{x}_{\text{test}}$ and DEBUGGABLE-LIN can be solved by GTA in linear time. The following theorem is straightforward.

Theorem 2. DEBUGGABLE-LIN *is linear time solvable for linear loss functions.*

Algorithm 1: Good Training-sample Assessment (GTA)

Input: Training data T, loss function $\mathcal{L}$, initial parameter $\mathbf{w}^{(0)}$, learning rate $\boldsymbol{\eta}$, threshold ε and test instance $(\mathbf{x}_{\text{test}}, y_{\text{test}})$.
Output: TRUE, iff $\text{SGD}_\Lambda(\mathcal{L}, \boldsymbol{\eta}, \varepsilon, T)$ is debuggable on $(\mathbf{x}_{\text{test}} y_{\text{test}})$.
$\mathbf{w} \leftarrow \mathbf{w}^{(0)}$;
for $(\mathbf{x}, y) \in T$ **do**
    if $y_{test}(\alpha y \boldsymbol{\eta} \otimes \mathbf{x})^{\top} \mathbf{x}_{test} > 0$ **then**
        $\mathbf{w} \leftarrow \mathbf{w} + \alpha y \boldsymbol{\eta} \otimes \mathbf{x}$;
    end
end
if $y_{test} \mathbf{w}^{\top} \mathbf{x}_{test} \geq 0$ **then**
    return TRUE;
end
return FALSE;

GTA is still effective for one-dimensional classifiers trained with hinge-like losses when $\beta \geq 0$.

Theorem 3. DEBUGGABLE-LIN *is linear time solvable for hinge-like loss functions, when $d = 1$ and $\beta \geq 0$.*

Proof. It suffices to prove that if $\exists T' \subseteq T$ such that $\text{SGD}_\Lambda(T', x_{\text{test}}) = y_{\text{test}}$, $\text{SGD}_\Lambda(T^*, x_{\text{test}}) = y_{\text{test}}$.
    a) Suppose all the data in T^* are activated, we have

$$y_{\text{test}} w^*(T^*) x_{\text{test}} = y_{\text{test}} w^{(0)} x_{\text{test}} + \sum_{(x,y) \in T^*} y_{\text{test}} \alpha y \eta x x_{\text{test}}$$

$$\geq y_{\text{test}} w^{(0)} x_{\text{test}} + \sum_{(x,y) \in T' \cap T^*} y_{\text{test}} \alpha y \eta x x_{\text{test}} + \sum_{(x,y) \in T' \setminus T^*} y_{\text{test}} \alpha y \eta x x_{\text{test}}$$

$$= y_{\text{test}} w^*(T') x_{\text{test}} \geq 0$$

b) Suppose $(x, y) \in T^*$ is the first inactivated data during the training phase, and w is the current parameter, we have $ywx > \beta$. Since $\alpha\eta \cdot (xy) \cdot (x_{\text{test}}y_{\text{test}}) \geq 0$, we have $(x_{\text{test}}y_{\text{test}}) \cdot w \geq 0$. Let T'' be the set of training data appeared before (x, y), we have $y_{\text{test}}w^*(T^*)x_{\text{test}} \geq y_{\text{test}}w^*(T'')x_{\text{test}} \geq 0$.

4.2   The Hard Case

The gradient of training data may not always be activated and could be affected by the training order. When the training order is adversarially chosen, the following theorem shows that DEBUGGABLE-LIN is NP-hard for all $d \geq 2$ and $\beta \in \mathbb{R}$.

Theorem 4. *If the training order is adversarially chosen and $d \geq 2$, DEBUGGABLE-LIN is NP-hard for each hinge-like loss function at every constant learning rate.*

Moreover, DEBUGGABLE-LIN is NP-hard even when $d = 1$ and $\beta < 0$.

Theorem 5. *If the training order is adversarially chosen and $d = 1$, DEBUGGABLE-LIN remains NP-hard for each hinge-like loss function with $\beta < 0$ at every constant learning rate.*

Remarks. The training order in this section can be arbitrary as long as the last three training samples are $(\mathbf{x}_c, y_c), (\mathbf{x}_b, y_b), (\mathbf{x}_a, y_a)$, respectively. All the training samples are "good" since for each $(\mathbf{x}, y) \in T$ we have $\mathbf{x}^\top \mathbf{x}_{\text{test}}yy_{\text{test}} > 0$. This implies that DEBUGGABLE-LIN is NP-hard even if all the training data are "good" training samples, and exemplifies why the GTA algorithm fails for higher dimensions.

5   Discussion and Conclusion

In this paper, we provided a comprehensive analysis on the complexity of DEBUGGABLE. We focus on the linear classifier that is trained using SGD, as it is a key component in the majority of popular models.

Since DEBUGGABLE is a special case of data debugging, the above results proved the intractability of data debugging and therefore gives a negative answer to Problem 1 declared in the introduction. The complexity results also demonstrated that it is not accurate to estimate the impact of subset of training data by summing up the score of each training samples in the subset, *as long as the scores can be calculated in polynomial time*.

In Sect. 4, a training sample is said to be "good" if it can help the resulting model to predict correctly on the test instance. That is, it can increase $y_{\text{test}}(\mathbf{w}^*)^\top \mathbf{x}_{\text{test}}$. However, in our proof we showed that DEBUGGABLE remains NP-hard even if all training samples are "good". This suggests that the quality of a training sample does not depend only on some properties of itself but also on the interaction between the rest of the training data, which should be taken into consideration when developing data cleaning approaches.

Moreover, the NP-hardness of DEBUGGABLE implies that, it is in general intractable to figure out the causality between even the prediction of a linear classifier and its training data. This may be seem surprising since linear classifiers have long been considered "inherently interpretable". As warned in [10], *a method being "inherently interpretable" needs to be verified before it can be trusted*, the concept of interpretability must be *rigorously defined*, or at least its boundaries specified.

Our results suggests the following directions for future research. Firstly, characterizing the training sample may be helpful in designing efficient algorithms for data debugging; Secondly, designing algorithms using CSP-solver is a potential way to solve data debugging more efficiently than the brute-force algorithms; Finally, developing random algorithms is a potential way to solve data debugging successfully with high probability.

Acknowledgments. This work was partially supported by the National Natural Science Foundation of China grant 62372138, Natural Science Foundation of Heilongjiang Province of China grant HSF20230095 and 2024ZXJ01A04.

References

1. Bae, J., Ng, N., Lo, A., Ghassemi, M., Grosse, R.: If influence functions are the answer, then what is the question? In: Proceedings of the 36th International Conference on Neural Information Processing Systems. NIPS '22, Curran Associates Inc., Red Hook, NY (2024)
2. Basu, S., You, X., Feizi, S.: On second-order group influence functions for blackbox predictions. In: Proceedings of the 37th International Conference on Machine Learning. ICML'20, JMLR.org (2020)
3. Brunet, M.E., Alkalay-Houlihan, C., Anderson, A., Zemel, R.: Understanding the origins of bias in word embeddings (2019)
4. Chang, E., Yeh, H.S., Demberg, V.: Does the order of training samples matter? improving neural data-to-text generation with curriculum learning. ArXiv **abs/2102.03554** (2021). https://api.semanticscholar.org/CorpusID:231846815
5. Demaine, E.D., Gasarch, W., Hajiaghayi, M.: Computational Intractability: A Guide to Algorithmic Lower Bounds. MIT Press (2024)
6. Deng, X., Papadimitriou, C.H.: On the complexity of cooperative solution concepts. Math. Oper. Res. **19**, 257–266 (1994).https://api.semanticscholar.org/CorpusID:12946448
7. Ghorbani, A., Zou, J.Y.: Data shapley: Equitable valuation of data for machine learning. ArXiv **abs/1904.02868** (2019). https://api.semanticscholar.org/CorpusID:102350503
8. Guo, H., Rajani, N., Hase, P., Bansal, M., Xiong, C.: FastIF: Scalable influence functions for efficient model interpretation and debugging. In: Moens, M.F., Huang, X., Specia, L., Yih, S.W.t. (eds.) Proceedings of the 2021 Conference on Empirical Methods in Natural Language Processing, pp. 10333–10350. Association for Computational Linguistics, Online and Punta Cana, Dominican Republic (2021). https://doi.org/10.18653/v1/2021.emnlp-main.808, https://aclanthology.org/2021.emnlp-main.808

9. Hara, S., Nitanda, A., Maehara, T.: Data Cleansing for Models Trained with SGD. Curran Associates Inc., Red Hook, NY (2019)

10. Jacovi, A., Goldberg, Y.: Towards faithfully interpretable NLP systems: How should we define and evaluate faithfulness? In: Annual Meeting of the Association for Computational Linguistics (2020). https://api.semanticscholar.org/CorpusID: 215416110

11. Jia, R., et al.: Towards efficient data valuation based on the shapley value. ArXiv **abs/1902.10275** (2019). https://api.semanticscholar.org/CorpusID:67855573

12. Jia, R., et al.: Efficient task-specific data valuation for nearest neighbor algorithms. Proc. VLDB Endow. **12**(11), 1610–1623 (2019).https://doi.org/10.14778/3342263. 3342637

13. Jia, R., et al.: Scalability vs. utility: Do we have to sacrifice one for the other in data importance quantification? In: 2021 IEEE/CVF Conference on Computer Vision and Pattern Recognition (CVPR), pp. 8235–8243 (2021).https://doi.org/ 10.1109/CVPR46437.2021.00814

14. Karlaš, B., et al.: Data debugging with shapley importance over end-to-end machine learning pipelines (2022)

15. Khanna, R., Kim, B., Ghosh, J., Koyejo, O.: Interpreting black box predictions using fisher kernels. In: International Conference on Artificial Intelligence and Statistics (2018). https://api.semanticscholar.org/CorpusID:53085397

16. Koh, P.W., Ang, K.S., Teo, H.H.K., Liang, P.: On the accuracy of influence functions for measuring group effects. In: Neural Information Processing Systems (2019). https://api.semanticscholar.org/CorpusID:173188850

17. Koh, P.W., Liang, P.: Understanding black-box predictions via influence functions. In: Proceedings of the 34th International Conference on Machine Learning - Volume 70, pp. 1885–1894. ICML'17, JMLR.org (2017)

18. Li, P., Rao, X., Blase, J., Zhang, Y., Chu, X., Zhang, C.: CleanML: A study for evaluating the impact of data cleaning on ml classification tasks. In: 2021 IEEE 37th International Conference on Data Engineering (ICDE), pp. 13–24 (2021).https:// doi.org/10.1109/ICDE51399.2021.00009

19. Liu, Y., Wu, W., Flokas, L., Wang, J., Wu, E.: Enabling SQL-based training data debugging for federated learning. In: Proceedings of the VLDB Endowment, Vol.15 ,pp. 388–400 (2022).https://doi.org/10.14778/3494124.3494125

20. Mange, J.: Effect of training data order for machine learning. In: 2019 International Conference on Computational Science and Computational Intelligence (CSCI), pp. 406–407 (2019).https://doi.org/10.1109/CSCI49370.2019.00078

21. Neutatz, F., Chen, B., Abedjan, Z., Wu, E.: From cleaning before ml to cleaning for ml. IEEE Data Eng. Bull. **44**, 24–41 (2021). https://api.semanticscholar.org/ CorpusID:237542697

22. Pradhan, R., Zhu, J., Glavic, B., Salimi, B.: Interpretable data-based explanations for fairness debugging. In: Proceedings of the 2022 International Conference on Management of Data. p. 247–261. SIGMOD '22, Association for Computing Machinery, New York, NY (2022).https://doi.org/10.1145/3514221.3517886

23. Wang, H., Ustun, B., Calmon, F.P.: Repairing without retraining: Avoiding disparate impact with counterfactual distributions (2019)

24. Wang, Q., Ma, Y., Zhao, K., Tian, Y.: A Comprehensive Survey of Loss Functions in Machine Learning. Annals of Data Science ,pp. 1–26 (2020). https://doi.org/10. 1007/s40745-020-00253-5

25. Wu, W., Flokas, L., Wu, E., Wang, J.: Complaint-driven training data debugging for query 2.0, pp. 1317–1334 (2020).https://doi.org/10.1145/3318464.3389696

Coresets for k-Median of Lines with Group Fairness Constraints

Ting Liang, Xiaoliang Wu, Junyu Huang, and Qilong Feng[✉]

School of Computer Science and Engineering, Central South University,
Changsha 410083, People's Republic of China
csufeng@mail.csu.edu.cn

Abstract. Clustering is a fundamental unsupervised learning problem with lots of applications in data mining, image classification, and other fields. Although clustering algorithms for k-means and k-median are widely used due to their simplicity and effectiveness on point-based data, they often perform poorly on structured data such as lines, graphs, and time series, where point-based representations fail to capture the underlying structure. Moreover, existing algorithms for structured data typically do not account for fairness constraints, which are increasingly important in modern applications involving sensitive attributes such as gender, race, or user groups. In this paper, we formally introduce the group fair k-median of lines problem (GF-k-ML), a new variant of k-median that integrates fairness constraints into the clustering of structured data represented by lines. Given a set L of n lines in $\mathbb{R}^d$, partitioned into t disjoint color groups $L_1, \ldots, L_t$, the goal of the GF-k-ML problem is to partition L into k clusters such that the proportion of lines from each color group in each cluster remains within a specified range, and the sum of distances over each line to its assigned center is minimized. We introduce a group-wise coreset construction algorithm that computes a separate coreset for each group obtained by partitioning the input lines according to sensitive attributes, and prove that these groups satisfy composability under fairness constraints. Our main result is a coreset that satisfies the fairness constraint and has a size of $O\left(\frac{td^2 k \log^2 k \log^2 n}{\varepsilon^2}\right)$ with an error parameter $\varepsilon \in (0, 1)$, which can be constructed in nearly linear time with respect to n.

Keywords: coresets · k-median of lines · fairness

1 Introduction

Clustering is a fundamental problem that has been extensively studied over the past few decades. The goal of clustering is to partition a given set of points

This work was supported by National Natural Science Foundation of China (62432016,62172446), and Central South University Research Program of Advanced Interdisciplinary Studies (2023QYJC023), and the Fundamental Research Funds for the Central Universities of Central South University (No.2024ZZTS0107).

F. V. Fomin and M. Xiao (Eds.): COCOON 2025, LNCS 15984, pp. 181–193, 2026.
https://doi.org/10.1007/978-981-95-0218-9_14

into several disjoint clusters such that similar points end up in the same cluster, and dissimilar points are separated into different clusters [19]. Several classic clustering models, including k-center, k-median, and k-means, have been widely explored [8,10,22], among which the k-median problem is one of the most frequently encountered. Given a set X of n points in $\mathbb{R}^d$ and an integer $k \in \mathbb{N}^{\geq 1}$, the goal of the k-median problem is to find a subset $S \subseteq \mathbb{R}^d$ of k centers, such that the sum of distances over each point $x \in X$ to its nearest center in S is minimized. The k-median problem is known to be NP-hard, even when the number k of centers or the dimension d is fixed [13,17], implying that exact algorithms are impractical for large-scale or high-dimensional datasets. This computational intractability becomes even more evident when the input consists of structured data, which often exhibit complex geometric or temporal dependencies. To overcome these challenges, several structured variants of the k-median problem have been proposed, such as k-median of lines [16], polygonal curves [3] and time series [20], each designed to capture the unique properties of structured or high-dimensional data.

In this paper, we study a practically important variant of the k-median problem, known as the k-median of lines problem (denoted as the k-ML problem), which has found applications in trajectory analysis [9], motion tracking [12], geometric data processing [24], and related domains. Given a set L of n lines in $\mathbb{R}^d$ and an integer $k \in \mathbb{N}^{\geq 1}$, and the goal is to find a subset S of k centers in $\mathbb{R}^d$, such that the sum of distances over each line $\ell \in L$ to its nearest center in S is minimized. For the k-ML problem, Ommer et al. [18] proposed the first heuristic algorithm, though it lacks theoretical guarantees, and provides no formal runtime analysis. More recently, research on the k-ML problem has primarily focused on the construction of coresets. Specifically, for a given error parameter $\varepsilon > 0$, Perets et al. [21] proved that the k-ML problem is solvable when k is a constant, whereas it becomes NP-complete when k is non-constant. Furthermore, they proposed a bi-criteria approximation algorithm in $\mathbb{R}^2$ based on coreset construction, achieving a $(1 + \varepsilon)$-approximation in $O(n(\log n/\varepsilon))^{O(k)}$ time. Marom et al. [16] constructed an ε-coreset of size $O(dk^{O(k)} \log(n)/\varepsilon^2)$ for the k-ML problem, with a total running time of $O(d^3 n \log(n)k \log k + (d/\varepsilon)^2 + ndk^{O(k)})$.

Despite these advances, little attention has been paid to incorporating fairness constraints into the k-ML problem, even though such constraints are crucial in real-world applications involving diverse groups. In contrast, fair clustering has recently received growing attention in the classical k-median setting, where the notion of group fairness has been introduced to ensure that each group is proportionally represented in the resulting clusters [1,2,4]. Specifically, Chierichetti et al. [4] introduced the definition of fairness with only two colors, requiring that the proportion of two colors has approximately equal representation in every cluster. Bercea et al. [2] and Bera et al. [1] proposed the notion of group fairness independently, where the difference is that colors are allowed to overlap in [1]. For the group fair k-median problem, Bercea et al. [2] gave a 4.675-approximation algorithm with 1 fairness violation. Bera et al. [1] presented a $(\rho + 2)$-approximation algorithm with $(4\delta + 3)$ violation, where ρ is the approximation ratio of any algorithm for the k-median problem, δ is the maximum number of colors a single point can belong to, respectively.

In this paper, we focus on the k-ML problem under the group fairness constraints, denoted as the group fair k-median of lines (GF-k-ML) problem. In the GF-k-ML problem, we are given a set L of n lines in $\mathbb{R}^d$, two fair vectors $\boldsymbol{\alpha} = \{\alpha_1, \ldots, \alpha_t\}$, $\boldsymbol{\beta} = \{\beta_1, \ldots, \beta_t\}$, where L comprises t disjoint groups $L_1, \ldots, L_t$, and the lines in L_h are colored with color h ($h \in \{1, \ldots, t\}$). The goal is to partition L into k clusters such that the proportion of lines with color h in each cluster is at least β_h and at most α_h, and the k-ML problem objective is minimized.

For the GF-k-ML problem, there exist some obstacles to construct coresets.

- Firstly, the triangle inequality is unsuitable for the GF-k-ML problem. It is known that existing coresets construction algorithms are heavily based on the triangle inequality between pair of points, which is not satisfied for the case of lines. To generalize these algorithms under the case of lines, the algorithms usually transform the point-to-line distance as corresponding point-to-point distance. However, the above idea is hard to extend for the GF-k-ML problem, since the algorithm needs to search for points in a large number of lines to construct the triangle inequality and satisfy the fairness constraints.
- Secondly, the definition of coreset for the k-ML problem is not workable for the group fairness constraints. For clustering of points, the definition of coreset is not universal with (or without) group fairness constraints [23]. Similar situations still exist for the GF-k-ML problem. Moreover, as pointed out in [15], the existing algorithms of coresets for lines do not support groups as input.

1.1 Our Contributions

In this paper, we have the following result for the GF-k-ML problem.

Theorem 1. *Given parameters $\varepsilon, \zeta \in (0, 1)$, there is an $O(tnd^2k \log k \log n + ndk^{O(k)})$ time randomized algorithm that with probability at least $1 - \zeta$, computes an ε-coreset of size $O(\frac{td^2 k \log^2 k \log^2 n + \log(1/\zeta)}{\varepsilon^2})$ for the GF-k-ML problem.*

To sum up, the key challenge for the GF-k-ML problem is how to ensure that the coreset construction algorithm, the triangle inequality, and the sensitivity sampling method are simultaneously applicable to line clustering and group fairness constraints. To overcome these challenges, our algorithm is based on two steps. We first construct a coreset separately for each group of lines with a distinct color, since each color group can be treated as an instance of the k-ML problem, thereby ensuring the validity of the triangle inequality. Then, we combine these group-wise coresets to form a unified coreset and prove that the coreset satisfies composability under fairness constraints, which allows the final coreset to be constructed without explicitly requiring the input to be partitioned into groups.

1.2   Other Related Work

Several variants of the k-ML problem have been extensively studied in the literature. Marom $et\ al.$ [16] introduced a streaming algorithm with communication complexity of $O(d^3 k^{O(k)} \log^2 n)$, which guarantees a $(1+\varepsilon)$-approximation. Lotan $et\ al.$ [15] considered the k-median problem on line-sets, and gave an ε-coreset of size $O(\log^2 n)$ with $O(n \log n)$ construction time. In addition to the k-ML problem, the k-center of lines problem has garnered significant attention, which aims to find k balls to cover a set of lines L such that the radius of the largest ball is minimized. Gao $et\ al.$ [6] studied the 1-line center problem in $\mathbb{R}^d$, and constructed an ε-coreset of size $O(1/\varepsilon)$ in time $O(nd\mathrm{poly}(1/\varepsilon))$, which is the first coreset construction algorithm for the k-center of lines problem. Gao $et\ al.$ [7] presented a $(2+\varepsilon)$-approximation algorithm for the cases $k = 2, 3$ of this problem in $\mathbb{R}^2$, with running time quasi-linear in the number of lines and the dimension of the ambient space. Extensive research has also been conducted on the k-ML problem of Δ-flats [6,14], where Δ is the dimension of affine subspaces.

2   Preliminaries

For any positive integer $m \in \mathbb{N}^{\geq 1}$, let $[m] = \{1, \ldots, m\}$. Given a set $X \subseteq \mathbb{R}^d$ of points, for any $p, q \in X$, the distance between p and q is denoted as $\tau(p, q) = \|p - q\|_2$, where $\|\cdot\|_2$ denotes the Euclidean norm. For any points $p, q, g \in X$ and a constant $\rho > 0$, the Euclidean distance satisfies the weak triangle inequality, i.e., $\tau(p, q) \leq \rho(\tau(p, g) + \tau(g, q))$. Given two set $X, Y \subseteq \mathbb{R}^d$ of points, for any point $p \in X$, the distance from p to Y is defined as the shortest distance from p to any point in Y, i.e., $\tau(p, Y) = \min_{y \in Y} \tau(p, y)$. Further, the distance between X and Y is defined as $\tau(X, Y) = \min_{x \in X, y \in Y} \tau(x, y)$. Given a line ℓ in $\mathbb{R}^d$ and a set $S \subseteq \mathbb{R}^d$ of points, let $\tau(\ell, S) = \min_{s \in S} \tau(\ell, s)$, where $\tau(\ell, s)$ is the shortest distance from any point on the line ℓ to s.

Definition 1. (the k-median of lines $(k$-ML$)$ problem) $Given\ a\ set\ L\ of\ n$ $lines\ in\ \mathbb{R}^d\ and\ an\ integer\ k,\ the\ goal\ is\ to\ find\ a\ subset\ S \subseteq \mathbb{R}^d\ of\ k\ centers$ $such\ that\ the\ cost$

$$\mathrm{cost}(L, S) = \sum_{\ell \in L} \tau(\ell, S)$$

$is\ minimized.$

Given an instance (L, d, k) of the k-ML problem, S is called a feasible solution of this instance if $S \subseteq \mathbb{R}^d$ is a set of k points.

Definition 2. (ε-Coreset for the k-ML problem [15]) $Given\ an\ instance$ $(L, d, k)\ of\ the\ k$-ML $problem\ and\ an\ error\ parameter\ \varepsilon \in (0, 1),\ a\ weighted$ $set\ C \subseteq L\ with\ non\text{-}negative\ weights\ u : C \to \mathbb{N}\ is\ an\ \varepsilon\text{-}coreset\ for\ the\ k$-ML $problem\ if\ for\ every\ set\ S \subseteq \mathbb{R}^d\ of\ k\ points,\ we\ have$

$$\mathrm{cost}_w(C, S) \in (1 \pm \varepsilon) \cdot \mathrm{cost}(L, S),$$

$where\ \mathrm{cost}_w(C, S) = \sum_{\ell \in C} u(\ell) \cdot \tau(\ell, S).$

Formally, in this paper, we consider the following group fair k-median of lines (GF-k-ML) problem.

Definition 3. (The GF-k-ML Problem). *Given a set L of n lines in $\mathbb{R}^d$, an integer k, a set of t colors $\mathcal{H} = \{1, \ldots, t\}$, t disjoint groups $\mathcal{L} = \{L_1, \ldots, L_t\}$ with $\cup_{h=1}^t L_h = L$, and two vector $\boldsymbol{\alpha} = (\alpha_1, \ldots, \alpha_t)$, $\boldsymbol{\beta} = (\beta_1, \ldots, \beta_t)$, where β_h, α_h are the lower and upper bounds on the proportions of group L_h in each cluster, respectively, the goal is to find a subset $S = \{s_1, \ldots, s_k\} \subseteq \mathbb{R}^d$ of k centers, and a mapping $\phi : L \to S$ such that the cost $\sum_{\ell \in L} \tau(\ell, \phi(\ell))$ is minimized, and ϕ satisfies the following group fairness constraints.*

$$\beta_h \leq \frac{|\{\ell \in L_h \mid \phi(\ell) = s_i\}|}{|\{\ell \in L \mid \phi(\ell) = s_i\}|} \leq \alpha_h, \forall i \in [k], h \in \mathcal{H}. \tag{1}$$

Given an instance $(L, d, k, \mathcal{H}, \mathcal{L}, \boldsymbol{\alpha}, \boldsymbol{\beta})$ of the GF-k-ML problem, if a mapping $\phi : L \to S$ satisfies constraint (1), we call ϕ a fair assignment. A pair (S, ϕ) is called a feasible solution if $S \subseteq \mathbb{R}^d$ is a set of k points, and $\phi : L \to S$ is a fair assignment. Further, let $\mathrm{colcost}(L, S)$ denote the minimum cost of a fair assignment of a set L of lines to a set S of k centers.

As pointed out in [23], due to the existence of the group fairness constraints, the Definition 2 of coreset is not workable for the GF-k-ML problem. Therefore, a stronger and complicated notion of coresets for the GF-k-ML problem is needed for satisfying the composability. Following the definition in [23], we now state the coreset satisfying the group fairness constraints. For a given a set $S = \{s_1, \ldots, s_k\}$ of k centers and a set $\mathcal{H} = \{1, \ldots, t\}$ of t colors, a color constraint is a $k \times t$ matrix K with non-negative integer entries, and let K_{ih} denote the entry of matrix K corresponding to row $i \in [k]$ and column $h \in [t]$. Given an instance $\mathcal{I} = (L, d, k, \mathcal{H}, \mathcal{L}, \boldsymbol{\alpha}, \boldsymbol{\beta})$ of the GF-k-ML problem, let $O = \{O_1, \ldots, O_k\}$ be a feasible partition of $\mathcal{I}$. For any $h \in [t]$, let O_i^h denote the set of points with color h in O_i. If for any $i \in [k]$ and $h \in [t]$, $|O_i^h| = K_{ih}$, then we call the partition O satisfies the color constraint K. Further, we redefine the cost of the GF-k-ML problem as $\mathrm{colcost}(L, K, S)$, which is the minimal cost of any clustering satisfying K. If no clustering satisfies K, $\mathrm{colcost}(L, K, S) := \infty$. We now give the formal definition of ε-coreset for the GF-k-ML problem as follows.

Definition 4. (ε-coreset for the GF-k-ML problem) *Given an instance $(L, d, k, \mathcal{H}, \mathcal{L}, \boldsymbol{\alpha}, \boldsymbol{\beta})$ of the GF-k-ML problem and an error parameter $\varepsilon \in (0, 1)$, a weighted set $C \subseteq L$ with non-negative weights $u : C \to \mathbb{N}$ is an ε-coreset for the GF-k-ML problem if for every set $S = \{s_1, \ldots, s_k\} \subseteq \mathbb{R}^d$ of k points and every coloring constraint K, we have*

$$\mathrm{colcost}_w(C, K, S) \in (1 \pm \varepsilon) \cdot \mathrm{colcost}(L, K, S),$$

where $\mathrm{colcost}_w(C, K, S)$ is the minimum value $\sum_{c \in C, s_i \in S} \psi(c, s_i) \cdot \tau(c, s_i)$ over all assignments $\psi : C \times S \to \mathbb{R}_{\geq 0}$ such that (1) For any $c \in C$, $\sum_{s_i \in S} \psi(c, s_i) = u(c)$; (2) For any $s_i \in S$ and $h \in [t]$, $\sum_{c \in L_h} \psi(c, s_i) = K_{ih}$.

We defer the proof of composability of ε-coreset for the GF-k-ML problem to Lemma 2.

Algorithm 1: COL-CORESET

Input: An instance $\mathcal{I} = (L, d, k, \mathcal{H}, \mathcal{L}, \boldsymbol{\alpha}, \boldsymbol{\beta})$ of the GF-k-ML problem,
    parameters c, m, ε

Output: An ε-coreset C of $\mathcal{I}$

1 $C \leftarrow \emptyset$, $\delta \leftarrow cdk \log k$;

2 **for** $h = 1$ **to** t **do**

3       $C_h \leftarrow \emptyset$;

4       $\sigma(\ell) \leftarrow$ SENSITIVITY(L_h, δ);

5   **for** $j = 1$ **to** m **do**

6   Sample a line ℓ from L_h with probability $\frac{\sigma(\ell)}{\sum_{\ell' \in L_h} \sigma(\ell')}$;

7           $u_\ell \leftarrow \frac{\sum_{\ell' \in L_h} \sigma(\ell')}{m\sigma(\ell)}$;

8           $C_h \leftarrow C_h \cup \{\ell\}$;

9       $C \leftarrow C \cup C_h$;

10 **return** C.

---

3   An ε-Coreset Algorithm for the GF-k-ML Problem

In this section, we propose an algorithm, called COL-CORESET, to construct a coreset for the GF-k-ML problem. Recall that the previous algorithms fail to generalize simultaneously to line clustering and group fairness constraints. Our improved strategy is to construct a fair coreset for each group of lines with different colors separately, which is based on an important observation, i.e., the union of these coresets is a coreset for the original instance. The high-level idea of COL-CORESET is as follows. Consider an instance $(L, d, k, \mathcal{H}, \mathcal{L}, \boldsymbol{\alpha}, \boldsymbol{\beta})$ of the GF-k-ML problem. Recall that L is partitioned into t disjoint $\{L_1, \ldots, L_t\}$ groups, where L_h is the set of lines with color $h \in [t]$. COL-CORESET begins by constructing a fair coreset on L_h for each $h \in [t]$. To obtain such a fair coreset, COL-CORESET employs a commonly used sensitivity sampling method [15,16,21], in which a coreset for the k-ML problem is provided. The framework computes the sensitivity for each line (step 4 of COL-CORESET), and try to sample some lines based on the obtained sensitivities, resulting in the final coreset of lines (steps 5-8 of COL-CORESET). Finally, COL-CORESET returns the union of these obtained coresets on each group L_h. Algorithm 1 details the above process.

3.1   The Sensitivity Sampling Method

In this section, we show how to use the sensitivity sampling method to construct coreset, which is based on the general framework in [15,16,21]. Since the sampling distribution of the sensitivity sampling method typically depends on the optimal solution, which is unknown and serves as the objective of clustering. Therefore, SENSITIVITY consists of two phases. In the first phase, SENSITIVITY computes a rough bi-criteria approximation by solving a robust median problem, which serves as a surrogate for sensitivity estimation. In the second phase, it calculates the sensitivities for a given set of lines based on the obtained bi-criteria

---

Algorithm 2: SENSITIVITY

Input: A set L lines in $\mathbb{R}^d$ and a parameter $\delta > 0$
Output: A sensitivity function $\sigma : L \to [0, \infty)$

1 $G \leftarrow \emptyset$, $M \leftarrow L$;
2 **repeat**
3 $\quad$ $E \leftarrow$ sample a set of δ lines from M randomly;
4 $\quad$ $G_E \leftarrow \emptyset$;
5 $\quad$ **for** each line $\ell \in E$ **do**
6 $\quad\quad$ $\lfloor$ $G_E \leftarrow G_E \cup \{p \in \ell \mid \tau(p, \ell') \le \tau(q, \ell'), \forall q \in \ell, \forall \ell' \in E \setminus \{\ell\}\}$;
7 $\quad$ $G \leftarrow G \cup G_E$;
8 $\quad$ $M \leftarrow$ remove the closest $\frac{2|M|}{3}$ lines in M to G_E;
9 **until** $|M| \le 1$;
10 **for** each $s \in G$ **do**
11 $\quad$ $\lfloor$ $L(s) \leftarrow \{\ell \in L \mid \tau(\ell, s) \le \tau(\ell, s'), \forall s' \in G\}$;
12 **for** each $s \in G$ **do**
13 $\quad$ $Q_s \leftarrow \emptyset$;
14 $\quad$ $\mathbb{B}_s \leftarrow \{x \in \mathbb{R}^d \mid \|x - s\| = 1\}$;
15 $\quad$ **for** each $\ell \in L(s)$ **do**
16 $\quad\quad$ $\ell' \leftarrow$ the line parallel to ℓ and passing through s;
17 $\quad\quad$ $p(\ell') \leftarrow$ an arbitrary point of $\ell' \cap \mathbb{B}_s$;
18 $\quad\quad$ $\lfloor$ $Q_s \leftarrow Q_s \cup \{p(\ell')\}$;
19 $\quad$ $\lfloor$ $\sigma \leftarrow$ compute the sensitivities of each line in $L(s)$ based on Q_s;
20 **return** σ.

---

approximate solution. Before presenting our main results, we provide the formal definitions of the bi-criteria approximation and robust median.

Definition 5. $((\kappa, \lambda)$-**approximate solution**) *Given an instance $\mathcal{I} = (L, d, k)$ of the k-ML problem, let $S^* \subseteq \mathbb{R}^d$ be the optimal solution of $\mathcal{I}$. Then, for given parameters $\kappa, \lambda \ge 0$, a set $S \subseteq \mathbb{R}^d$ is called a (κ, λ)-approximate solution of $\mathcal{I}$ if $|S| = \lambda k$ and $\mathrm{cost}(L, S) \le \kappa \cdot \mathrm{cost}(L, S^*)$.*

Definition 6. (robust median *[5]*) *Given a set L of n lines in $\mathbb{R}^d$, a parameter $\gamma \in (0, 1]$, define $\mathrm{cost}^*(L, \gamma) = \min_{c \in X} \sum_{\ell \in closest(L, \{c\}, \gamma)} \tau(\ell, c)$. For $\varepsilon \in (0, 1]$ and $\kappa, \lambda > 0$, a set $Y \subseteq X$ is a $(\gamma, \varepsilon, \kappa, \lambda)$-median of L if $|Y| = \lambda$, and $\sum_{\ell \in closest(L, Y, (1-\varepsilon)\gamma)} \mathrm{cost}(\ell, Y) \le \kappa \cdot d^*(L, \gamma)$.*

3.1.1 (κ, λ)-Approximate Solution

In this section, we prove that the set G returned by step 7 of SENSITIVITY is a (κ, λ)-approximate solution for the given k-ML problem instance $\mathcal{I} = (L, d, k)$. Recall that in the first phase, given a set L lines in $\mathbb{R}^d$ and a parameter $\delta > 0$, SENSITIVITY computes an approximate solution G across multiple refinement iterations. In each iteration, SENSITIVITY randomly samples a subset E of δ

lines from L, and determines a set G_E of points from the lines in E. Here, we introduce a notion about the property of E. Given a set L of lines in $\mathbb{R}^d$, a subset $S \subseteq \mathbb{R}^d$ of k points, and an integer $z > 0$, let $closest(L, S, z)$ denote the set of z closest lines from L to S. Further, we have the following property of E.

Lemma 1. *Given an instance (L, d, k) of the k-ML problem, in one iteration of steps 2–9 of* SENSITIVITY, *let E be the set of lines sampled from L, and G_E be the set of points obtained by E, respectively. Then, for any integer $z \in [n-1]$, and a set $S \subseteq \mathbb{R}^d$ of size k, there exists a set of $S'' \subseteq G_E$ of k points such that*

$$cost(closest(E, S'', z), S'') \le 4\rho^2 \cdot cost(closest(E, S, z), S).$$

Proof. For each line $\ell \in E$, let s be the closest point in S to ℓ, and ℓ' be the closest line in E to s, respectively. Denote by $\pi_\ell(v)$ the projection of a point $v \in \mathbb{R}^d$ onto the line ℓ. We now bound the distance from ℓ to S''. Assume that s'' is the closest point to ℓ in G_E, we have $\tau(\ell, S'') \le \tau(\ell, s'')$. By the result in [16], for any two lines $\ell_1, \ell_2 \in \mathbb{R}^d$ that are not parallel, there exists an algorithm that returns a point $v \in \mathbb{R}^d$ such that for any $p \in \ell_1$, $\tau(p, \ell_2) = w \cdot \tau(p, v)$, where $p \in \ell_1$, and w is a given constant, respectively. By applying this result, we can obtain $\tau(\ell, s'') = w \cdot \tau(v, s'')$, where $v \in \mathbb{R}^d$ and w is a constant, respectively. By the definition of $\pi_\ell(v)$, the weak triangle inequality and the result in [16], we have

$$
\begin{aligned}
\tau(v, s'') &\le \rho(\tau(v, \pi_{\ell'}(s)) + \tau(s'', \pi_{\ell'}(s))) \\
&\le \rho(\tau(v, \pi_{\ell'}(s)) + \tau(\pi_{\ell'}(s), \pi_{\ell'}(v))) \\
&\le 2\rho \cdot \tau(v, \pi_{\ell'}(s)) = 2\rho/w \cdot \tau(\ell, \pi_{\ell'}(s)),
\end{aligned}
$$

and

$$
\begin{aligned}
\tau(\ell, \pi_{\ell'}(s)) = \tau(\pi_\ell(\pi_{\ell'}(s)), \pi_{\ell'}(s)) &\le \tau(\pi_\ell(s), \pi_{\ell'}(s)) \\
&\le \rho(\tau(\pi_\ell(s), s) + \tau(\pi_{\ell'}(s), s)) \\
&\le 2\rho\tau(\ell, s) = 2\rho\tau(\ell, S).
\end{aligned}
$$

Thus, we have $\tau(\ell, S'') \le 4\rho^2 \tau(\ell, S)$. Then, based on the above inequalities and the theorem in [11], we get

$$cost(closest(E, S'', z), S'') \le 4\rho^2 \cdot cost(closest(E, S, z), S).$$

$\square$

Based on Lemma 1 and Definition 6, we can easily observe that for the case of $k = 1$ and $z = \frac{11|E|}{12}$, there exists a set $S'' = \{s'\} \subseteq G_E$ that is a $((1 - \varepsilon)\gamma, \varepsilon, \kappa, 1)$-median of E, where $\gamma = 1$, $\varepsilon = \frac{1}{12}$, and $\kappa = 4\rho^2$, respectively. Then, by the property of robust median problem, we can obtain such an approximate solution based on the general framework of sensitivity sampling [15,16,21]. We now introduce the property of robust median as follows.

Theorem 2. ([15,16]) *Consider a set L of n lines in $\mathbb{R}^d$, an integer $k \geq 1$, parameters $\gamma \in (0,1]$, $\varepsilon, \zeta \in (0, 1/10)$, and $\kappa > 0$. Let $\mathcal{Q} = \{S \subseteq \mathbb{R}^d \mid |S| = k\}$ as the family of all set of k points, and $(L, \mathcal{Q})$ be a lines clustering query space of dimension d_{VC}. Then, there exists an algorithm that samples a set $P \subseteq L$ with a size of $|P| = \frac{c(d_{VC} + \log(\frac{1}{\zeta}))}{\varepsilon^4 \gamma^2}$, and returns a point set $\mathcal{R}(P)$, where c is a sufficiently large constant. Suppose that there exists a point $x \in \mathcal{R}(P)$ being of $((1 - \varepsilon)\gamma, \varepsilon, \kappa, 1)$-median of P, then, with probability at least $1 - \zeta$, x is a $(\gamma, 4\varepsilon, \kappa, \lambda)$-median of L.*

By Theorem 2, we get that with probability at least $1 - \zeta$, G_E is a $(\gamma, 4\varepsilon, \kappa, \lambda)$-median of M, where $\gamma = 1$, $\varepsilon = \frac{1}{3}$, $\kappa = 4\rho^2$ and $\lambda = |\delta|^2$, respectively. Note that δ is given to be $cdk \log k$, since d_{VC} has already been proven to be $dk \log k$ in [16]. By this conclusion, we note that G_E is very close to (κ, λ)-approximate solution, since a (κ, λ)-bicriteria approximation is precisely a $(1, 0, \kappa, \lambda)$-median. However, G_E is not consistent with a (κ, λ)-approximation solution of L, since the clustering result based on G_E contains outliers with a factor of 4ε, where the number of outliers is $4\varepsilon \cdot |L| = \frac{1}{3}|L|$. Therefore, in step 8 of SENSITIVITY, it remains $\frac{2}{3}|M|$ lines for G_E, which ensures that the final set of G constitutes a (κ, λ)-approximation solution of L.

3.1.2 Sensitivity

In this section, we show how to compute the sensitivity of each lines based on G, which is presented in [16] for solving the k-ML problem. Here, we briefly describe the process of computing the sensitivity for each line. Given an instance $\mathcal{I} = (L, d, k)$ of the k-ML problem, SENSITIVITY first assigns each line in L to its closest point in G, and obtains a partition $\{L(s) \mid s \in G\}$ of L, where $L(s)$ denotes the set of lines in L closest to point $s \in G$. Then, for each $s \in G$, SENSITIVITY constructs a unit sphere $\mathbb{B}_s$ centered at s. Further, for each line $\ell \in L(s)$, SENSITIVITY computes the line ℓ' that parallels ℓ and intersects s, and builds a set Q_s to store any intersection of ℓ' and $\mathbb{B}_s$. Finally, SENSITIVITY computes the sensitivities of each line in $L(s)$ based on Q_s, and returns the sensitivity function $\sigma : L \to [0, \infty)$. Based on the above discussion, we have the following theorem about the bound of the total sensitivity.

Theorem 3. ([16]) *Given a set $L = \{\ell_1, \ldots, \ell_n\}$ of n lines in $\mathbb{R}^d$, an integer $k \geq 1$, parameters $\kappa, \lambda > 0$, $\zeta \in (0, 1)$, and let $G \subseteq \mathbb{R}^d$ be a (κ, λ)-approximation of the instance $\mathcal{I} = (L, d, k)$. Then, given G, with a probability at least $1 - \zeta$, the total sensitivity bound is $\kappa \lambda k^{O(k)} \log n$.*

3.1.3 Coreset Construction

In this section, we show how to construct the coreset based on the obtained sensitivities. Here, we briefly describe the process of sensitivity sampling. Given an instance $\mathcal{I} = (L, d, k)$ for the k-ML problem, SENSITIVITY samples some lines with a certain probability that is positively correlated with itself sensitivity, and assigns these lines a weight that is negatively correlated with itself sensitivity. Further, we have the following results.

Theorem 4. ([15,16]) *Consider an instance* (L, d, k) *of the* k-ML *problem, parameters* $\varepsilon, \zeta > 0$ *and let* C_h *be the coreset returned by step 8 of* COL-CORESET. *Then, the size of* C_h *is* $O(\frac{d^2 k \log^2 k \log^2 n + \log(1/\zeta)}{\varepsilon^2})$.

3.2   The Proof of Composability

We now prove that Definition 4 satisfies the composability for the GF-k-ML problem, which is similar to the proof in [23].

Lemma 2. (Composability) *Given two lines set* $L_1, L_2 \subseteq \mathbb{R}^d$, *let* C_1 *and* C_2 *be the* ε-*coresets of* L_1 *and* L_2 *that satisfy Definition 4, respectively. Then,* $C = C_1 \cup C_2$ *is an* ε-*coreset of* $L = L_1 \cup L_2$ *satisfying Definition 4.*

Proof. Let $S \subseteq \mathbb{R}^d$ be a set of k points, and $K \in \mathbb{N}^{k \times t}$ be an coloring constraint, and let $\psi : L \to S$ be a mapping that minimizes the cost colcost(L, S) among all assignments satisfying K, respectively. For $L_1, L_2 \subseteq L$, we define two mappings $\psi_1 : L_1 \to S$ and $\psi_2 : L_2 \to S$, and two coloring constraints K_1 and K_2 by counting the number of lines of each color at each point in S that belongs to L_1 and L_2, respectively. Since ψ satisfies K, that is, the number of lines of color h ($h \in [t]$) assigned to each point in S is exactly K_{it} for $i \in [k]$, we have $K = K_1 + K_2$. Moreover, we have $\psi_1(s) = \psi(s)$ for all $\ell_1 \in L_1$, and $\psi_2(s) = \psi(s)$ for all $\ell_2 \in L_2$.

First, we prove that

$$\text{colcost}(L, S, K) = \text{colcost}(L_1, S, K_1) + \text{colcost}(L_2, S, K_2).$$

Since ψ_1 and ψ_2 satisfy the constraints K_1 and K_2, respectively, and they result in the same cost for each line as ψ, we have

$$\text{colcost}(L, S, K) \leq \text{colcost}(L_1, S, K_1) + \text{colcost}(L_2, S, K_2).$$

Then, assume that there exists an assignment ψ_1' that satisfies

$$\sum_{\ell \in L_1} \tau(\ell, \psi_1'(\ell)) < \text{colcost}(L_1, S, K_1).$$

In this case, replacing ψ_1 with ψ_1' for the lines in L_1 will result in a lower cost, whereas this contradicts with the optimality of ψ. A similar argument holds for ψ_2. Thus, we get

$$\text{colcost}(L, S, K) = \text{colcost}(L_1, S, K_1) + \text{colcost}(L_2, S, K_2).$$

Second, we prove that

$$\text{colcost}_w(C, S, K) \leq (1 + \varepsilon) \cdot \text{colcost}(L, S, K).$$

Since both C_1 and C_2 are coresets for L_1 and L_2 satisfying Definition 4, respectively. Then, we have

$$\text{colcost}_w(C_1, S, K_1) + \text{colcost}_w(C_2, S, K_2)$$
$$\in (1 \pm \varepsilon) \cdot \text{colcost}(L_1, S, K_1) + (1 \pm \varepsilon) \cdot \text{colcost}(L_2, S, K_2)$$
$$\in (1 \pm \varepsilon) \cdot \text{colcost}(L, S, K).$$

Since the optimal assignments for C_1 and C_2 can be combined to form an assignment for C, it follows that

$$\text{colcost}_w(L, S, K) \leq \text{colcost}_w(L_2, S, K_2) + \text{colcost}_w(L_1, S, K_1).$$

Finally, we prove that

$$\text{colcost}_w(C, S, K) \geq (1 - \varepsilon) \cdot \text{colcost}(L, S, K).$$

Let $\psi' : C \to S$ be a mapping that satisfies K with cost $\text{colcost}_w(C, S, K)$. Similarly, let $\psi_1' : L_1 \to S$ and $\psi_2' : L_2 \to S$ be the result of translating ψ' to L_1 and L_2, and split K into K_1' and K_2' according to ψ'. Then, by the same argumentation as above, we get

$$\text{colcost}_w(C, S, K) = \text{colcost}_w(C_1, S, K_1') + \text{colcost}_w(C_2, S, K_2').$$

Furthermore,

$$\text{colcost}_w(C, S, K) = \text{colcost}_w(C_1, S, K_1') + \text{colcost}_w(C_2, S, K_2')$$
$$\geq (1 - \varepsilon)\text{colcost}(P_1, C, K_1') + (1 - \varepsilon)\text{colcost}(P_2, C, K_2')$$
$$\geq (1 - \varepsilon)\text{colcost}(P, C, K).$$

$\square$

Lemma 6 implies that Definition 4 is a suitable definition of coresets for the GF-k-ML problem.

Theorem 5. *Given an instance* $(L, d, k, \mathcal{H}, \mathcal{L}, \boldsymbol{\alpha}, \boldsymbol{\beta})$ *of the* GF-k-ML *problem and parameters* $\varepsilon, \zeta > 0$, *Let* $C = \sum_{h \in [t]} C_h$ *be the set returned by step 9 of* COL-CORESET. *Then, the size of* C *is* $O(\frac{tkd^2 \log^2 k \log^2 n + \log(1/\zeta)}{\varepsilon^2})$.

Theorem 6. *The time complexity of* COL-CORESET *is* $O(tnd^2 k \log k \log n + ndk^{O(k)})$.

Proof. For any color group L_h ($h \in [t]$) with size n_h, we first bound the time complexity of SENSITIVITY. It is easy to get that step 1 and step 4 can be executed in time $O(1)$, step 3 can be executed in time $O(n_h)$, steps 5–6 can be executed in time $O(\delta^2 d^2)$, and step 7 can be executed in time $O(n_h \log n_h)$, respectively. Thus, the first phase of SENSITIVITY can be executed in time $O(\log n_h(\delta^2 d^2 + n_h + n_h \log n_h))$, since there are at most $O(\log n_h)$ iterations.

Steps 10–11 can be executed in time $O(d^2 k \log k n_h \log n_h)$, since the distance between a point and a line can computed with $O(d)$ time. By [16], a sensitivity bound can be computed in time $O(dn_h k^{O(k)})$. Therefore, the time complexity of SENSITIVITY is $O(d^2 k \log k n_h \log n_h + dn_h k^{O(k)})$. The time complexity of COL-CORESET is $O(ktd^2 n \log k \log n + ndk^{O(k)})$. □

By the above discussion, Theorem 1 can be proved.

4   Conclusions

In this paper, we introduce the group fair k-median of lines (GF-k-ML) problem, a new extension of the k-median of lines problem that incorporates group fairness constraints. To ensure fairness, we construct coresets separately for each group defined by sensitive attributes, and show that they can be combined into a unified coreset while preserving fairness guarantees. The algorithm begins by estimating the sensitivity of each line through a bi-criteria approximation derived from a robust median, and uses these sensitivities to sample a weighted subset of lines that constitutes the final coreset. The resulting coreset satisfies fairness constraints and has a size of $O\left(\frac{td^2 k \log^2 k \log^2 n}{\varepsilon^2}\right)$ with an error parameter $\varepsilon \in (0, 1)$, which can be constructed in nearly linear time with respect to n. An important direction for future work is to extend our framework to other structured clustering settings, such as curves, trajectories, or higher-order subspaces, under fairness constraints.

References

1. Bera, S.K., Chakrabarty, D., Flores, N.J., Negahbani, M.: Fair algorithms for clustering. In: Proceedings of the 33rd International Conference on Neural Information Processing Systems, pp. 4954–4965 (2019)
2. Bercea, I.O., et al.: On the cost of essentially fair clusterings. In: Proceedings of the 22nd International Conference on Approximation Algorithms for Combinatorial Optimization Problems and 23rd International Conference on Randomization and Computation, pp. 18:1–18:22 (2019)
3. Buchin, M., Driemel, A., Rohde, D.: Approximating (k, ℓ)-median clustering for polygonal curves. ACM Trans. Algorithms **19**(1), 1–32 (2023)
4. Chierichetti, F., Kumar, R., Lattanzi, S., Vassilvitskii, S.: Fair clustering through fairlets. In: Proceedings of the 31st International Conference on Neural Information Processing Systems, pp. 5036–5044 (2017)
5. Feldman, D., Schulman, L.J.: Data reduction for weighted and outlier-resistant clustering. In: Proceedings of the 23rd Annual ACM-SIAM Symposium on Discrete Algorithms, pp. 1343–1354 (2012)
6. Gao, J., Langberg, M., Schulman, L.J.: Analysis of incomplete data and an intrinsic-dimension helly theorem. Discrete Comput. Geom. **40**, 537–560 (2008)
7. Gao, J., Langberg, M., Schulman, L.J.: Clustering lines in high-dimensional space: Classification of incomplete data. ACM Trans. Algorithms **7**(1), 1–26 (2010)

8. Hansen, P., Brimberg, J., Urošević, D., Mladenović, N.: Solving large p-median clustering problems by primal-dual variable neighborhood search. Data Min. Knowl. Disc. **19**, 351–375 (2009)
9. Hirano, S., Tsumoto, S.: Cluster analysis of time-series medical data based on the trajectory representation and multiscale comparison techniques. In: Proceedings of the 6th International Conference on Data Mining, pp. 896–901 (2006)
10. Ho, G.T., Ip, W., Lee, C.K., Mou, W.: Customer grouping for better resources allocation using GA based clustering technique. Expert Syst. Appl. **39**(2), 1979–1987 (2012)
11. Jubran, I., Feldman, D.: Minimizing sum of non-convex but piecewise log-lipschitz functions using coresets. arXiv preprint arXiv:1807.08446 (2018)
12. Keuper, M., Tang, S., Andres, B., Brox, T., Schiele, B.: Motion segmentation and multiple object tracking by correlation co-clustering. IEEE Trans. Pattern Anal. Mach. Intell. **42**(1), 140–153 (2018)
13. Kumar, A., Sabharwal, Y., Sen, S.: Linear-time approximation schemes for clustering problems in any dimensions. J. ACM **57**(2), 1–32 (2010)
14. Lee, E., Schulman, L.J.: Clustering affine subspaces: hardness and algorithms. In: Proceedings of the 24th Annual ACM-SIAM Symposium on Discrete Algorithms, pp. 810–827 (2013)
15. Lotan, S., Shayda, E.E.S., Feldman, D.: Coreset for line-sets clustering. In: Proceedings of the 36th International Conference on Neural Information Processing Systems, pp. 37363–37375 (2022)
16. Marom, Y., Feldman, D.: k-means clustering of lines for big data. In: Proceedings of the 33rd International Conference on Neural Information Processing Systems, pp. 12817–12826 (2019)
17. Matoušek, J.: On approximate geometric k-clustering. Discrete Comput. Geom. **24**(1), 61–84 (2000)
18. Ommer, B., Malik, J.: Multi-scale object detection by clustering lines. In: Proceedings of the 12th International Conference on Computer Vision, pp. 484–491 (2009)
19. Pan, W., Shen, X., Liu, B.: Cluster analysis: unsupervised learning via supervised learning with a non-convex penalty. J. Mach. Learn. Res. **14**(1), 1865–1889 (2013)
20. Paparrizos, J., Gravano, L.: k-shape: Efficient and accurate clustering of time series. In: Proceedings of the 2015 ACM SIGMOD International Conference on Management of Data, pp. 1855–1870 (2015)
21. Perets, T.: Clustering of lines. Open University of Israel Ra'anana, Israel (2011)
22. Rollet, R., Benie, G., Li, W., Wang, S., Boucher, J.: Image classification algorithm based on the RBF neural network and k-means. Int. J. Remote Sens. **19**(15), 3003–3009 (1998)
23. Schmidt, M., Schwiegelshohn, C., Sohler, C.: Fair coresets and streaming algorithms for fair k-means. In: Proceedings of the 17th International Workshop on Approximation and Online Algorithms, pp. 232–251 (2020)
24. Shi, X., Liang, C., Wang, H.: Multiview robust graph-based clustering for cancer subtype identification. IEEE/ACM Trans. Comput. Biol. Bioinf. **20**(1), 544–556 (2022)

Redefining Entity Integration: Theoretical Insights for GNN-Based Recommender Systems

Yifei Wang[1,2,3], Jiayan Zhu[3], Yao Xu[3], Xin Li[3], and Jiamou Liu[2(✉)]

[1] University of Electronic Science and Technology of China, Chengdu, China
wany107@aucklanduni.ac.nz
[2] School of Computer Science, The University of Auckland, Auckland, New Zealand
jiamou.liu@auckland.ac.nz
[3] Gaode Map, Alibaba Group, Beijing, China
{tongyan.zjy,xuenuo.x,beilai.bl}@alibaba-inc.com

Abstract. Recommender systems utilize Graph Neural Networks (GNNs) to learn vectorized representations of users and items from user-item interactions for predicting recommendations. Recent methods improve recommendations by incorporating item-related entities through a technique known as the Collaborative Knowledge Graph (CKG). However, the theoretical foundation of entity integration remains underexplored, leading to unresolved challenges in maintaining two critical properties for GNN-based recommender models: Local Consistency and the Inclusion of Indispensable Entities. This paper addresses two key research questions: (1) Do CKG-based models align well with these requirements? (2) Can an alternative graph structure better integrate entities into recommender systems? To answer these questions, we analyze CKG-based models and prove their fundamental limitation: they fail to simultaneously satisfy both properties. To resolve this issue, we propose a novel graph structure, the Fusion Graph (FG). We prove a theorem that demonstrates FG-based models meet the requirements of recommender systems. The source code is available at https://github.com/wangyifeibeijing/FGN.

Keywords: Graph-based Recommendation · Graph Neural Networks

1 Introduction

Recommender systems process vast volumes of online content to provide personalized suggestions to users. They play a crucial role in web-based shopping and service platforms such as Amazon and Netflix [20,22]. Graph Neural Network (GNN) models have emerged as the predominant approach for recommender systems [4,17]. These models represent users and items through graph-based modeling of past interactions, generating vectorized representations, also known as embeddings, of nodes (see Sect. 2). Recently, knowledge graphs (KGs), which

F. V. Fomin and M. Xiao (Eds.): COCOON 2025, LNCS 15984, pp. 194–208, 2026.
https://doi.org/10.1007/978-981-95-0218-9_15

contain entities linked to relevant items, have been leveraged as an additional source of information in recommender systems. By incorporating the relationships between items and entities from a KG, GNNs can adjust their learned embeddings based on the relationships between items and entities [14,15].

To achieve this, a standard method constructs a graph-structured data model comprising user, item, and entity nodes. The edges of this graph represent user-item interactions and item-entity relationships, forming what is known as a collaborative knowledge graph (CKG) [16]. This approach is believed to meet the demands of modern recommender systems [2,10,18].

To capture the structure of graph data, a model should ideally assign embedding vectors to nodes that reflect their local structure. Simultaneously, the inclusion of entities aims to introduce additional knowledge that affects the learning process by aligning items that lack direct user interactions. We formalize two desirable properties for GNN models in recommender systems: *Local Consistency* (Definition 1) and *Indispensable Entities* (Definition 2). These properties serve as criteria to evaluate whether a GNN model meets the requirements of recommender systems (see Sect. 2)

Then, a key challenge in this research direction, which this study aims to address, is: What is the optimal way to integrate entity-item relationships into the design of GNN-based recommender systems?

First, it remains unclear how well the CKG-based methods align with the requirements of recommender systems. Specifically, it is unclear how the embeddings learned by CKG-based models preserve local structures and how the presence of entities influences the learned representations. Our **first research question** examines the extent to which CKG-based GNN models align with the objectives of the recommendation tasks. Second, in a CKG, entities are only associated with items, meaning that entity embeddings learned by the model do not explicitly capture entity-user relationships. More fundamentally, the question arises as to whether the CKG structure is the most appropriate framework for integrating knowledge into a data-driven recommendation model. Our **second research question** explores whether a different graph structure can more effectively incorporate knowledge into recommender systems.

To address these research questions, this paper makes the following contributions: **Identifying Limitations of CKG-Based Models.** We investigate the properties of CKG-based models and uncover a critical limitation: These models fail to simultaneously satisfy both local consistency and indispensable entity properties. Our **first contribution** is the theoretical analysis of this issue (Theorem 1), demonstrating why CKG-based models struggle to balance these two properties (see Sect. 4). **Introducing a Novel Graph Structure: Fusion Graph (FG).** To address the limitations of CKG, we propose a new graph-structured data model called the *Fusion Graph (FG)*. Unlike CKG, FG includes edges not only between items and entities (as in CKG) but also between users and entities. Our **second contribution** is the theoretical validation of FG, showing that it enables GNN models to satisfy the fundamental requirements of recommender systems (Theorem 2, see Sect. 5). In addition to our theoretical analysis, we conduct extensive experiments on four real-world datasets, demonstrating that FG-based models outperform CKG-based models in real-world recommendation scenarios (see Sect. 6).

Paper Organization. This paper is structured as follows. Section 2 presents the preliminaries. Section 3 defines the desired properties of GNN-based recommender models. In Sect. 4, we analyze the widely used CKG-based approach and prove Theorem 1, which highlights its limitations. Section 5 introduces a novel graph structure, FG, and presents Theorem 2, demonstrating its suitability for recommender systems. Section 6 provides experimental results that support the effectiveness of FG. Section 7 discusses related work, and finally, Sect. 8 concludes the paper.

2   Preliminaries

The Recommendation Task. In a recommender system, the core task is to analyze each user's interactions with items. These interactions can be represented by a graph. Let $\mathcal{V}_U$ denote the set of vertices corresponding to users, $\mathcal{V}_I$ represent the set of vertices corresponding to items, and $\mathcal{E}^{\text{all}} \subseteq \{\{u, i\} \mid u \in \mathcal{V}_U, i \in \mathcal{V}_I\}$ denote the set of undirected edges representing all the user item interactions. However, only part of these edges is known, I.e., the historical ones, denoted by $\mathcal{E}' \subseteq \mathcal{E}^{\text{all}}$. Thus, the recommendation task is a task to search for a function to model the latent distribution and thus to predict the unknown user-item interactions $\mathcal{E}^{\text{all}} \setminus \mathcal{E}'$. Let $d \in \mathbb{N}^+$ denote the pre-defined vector dimension. This task is done by learning a function $\mathcal{F} : \mathcal{V}_U \cup \mathcal{V}_I \to \mathbb{R}^d$ that maps each user $u \in \mathcal{V}_U$ and item $i \in \mathcal{V}_I$ to a d-dimensional vector, referred to as the *embedding* of the corresponding node. These embeddings are then used to predict user preferences for items. Then the probability of interaction $p(\{u, i\} \in \mathcal{E}^{\text{all}})$ is approximated with a transform function, the activation function, denoted by $\sigma : \mathbb{R} \to [0, 1]$:

$$q(\{u, i\} \in \mathcal{E}^{\text{all}}) := \sigma(\langle \mathcal{F}(u), \mathcal{F}(i) \rangle). \tag{1}$$

Various methods exist for the input of this activation function. In equation (1), we use the inner product as a representative example. Other similar inputs can be applied without affecting the subsequent discussion. Then, let $\mathcal{C}$ be the set of functions denoting the search scope, which is known as the concept set. The learning of $\mathcal{F}$ in recommendation task is formed as:

$$\mathcal{F} = \arg\max_{\mathcal{F}' \in \mathcal{C}} \sum_{\{u,i\} \in \mathcal{E}'} \log q(\{u, i\} \mid \mathcal{F}') = \arg\max_{\mathcal{F}' \in \mathcal{C}} \sum_{\{u,i\} \in \mathcal{E}'} \log \sigma(\langle \mathcal{F}(u), \mathcal{F}(i) \rangle).$$

In addition to the users' interactions with items, modern recommender systems incorporate additional item-related information [2, 10, 14, 15, 18]. This information can include multiple entities, such as the actors or actresses (i.e., entities) who have starred in a movie (i.e., the item) in a movie recommender system. Let $\mathcal{V}_E$ denote the set of vertices corresponding to entities, and $\mathcal{E}^* \subseteq \{\{i, e\} \mid i \in \mathcal{V}_I, e \in \mathcal{V}_E\}$ denote the set of undirected edges between items and their corresponding entities. Thus, the additional information can be represented by $\mathcal{E}^*$.

Graph Neural Networks. In general, the search for an appropriate function $\mathcal{F}$ is restricted to a specific class of functions. A widely used class is Graph Neural Networks (GNNs). For a graph $\mathcal{G} := (\mathcal{V}, \mathcal{E})$, where $\mathcal{V}$ is the set of nodes and $\mathcal{E} \subseteq \mathcal{V} \times \mathcal{V}$ is the set of edges, a GNN model $f : \mathcal{V} \to \mathbb{R}^d$ maps each node $v \in \mathcal{V}$ to a d-dimensional vector, i.e., node embedding of v. The implementation of a GNN model is formally defined as follows:

Graph Attention. Let $A \in \{0, 1\}^{n \times n}$ be the adjacency matrix of the input graph $\mathcal{G}$, where $A_{jk} = 1$ if $(v_j, v_k) \in \mathcal{E}$, and $A_{jk} = 0$ otherwise. The GNN identifies an attention score α_{jk} between nodes v_j and v_k. These scores form the attention matrix $\mathcal{A} \in \mathbb{R}^{n \times n}$, with the (j, k)-th entry being α_{jk}. Intuitively, the attention score α_{jk} indicates how much information from v_k should contribute to the embedding of v_j. There are various methods for defining attention scores, typically based on the existence of an edge between v_j and v_k, along with any trainable parameters. For further details, we refer the reader to the description of graph attention mechanisms (e.g., [12]). Notably, all definitions of attention scores so far satisfy the following conditions:

$$\alpha_{jk} \geq 0 \text{ where equality holds iff } j \neq k \,\&\, A_{jk} = 0,$$
$$\sum_{k=1}^{n} \alpha_{jk} = 1 \text{ for any } j \in \{1, \ldots, n\}. \tag{2}$$

Layered Aggregation. A layered architecture is used to iteratively derive embeddings. Let $\ell \in \mathbb{N}$ denote the predefined number of layers. Let $j \in [1, |\mathcal{V}|]$ be the index of a node, and $h_j^0 \in \mathbb{R}^{d_0}$ represent a trainable vector, which is initialized randomly. For each $t \in [1, \ell]$, given h_j^t at layer t, h_j^{t+1} is computed in two steps:

(1) First, each h_j^t is transformed using a learnable weight matrix $W^t \in \mathbb{R}^{d_t \times d_{t+1}}$ and a non-linear map $\sigma : \mathbb{R}^{d_{t+1}} \to \mathbb{R}^{d_{t+1}}$ to produce a transform vector g_j^t, which represents the message about v_j to be passed to adjacent nodes:

$$g_j^t := \sigma \left(h_j^t W^t \right). \tag{3}$$

(2) Then, the transform vectors are aggregated using the attention scores as:

$$h_j^{t+1} := \alpha_{jj} g_j^t + \sum_{1 \leq i \leq n} \alpha_{ij} g_i^t. \tag{4}$$

Note that by the attention matrix conditions, the second term aggregates the transform vectors of all nodes adjacent to v_j. In this manner, h_j^{t+1} combines information from v_j as well as all nodes in its 1-hop locality.

The final node embedding learned by the GNN is $f(v_j) = h_j^\ell$. For convenience, we define $H^t := (h_1^t, \ldots, h_n^t)$ and $G^t := (g_1^t, \ldots, g_n^t)$ as the GNN-learned vectors if they satisfy the above transformations (3) and (4).

3   Demanded Properties of GNN-Based Recommneder

When a GNN model is applied as the embedding function in the recommendation task, the embeddings of a user u and an item i are denoted as h_u and h_i, respectively. As discussed in Sect. 2, in a recommender system, the inner product $\langle h_u, h_i \rangle$ plays a crucial role in determining recommendation outcomes. A higher $\langle h_u, h_i \rangle$ score indicates a greater likelihood that item i will be recommended to user u. Therefore, generating appropriate embeddings is crucial for the recommender model. By incorporating additional entities into the user-item interaction graph, the GNN model has two objectives: (1) to learn user embeddings such that their inner products with items reflect their interactions, and (2) to learn entity embeddings that are indispensable in increasing the inner products of interacted user-item pairs.

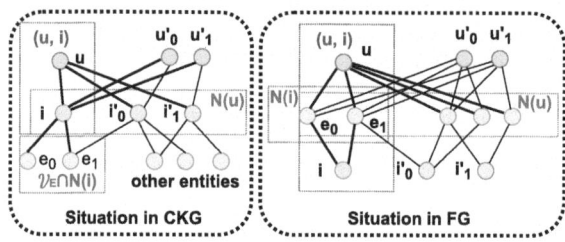

Fig. 1. An example of a (u, i) pair and their corresponding localities in CKG and FG is shown, where u and i represent the target user-item pair, u' denotes the users in the 1-hop neighborhood of i, and i' denotes the items in the 1-hop neighborhood of u. The node e and other yellow nodes represent the entities included in the 1-hop localities of the aforementioned nodes, which influence the learned embeddings of users and items. (Color figure online)

To simplify our analysis and avoid unnecessary complexity, we focus on a simplified and idealized GNN model. This model, although simplified, captures the core functionality of the original GNN and offers valuable insights into the learning mechanism. Here, we examine the process of learning representations for a single user-item pair $(v_u, v_i) \in \mathcal{V}_U \times \mathcal{V}_I$ within a single layer t. For convenience, we drop the superscripts and denote g_j as g_j^{t-1} and h_j as h_j^t.

Assume the GNN model has been optimized to produce h_j and g_j for each node v_j. We define two desirable properties of these vectors. The first property pertains to the optimization of g_j. Since the GNN learns the embedding of each node v_j by aggregating information from its 1-hop neighbors, we expect g_j to resemble the aggregated transformed embeddings of the neighboring nodes. Specifically, let $N(j) := \{k \mid (v_k, v_j) \in \mathcal{E}\}$ represent the 1-hop neighborhood of node v_j. Then, define $\psi_j := \sum_{k \in N(j)} \alpha_{kj} g_k$, which expresses the aggregation of v_j's local neighborhood. The first property stipulates that, in the embedding space, the transformed embedding of v_j should be the closest to ψ_j.

Definition 1 (Locally Consistent). *Let* $H = (h_1, \ldots, h_n)$ *and* $G = (g_1, \ldots, g_n)$ *be the GNN-learned embedding vectors. These vectors are said to be* locally consistent *if, for any node* $v_j \in \mathcal{V}$ *and* $j' \neq j$, *the following condition holds:*

$$\langle \psi_j, g_j \rangle \geq \langle \psi_j, g_{j'} \rangle.$$

The second property asserts that the entities make a non-trivial contribution to learning the embedding of v_i. By applying Eqs. (4) and (2), we obtain the following expressions for h_i and h_u:

$$
\begin{aligned}
h_i &= \alpha_{ii} g_i + \sum_{u' \in N(i) \cap \mathcal{V}_U} \alpha_{u'i} g_{u'} + \sum_{e \in N(i) \cap \mathcal{V}_E} \alpha_{ei} g_e, \\
h_u &= \alpha_{uu} g_u + \sum_{i' \in N(u) \cap \mathcal{V}_I} \alpha_{i'u} g_{i'} + \sum_{e \in N(u) \cap \mathcal{V}_E} \alpha_{eu} g_e.
\end{aligned}
\tag{5}
$$

Thus, by treating the attention scores as constants, we can express h_i and h_u as weighted sums of g_i, g_u' (for $u' \in \mathcal{V}_U$), and g_e (for $e \in \mathcal{V}_E$), where the leading coefficients sum to 1 (as shown in (2)). For a CKG-based model, since no term in h_u contains g_e, the last term in h_i represents the "contribution" of entities. The following definition formalizes the idea that entities are crucial for maximizing objective $\langle h_u, h_i \rangle$, which quantifies likelihood of interaction between u and i.

Definition 2 (Having Indispensable Entities). *Let* $H = (h_1, \ldots, h_n)$ *and* $G = (g_1, \ldots, g_n)$ *be GNN-learned embedding vectors over a graph. We say that the vectors* have indispensable entities *if, for any* $c_u, c_i, d_u, d_i \in \mathbb{R}^+$ *satisfying:*

$$c_u + \sum_{i' \in N(u)} c_{i'} = d_i + \sum_{u' \in N(i)} c_{u'} = 1,$$

the following condition holds:

$$\langle h_u, h_i \rangle > \left\langle \left(c_u g_u + \sum_{i' \in N(u) \cap \mathcal{V}_I} c_{i'} g_{i'} \right), \left(d_i g_i + \sum_{u' \in N(i) \cap \mathcal{V}_U} c_{u'} g_{u'} \right) \right\rangle. \tag{6}$$

4   Limitation of CKG-Based Models

A *collaborative knowledge graph* (CKG) combines the interaction set $\mathcal{E}'$ and the KG edges set $\mathcal{E}^*$ to form a graph:

Definition 3 (Collaborative Knowledge Graph [16]). *A Collaborative Knowledge Graph is the undirected graph:* $\mathcal{G}^\dagger := (\mathcal{V}_U \cup \mathcal{V}_I \cup \mathcal{V}_E, \mathcal{E}' \cup \mathcal{E}^*)$.

CKGs are widely applied in various recommender systems [2,10,14,15,18]. However, there is a lack of theoretical analysis regarding their properties. By combining the definition of CKG with the workflow of GNNs, we observe that a CKG-based GNN model might fail to maintain both local consistency and the presence of indispensable entities:

Theorem 1. *For a CKG, no GNN input with learned embeddings H and G can satisfy both local consistency and the presence of indispensable entities.*

Proof. Notice that for a CKG:

$$N(u) \cap \mathcal{V}_E = \emptyset. \tag{7}$$

From (5) and (7), the CKG-input GNN-learned matrices satisfy:

$$
\begin{aligned}
h_i &= \alpha_{ii} g_i + \sum_{u' \in N(i) \cap \mathcal{V}_U} \alpha_{u'i} g_{u'} + \sum_{e \in N(i) \cap \mathcal{V}_E} \alpha_{e,i} g_e, \\
h_u &= \alpha_{uu} g_u + \sum_{i' \in N(u) \cap \mathcal{V}_I} \alpha_{i'u} g_{i'}.
\end{aligned}
\tag{8}
$$

Assume, for the sake of contradiction, that a certain GNN model with CKG input is both locally consistent and contains indispensable entities. Then, by local consistency, for any entity $e \in \mathcal{V}_E$:

$$\langle \psi_u, g_u \rangle = \sum_{i' \in N(u)} \alpha_{i'u} \langle g_{i'}, g_u \rangle \geq \langle \psi_u, g_e \rangle = \sum_{i' \in N(u)} \alpha_{i'u} \langle g_{i'}, g_e \rangle.$$

At the same time, it is evident that $u \in N(i)$. Since $|g_v| = 1$ for all $v \in \mathcal{V}$, for u, we have: $\langle g_u, g_u \rangle \geq \langle g_u, g_e \rangle$. It then follows that:

$$
\begin{aligned}
\langle h_u, g_u \rangle &= \left(\alpha_{uu} (g_u)^{\mathrm{T}} + \sum_{i' \in N_u} \alpha_{i'u} g_{i'}^{\mathrm{T}} \right) \cdot g_u \\
&\geq \left(\alpha_{uu} (g_u)^{\mathrm{T}} + \sum_{i' \in N(u)} \alpha_{i'u} g_{i'}^{\mathrm{T}} \right) \cdot g_e = \langle h_u, g_e \rangle. \tag{9}
\end{aligned}
$$

By (5), $\langle h_u, h_i \rangle = (h_u)^{\mathrm{T}} \cdot h_i$ can be expressed as the sum:

$$
\begin{aligned}
&\alpha_{ii} \langle h_u, g_i \rangle + \sum_{u' \in \mathcal{V}_U} \alpha_{u'i} \langle h_u, g_{u'} \rangle + \sum_{e \in \mathcal{V}_E} \alpha_{ei} \langle h_u, g_e \rangle \\
&= \alpha_{ii} \langle h_u, g_i \rangle + \alpha_{ui} \langle h_u, g_u \rangle + \sum_{u' \in (\mathcal{V}_U \setminus \{u\})} \alpha_{u'i} \langle h_u, g_{u'} \rangle + \sum_{e \in \mathcal{V}_E} \alpha_{ei} \langle h_u, g_e \rangle.
\end{aligned}
\tag{10}
$$

Consider $\hat{h}_i := \alpha_{ii} g_i + \hat{\alpha}_{ui} g_u + \sum_{u' \in (\mathcal{V}_U \setminus \{u\})} \alpha_{u'i} g_{u'}$, where $\hat{\alpha}_{ui} = \alpha_{ui} + \sum_{u' \in (\mathcal{V}_U \setminus \{u\})} \alpha_{u'i} + \sum_{e \in \mathcal{V}_E} \alpha_{ei}$. We have:

$$\langle h_u, \hat{h}_i \rangle = \hat{\alpha}_{ii} \langle h_u, g_i \rangle + \hat{\alpha}_{ui} \langle h_u, g_u \rangle + \sum_{u' \in (\mathcal{V}_U \setminus \{u\})} \alpha_{u'i} \langle h_u, g_{u'} \rangle. \tag{11}$$

Equation (11) can be transformed into:

$$
\begin{aligned}
&\hat{\alpha}_{ii} \langle h_u, g_i \rangle + \hat{\alpha}_{ui} \langle h_u, g_u \rangle + \sum_{u' \in (\mathcal{V}_U \setminus \{u\})} \alpha_{u'i} \langle h_u, g_{u'} \rangle \\
&= \alpha_{ii} \langle h_u, g_i \rangle + \alpha_{ui} \langle h_u, g_u \rangle + \sum_{u' \in (\mathcal{V}_U \setminus \{u\})} \alpha_{u'i} \langle h_u, g_{u'} \rangle + \sum_{e \in \mathcal{V}_E} \alpha_{ei} \langle h_u, g_u \rangle.
\end{aligned}
\tag{12}
$$

By (9), (10), (11), and (12), we have: $\langle h_u, h_i \rangle \leq \langle h_u, \hat{h}_i \rangle$. In other words, the model does not have indispensable entities. This contradicts the assumption that the model has indispensable entities. Thus, for a CKG, no GNN input with learned embeddings H and G can satisfy both local consistency and the presence of indispensable entities. □

The theorem above highlights the limitation of using CKG as the GNN input: there is an inherent conflict between achieving local consistency in a CKG and maintaining indispensable entities (under certain reasonable assumptions). Specifically, if one were to achieve local consistency for the users—that is, ensuring that the (transformed) embedding vector g_u of a user v_u resembles the items that v_u interacts with—then it becomes necessary for the entities to be dispensable in recommendation tasks. To address this limitation, it is crucial to resolve this conflict. In the next section, we present an alternative method for combining the provided $\mathcal{E}'$ and $\mathcal{E}^*$ so that local consistency and the presence of indispensable entities are not mutually exclusive.

5   Fusion Graph

We make the following observation: the local consistency condition (Def. 1) in a CKG-based GNN model implies aligning the embedding of a user v_u with the embeddings of items with which v_u has interacted. In other words, this model does not explicitly explore the relations between the user set $\mathcal{V}_U$ and the entity set $\mathcal{V}_E$, leading to underutilization of $\mathcal{V}_E$. A potential solution to this limitation is to build node representations that explicitly capture the relations between users and entities. For any user $v_u \in \mathcal{V}_U$, let $\mathcal{N}_{\mathcal{E}'}(u) := \{i \mid (v_u, v_i) \in \mathcal{E}'\}$ denote the set of items that v_u interacts with in the training set. Similarly, for any item $v_i \in \mathcal{V}_I$, let $\mathcal{N}_{\mathcal{E}^*}(i) := \{e \mid (v_i, v_e) \in \mathcal{E}^*\}$ denote the set of entities associated with the item v_i. Now, define the following set:

$$\mathcal{N}^{\ddagger}(u) := \bigcup_{i \in \mathcal{N}_{\mathcal{E}'}(u)} \mathcal{N}_{\mathcal{E}^*}(i).$$

Any entity e such that $e \in \mathcal{N}^{\ddagger}(u)$ is called *relevant* to u. Intuitively, if $e' \notin \mathcal{N}^{\ddagger}(u)$, then no item that has been interacted with by u is associated with the attribute e', and thus e' is irrelevant to user u. The relevance relation between users and entities provides an alternative way to integrate $\mathcal{E}'$ with $\mathcal{E}^*$ in the form of a graph structure[1].

Definition 4 (Fusion Graph). *Construct undirected edges between each user and each of the user's relevant entities, denoted by:* $\mathcal{E}^{\ddagger} := \{\{u, e\} \mid u \in \mathcal{V}_U, e \in \mathcal{N}^{\ddagger}(u)\}$. *A Fusion Graph (FG) is defined as* $\mathcal{G}^{\ddagger} := (\mathcal{V}_U \cup \mathcal{V}_I \cup \mathcal{V}_E, \mathcal{E}^* \cup \mathcal{E}^{\ddagger})$.

Next, to analyze the learning outcome in a GNN model with FG as input, we adopt the same setting as in Sect. 4, focusing on the representations of user-item

---

[1] We introduce a virtual entity for each item that has no related entities, thereby preventing isolated nodes.

interactions $(v_u, v_i) \in \mathcal{V}_U \times \mathcal{V}_I$ (as shown in Fig. 1) at layer t. The following theorem highlights the key difference between CKG- and FG-based models.

Theorem 2. *Over FG, there exist GNN-learned embedding vectors H and G that are both locally consistent and contain indispensable entities.*

Proof. In FG, observe that:

$$N(i) \cap \mathcal{V}_U = N(u) \cap \mathcal{V}_I = \emptyset. \tag{13}$$

From (5) and (13), the GNN-learned matrices over FG satisfy:

$$
\begin{aligned}
h_i &= \alpha_{ii} g_i + \sum_{e \in N(i) \cap \mathcal{V}_E} \alpha_{ei} g_e, \\
h_u &= \alpha_{uu} g_u + \sum_{e \in N(u) \cap \mathcal{V}_E} \alpha_{eu} g_e.
\end{aligned}
\tag{14}
$$

Thus, by (14), GNN-learned embeddings over FG having indispensable entities, as denoted in (6), is equivalent to:

$$\langle h_u, h_i \rangle > \langle g_u, g_i \rangle.$$

We now provide an example of GNN-learned embeddings H and G over FG that are both locally consistent and contain indispensable entities. Let S be the matrix $A + I$, where A is the adjacency matrix of the FG. S is a positive semidefinite symmetric real matrix. Applying Cholesky's decomposition [5], we obtain $Q \in \mathbb{R}^{n \times n}$ such that $QQ^\mathrm{T} = S$. Let g_j denote the row vector $Q_{(j \cdot)}$ of Q for each $1 \leq j \leq n$, and let h_j denote the corresponding embedding vectors obtained by applying (4) to the g_js. Let $H = (h_1, \ldots, h_n)$ and $G = (g_1, \ldots, g_n)$. It is clear that H and G are GNN-learned embeddings over FG. Since $QQ^\mathrm{T} = S$, the inner product between any two nodes v_j, v_k satisfies:

$$\langle g_j, g_k \rangle = Q_{(j \cdot)} \cdot Q_{(k \cdot)}^\mathrm{T} = S_{jk}. \tag{15}$$

We now analyze the properties of these embeddings.

(1) Local consistency. For any node v_j, let $\psi_j = \sum_{k \in N(j)} \alpha_{kj} g_k$. By (15), we have:

$$\langle \psi_j, g_j \rangle = \sum_{k \in N(j)} \alpha_{kj} \langle g_k, g_j \rangle = \sum_{k \in N(j)} \alpha_{kj} S_{kj} = \sum_{k \in N(j)} \alpha_{kj}.$$

For a different node $v_{j'}$ where $j' \neq j$, we have two cases:

Case 1: If $N(j) \setminus N(j') = \emptyset$, then for all $k \in N(j)$, $S_{kj'} = 1$. Thus, we obtain:

$$\langle \psi_j, g_{j'} \rangle = \sum_{k \in N(j)} \alpha_{kj} \langle g_k, g_{j'} \rangle = \sum_{k \in N(j)} \alpha_{kj} S_{kj'} = \sum_{k \in N(j)} \alpha_{kj} = \langle \psi_j, g_j \rangle.$$

Case 2: If $N(j) \setminus N(j') \neq \emptyset$, then for $l \in N(j)/N(j')$, $S_{lj'} = 0$. Thus:

$$
\begin{aligned}
\langle \psi_j, g_{j'} \rangle &= \sum_{k \in N(j)} \alpha_{kj} \langle g_k, g_{j'} \rangle \\
&= \sum_{k \in N(j) \cap N(j')} \alpha_{kj} = \langle \psi_j, g_j \rangle - \sum_{l \in N(j)/N(j')} \alpha_{lj} \leq \langle \psi_j, g_j \rangle.
\end{aligned}
$$

Therefore, for any node $v_j \in \mathcal{V}$ and $j' \neq j$, we have: $\langle \psi_j, g_j \rangle \geq \langle \psi_j, g_{j'} \rangle$. Thus, the model is locally consistent.

(2) Indispensable entities. By Definition 4, there is no edge between u and i. Therefore, when we assign $Q_{(j \cdot)}$ as the embedding of v_j for u and j, we have: $\langle g_u, g_i \rangle = Q_{(u \cdot)} \cdot Q_{(i \cdot)}^{\mathrm{T}} = S_{ui} = 0$. Simultaneously, by (14), we have:

$$
\begin{aligned}
\langle h_u, h_i \rangle = \alpha_{ii} \alpha_{uu} \langle g_u, g_i \rangle + \alpha_{uu} \sum_{e \in N(i)} \alpha_{ei} \langle g_u, g_e \rangle \\
+ \alpha_{ii} \sum_{e \in \mathcal{N}^{\ddagger}(u)} \alpha_{ui} \langle g_i, g_e \rangle + \langle \sum_{e \in N(i)} \alpha_{ei} g_e, \sum_{e \in \mathcal{N}^{\ddagger}(u)} \alpha_{ui} g_e \rangle.
\end{aligned}
\tag{16}
$$

As there is no edge between u and i, we get:

$$
\begin{aligned}
\langle h_u, h_i \rangle = \alpha_{uu} \sum_{e \in N(i)} \alpha_{ei} \langle g_u, g_e \rangle \\
+ \alpha_{ii} \sum_{e \in \mathcal{N}^{\ddagger}(u)} \alpha_{ui} \langle g_i, g_e \rangle + \langle \sum_{e \in N(i)} \alpha_{ei} g_e, \sum_{e \in \mathcal{N}^{\ddagger}(u)} \alpha_{ui} g_e \rangle.
\end{aligned}
\tag{17}
$$

Since there is no edge between different entities and according to the definition of $\mathcal{E}_{\mathsf{UE}}(u)$, we have $N(i) \subseteq \mathcal{E}_{\mathsf{UE}}(u)$. Therefore:

$$
\langle h_u, h_i \rangle = \alpha_{uu} \sum_{e \in N(i)} \alpha_{ei} S_{ue} + \alpha_{ii} \sum_{e \in N(i)} \alpha_{ui} S_{ie} + \sum_{e \in N(i)} \alpha_{ei} \alpha_{eu} S_{ee}.
$$

Thus, by (2): $\langle h_u, h_i \rangle = \sum_{e \in N(i)} (\alpha_{uu} \alpha_{ei} + \alpha_{ii} \alpha_{ui} + \alpha_{ei} \alpha_{eu}) > 0 = \langle g_u, g_i \rangle$. Therefore, the model contains indispensable entities. $\square$

The theorem above shows that, when FG is used in GNN, local consistency and the presence of indispensable entities are no longer mutually exclusive.

6 Experimental Evaluation

We also present an experimental evaluation on real-world recommendation datasets to assess the improvements achieved by the Fusion Graph (FG) approach.

Table 1. Statistics of datasets.

	Amazon-book	LastFM	Yelp-2018	MovieLens
#Users	70,679	23,566	45,919	37,385
#Items	24,915	48,123	45,538	6,182
#Entities	113,487	106,389	136,499	24,536
#Interactions	846,434	1,712,638	1,183,610	237,155

Datasets. We utilize the 4 datasets: **Amazon-book** (1M Amazon book reviews), **LastFM** (20K music listening information), **Yelp2018** (dataset of local business activities), and **MovieLens** (30K user's movie ratings). The statistics are summarized in Table 1. For all datasets, We adopt the same train/test separation as [10]: randomly split 80% of user-item interactions as train set for each user, and for the remaining 20% as test set. We adopt two widely-used metrics to evaluate the performance of top-K recommendation and preference ranking [16]: recall@K and ndcg@K. K is set to 20. The average metrics for all users in test set are reported. All experiments are reported with the average of 5 runs.

Baselines. We compare our proposed FGN with 6 baselines: **Top-popular**, a traditional solution that recommends the most popular items to every user; **BPR-MF** [11], a CF method that factorsizes the user-item rating matrix; **DICE** [21], a de-bias causal model that learns representations with interest and conformity structurally disentangled; **KGCN** [15], an IF-based method utlize GCN encoders on the KG and user-item interaction graph separately; **KGNN-LS** [14], an improved version of KGCN that incorporates a regularization term; **KGAT** [16], a UF-based method firstly introduced the CKG; **Simple-HGN** [10], a CKG-based method with residual connection and multi-relational attention. Table 2 summarizes the results. We observe that FGN consistently achieves the best performance across all four datasets. This demonstrates that by addressing the limitation of CKG, FGN can more effectively improve performance.

To simulate situation when user-item interaction information is limited, we constructed sparse datasets by: the user interaction information in the training

Table 2. Overall Performance Comparison.

	Amazon-book		LastFM		Yelp-2018		MovieLens	
	recall@20	ndcg@20	recall@20	ndcg@20	recall@20	ndcg@20	recall@20	ndcg@20
Top-popular	0.0287	0.0123	0.0121	0.0106	0.0174	0.0110	0.2498	0.1630
DICE	0.1342	0.0731	0.0876	0.08305	0.0598	0.0384	0.3329	0.2124
BPR-MF	0.1300	0.0679	0.0724	0.0617	0.0627	0.0393	0.3992	0.2559
KGCN	0.1464	0.0769	0.0819	0.0705	0.0683	0.0431	0.4237	0.2753
KGNN-LS	0.1448	0.0759	0.0806	0.0695	0.0671	0.0422	0.4218	0.2741
KGAT	0.1507	0.0802	0.0877	0.0749	0.0697	0.0450	0.4532	0.3007
Simple-HGN	0.1587	0.0854	0.0917	0.0797	0.0732	0.0466	0.4618	0.3090
FGN	**0.1670**	**0.0901**	**0.0964**	**0.08357**	**0.0757**	**0.0484**	**0.4661**	**0.3147**

Table 3. Performance Comparison on sparse data. The drop rate is the percentage of user-item edges we cut off from the original dataset.

drop rate	recall@20 on Amazon-book					recall@20 on LastFM				
	10%	30%	50%	70%	90%	10%	30%	50%	70%	90%
KGAT	0.1176	0.1039	0.0989	0.0949	0.0944	0.071	0.0662	0.0639	0.0605	0.0563
Simple-HGN	0.1181	0.1018	0.0926	0.0906	0.0886	0.0754	0.0664	0.0622	0.0586	0.0529
FGN	**0.1283**	**0.1173**	**0.1160**	**0.1092**	**0.1103**	**0.0799**	**0.0691**	**0.0644**	**0.0626**	**0.0598**
%Improv	**8.63%**	**12.9%**	**17.3%**	**15.07%**	**16.84%**	**5.97%**	**4.07%**	**0.7%**	**3.47%**	**6.2%**
drop rate	ndcg@20 on Amazon-book					ndcg@20 on LastFM				
	10%	30%	50%	70%	90%	10%	30%	50%	70%	90%
KGAT	0.0614	0.0535	0.0506	0.0482	0.0478	0.0617	0.0571	0.0548	0.0522	0.0489
Simple-HGN	0.0615	0.0515	0.0467	0.0454	0.0445	0.0657	0.0579	0.054	0.05	0.0454
FGN	**0.0674**	**0.0606**	**0.0600**	**0.0561**	**0.0570**	**0.0696**	**0.0609**	**0.0563**	**0.0538**	**0.0518**
%Improv	**9.59%**	**13.27%**	**18.58%**	**16.39%**	**19.25%**	**5.94%**	**5.18%**	**2.74%**	**3.07%**	**5.93%**
drop rate	recall@20 on yelp2018					recall@20 on Movie-lens				
	10%	30%	50%	70%	90%	10%	30%	50%	70%	90%
KGAT	0.0650	0.0605	0.0571	0.0511	0.0457	0.4304	0.3793	0.32769	0.2437	0.1967
Simple-HGN	0.0665	0.0593	0.0553	0.0478	0.0421	0.4386	0.3870	0.3357	0.2583	0.2481
FGN	**0.0705**	**0.0654**	**0.0608**	**0.0531**	**0.0484**	**0.4412**	**0.3943**	**0.3620**	**0.3226**	**0.3044**
%Improv	**5.90%**	**8.07%**	**6.42%**	**3.99%**	**5.97%**	**0.60%**	**1.87%**	**7.84%**	**24.89%**	**22.68%**
drop rate	ndcg@20 on yelp2018					ndcg@20 on Movie-lens				
	10%	30%	50%	70%	90%	10%	30%	50%	70%	90%
KGAT	0.0419	0.0387	0.0364	0.0321	0.0285	0.2857	0.2513	0.2160	0.1639	0.1259
Simple-HGN	0.0419	0.0376	0.0346	0.0299	0.0265	0.2943	0.2582	0.2214	0.1646	0.1480
FGN	**0.0444**	**0.0414**	**0.0382**	**0.0340**	**0.0305**	**0.2986**	**0.2646**	**0.2240**	**0.2038**	**0.1822**
%Improv	**5.98%**	**7.13%**	**5.16%**	**5.95%**	**6.93%**	**1.48%**	**2.49%**	**1.17%**	**23.78%**	**23.07%**

set is randomly discarded by a certain percentage, and the test dataset remains unchanged. We train FGN and CKG baselines (KGAT, Simple-HGN) on these sparse datasets, and the experimental results are shown in Table. 3. FGN continues to outperform baselines on sparse datasets. As user interaction information decreases, FGN's advantages over baselines expand. This shows that FGN can indeed make better use of KG, which play a greater role facing data sparsity. This is also consistent with the conclusion of Theorem 1.

We also visualize the embeddings of entities of CKG-based and FG-based models. Taking the MoiveLens dataset as an example, we randomly select a user u, and then select 10 u-preferred entities and 10 other entities (noises) from $\mathcal{E}_{\mathsf{UE}}(u)$ according to $p^-(u, e)$. The embeddings of these entities in Simple-HGN (baseline with best performance) and FGN are visualized via T-SNE(Fig. 2). The embeddings of u-preferred entities and others are mixed together in the Simple-HGN model, while in FGN, they can be clearly distinguished. It also indicates that FGN can make entities play a more important role in recommendations.

 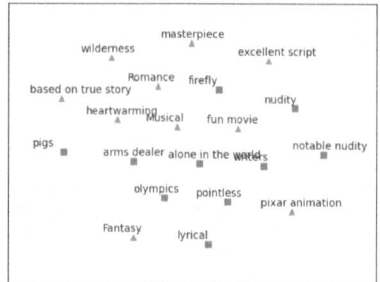

(a) Embeddings learned by Simple-HGN   (b) Embeddings learned by FGN

Fig. 2. The embeddings of entities learned by Simple-HGN and FGN on MoiveLens. The squares represent the u-preferred entities, and the triangles represent others.

7   Related Works

Graph Neural Networks (GNN) and Recommendation. [9] introduced graph into recommendation, by using matrix factorization to project users and items to vector representations, and taking the inner products of these vectors to predict user-item interactions. Subsequently deep learning techniques were introduced to capture nonlinear user-item correlations [4,6,13]. *Graph neural networks* (GNN) emerged in the early 2010s as an effect GRL approach to acquire multi-hop relations between nodes [1,3,8]. These techniques were first introduced to the recommendation task by [17], which applied graph convolutional network (GCN), a typical form of GNN, with CF to process user-item interactions. Other notable work include [7,19].

Collaborative Knowledge Graphs. Recent developments that integrate KG in CF can be classified into two categories, *independent framework* (IF)-based and *unifying framework* (UF)-based: The former handle the user-item interaction graph and the KG separately as two graphs [2,14,15]. The latter encode users, items, and entities in a single CKG to capture high-order interactions. For example, [10,16] used graph attention networks (GAT) [12] on CKG to capture high order interactions between users and items. [16] demonstrates that UF-based methods significantly outperform the IF-based counterparts. In this paper, we aim to investigate the usage of entities in the UF-based methods.

8   Conclusion

In this paper, we provide a theoretical analysis of how to incorporate entity-item information into recommender systems. We prove the limitation of state-of-the-art CKG-based approach. To address this, we propose a new approach, the FG, to overcome this limitation. Theoretical analysis demonstrates the feasibility of FG-based GNN models. A promising direction for future work is to explore more intrinsic goals of recommender systems and formalize them, providing guidance for the development of machine learning-based methods.

Acknowledgements. Yifei Wang is supported by a PhD scholarship from China Scholarship Council.

References

1. Bruna, J., Zaremba, W., Szlam, A., LeCun, Y.: Spectral networks and deep locally connected networks on graphs. In: ICLR (2014)
2. Cao, Y., Wang, X., He, X., Hu, Z., Chua, T.S.: Unifying knowledge graph learning and recommendation: towards a better understanding of user preferences. In: WWW (2019)
3. Hammond, D.K., Vandergheynst, P., Gribonval, R.: Wavelets on graphs via spectral graph theory. Appl. Comput. Harm. Anal. (2011)
4. He, X., Liao, L., Zhang, H., Nie, L., Hu, X., Chua, T.S.: Neural collaborative filtering. In: WWW (2017)
5. Horn, R.A., Johnson, C.R.: Matrix Analysis. Cambridge university press, Cambridge (2012)
6. Hsieh, C.K., Yang, L., Cui, Y., Lin, T.Y., Belongie, S., Estrin, D.: Collaborative metric learning. In: WWW (2017)
7. Jin, B., Gao, C., He, X., Jin, D., Li, Y.: Multi-behavior recommendation with graph convolutional networks. In: SIGIR (2020)
8. Kipf, T.N., Welling, M.: Semi-supervised classification with graph convolutional networks (2016)
9. Koren, Y., Bell, R., Volinsky, C.: Matrix factorization techniques for recommender systems. Computer (2009)
10. Lv, Q., et al.: Are we really making much progress? Revisiting, benchmarking, and refining heterogeneous graph neural networks (2021)
11. Rendle, S., Freudenthaler, C., Gantner, Z., Schmidt-Thieme, L.B.: Bayesian personalized ranking from implicit feedback. In: Proceedings of Uncertainty in Artificial Intelligence (2014)
12. Veličković, P., Cucurull, G., Casanova, A., Romero, A., Liò, P., Bengio, Y.: Graph attention networks. In: ICLR (2018)
13. Wang, H., Wang, N., Yeung, D.Y.: Collaborative deep learning for recommender systems. In: KDD (2015)
14. Wang, H., et al.: Knowledge-aware graph neural networks with label smoothness regularization for recommender systems. In: KDD (2019)
15. Wang, H., Zhao, M., Xie, X., Li, W., Guo, M.: Knowledge graph convolutional networks for recommender systems. In: WWW (2019)
16. Wang, X., He, X., Cao, Y., Liu, M., Chua, T.S.: KGAT: knowledge graph attention network for recommendation. In: KDD (2019)
17. Wang, X., He, X., Wang, M., Feng, F., Chua, T.S.: Neural graph collaborative filtering. In: SIGIR (2019)
18. Wang, X., Wang, D., Xu, C., He, X., Cao, Y., Chua, T.S.: Explainable reasoning over knowledge graphs for recommendation. In: AAAI (2019)
19. Ying, R., He, R., Chen, K., Eksombatchai, P., Hamilton, W.L., Leskovec, J.: Graph convolutional neural networks for web-scale recommender systems. In: KDD (2018)

20. Zheng, G., et al.: DRN: a deep reinforcement learning framework for news recommendation. In: WWW (2018)
21. Zheng, Y., Gao, C., Li, X., He, X., Li, Y., Jin, D.: Disentangling user interest and conformity for recommendation with causal embedding. In: Proceedings of the Web Conference 2021, pp. 2980–2991 (2021)
22. Zhou, G., et al.: Deep interest network for click-through rate prediction. In: KDD (2018)

Tight Gap-Dependent Memory-Regret Trade-Off for Single-Pass Streaming Stochastic Multi-Armed Bandits

Zichun Ye[1], Chihao Zhang[1(✉)], and Jiahao Zhao[2]

[1] Shanghai Jiao Tong University, Shanghai, China
{alchemist,chihao}@sjtu.edu.cn
[2] The University of Hong Kong, Hong Kong, China
zjiahao@connect.hku.hk

Abstract. We study the problem of minimizing gap-dependent regret for single-pass streaming stochastic multi-armed bandits (MAB). In this problem, the n arms are present in a stream, and at most $m < n$ arms and their statistics can be stored in the memory. We establish tight *non-asymptotic* regret bounds regarding all relevant parameters, including the number of arms n, the memory size m, the number of rounds T and $(\Delta_i)_{i \in [n]}$ where Δ_i is the reward mean gap between the best arm and the i-th arm. These gaps are *not* known in advance by the player. Specifically, for any constant $\alpha \geq 1$, we present two algorithms: one applicable for $m \geq \frac{2}{3}n$ with regret at most $\mathcal{O}_\alpha\left(\frac{(n-m)T^{\frac{1}{\alpha+1}}}{n^{1+\frac{1}{\alpha+1}}} \sum_{i:\Delta_i>0} \Delta_i^{1-2\alpha} \right)$[1] and another applicable for $m < \frac{2}{3}n$ with regret at most $\mathcal{O}_\alpha\left(\frac{T^{\frac{1}{\alpha+1}}}{m^{\frac{1}{\alpha+1}}} \sum_{i:\Delta_i>0} \Delta_i^{1-2\alpha} \right)$. We also prove matching lower bounds for both cases by showing that for any constant $\alpha \geq 1$ and any $m \leq k < n$, there exists a set of hard instances on which the regret of any algorithm is $\Omega_\alpha\left(\frac{(k-m+1)T^{\frac{1}{\alpha+1}}}{k^{1+\frac{1}{\alpha+1}}} \sum_{i:\Delta_i>0} \Delta_i^{1-2\alpha} \right)$. This is the first tight gap-dependent regret bound for streaming MAB. Prior to our work, an $\mathcal{O}\left(\sum_{i:\ \Delta_i>0} \frac{\sqrt{T}\log T}{\Delta_i} \right)$ upper bound for the special case of $\alpha = 1$ and $m = \mathcal{O}(1)$ was established by Agarwal, Khanna and Patil (COLT'22). In contrast, our results provide the correct order of regret as $\Theta\left(\frac{1}{\sqrt{m}} \sum_{i:\ \Delta_i>0} \frac{\sqrt{T}}{\Delta_i} \right)$(1 In this paper, the notations $\mathcal{O}_\alpha, \Omega_\alpha, \Theta_\alpha$ subsume a multiplicative factor depending only on α. This is fine since we usually take α to be a constant.).

Keywords: Multi-Armed Bandits · Online Learning · Streaming Algorithms

The full version of the paper is available at https://arxiv.org/abs/2503.02428.

F. V. Fomin and M. Xiao (Eds.): COCOON 2025, LNCS 15984, pp. 209–222, 2026.
https://doi.org/10.1007/978-981-95-0218-9_16

1   Introduction

The stochastic multi-armed bandits (MAB) is a popular T-round game that has been widely studied in online learning. In the game, one player faces n arms. In each round $t \in [T]$, the player chooses an arm A_t among the n arms and gets a reward drawn from a predetermined reward distribution with mean $\mu_{A_t} \in [0,1]$. The arm with the largest reward mean μ_* is called the best arm. The total expected regret is defined as $\mathbf{E}\left[R(T)\right] = \mathbf{E}\left[\sum_{t=1}^{T} \mu_* - \mu_{A_t}\right]$, where the expectation is over the randomness from the player's strategy, and the aim is to minimize $\mathbf{E}\left[R(T)\right]$.

The classic MAB problem defined above has been thoroughly studied. It is known that the expected minimax regret, namely the regret of the best algorithm against the worst input, is $\Theta\left(\sqrt{nT}\right)$ via the *Upper Confidence Bounds* (UCB) algorithm and its variants (see e.g. [3,4,7]). As for the *gap-dependent* regret, the mean gap $\Delta_i := \mu_* - \mu_i$ for each $i \in [n]$ is involved in such bound. The UCB algorithm provides an regret upper bound $\mathcal{O}\left(\sum_{i\in[n]:\,\Delta_i>0} \frac{\log T}{\Delta_i}\right)$ (see e.g. [7,10, 11]).

A recent line of research modeled the MAB problem in the streaming setting to incorporate the situation where the number of arms is huge and cannot be stored in the memory at the same time. In this model, the n arms arrive one by one in a stream, and only $m < n$ arms and their statistics can be stored at the same time. When an arm is read into memory and stored, it can be pulled and its corresponding statistics can be stored. Once an arm is discarded from the memory, all of its information will be forgotten and can never be pulled again. The minimax regret of the problem has recently been settled in [8,15], which is $\Theta\left(\frac{n-m}{n^{\frac{2}{3}}} \cdot T^{\frac{2}{3}}\right)$ for any $2 \leq m \leq n-1$.

However, the gap-dependent regret in this setting has not been fully explored. In fact, the minimax regret bound is derived by considering the worst cases over all choices of possible gaps, and might be much worse compared to the case where the actual gap values are explicitly taken into account. To the best of our knowledge, the only known upper bound is $\mathcal{O}\left(\sum_{i:\Delta_i>0} \frac{\sqrt{T}\log T}{\Delta_i}\right)$, proved in [1], which holds only for $m = O(1)$. This upper bound already suggests that the gap-dependent regret bound can be superior to the minimax regret bound ($\widetilde{\mathcal{O}}(\sqrt{T})$ v.s. $\mathcal{O}\left(T^{\frac{2}{3}}\right)$). However, as mentioned before, it is well known that when no memory constraint is considered, the dependency on T in the gap-dependent bound can be as low as $\log T$. Therefore, it is natural to ask what is the correct regret bound in the memory-constrained setting, and particularly how the memory affects the bound.

On the other hand, since the mean gap Δ_i's are part of the input instance and might depend on T, there might be a trade-off between the dependency on $\frac{1}{\Delta_i}$ and T in the regret bound. To capture this trade-off, we introduce a new parameter $\alpha \geq 1$ and aim at establishing the regret bounds of the form $f(\alpha, m, n, T) \cdot \sum_{i:\,\Delta_i>0} \Delta_i^{1-2\alpha}$ for some function f. Therefore previous gap-dependent bounds, either with or without memory constraint, correspond to the case $\alpha = 1$.

1.1 Our Results

In this work, we design new algorithms and prove matching regret lower bounds for the problem, confirming the gap-dependent regret is

$$\Theta_\alpha \left(\frac{(n-m)T^{\frac{1}{\alpha+1}}}{n^{1+\frac{1}{\alpha+1}}} \sum_{i:\Delta_i>0} \Delta_i^{1-2\alpha} \right) \text{ when } m \geq \tfrac{2}{3}n \text{ and } \Theta_\alpha \left(\frac{T^{\frac{1}{\alpha+1}}}{m^{\frac{1}{\alpha+1}}} \sum_{i:\Delta_i>0} \Delta_i^{1-2\alpha} \right)$$

when $2 \leq m < \tfrac{2}{3}n$ for any constant $\alpha \geq 1$. Our results are summarized in Table 1.

Table 1. Summary of results for streaming MAB

	Regret Bounds	Memory
[15]	$\mathcal{O}\left(n^{\frac{1}{3}} T^{\frac{2}{3}} \right)$	$m = \Theta(\log^* n)$
	$\Omega\left(n^{\frac{1}{3}} T^{\frac{2}{3}} \right)$	$m \leq \frac{n}{20}$
[8]	$\Theta\left(\frac{n-m}{n^{\frac{2}{3}}} \cdot T^{\frac{2}{3}} \right)$	$2 \leq m < n$
[1]	$\mathcal{O}\left(\sum_{i:\Delta_i>0} \frac{\sqrt{T}\log T}{\Delta_i} \right)$	$m = O(1)$
[This work]	$\alpha = 1, \Theta\left(\frac{(n-m)\sqrt{T}}{n^{\frac{3}{2}}} \sum_{i:\Delta_i>0} \frac{1}{\Delta_i} \right)$	$m \geq \frac{2}{3}n$
	$\alpha = 1, \Theta\left(\frac{\sqrt{T}}{m^{\frac{1}{2}}} \sum_{i:\Delta_i>0} \frac{1}{\Delta_i} \right)$	$2 \leq m < \frac{2}{3}n$
[This work]	$\forall \alpha \geq 1, \Theta_\alpha \left(\frac{(n-m)T^{\frac{1}{\alpha+1}}}{n^{1+\frac{1}{\alpha+1}}} \sum_{i:\Delta_i>0} \Delta_i^{1-2\alpha} \right)$	$m \geq \frac{2}{3}n$
	$\forall \alpha \geq 1, \Theta_\alpha \left(\frac{T^{\frac{1}{\alpha+1}}}{m^{\frac{1}{\alpha+1}}} \sum_{i:\Delta_i>0} \Delta_i^{1-2\alpha} \right)$	$2 \leq m < \frac{2}{3}n$

Similar to the minimax regret case, the algorithm for the large memory case ($m \geq \frac{2}{3}n$) differs from that of the small memory case ($m < \frac{2}{3}n$). The reason is that the player continually faces the task of determining which arm to discard from the memory during the game. The task is called the *best arm retention* (BAR) problem and has been recently thoroughly studied [6,8]. It is known that the complexity of the problem is the same as that of the best arm identification (BAI) problem when the memory is small while a more efficient algorithm exists when the memory is large. Therefore, algorithms tailored for both small and large memory are necessary.

Our bounds suggest many interesting behaviors of the model. Taking $\alpha = 1$, our results reveal that with $m = n - 1$, namely with only one unit less memory, the dependency on T significantly increases from $\log T$ to $\sqrt{T}$. Another interesting phenomenon shown by our results is that the regret is not "smooth" in m, as shown in Fig. 1 when $\alpha = 1$[1]. This is in sharp contrast with the minimax regret $\Theta\left(\frac{n-m}{n^{\frac{2}{3}}} \cdot T^{\frac{2}{3}}\right)$ which is smooth with respect to m for all $2 \leq m < n$. Such non-smoothness appears in our lower bound proof in the following manner. We essentially show that give any $\alpha \geq 1$ and any $m \leq k < n$, there exists a set of hard instances on which any algorithm incurs regret $\Omega\left(16^{-\alpha} \cdot \frac{(k-m+1)T^{\frac{1}{\alpha+1}}}{k^{1+\frac{1}{\alpha+1}}} \sum_{i:\Delta_i>0} \Delta_i^{1-2\alpha}\right)$. Therefore, the best lower bound is obtained by optimizing k. For $m < \frac{2}{3}n$, the best choice is $k = \frac{3}{2}m$ while for $m \geq \frac{2}{3}n$, the best choice is $k = n - 1$.

It is helpful to compare our results with the previous ones for $\alpha = 1$. In this case, our bound is $\Theta\left(\frac{n-m}{n^{\frac{3}{2}}} \sum_{i:\,\Delta_i>0} \frac{\sqrt{T}}{\Delta_i}\right)$ when $m \geq \frac{2}{3}n$ and $\Theta\left(\frac{1}{\sqrt{m}} \sum_{i:\,\Delta_i>0} \frac{\sqrt{T}}{\Delta_i}\right)$ when $2 \leq m < \frac{2}{3}n$. It improves the previous best bound in [1], which only holds for $m = \mathcal{O}(1)$, and also fills the blank space in the large memory setting.

1.2   Related Work

The MAB problem was first introduced in [14]. The work [3] proved an optimal regret bound of $\Theta(\sqrt{nT})$ for both stochastic and adversarial cases. Then [12] first took the streaming MAB problem into consideration and obtained an instance-dependent upper bound using $\mathcal{O}(\log T)$ passes and $\mathcal{O}(1)$ memory. The work of [5] gave a generalized upper bound of $\mathcal{O}\left(\frac{n^{\frac{3}{2}}}{m}\sqrt{T\log\frac{T}{nm}}\right)$ for $2 \leq m < n$ in $O(\log T)$ passes. And [2,9] further explores the sample-pass trade-offs. Subsequently, the works of [13] and [15] studied the single-pass scenario and [15] gave tight regret bounds of $\Theta\left(n^{\frac{1}{3}}T^{\frac{2}{3}}\right)$ when $\log^* n \leq m \leq \frac{n}{20}$. The work of [1] provided the first minimax regret lower bound with regard to the number of passes P. They also gave an instance-dependent upper bound of $\mathcal{O}\left(\sum_{i:\Delta_i>0} \frac{\sqrt{T}\log T}{\Delta_i}\right)$ when $m = \mathcal{O}(1)$ and $P = 1$. Very recently, [8] studied the minimax regret bound in the multi-pass setting and obtained tight bounds in terms of m, n, T and P. In particular, they obtained a tight bound of $\Theta\left(\frac{n-m}{n^{\frac{2}{3}}} \cdot T^{\frac{2}{3}}\right)$ in a single pass.

---

[1] Since we only provide non-asymptotic bounds, the curve in Fig. 1 demonstrates the regret bound qualitatively. In fact, the threshold $m = \frac{2}{3}n$ can be replaced by any $m = c \cdot n$ for constant $c \in [\frac{2}{3}, 1)$.

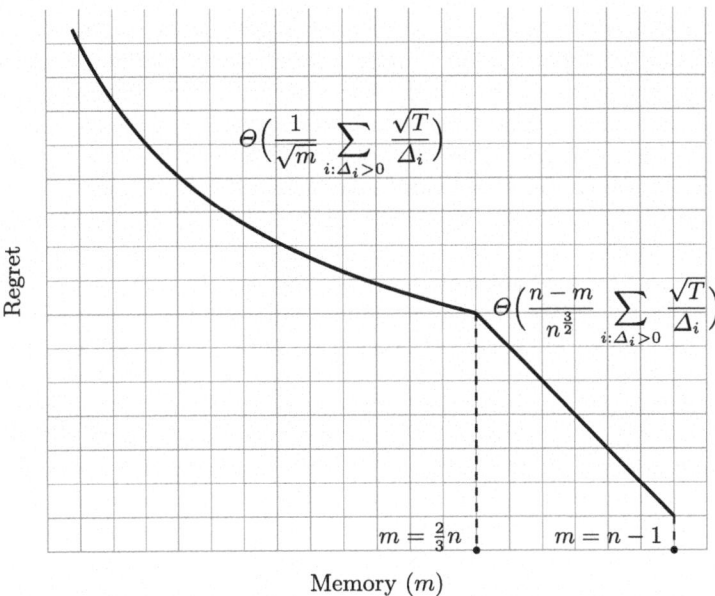

Fig. 1. Regret with respect to the memory size m.

2 Preliminaries

Multi-armed bandit (MAB) We defined the problem of MAB at the beginning of the introduction. Here we introduce some further notations and define the game in detail. We use a mean vector $\nu = (\mu_1, \mu_2, \cdots, \mu_n)$ to denote an instance of MAB in which the i-th arm has a Bernoulli reward $Ber(\mu_i)$. We use $i^* = \arg\max_{i \in [n]} \mu_i$ to denote the index of the arm with maximum mean reward. We assume the choice of i^* is unique. We will sometimes call the i-th arm arm_i and write arm_{i^*} as arm_*. We will also write μ_* instead of μ_{i^*} for convenience.

For every $i \in [n]$, we use a random variable $T_i := \sum_{t=1}^{T} \mathbb{1}\left[A_t = i\right]$ to denote the number of times the i-th arm has been pulled during the game. We use $\mathbb{P}_\nu [\cdot]$ and $\mathbf{E}_\nu[\cdot]$ to denote the probability and expectation of the algorithm running on instance ν.

In the MAB game, the player is given a set of n arms, denoted as $[n]$. Each arm has a reward mean μ_i from a fixed distribution $\mathcal{D}_i$. A T-round decision game starts as follows: in each round $t \in [T]$, the player first selects an arm $A_t \in [n]$ to pull based on the information observed in previous rounds; then the player observes and gains reward of $r_t(A_t) \sim \mathcal{D}_{A_t}$. The player's objective is to minimize the difference between the cumulative reward of the best arm and the player's own cumulative reward. That is, the player aims to design an algorithm $\mathcal{A}$ to minimize the expected regret $\mathbf{E}\left[R(T)\right] = \mathbf{E}\left[\sum_{t=1}^{T} r_t(i^*) - r_t(A_t)\right] = \mathbf{E}\left[\sum_{t=1}^{T} \mu_* - \mu_{A_t}\right]$.

Streaming Stochastic MAB. The streaming MAB model was first formalized in [12]. In this model, the n arm arrives sequentially in the stream and the number of arms that can be stored at the same time is substantially smaller than n. A single pass means every arm in the stream only comes once, that is, if an arm is not stored yet or has been discarded from the memory, it cannot be retrieved later. We consider the worst-case order of the stream.

In each round $t \in [T]$, the player acts in two stages, which include manipulating arm storage (discard arms in the memory, read new arms in the stream) and pulling arm (choose an arm A_t in the memory to pull, observe and gain rewards) respectively. We emphasize that in one round, the player can discard and read any number of arms (including zero) but can only pull exactly once.

Best Arm Retention. To obtain the lower bound, we observe that it is important for an algorithm to pull an arm enough times before discarding it. Otherwise, it can probably discard all good arms. The work of [6] modeled and called it the *best arm retention* (BAR) problem. The BAR problem is to retain m arms out of all n arms after T rounds and make sure that the best arm is not discarded with high probability. In other words, the player has to discard $n - m$ relatively bad arms during the game. A (ε, δ)-PAC algorithm for BAR problem satisfies that for any fixed parameter $\varepsilon, \delta \in (0, 1)$, it retains an ε-best arm in the m arms with probability at least $1 - \delta$.

Gap-Dependent Bound. This paper focuses on gap-dependent regret bounds. Define the gap vector of a certain instance ν as $S(\nu) = (\Delta_i(\nu))_{i \in [n]}$ where $\Delta_i(\nu)$ is the mean gap between the best arm and the i-th arm in the instance ν. For an instance ν of MAB, the gap-dependent regret ideally involves parameters m, n, T and the gap vector $S(\nu)$. In this work, we define the *lower bound* for the gap-dependent regret in the following sense.

Definition 1. *Let m, n be fixed. Denote Π as the set of all possible algorithms with memory size m, and $\mathscr{I}$ as the set of all possible instances with n arms. We say $R(m, n, T, S)$ is a non-asymptotic gap-dependent regret lower bound, if and only if for any algorithm $\mathcal{A} \in \Pi$, there exists $\nu \in \mathscr{I}$ such that $R^{\mathcal{A}, \nu}(T) \geq R(m, n, T, S(\nu))$, for sufficiently large T, where $R^{\mathcal{A}, \nu}(T)$ represents the regret incurred by running $\mathcal{A}$ on ν after T rounds.*

The *non-asymptotic bounud* means that T is a part of the input and therefore Δ_i might depend on T. Compared to the minimax lower bound, $R(m, n, T, S)$ is a functional depending on the gap-vector $S(\nu) = (\Delta_i(\nu))_{i \in [n]}$ where each $\Delta_i(\cdot)$ is a function on the instance. Nevertheless, we will use $R(T)$ as a shorthand for $R(m, n, T, S)$ when no ambiguity arises.

We will use the UCB algorithm as a black box and place the details of it in our full version. We will use the following important property of the UCB algorithm in our analysis.

Lemma 1 ([4]). *Giving n arms which have Bernoulli rewards in the memory and running UCB on them for T rounds, then for any arm $_i$ with $\Delta_i > 0$: $\mathbf{E}[T_i] \leq \frac{8 \log T}{\Delta_i^2}$.*

We will also use the following technical lemma, which is a simple consequence of the Hoeffding's inequality. Proofs of these two lemma are provided in our full version.

Lemma 2. *Let arm_1 and arm_2 be two different arms with reward means of μ and $\mu + \Delta, \Delta > 0$ respectively. Suppose we sample each arm L times and obtain empirical mean of $\widehat{\mu}_1$ and $\widehat{\mu}_2$ respectively, then $\mathbb{p}[\widehat{\mu}_1 \geq \widehat{\mu}_2] \leq e^{-\frac{L\Delta^2}{2}}$.*

3   Gap-Dependent Regret Upper Bound

In this section, we will propose two algorithms, which apply to large m and small m respectively. The philosophy behind the two algorithms is the same: one tries to retain the good arm in the memory and therefore solves a BAR problem in the streaming setting. However, when the memory is large, this can be done more efficiently.

3.1   Large Memory Case ($\frac{2}{3}n \leq m \leq n - 1$)

When $m \geq \frac{2}{3}n$, we can simply read the first m arms into the memory and have enough room to replace parts of them by rest arms in one batch. We simply pick $n - m$ pairs out of the m arms, compare them pairwise, and replace the $n - m$ worst arms with the remaining $n - m$ fresh arms in the stream. After the manipulation of the memory, we apply a standard UCB algorithm for arms in the memory in the remaining rounds.

Algorithm 1. Single-pass algorithm for MAB when $\frac{2}{3}n \leq m \leq n - 1$

Input: Time horizon T, memory size m, number of arms n and a constant $\alpha \geq 1$.

1: Let $L \leftarrow \left(\frac{2\alpha}{e}\right)^{\frac{\alpha}{\alpha+1}} \cdot \left(\frac{T}{n}\right)^{\frac{1}{\alpha+1}}$, $c \leftarrow n - m$;
2: Read in the first m arms;
3: Choose $2c$ arms from the memory u.a.r and denote them by $S = \{s_1, s_2, \ldots, s_{2c}\}$;
4: **for** $i = 1, 2, \ldots, c$ **do**
5:     Pull s_{2i-1}, s_{2i} each L times, calculate their empirical means respectively;
6:     Discard the one with less empirical mean;
7: Read in the remaining c arms in the stream, denote all the m arms in memory as M;
8: Run UCB on M until the game ends;   ▷ *The Exploitation Phase*

The strategy for the analysis of the algorithm is as follows. If the best arm arm_* does not show up in the first m arms, then it must belong to the last M, and thus the UCB algorithm will take care of everything. Otherwise, arm_* has a probability of $\frac{2c}{m}$ to be chosen into S. Then we carefully analyze its probability of being beaten by another sub-optimal arm in L rounds and deduce the bound for regret incurred by this bad event. We obtain the following theorem. The proof is in our full version.

Theorem 1. *Given any $\alpha \geq 1$ and any input instance, assuming T is sufficiently large, Algorithm 1 uses the memory of m arms with expected regret*

$$\mathbf{E}\left[R(T)\right] = \mathcal{O}\left(\left(\frac{2\alpha}{e}\right)^{\frac{\alpha}{\alpha+1}} \frac{(n-m)T^{\frac{1}{\alpha+1}}}{n^{1+\frac{1}{\alpha+1}}} \sum_{i:\Delta_i>0} \Delta_i^{1-2\alpha}\right).$$

3.2   Small Memory Case ($2 \leq m < \frac{2}{3}n$)

When $m < \frac{2}{3}n$, Algorithm 1 is not feasible since the rest $n - m$ arms cannot be read into memory at once. We design Algorithm 2 to deal with this case.

Algorithm 2. Single-pass algorithm for MAB when $2 \leq m < \frac{2}{3}n$

Input: Time horizon T, memory size m, number of arms n and a constant $\alpha \geq 1$.

1: Let $L \leftarrow \left(\frac{2\alpha}{e}\right)^{\frac{\alpha}{\alpha+1}} \cdot \left(\frac{T}{m}\right)^{\frac{1}{\alpha+1}}$;
2: Read the first $m - 1$ arms into memory;
3: Pull each of them L times and calculate their empirical means respectively;
4: **for** each arriving arm arm_i **do**
5:     Choose an arm arm_j u.a.r. in the memory;
6:     Read arm_i into memory;
7:     Pull arm_i L times and calculate its empirical mean $\widehat{\mu}_i$;
8:     **if** $\widehat{\mu}_i > \widehat{\mu}_j$ **then**
9:         Discard arm_j;
10:    **else**
11:        Discard arm_i;
12: Run UCB on all arms in the memory until the game ends; ▷ *The Exploitation Phase*

Algorithm 2 will pull each arm L times before the exploitation phase. Every incoming arm will be compared with a uniformly and randomly chosen arm in the memory. The worse of the two will be discarded to incorporate future new arms. This operation lasts until the end of the stream. Finally, we apply a standard UCB algorithm for arms in the memory until the end. The proof is in our full version.

Theorem 2. *Given any $\alpha \geq 1$ and any input instance, assuming T is sufficiently large, Algorithm 2 uses the memory of m arms with expected regret*

$$\mathbf{E}\left[R(T)\right] = \mathcal{O}\left(\left(\frac{2\alpha}{e}\right)^{\frac{\alpha}{\alpha+1}} \frac{T^{\frac{1}{\alpha+1}}}{m^{\frac{1}{\alpha+1}}} \sum_{i:\Delta_i>0} \Delta_i^{1-2\alpha}\right).$$

4   Gap-Dependent Regret Lower Bounds

In this section, we will prove the regret lower bounds for every $2 \leq m < n$. We first provide our construction of the hard instances in Sect. 4.1. Then we reduce the problem of minimizing regret on these instances to the problem of best arm retention, whose sample complexity lower bounds have recently been established. Here we propose our lower bound of regret.

Theorem 3. *Given any $\alpha \geq 1$, for any integer k satisfying $m \leq k < n$ and any algorithm $\mathcal{A}$, assuming T is sufficiently large, there exists a set of hard instances on which the expected regret of $\mathcal{A}$ is $\Omega\left(16^{-\alpha} \cdot \frac{(k-m+1)T^{\frac{1}{\alpha+1}}}{k^{1+\frac{1}{\alpha+1}}} \sum_{i:\Delta_i>0} \Delta_i^{1-2\alpha}\right)$.*

By picking $k = \frac{3}{2}m$ when $m < \frac{2}{3}n$ and $k = n - 1$ otherwise, we obtain

Corollary 1. *Given any $\alpha \geq 1$, for any algorithm $\mathcal{A}$ using $2 \leq m < n$ memory on n arms, assuming T is sufficiently large, there always exists a set of hard instances on which the expected regret of $\mathcal{A}$ is $\Omega\left(16^{-\alpha} \cdot \frac{T^{\frac{1}{\alpha+1}}}{m^{\frac{1}{\alpha+1}}} \sum_{i:\Delta_i>0} \Delta_i^{1-2\alpha}\right)$ when $m < \frac{2}{3}n$ and $\Omega\left(16^{-\alpha} \cdot \frac{(n-m)T^{\frac{1}{\alpha+1}}}{k^{1+\frac{1}{\alpha+1}}} \sum_{i:\Delta_i>0} \Delta_i^{1-2\alpha}\right)$ when $m \geq \frac{2}{3}n$.*

4.1 The Construction of the Hard Instances

In this section, we provide the family of hard instances and prove in the next section that any algorithm exhibits large regret on at least *one of* the instances in the family.

Given a constant $\alpha \geq 1$, let T be sufficiently large and $\varepsilon = \frac{1}{4} \cdot \left(\frac{k}{T}\right)^{\frac{1}{2+2\alpha}}$. For every $m \leq k < n$, the hard instances for our problem are as follows:

$$\mathscr{I}: \begin{cases} \nu_1 = \left(\underbrace{\frac{1}{2} + n\varepsilon, \frac{1}{2} + (n-1)\varepsilon, \frac{1}{2} + (n-1)\varepsilon, \cdots, \frac{1}{2} + (n-1)\varepsilon}_{k \text{ arms}}, \frac{1}{2}, \cdots, \frac{1}{2}\right) \\ \nu_2 = \left(\frac{1}{2} + n\varepsilon, \frac{1}{2} + (n+1)\varepsilon, \frac{1}{2} + (n-1)\varepsilon, \cdots, \frac{1}{2} + (n-1)\varepsilon, \frac{1}{2}, \cdots, \frac{1}{2}\right) \\ \quad\vdots \\ \nu_k = \left(\frac{1}{2} + n\varepsilon, \frac{1}{2} + (n-1)\varepsilon, \cdots, \frac{1}{2} + (n-1)\varepsilon, \frac{1}{2} + (n+1)\varepsilon, \frac{1}{2}, \cdots, \frac{1}{2}\right) \end{cases}$$

$$\mathscr{I}': \begin{cases} \nu_1' = \left(\underbrace{\frac{1}{2} + n\varepsilon, \frac{1}{2} + (n-1)\varepsilon, \frac{1}{2} + (n-1)\varepsilon, \cdots, \frac{1}{2} + (n-1)\varepsilon}_{k \text{ arms}}, 1, \cdots, 1\right) \\ \nu_2' = \left(\frac{1}{2} + n\varepsilon, \frac{1}{2} + (n+1)\varepsilon, \frac{1}{2} + (n-1)\varepsilon, \cdots, \frac{1}{2} + (n-1)\varepsilon, 1, \cdots, 1\right) \\ \quad\vdots \\ \nu_k' = \left(\frac{1}{2} + n\varepsilon, \frac{1}{2} + (n-1)\varepsilon, \cdots, \frac{1}{2} + (n-1)\varepsilon, \frac{1}{2} + (n+1)\varepsilon, 1, \cdots, 1\right) \end{cases}$$

In other words, we have two families of instances, $\mathscr{I}$ and $\mathscr{I}'$, each consisting of k instances. The first arm of each instance has $\mathsf{Ber}\left(\frac{1}{2} + n\varepsilon\right)$ reward. For each $2 \leq i \leq k$, the i-th arm of ν_i and ν_i' has reward $\mathsf{Ber}\left(\frac{1}{2} + (n+1)\varepsilon\right)$ and the remaining arms among arm_2 to arm_k has reward $\mathsf{Ber}\left(\frac{1}{2} + (n-1)\varepsilon\right)$. The difference between ν_i and ν_i' is the last $n - k$ arms where in ν_i they all have reward $\mathsf{Ber}\left(\frac{1}{2}\right)$ while in ν_i' the rewards are $\mathsf{Ber}(1)$.

We will derive lower bounds for our streaming algorithm from the sample complexity lower bounds for the BAR problem. We also specify our hard instances for the BAR problem. Let $\nu_i[1:k]$ represent the first k arms in ν_i and let $\rho_i = \nu_i[1:k]$, that is

$$\rho_i = \Big(\frac{1}{2} + n\varepsilon, \frac{1}{2} + (n-1)\varepsilon, \cdots, \underbrace{\frac{1}{2} + (n+1)\varepsilon}_{\text{the } i\text{-th arm}}, \cdots, \frac{1}{2} + (n-1)\varepsilon\Big).$$

We construct the set of instances $\mathscr{I}_0 = \{\rho_i = \nu_i[1:k], i \in [k]\}$. Then we have the following sample complexity for BAR on $\mathscr{I}_0$.

Lemma 3 (implicitly in[6]). *For any (ε, δ)-PAC algorithm for the BAR problem on instances in $\mathscr{I}_0$ such that $\varepsilon \leq \frac{1}{8(n-1)}$ and $\delta \leq \frac{k-m}{k}(1-\beta)$, where $\beta \in (0,1)$ is an arbitrary constant, its sample times T on the input ρ_1 satisfies $\mathbf{E}_{\rho_1}[T] \geq \frac{\beta}{32} \cdot \frac{k-m-\delta}{\varepsilon^2} \log \frac{k-m-\delta}{(k-1)\delta}$.*

The same sample complexity lower bound has been proved in [6] for a similar hard instance family and their proof can be adapted to our hard instances. We provide a proof of Lemma 3 in our full version for the sake of self-containment.

4.2   The Proof of the Lower Bounds

For every instance $\nu \in \mathscr{I} \cup \mathscr{I}'$, the execution of an algorithm can be divided into two stages: rounds before reading the $(k+1)$-th arm and the rounds of and after reading the $(k+1)$-th arm. We call them stage one and stage two respectively. We use L_1 and L_2 To Denote The number of rounds for the two stages and use R_1 and R_2 to denote the regret incurred in the two stages respectively. Therefore $L_1 + L_2 = T$. Do not confuse the notations here with ones in the upper bound proof since we are dealing with any algorithm here.

Our lower bounds proof is by formalizing the following dilemma faced by each algorithm: For instances in $\mathscr{I}$,

- L_1 cannot be too large, otherwise, their counterparts in $\mathscr{I}'$ will incur large R_1,
- and L_1 cannot be too small either since otherwise the algorithm cannot identify the best arm among the first k arms, which will cause R_2 large for some instances.

For the second point above, we reduce the lower bound to the sample complexity of the best arm retention problem.

Let $f(k,m) = \frac{2}{16^{\alpha+1}} \cdot \frac{k-m+1}{k^{\frac{1}{\alpha+1}}}$. If there exists an algorithm $\mathcal{A}$ with regret at

most $\frac{2}{16^{\alpha+1}} \cdot \frac{(k-m+1)T^{\frac{1}{\alpha+1}}}{k^{1+\frac{1}{\alpha+1}}} \sum_{i:\ \Delta_i>0} \Delta_i^{1-2\alpha}$, since $\alpha \geq 1$, then it holds that

$$\mathbf{E}_{\nu_1}[R(T)] \leq \frac{2}{16^{\alpha+1}} \cdot \frac{(k-m+1)T^{\frac{1}{\alpha+1}}}{k^{1+\frac{1}{\alpha+1}}} \left((k-1)\varepsilon^{1-2\alpha} + (n-k)(n\varepsilon)^{1-2\alpha}\right)$$

$$\leq \frac{2}{16^{\alpha+1}} \cdot \frac{(k-m+1)T^{\frac{1}{\alpha+1}}}{k^{1+\frac{1}{\alpha+1}}} \cdot k\varepsilon^{1-2\alpha} = f(k,m) \cdot T^{\frac{1}{\alpha+1}} \cdot \varepsilon^{1-2\alpha},$$

and for any $2 \leq i \leq k$,

$$\mathbf{E}_{\nu_i}[R(T))] \leq \frac{2}{16^{\alpha+1}} \cdot \frac{(k-m+1)T^{\frac{1}{\alpha+1}}}{k^{1+\frac{1}{\alpha+1}}} \left(\varepsilon^{1-2\alpha} + (k-2)(2\varepsilon)^{1-2\alpha}\right.$$

$$\left. + (n-k)((n+1)\varepsilon)^{1-2\alpha}\right)$$

$$\leq \frac{2}{16^{\alpha+1}} \cdot \frac{(k-m+1)T^{\frac{1}{\alpha+1}}}{k^{1+\frac{1}{\alpha+1}}} \cdot k\varepsilon^{1-2\alpha} = f(k,m) \cdot T^{\frac{1}{\alpha+1}} \cdot \varepsilon^{1-2\alpha},$$

Similarly for any $\nu' \in \mathscr{I}'$ and sufficiently large T,

$$\mathbf{E}_{\nu'}[R(T)] \leq \frac{2}{16^{\alpha+1}} \cdot \frac{(k-m+1)T^{\frac{1}{\alpha+1}}}{k^{1+\frac{1}{\alpha+1}}} \cdot k\left(\frac{1}{2} - (n+1)\varepsilon\right)^{1-2\alpha}$$

$$\leq \frac{2}{16^{\alpha+1}} \cdot \frac{(k-m+1)T^{\frac{1}{\alpha+1}}}{k^{\frac{1}{\alpha+1}}} \cdot \left(\frac{1}{4}\right)^{1-2\alpha} = \frac{1}{32} \cdot f(k,m) \cdot T^{\frac{1}{\alpha+1}}.$$

The above discussions are summarized below.

Lemma 4. *If there exists an algorithm $\mathcal{A}$ with regret at most*

$$\frac{2}{16^{\alpha+1}} \cdot \frac{(k-m+1)T^{\frac{1}{\alpha+1}}}{k^{1+\frac{1}{\alpha+1}}} \sum_{i:\ \Delta_i>0} \Delta_i^{1-2\alpha}$$

on each instance in $\mathscr{I} \cup \mathscr{I}'$, then for every $\nu \in \mathscr{I}$, $\nu' \in \mathscr{I}'$,

$$\mathbf{E}_\nu[R(T)] \leq f(k,m)T^{\frac{1}{\alpha+1}}\varepsilon^{1-2\alpha}, \quad \mathbf{E}_{\nu'}[R(T)] \leq \frac{1}{32} \cdot f(k,m) \cdot T^{\frac{1}{\alpha+1}} \quad (1)$$

where the randomness in the expectation is from both the (possible) randomness of $\mathcal{A}$ and the randomness of the instance.

Provided the upper bound on the regret, we now show that L_1 cannot be too large.

Lemma 5. *If there exists an algorithm $\mathcal{A}$ satisfying Eq. (1), then for every $i \in [k]$, $\mathbf{E}_{\nu_i}[L_1] \leq \frac{1}{8}f(k,m)T^{\frac{1}{\alpha+1}}$.*

Proof. Note that the random variable L_1 only depends on the performance of the first k arms in the stream, and the first k arms of ν_i are the same as that of ν_i'. Therefore $\mathbf{E}_{\nu_i'}[L_1] = \mathbf{E}_{\nu_i}[L_1]$. Each pull of the first k arms in ν_i' incurs a regret at least $\frac{1}{2} - (n+1)\varepsilon$. So we have

$$\mathbf{E}_{\nu_i'}[L_1] \cdot \left(\frac{1}{2} - (n+1)\varepsilon\right) \leq \mathbf{E}_{\nu_i'}[R] \leq \frac{1}{32} \cdot f(k,m) \cdot T^{\frac{1}{\alpha+1}}.$$

This implies that

$$\mathbf{E}_{\nu_i}[L_1] = \mathbf{E}_{\nu_i'}[L_1] \leq \frac{\frac{1}{32}f(k,m)T^{\frac{1}{\alpha+1}}}{(1/2 - (n+1)\varepsilon)} \leq \frac{1}{8} \cdot f(k,m) \cdot T^{\frac{1}{\alpha+1}},$$

for sufficiently large T.

On the other hand, we prove the following lemma, justifying that for some ν_i, $\mathbf{E}_{\nu_i}[L_1]$ cannot be small provided the algorithm has small regret.

Lemma 6. *If there exists an algorithm $\mathcal{A}$ satisfying Eq. (1), then, $\mathbf{E}_{\nu_1}[L_1] > 8 \cdot f(k,m) \cdot T^{\frac{1}{\alpha+1}}$.*

Clearly Theorem 3 holds by combining Lemma 4, Lemma 5 and Lemma 6.

The remaining part of the section devotes to a proof of Lemma 6. For every $i \in [k]$, we let τ_i be the event that "the arm i is not in the memory at the beginning of stage two". We now show that for every $\nu_i \in \mathscr{I}$, provided the algorithm $\mathcal{A}$ has small regret, $\mathbb{P}_{\nu_i}[\tau_i]$ is small.

Lemma 7. *For every $i \in [k]$ and sufficiently large T, it holds that $\mathbb{P}_{\nu_i}[\tau_i] \leq \frac{1}{2} \cdot \frac{k-m+1}{k}$.*

Proof. For every $i \in [k]$, it follows from Lemma 5 that

$$\mathbf{E}_{\nu_i}[L_2] = T - \mathbf{E}_{\nu_i}[L_1] \geq T - \frac{1}{8}f(k,m)T^{\frac{1}{\alpha+1}} \geq \frac{T}{4}.$$

Note that once τ_i happens on ν_i, each pull in stage two incurs a regret at least ε. Therefore, for every $i \in [k]$,

$$\mathbf{E}_{\nu_i}[R] \geq \mathbb{P}_{\nu_i}[\tau_i]\,\mathbf{E}_{\nu_i}[R_2|\tau_i] \geq \mathbb{P}_{\nu_i}[\tau_i] \cdot \frac{T}{4} \cdot \varepsilon.$$

It then follows from Eq. (1) that for sufficiently large T,

$$\mathbb{P}_{\nu_i}[\tau_i] \leq 4f(k,m)T^{-\frac{\alpha}{\alpha+1}}\varepsilon^{-2\alpha} = \frac{1}{2} \cdot \frac{k-m+1}{k}.$$

Now we claim that the algorithm $\mathcal{A}$ can be used to solve the best arm retention problem with instances in $\mathscr{I}_0$ that retain $m-1$ arms out of k arms: Given an instance for BAR in $\mathscr{I}_0$, we put them in a stream and call the algorithm $\mathcal{A}$.

At the end of stage one, we output all the arms in the memory. Note that the memory is $m - 1$ instead of m here since we do not count the one retaining the $(k + 1)$-th arm in the stream.

Clearly, the algorithm above is a (δ, ε)-PAC algorithm with $\delta = \mathbb{P}_{\nu_i}[\tau_i] \leq \frac{1}{2} \cdot \frac{(k-m+1)}{k}$ on these instances. On the other hand, by picking $\beta = \frac{1}{2}$, it follows from Lemma 3 that

$$
\begin{aligned}
\mathbf{E}_{\nu_1}[L_1] &\geq \frac{k - m + 1 - \delta}{64\varepsilon^2} \log \frac{k - m + 1 - \delta}{(k-1)\delta} \\
&\geq \frac{(k - m + 1) - \frac{(k-m+1)}{2k}}{64\varepsilon^2} \log \frac{(k - m + 1) - \frac{(k-m+1)}{2k}}{(k - 1) \cdot \frac{(k-m+1)}{2k}} \\
&= \left(\frac{2k - 1}{k} \log \frac{2k - 1}{k - 1} \cdot 16^\alpha \right) f(k, m) T^{\frac{1}{\alpha+1}} > 8 \cdot f(k, m) \cdot T^{\frac{1}{\alpha+1}}.
\end{aligned}
$$

Acknowledgement. The authors would like to thank Yuchen He for the help at various stages of the work.

References

1. Agarwal, A., Khanna, S., Patil, P.: A sharp memory-regret trade-off for multi-pass streaming bandits. In: Conference on Learning Theory (COLT 2022), pp. 1423–1462 (2022)
2. Assadi, S., Wang, C.: The best arm evades: near-optimal multi-pass streaming lower bounds for pure exploration in multi-armed bandits. In: Conference on Learning Theory (COLT 2024), pp. 311–358 (2024)
3. Audibert, J.Y., Bubeck, S.: Minimax policies for adversarial and stochastic bandits. In: Conference on Learning Theory (COLT 2009), pp. 217–226 (2009)
4. Auer, P.: Finite-Time Analysis of the Multiarmed Bandit Problem. Kluwer Academic Publishers (2002)
5. Chaudhuri, A.R., Kalyanakrishnan, S.: Regret minimisation in multi-armed bandits using bounded arm memory. In: AAAI Conference on Artificial Intelligence (AAAI 2020), pp. 10085–10092 (2020)
6. Chen, H., He, Y., Zhang, C.: On the problem of best arm retention. In: International Joint Conference on Theoretical Computer Science – Frontier of Algorithmic Wisdom (IJTCS-FAW 2024), pp. 1–20 (2024)
7. Garivier, A., Cappé, O.: The KL-UCB algorithm for bounded stochastic bandits and beyond. In: Conference on Learning Theory (COLT 2011), pp. 359–376 (2011)
8. He, Y., Ye, Z., Zhang, C.: Understanding memory-regret trade-off for streaming stochastic multi-armed bandits. In: Proceedings of the ACM-SIAM Symposium on Discrete Algorithms (SODA 2025), pp. 3450–3485 (2025)
9. Karpov, N., Wang, C.: Nearly tight bounds for exploration in streaming multi-armed bandits with known optimality gap. In: AAAI Conference on Artificial Intelligence (AAAI 2025) (to appear)
10. Kaufmann, E., Cappé, O., Garivier, A.: On Bayesian upper confidence bounds for bandit problems. In: International Conference on Artificial Intelligence and Statistics (AISTATS 2012), pp. 592–600 (2012)

11. Lattimore, T., Szepesvári, C.: Bandit Algorithms. Cambridge University Press (2020)
12. Liau, D., Song, Z., Price, E., Yang, G.: Stochastic multi-armed bandits in constant space. In: International Conference on Artificial Intelligence and Statistics (AISTATS 2018), pp. 386–394 (2018)
13. Maiti, A., Patil, V., Khan, A.: Multi-armed bandits with bounded arm-memory: near-optimal guarantees for best-arm identification and regret minimization. Adv. Neural. Inf. Process. Syst. **34**, 19553–19565 (2021)
14. Robbins, H.: Some aspects of the sequential design of experiments. Bull. Am. Math. Soc. **58**, 527–535 (1952)
15. Wang, C.: Tight regret bounds for single-pass streaming multi-armed bandits. In: International Conference on Machine Learning (ICML 2023), pp. 35525–35547 (2023)

A Robust Distributed Minimax Learning Method Against Model Poisoning Attacks

Tingting Zhang[1], Yuan Yuan[2](✉), Xiao Zhang[1], Yifei Zou[1], Zhipeng Cai[3], and Dongxiao Yu[1]

[1] School of Computer Science and Technology, Shandong University, Qingdao 266237, China
zhang_tingting@mail.sdu.edu.cn, {xiaozhang,yfzou,dxyu}@sdu.edu.cn
[2] School of Software and Joint SDU-NTU Centre for Artificial Intelligence Research (C-FAIR), Shandong University, Jinan 250100, China
yyuan@sdu.edu.cn
[3] Department of Computer Science, Georgia State University, Atlanta, GA 30303, USA
zcai@gsu.edu

Abstract. Distributed minimax optimization plays a crucial role in modern machine learning. However, this framework is highly susceptible to adversarial attacks, especially model poisoning attacks, where malicious participants inject harmful updates to compromise the learning process. Existing defense mechanisms primarily focus on traditional distributed learning in minimization problems. To address this challenge, we propose Robust-LSGDA, a novel distributed minimax learning algorithm designed to defend against model poisoning attacks. Our algorithm achieves an asymptotically optimal convergence rate of $\mathcal{O}\left(\frac{1}{\sqrt{KT}}\right)$, where K is the number of clients and T is the global maximum number of iterations. Furthermore, we establish a robustness guarantee for Robust-LSGDA by analyzing its certified radius, providing theoretical assurance of its defense capabilities.

Keywords: Distributed Minimax Optimization · Model Poisoning · Certified Radius

1 Introduction

Distributed minimax optimization has been attracting increasing attention since numerous modern machine learning applications can be formulated as a minimax

This work was supported in part by Joint Key Funds of National Natural Science Foundation of China under Grant U23A20302 and U24A20244, National Natural Science Foundation of China under Grant 62302247 and 62202273, Shandong Natural Science Foundation under Grant ZR2022QF140, Shandong Science Foundation for Excellent Young Scholars under Grant 2023HWYQ-007, Postdoctoral Fellowship Program of CPSF under Grant GZC20231460, China Postdoctoral Science Foundation under Grant 2024M761806.

F. V. Fomin and M. Xiao (Eds.): COCOON 2025, LNCS 15984, pp. 223–236, 2026.
https://doi.org/10.1007/978-981-95-0218-9_17

optimization problem, including Generative Adversarial Networks (GANs) [8,9], reinforcement learning [2], and robust optimization [13,14]. Their objective is to solve optimization problems involving two competing agents: a minimizer and a maximizer. In a distributed setting, the optimization process is carried out across multiple clients, each possessing a portion of the data, without requiring direct data sharing, as shown in Fig. 1. Formally, a general distributed minimax optimization problem can be formulated as:

$$\min_{x \in \mathbb{R}^{d_x}} \max_{y \in \mathbb{R}^{d_y}} \left\{ F(x,y) := \frac{1}{K} \sum_{k=1}^{k} f^k(x,y) \right\}, \tag{1}$$

where $F(x,y)$ is a function that depends on variables x and y, representing the objectives of the minimizer and maximizer, respectively. And $f^k(x,y)$ represents the client k local optimization function. The local optimization function is defined as $f^k(x,y) = \mathbb{E}_{\xi^k \sim \mathcal{D}^k} \left[L\left(x,y;\xi^k\right) \right]$, where $\mathcal{D}^k$ is the local data distribution of k-th client and L is the loss function for the data point ξ^k. We assume $f^k(x,y)$ is nonconvex regarding x and concave regarding y.

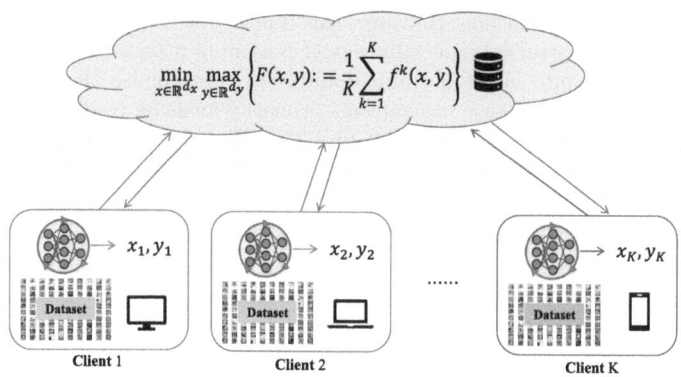

Fig. 1. Framework of distributed minimax optimization.

In real-world scenarios, distributed minimax optimization is particularly vulnerable in settings where multiple parties contribute to model training without centralized oversight [22]. One of the critical security threats in distributed minimax optimization is model poisoning attacks [5], where adversaries manipulate the learning process by injecting malicious updates. In fact, the existence of the attack poses significant challenges to the design of distributed min-max optimization algorithms. 1) **High communication cost:** To defend against model poisoning attacks, the server usually needs to introduce defense mechanisms. These mechanisms typically involve additional data exchanges, thereby increasing the communication burden [6]. 2) **Hierarchical dependency:** The hierarchical dependency between the minimization and maximization optimizations enables malicious manipulations in the maximization process to indirectly influence the minimization optimizations [7]. **3) Convergence guarantee:** Attacks may undermine the stability of optimization, making the optimization process difficult to converge or leading it to suboptimal solutions. For example, malicious clients can manipulate gradient information, causing the optimization algorithm to oscillate or deviate from the optimal solution [1].

Indeed, existing defense methods primarily address traditional model poisoning attacks in distributed learning, typically approaching defense from three perspectives: pre-aggregation, in-aggregation, and post-aggregation. AUROR [20] is an automated defense that filters out malicious users based only on their masked features, that does not require the availability of the entire training data. RFA [17] utilizes a robust aggregation approach by replacing the weighted arithmetic mean with an approximate geometric median, so as to minimize the impacts of "outlier" updates. Wu et al. [23] introduce a method to counteract model poisoning attacks by developing a pruning method to remove redundant neurons in the network and then adjusting the model's extreme weight values after the training phase. However, it only provides protection for the minimum optimization. Therefore, *how to design a robust defense mechanism tailored to the unique characteristics of distributed minimax optimization is an urgent and necessary problem.*

To solve the above problem, we design a robust framework of distributed minimax learning method against model poisoning attacks. Firstly, to mitigate the communication bottleneck, we perform multiple local model updates using stochastic gradient descent ascent (SGDA) and periodically aggregate them by averaging. Through the above operation, we reduce the number of communication rounds from T to S, where $S = T/\tau$. Secondly, to eliminate the impact of attacks, we introduce a regularization term to optimally perturb the Hessian matrix of local gradients and local models. The goal is to make the Hessian matrix as close to the identity matrix as possible, thereby reducing the effect of attacks. Finally, we conducted convergence analysis and certified radius analysis to verify the effectiveness of the proposed algorithm. Meanwhile, the certified radius quantifies the distance between benign and poisoned models. Specifically, we make the following contributions:

- To the best of our knowledge, Robust-LSGDA is the first method designed to effectively defend against model poisoning attacks in distributed minimax optimization.
- We demonstrate the effectiveness of the proposed algorithm through convergence analysis. Specifically, we prove that Robust-LSGDA achieves an asymptotically optimal convergence rate of $\frac{1}{\sqrt{KT}}$, where K is the number of clients and T is the global maximum iteration number.
- We minimize the certified radius under two types of model poisoning attacks to demonstrate the robustness of the algorithm, which serves as an upper bound on the distance between malicious and benign models.

2   Related Work

In this section, we introduce the related work on model poisoning and minimax optimization.

2.1   Defenses for Poisoning Attacks in Federated Learning

In Federated Learning (FL), model poisoning attacks pose a significant security threat. FL enables distributed model training across multiple clients without sharing raw data, but its distributed framework and data heterogeneity create vulnerabilities. Model poisoning attacks embed adversarial triggers in training data, causing models to misclassify triggered test samples. Recently, several studies have proposed methods to train FL models while mitigating specific types of attacks. AUROR [20] is a defense against poisoning attacks in the direct collaborative learning setting that can detect malicious users with high accuracy, thereby building a robust global model. RFA [17] is a robust aggregation method that uses the geometric median and the smoothed Weiszfeld algorithm to efficiently compute updates, enhancing FL's resilience against scenarios where some devices may send corrupted updates to the central server. Wu et al. [23] simplifies model architectures via federated neuron pruning while preserving model performance on intended tasks. Adjusting extreme weights in the pruned model significantly reduces the success rate of model poisoning attacks. LeadFL [24] perturbs the local model updates by adding a novel regularization term. FoolsGold [6] is a defense mechanism based on party update diversity, with no restriction on the number of adversarial parties. By calculating cosine similarity among participants, it assigns individual global learning rates. Parties with similar update vectors receive lower learning rates, effectively protecting the global model from model poisoning attacks. However, the current algorithms mainly focus on the single layer, do not consider the double layer structure, and there is no defense method designed for the distributed minimax optimization.

2.2   Distributed Minimax Optimization

Most previous work on minimax optimization has primarily focused on centralized settings. [4] proves the linear convergence of the primal-dual gradient method for a class of convex-concave functions. Some studies [11, 15] also conduct trials on nonconvex-concave minimax optimization. [11] studies single-machine SGDA in the nonconvex-concave case. In recent years, studies have also been conducted on distributed minimax learning. A few recent studies are devoted to distributed minimax optimization. The work [12] analyzes the convergence of networked optimistic stochastic gradient descent ascent (OSGDA) on the nonconvex-nonconcave setting. [19] studies a variant of local SGDA. FedGAN [18] is a method to solve large-scale and privacy-preserving minimax problems. However, the current research primarily focuses on distributed minimax optimization, and there is little work on defense against model poisoning attacks in distributed minimax problems.

3 Model Poisoning Attacks in Distributed Minimax Optimization

In this section, we provide a detailed definition of the model poisoning attack scenario within the context of minimax optimization and highlight the significance of such attacks in this framework.

3.1 Distributed Minimax Optimization

In distributed learning, a general min-max optimization problem is defined as Eq. (1). The server sends the global model to selected clients, these clients utilize their local data to train the received model. Client k updates local model w_t^k as follows:

$$w_{t+1}^k = w_t^k - \eta \nabla f^k(w_t). \tag{2}$$

Then these clients send the trained model back to the server. The server aggregates collected models to obtain a new global model. The above steps form one round, which is repeated until the stop condition is satisfied.

3.2 Poisoning Attack

During distributed training, malicious clients may participate in the process. Suppose there are a total of K clients, where the set of benign clients is denoted as $\mathcal{G}$ and the set of malicious clients as $\mathcal{M}$, satisfying $|\mathcal{G}| + |\mathcal{M}| = K$. We assume that malicious clients have similar computational capabilities as benign clients and cannot directly access the weights or data of other clients. Their goal is either to degrade the model's performance on specific tasks, known as targeted attacks [5] or to manipulate the global model into making incorrect predictions on inputs containing predefined triggers, referred to as model poisoning attacks [23], all while maintaining the model's overall accuracy.

To obtain such a poisoned model, malicious clients train their local models on malicious data by minimizing the malicious loss function f_M. The malicious client k optimization process is formulated as follows:

$$w_{t+1}^k \leftarrow w_t^k - \eta_t \left[\pi \nabla f(w_t^k) + (1 - \pi) \nabla f_M(w_t^k) \right]. \tag{3}$$

Here, π controls the balance between the normal gradient $\nabla f(w_t^k)$ and the malicious gradient $\nabla f_M(w_t^k)$. Through this optimization strategy, the updates generated by malicious clients influence the global model's parameters during model aggregation.

It is important to note that the malicious dataset is assumed to follow the same distribution as the benign training data. The key distinction lies in the attack strategy: for targeted attacks, the labels of certain samples are modified to a specific target class, while for model poisoning attacks, samples containing predefined patterns are introduced into the dataset.

3.3   Attack Effect

We formalize the impact of the attack in round t as $\delta_t = w_t - w_t^M$ based on the work [21], where w_t is the global model parameters in round t without attacks by malicious clients and w_t^M is the malicious clients' model parameters. The transmission of the attack effect can be estimated as follows:

$$\hat{\delta}_t = \left[\frac{1}{|\mathcal{C}_t|} \sum_{k \in \mathcal{C}_t} \prod_{i=0}^{I-1} (I - \eta_t H_{t,i}^k) \right] \hat{\delta}_{t-1}, \tag{4}$$

where $\mathcal{C}_t$ is the set of selected clients in global round t. $H_{t,i}^k = \nabla^2 f(w_{t,i}^k)$ represents the Hessian matrix at local iteration i of global round t and I is the identify matrix.

During the training process, the Hessian matrix is highly sparse in both benign and malicious clients. So in Eq. (4), the weight of the attack at time $t-1$ is close to $\sum_{k \in \mathcal{C}_t} \prod_{i=0}^{I-1} (I)$. As a result, the influence of $\hat{\delta}_{t-1}$ will persistently propagate and extend to $\hat{\delta}_t$. The challenge of minimax optimization in defending against model poisoning attacks is to mitigate the mutual influence between maximal and minimal parameters attacks while maintaining optimization communication efficiency and global model accuracy.

4   Algorithm

In this section, we first present the main idea of the proposed Robust-LSGDA, then describe how each component of Robust-LSGDA is designed.

4.1   Local SGDA

Local Stochastic Gradient Descent Ascent [3] is a simple extension of the centralized algorithm SGDA [11]. Let S denote the total number of communication rounds between the server and clients, and let τ represent the number of local iterations executed by clients between two successive communication rounds, so the number of global iterations is $T = S \times \tau$. Each client updates their local parameters $\{x_t^k, y_t^k\}$ at time t as follows:

$$x_{t+1}^k = x_t^k - \eta_x \nabla_x f^k(x_t^k, y_t^k; \xi_t^k) \tag{5}$$

$$y_{t+1}^k = y_t^k + \eta_y \nabla_y f^k(x_t^k, y_t^k; \xi_t^k), \tag{6}$$

where ξ_t^k is the data sample by the client k at time t. After clients update the τ local iterations, the server collects the parameters $x_{s\tau}^k$ and $y_{s\tau}^k$ from clients to aggregate minimax parameters: $x_{s\tau} = \frac{1}{K} \sum_{k=1}^{K} x_{s\tau}^k$ and $y_{s\tau} = \frac{1}{K} \sum_{k=1}^{K} y_{s\tau}^k$. Then the server sends the updated models back to the local clients.

4.2   Client Defense Mechanism

In order to defend against poisoning attacks, we need to add perturbation to the Hessian matrix such that this coefficient term $(I - \eta_t H^k_{t,i})$ vanishes. We show that this is equivalent to adding the same amount of perturbation in model updates. However, in the minimax optimization $\min_{x \in \mathbb{R}^{d_x}} \max_{y \in \mathbb{R}^{d_y}} F(x, y)$, the model update is related to parameters x and y, so we need to adjust the model update method. We first summarize the proposed regularized x and y update protocol as follows:

$$\widetilde{x}^k_t \leftarrow x^k_t - \eta_x \nabla_x f^k(x^k_t, y^k_t; \xi^k_t) \tag{7}$$

$$x^k_{t+1} \leftarrow \widetilde{x}^k_t - \eta_x \alpha \text{clip}\left(\nabla(I - \eta_x \widetilde{HE}^k_{t,x}), q_x\right) \tag{8}$$

$$\widetilde{y}^k_t \leftarrow y^k_t + \eta_y \nabla_y f^k(x^k_t, y^k_t; \xi^k_t) \tag{9}$$

$$y^k_{t+1} \leftarrow \widetilde{y}^k_t + \eta_y \alpha \text{clip}\left(\nabla(I - \eta_y \widetilde{HE}^k_{t,y}), q_y\right) \tag{10}$$

where $\widetilde{x}^k_t$ and $\widetilde{y}^k_t$ are the intermediate parameters x and y in local iteration t of client k. α serves as a hyperparameter that regulates the strength of the regularization term. $\text{clip}(\cdot, q)$ denotes the process of restricting the regularization term to a threshold q in order to guarantee convergence. $\widetilde{HE}^k_{t,x}$ and $\widetilde{HE}^k_{t,y}$ are the estimations of the Hessian matrices of the parameters x and y before adding the regularization term.

Hessian Matrix Estimation. Based on [10], we employ the estimated Hessian matrix, with a particular focus on its diagonal terms. In minimax optimization, we rewrite the estimate of the Hessian matrix as the change of x and y parameters:

$$\widetilde{HE}^k_{t,x} = (\widetilde{x}^k_t - x^k_t - \Delta x^k_t)/\eta_x \tag{11}$$

$$\widetilde{HE}^k_{t,y} = (\widetilde{y}^k_t - y^k_t - \Delta y^k_t)/\eta_y, \tag{12}$$

where $\Delta x^k_t = x^k_t - x^k_{t-1}$ and $\Delta y^k_t = y^k_t - y^k_{t-1}$, we use the change in x and y parameters during local iterations to approximate the change in gradient.

Gradient Clipping. To guarantee convergence of the model following the introduction of the regularization term, the gradients are constrained using a threshold q during local training. The clipping function for the x parameter can be specified as:

$$\text{clip}\left(\nabla(I - \eta_x \widetilde{HE}^k_{t,x}), q_x\right)_{r,c} = \begin{cases} \nabla(I - \eta_x \widetilde{HE}^k_{t,x})_{r,c}, & \text{if } \left|\nabla(I - \eta_x \widetilde{HE}^k_{t,x})_{r,c}\right| \leq q_x \\ q_x, & \text{otherwise} \end{cases} \tag{13}$$

where r and c are the indices of rows and columns of the matrix. It is a similar clipping function for the y parameter.

4.3   Entire Procedure

Integrated with the above strategies, Robust-LSGDA is described in Algorithm 1. Similar to traditional FL, for each time t, each client sends its parameters x and y to the server at the beginning (Line 2). The server aggregates collected local parameters to derive new global parameters (Line 2) and sends the updated parameters to selected clients (Line 3). Clients start local training (Lines 9–18). Client first updates the parameter x, reaching an intermediate state of the parameter x (Line 11). Then calculates the estimate of the Hessian matrix of x to compute the regularization term. After that, allowing the regularization loss to backpropagate (Lines 12–14). For parameter y, the optimization approach corresponds to that for parameter x, but follows a stochastic gradient ascent algorithm (Lines 15–18).

Algorithm 1: Robust-LSGDA

Input: number of communication rounds S; number of local iterations τ; number of global iterations $T = S\tau$; the number of nodes K; local learning rates η_x and η_y; regularization rate α; clipping bound q_x for x and q_y for y.

1   **for** $s = 0, 1, 2, ..., S - 1$ **do**

2   Each client sends $x_{s\tau-1}, y_{s\tau-1}$ to server, and server aggregates models: $x_{s\tau} = \frac{1}{K}\sum_{k=1}^{K} x_{s\tau-1}^k$, $y_{s\tau} = \frac{1}{K}\sum_{k=1}^{K} y_{s\tau-1}^k$;

3   Server sends $x_{s\tau}, y_{s\tau}$ to nodes;

4   **for** *each client in parallel* **do**

5   Synchronize $x_{s\tau}^k, y_{s\tau}^k$ with the global $x_{s\tau}, y_{s\tau}$ from the previous round: $x_{s\tau}^k \leftarrow x_{s\tau}, y_{s\tau}^k \leftarrow y_{s\tau}$;

6       $x_{(s+1)\tau-1}^k, y_{(s+1)\tau-1}^k \leftarrow$ **RunClient** $(x_{s\tau}, y_{s\tau})$;

7   **end**

8   **end**

9   **RunClient** $(x_{s\tau}, y_{s\tau})$:

10 **for** $t = s\tau, ..., (s + 1)\tau - 1$ **do**

11   Calculate gradients and update the x: $\widetilde{x}_t^k = x_t^k - \eta_x \nabla_x f^k(x_t^k, y_t^k; \xi_t^k)$;

12   Approximate the Hessian matrix of x: $\widetilde{HE}_{t,x}^k = (\widetilde{x}_t^k - x_t^k - \Delta x_t^k)/\eta_x$;

13   Calculate and clip the gradients of the regularization term of x:
$R_{t,x}^k = \mathrm{clip}(\triangledown(I - \eta_x \widetilde{HE}_{t,x}^k), q_x)$;

14   Update $x_{t+1}^k = \widetilde{x}_t^k - \eta_x \alpha R_{t,x}^k$;

15   Calculate gradients and update the y: $\widetilde{y}_t^k = y_t^k + \eta_y \nabla_y f^k(x_t^k, y_t^k; \xi_t^k)$;

16   Approximate the Hessian matrix of y: $\widetilde{HE}_{t,y}^k = (\widetilde{y}_t^k - y_t^k - \Delta y_t^k)/\eta_y$;

17   Calculate and clip the gradients of the regularization term of y:
$R_{t,y}^k = \mathrm{clip}(\nabla(I - \eta_y \widetilde{HE}_{t,y}^k), q_y)$;

18   Update $y_{t+1}^k = \widetilde{y}_t^k + \eta_y \alpha R_{t,y}^k$;

19 **end**

20 **return** x_{t+1}^k, y_{t+1}^k;

---

5   Main Result

In this section, we provide the theoretical results of our algorithm Robust-LSGDA. We first give some preliminary Assumptions, Definitions, and relevant Lemmas.

Assumption 1 (Smoothness). *The local function f^k has Lipschitz continuous gradients, i.e., there exists a constant $L_f > 0$, such that each client $k \in [K]$, for all $x, x' \in \mathbb{R}^{d_x}$ and $y, y' \in \mathbb{R}^{d_y}$,*

$$\left\| \nabla f^k(x, y) - \nabla f^k(x', y') \right\| \leq L_f \left\| (x, y) - (x', y') \right\|.$$

Assumption 2 (Bounded Variance). *The variance of the stochastic gradient at each client $k \in [K]$ is bounded as follows.*

$$\mathbb{E}_{\xi_t^k}[\nabla f^k(x, y; \xi_t^k)] = \nabla f^k(x, y), \quad \mathbb{E}_{\xi_t^k} \|\nabla f^k(x, y; \xi_t^k) - \nabla f^k(x, y)\|^2 \leq \sigma^2.$$

Assumption 3 (Bounded Heterogeneity). *To quantify the variability of the local functions $f^k(x, y)$ across different clients, the bounded heterogeneity is defined as follows:*

$$\varsigma_x^2 = \sup_{x \in \mathbb{R}^{d_x}, y \in \mathbb{R}^{d_y}} \frac{1}{K} \sum_{k=1}^{K} \left\| \nabla_x f^k(x, y) - \nabla_x f(x, y) \right\|^2,$$

$$\varsigma_y^2 = \sup_{x \in \mathbb{R}^{d_x}, y \in \mathbb{R}^{d_y}} \frac{1}{K} \sum_{k=1}^{K} \left\| \nabla_y f^k(x, y) - \nabla_y f(x, y) \right\|^2.$$

We assume that ς_x and ς_y are bounded.

Assumption 4 (Polyak Łojasiewicz (PL) Condition in y). *The function f fulfills μ-PL condition in y ($\mu > 0$), if for any fixed x: 1) $\max_{y'} f(x, y')$ has a nonempty solution set; 2) for all y, $\|\nabla_y f(x, y)\|^2 \geq 2\mu(\max_{y'} f(x, y') - f(x, y))$.*

Assumption 5 (Coordinate-wise Lipschitz). *If for any models $\theta_t, \theta_t^* \in \mathcal{M}$, communication round $s \in [S]$, and a dataset D, the function $f(\mathcal{G}, \mathcal{A}, \eta)$ is c-coordinate-wise Lipschitz, so the outputs of the gradient oracle on any coordinate cannot drift too much farther apart. Specifically, for any coordinate index $i \in [d]$*

$$|\mathcal{G}(\theta_t^*, D)[i] - \mathcal{G}(\theta_t, D)[i]| \leq c \cdot |\theta_t^* - \theta_t|_1 \tag{14}$$

Definition 1. *(Poisoning Attack). For a protocol $f = (\mathcal{G}, \mathcal{A}, \eta)$ we define the set of poisoned protocols $F(\rho)$ as all protocols $f^* = (\mathcal{G}^*, \mathcal{A}, \eta)$ that are identical to f except for the gradient oracle $\mathcal{G}^*$, which is replaced by a ρ-corrupted version of $\mathcal{G}$. Specifically, for any round t, any model θ_t and any dataset D, for some ϵ with ϵ with $\|\epsilon\|_1$, the following holds: $\mathcal{G}^*(\theta_t, D) = \mathcal{G}(\theta_t, D) + \epsilon$.*

Lemma 1. *If the function $f(x, \cdot)$ satisfies Assumptions 1, 4, the $\Phi(x)$ is L_Φ-smooth with $L_\Phi = \kappa L/2 + L$, where $\kappa = L/\mu$ is the condition number.*

Lemma 2. *Assume the local loss function f^k satisfy Assumptions 1, 4 and the stochastic oracles for local functions satisfy Assumption 2.*

$$\mathbb{E}\left[\Phi(x_{t+1})\right] \leq \mathbb{E}\left[\Phi(x_t)\right] - \frac{7\eta_x}{16}\mathbb{E}\left\|\Phi(x_t)\right\|^2 + \left(3L_\Phi\eta_x^2 - \frac{\eta_x}{2}\right)\mathbb{E}\left[\left\|\frac{1}{K}\sum_{k=1}^{K}\nabla_x f^k(x_t, y_t)\right\|^2\right]$$

$$+ \frac{\eta_x L_f^2}{\mu}\mathbb{E}\left[\Phi(x_t) - f(x_t, y_t)\right] + \left(3L_\Phi\eta_x^2 + 8\eta_x\right)L_f^2\Delta_t^{x,y}$$

$$+ \left(8\eta_x + L_\Phi\eta_x^2\right)\alpha^2\mathbb{E}\left[\left\|\frac{1}{K}\sum_{k=1}^{K}R_{t,x}^k\right\|^2\right] + \frac{3L_\Phi\eta_x^2\sigma^2}{K},$$

where $\Delta_t^{x,y} \triangleq \frac{1}{K}\sum_{k=1}^{K}\mathbb{E}\left(\left\|x_t^k - x_t\right\|^2 + \left\|y_t^k - y_t\right\|^2\right)$ is defined as the synchronization error.

Lemma 3. *Assume the local loss function f^k satisfies Assumptions 1, 3, and the stochastic oracles for local satisfy Assumption 2. The learning rates η_x, η_y satisfy $\eta_y \leq 1/\mu$, $\frac{\eta_x}{\eta_y} \leq 2(3+2\alpha)\kappa^2$. We have*

$$\frac{1}{T}\sum_{t=0}^{T-1}\mathbb{E}\left(\Phi(x_t) - f(x_t, y_t)\right) \leq \frac{2\left(\Phi(x_0) - f(x_0, y_0)\right)}{\eta_y\mu T} + \frac{2\eta_x(1-\eta_y\mu)(\frac{9}{16}+\alpha)}{\mu\eta_y T}\sum_{t=0}^{T-1}\mathbb{E}\left\|\nabla\Phi(x_t)\right\|^2$$

$$+ \frac{2\eta_x^2}{\mu\eta_y}\left[(1-\eta_y\mu)(3L_\Phi + L_f) + 3\eta_y L_f^2\right]\frac{1}{T}\sum_{t=0}^{T-1}\mathbb{E}\left\|\frac{1}{K}\sum_{k=1}^{K}\nabla_x f^k(x_t^k, y_t^k)\right\|^2$$

$$+ \frac{2L_f^2}{\mu\eta_y}\left[(1-\eta_y\mu)(3L_\Phi\eta_x^2 + 8\eta_x) + \frac{3}{2}\eta_y\right]\frac{1}{T}\sum_{t=0}^{T-1}\Delta_t^{x,y} + \frac{L_f\sigma^2(6\eta_x^2 L_f + \eta_y)}{\mu K}$$

$$+ \frac{2\alpha\eta_x}{\mu\eta_y}\left[(1-\eta_y\mu)(1 + \alpha\eta_x L_f + 8\alpha + \alpha\eta_x L_\Phi) + 3L_f^2\alpha\eta_x\eta_y\right]\frac{1}{T}\sum_{t=0}^{T-1}\mathbb{E}\left\|\frac{1}{K}\sum_{k=1}^{K}R_{t,x}^k\right\|^2$$

$$+ \frac{2}{\mu}\left[\alpha^2(\eta_y L_f + \frac{1}{2})\right]\frac{1}{T}\sum_{t=0}^{T-1}\mathbb{E}\left\|\frac{1}{K}\sum_{k=1}^{K}R_{t,y}^k\right\|^2 + \frac{2(1-\eta_y\mu)(L_f + 3L_\Phi)\eta_x^2\sigma^2}{\mu\eta_y K}.$$

Lemma 4. *Assume the local loss function f^k satisfy Assumptions 1, 3, and the local stochastic oracles satisfy Assumption 2. The learning rates $\eta_x, \eta_y \leq \frac{1}{2(3+2\alpha)\tau L_f}$. Then the iterates $\{x_t^k, y_t^k\}$ satisfy*

$$\frac{1}{T}\sum_{t=0}^{T-1}\Delta_t^{x,y} \triangleq \frac{1}{T}\sum_{t=0}^{T-1}\frac{1}{K}\sum_{k=1}^{K}\mathbb{E}\left(\left\|x_t^k - x_t\right\|^2 + \left\|y_t^k - y_t\right\|^2\right)$$

$$\leq 10(\tau-1)^2\left[\left(\eta_x^2 + \eta_y^2\right)\sigma^2\left(1 + \frac{1}{K}\right) + 3\left(\eta_x^2\varsigma_x^2 + \eta_y^2\varsigma_y^2\right)\right]$$

$$+ \frac{10\alpha^2\left(\eta_x^2 + \eta_y^2\right)(\tau-1)^2}{\tau K}\sum_{t=s\tau+1}^{(s+1)\tau-1}\sum_{k=1}^{K}\mathbb{E}\left[\left\|R_{i,x}^k - \frac{1}{K}\sum_{j=1}^{K}R_{i,x}^j\right\|^2 + \left\|R_{i,y}^k - \frac{1}{K}\sum_{j=1}^{K}R_{i,y}^j\right\|^2\right].$$

5.1   Convergence Result

Theorem 1. *Based on the Assumptions 1 to 4. Suppose the step-sizes η_x, η_y are chosen such that $\eta_y \leq \frac{1}{2(3+2\alpha)L_f\tau}$, $\frac{\eta_x}{\eta_y} = \frac{1}{2(3+2\alpha)\kappa^2}$, where $\kappa = \frac{L_f}{\mu}$ is the condition number, $\Phi(x) \triangleq \max_y f(x,y)$ is the envelope function, and $\Delta_\Phi \triangleq \Phi(x_0) - \min_x \Phi(x)$. Specifically, using $\eta_y = \sqrt{\frac{K}{L_fT}}$ and $\eta_x = \frac{1}{2(3+2\alpha)\kappa^2}\sqrt{\frac{K}{L_fT}}$, we can obtain the convergence result of Robust-LSGDA as follows:*

$$
\mathbb{E}\left\|\nabla\Phi(\bar{x}_T)\right\|^2 \leq \mathcal{O}\left(\frac{\kappa^2\left(\Delta_\Phi + \sigma^2\right)}{\sqrt{KT}}\right)
$$

$$
+ \mathcal{O}\left(\left(\frac{1}{\alpha\kappa^2}\sqrt{\frac{K}{T}} + \kappa^2\right)(\tau-1)^2\frac{\sigma^2\left(\alpha^2\kappa^4+1\right)(K+1) + K\left(\zeta_x^2 + \zeta_y^2\alpha^2\kappa^4\right)}{\alpha^2\kappa^4T}\right)
$$

$$
+ \mathcal{O}\left(\left(\frac{1}{\alpha\kappa^2}\sqrt{\frac{K}{T}} + \kappa^2\right)(\tau-1)^2\frac{K\left(\alpha^2\kappa^4+1\right)}{\kappa^4T}(d_x^2q_x^2 + d_y^2q_y^2)\beta\right)
$$

$$
+ \left(64\alpha^2 + \frac{8\alpha(1+8\alpha)}{2+3\alpha} + \frac{4\alpha^2L_\Phi\sqrt{\frac{K}{L_fT}}}{(2+3\alpha)\kappa^2} + \frac{8\alpha L_\Phi\sqrt{\frac{K}{L_fT}}}{(2+3\alpha)^2\kappa^2} + \frac{12L_fK\alpha^2}{(2+3\alpha)^2\kappa^2T}\right)
$$

$$
\frac{1}{T}\sum_{t=0}^{T-1}\mathbb{E}\left\|\frac{1}{K}\sum_{k=1}^{K}R_{t,x}^k\right\|^2 + \left[8\kappa^2\alpha^2(2\sqrt{\frac{KL_f}{T}}+1)\right]\frac{1}{T}\sum_{t=0}^{T-1}\mathbb{E}\left\|\frac{1}{K}\sum_{k=1}^{K}R_{t,y}^k\right\|^2
$$

$$\tag{15}$$

Remark 1. **Convergence Result** From the result of Theorem 1, we can see that our convergence is controlled by the term of $\mathcal{O}\left(\frac{1}{\sqrt{KT}}\right)$. The first term in Theorem 1 corresponds to the optimization error of a fully synchronous algorithm ($\tau = 1$), where local models are aggregated after each update. The second term emerges as a result of clients performing multiple local updates ($\tau > 1$) between consecutive communication rounds. This term is influenced by the degree of data heterogeneity across clients, denoted by ζ_x and ζ_y. Since the dependence on the learning rates η_x and η_y is quadratic, selecting sufficiently small values for η_x and η_y, along with a carefully chosen number of local updates τ, ensures that performing multiple local updates does not affect the asymptotic convergence rate. The third term is due to the defensive measure and is affected by the dimensions and the clipping values.

Remark 2. **Impact of clipping term** We assume the clipping value is relatively small, and suppose the clipping term is relatively small. In practical scenarios, this is achieved because most clipping strategies are inclined to focus on restraining weights, thereby ensuring that $R_{t,x}^k$ and $R_{t,y}^k$ remain small. In our convergence analysis, $\left(64\alpha^2 + \frac{8\alpha(1+8\alpha)}{2+3\alpha} + \frac{4\alpha^2L_\Phi\sqrt{\frac{K}{L_fT}}}{(2+3\alpha)\kappa^2} + \frac{8\alpha L_\Phi\sqrt{\frac{K}{L_fT}}}{(2+3\alpha)^2\kappa^2} + \frac{12L_fK\alpha^2}{(2+3\alpha)^2\kappa^2T}\right)$

$$\frac{1}{T}\sum_{t=0}^{T-1}\mathbb{E}\left\|\frac{1}{K}\sum_{k=1}^{K}R_{t,x}^{k}\right\|^{2} \quad + \quad \left[8\kappa^{2}\alpha^{2}(2\sqrt{\frac{KL_{f}}{T}}+1)\right]\frac{1}{T}\sum_{t=0}^{T-1}\mathbb{E}\left\|\frac{1}{K}\sum_{k=1}^{K}\right.$$

$R_{t,y}^{k}\big\|^{2}$, clipping will incur an error term, which is affected by the regularization rate α. As the number of communication rounds increases, the error term decreases gradually.

Remark 3. (Nonconvex-Strongly-Concave Problems). Since the PL condition is a more general assumption than strong concavity, our results also extend to Nonconvex-Strongly-Concave minimax problems.

5.2 Robustness Analysis

We analyze the robustness of Robust-LSGDA using the certified radius framework. We examine two threat models: periodic poisoned model submissions and bursty poisoned model submissions. The certified radius represents the maximum distance between a poisoned model and a benign model, and minimizing this radius enhances robustness, as models closer to each other are more likely to yield similar predictions. Building on these assumptions, [16] proposed the certified radius for general protocols.

Theorem 2. *On a dataset D, f is a c-coordinatewise-Lipschitz protocol. The certified radius $R(\rho)$ for f is defined as $R(\rho) = \Lambda(T)(1+dc)^{\Lambda(T)}\rho$, where $\Lambda(t) = \sum_{t=0}^{T-1}\eta_t$ is the cumulative step size, and d is the dimension of model parameters.*

Scenario I. Scenario I presumes that a malicious client periodically submits harmful updates at specific intervals during the global training rounds. More precisely, this malicious client introduces poisonous updates in the global round designated as T_A. Subsequently, from the T_A round until the $T-1$ round, no further malicious updates are submitted. This setup represents a streamlined version of a burst adversarial model. This scenario is characterized by its suddenness and unpredictability, making it particularly apt for assessing the robustness of defense algorithms under extreme conditions.

Theorem 3. *(Certified Radius in Scenario I) Based on Assumption 5 holding on T_A, let c be as defined therein. We can obtain the certified radius as follows:*

$$R_x(\rho) = \left(\frac{1}{|\mathcal{C}_t|}\right)^{T-T_A}\left|\prod_{t=T_A}^{T}\left[\sum_{k\in\mathcal{C}_t}\prod_{i=0}^{I-1}\left(I-\eta_{t,x}\widetilde{HE}_{t,i,x}^{k}\right)\right]\right|$$
$$\cdot\sum_{t=0}^{T_A-1}\eta_{t,x}(1+d_xc)^{\sum_{t=0}^{T_A-1}\eta_{t,x}}\rho \tag{16}$$

$$R_y(\rho) = \left(\frac{1}{|\mathcal{C}_t|}\right)^{T-T_A}\left|\prod_{t=T_A}^{T}\left[\sum_{k\in\mathcal{C}_t}\prod_{i=0}^{I-1}\left(I-\eta_{t,y}\widetilde{HE}_{t,i,y}^{k}\right)\right]\right|$$
$$\cdot\sum_{t=0}^{T_A-1}\eta_{t,y}(1+d_yc)^{\sum_{t=0}^{T_A-1}\eta_{t,y}}\rho. \tag{17}$$

Scenario II. Here, we consider a more general threat model where the number of malicious clients fluctuates from round to round, leading to bursty adversarial patterns. Specifically, we presume that clients are chosen at random. For attack, we can derive the certified radius of Robust-LSGDA under bursty adversarial patterns as follows:

Theorem 4. *(Certified Radius in Scenario II). Based on Assumption 5 holding on T_A. The certified radius of the threat model can be defined as follows:*

$$R_x(\rho) = |x_T - x_T^*| = (1 + d_x c)^{\sum_{t=0}^T \eta_{t,x}} \rho T \sum_{t=0}^T \eta_{t,x} \tag{18}$$

$$R_y(\rho) = |y_T - y_T^*| = (1 + d_y c)^{\sum_{t=0}^T \eta_{t,y}} \rho T \sum_{t=0}^T \eta_{t,y}. \tag{19}$$

6   Conclusion

In this paper, we propose Robust-LSGDA, a novel framework that ensures robustness against model poisoning attacks. This framework not only enhances the security of distributed minimax learning but also provides a foundation for future research in robust distributed optimization. In the future, it deserves to explore the impact of heterogeneous and non-IID settings for distributed minimax learning.

References

1. Bhagoji, A.N., Chakraborty, S., Mittal, P., Calo, S.: Analyzing federated learning through an adversarial lens. In: International Conference on Machine Learning, pp. 634–643 (2019)
2. Dai, B., et al.: SBEED: convergent reinforcement learning with nonlinear function approximation. In: International Conference on Machine Learning, pp. 1125–1134 (2018)
3. Deng, Y., Mahdavi, M.: Local stochastic gradient descent ascent: convergence analysis and communication efficiency. In: International Conference on Artificial Intelligence and Statistics, pp. 1387–1395 (2021)
4. Du, S.S., Hu, W.: Linear convergence of the primal-dual gradient method for convex-concave saddle point problems without strong convexity. In: The 22nd International Conference on Artificial Intelligence and Statistics, pp. 196–205 (2019)
5. Fang, M., Cao, X., Jia, J., Gong, N.: Local model poisoning attacks to {Byzantine-Robust} federated learning. In: 29th USENIX Security Symposium (USENIX Security 2020), pp. 1605–1622 (2020)
6. Fung, C., Yoon, C.J., Beschastnikh, I.: The limitations of federated learning in sybil settings. In: 23rd International Symposium on Research in Attacks, Intrusions and Defenses (RAID 2020), pp. 301–316 (2020)
7. García Trillos, N., Akash, A.K., Li, S., Riedl, K., Zhu, Y.: Defending against diverse attacks in federated learning through consensus-based bi-level optimization. arXiv e-prints pp. arXiv–2412 (2024)

8. Goodfellow, I., et al.: Generative adversarial nets. In: Advances in Neural Information Processing Systems, vol. 27 (2014)
9. Gulrajani, I., Ahmed, F., Arjovsky, M., Dumoulin, V., Courville, A.C.: Improved training of Wasserstein GANs. In: Advances in Neural Information Processing Systems, vol. 30 (2017)
10. Le Cun, Y., Denker, J., Solla, S.: Optimal brain damage, advances in neural information processing systems. Denver 1989, Ed. D. Touretzsky, Morgan Kaufmann **598**, 605 (1990)
11. Lin, T., Jin, C., Jordan, M.: On gradient descent ascent for nonconvex-concave minimax problems. In: International Conference on Machine Learning, pp. 6083–6093 (2020)
12. Liu, M.L., Mroueh, Y., Zhang, W., Cui, X., Ross, J., Das, P.: Decentralized parallel algorithm for training generative adversarial nets. In: Advances in Neural Information Processing Systems, vol. 33, pp. 11056–11070 (2020)
13. Mohri, M., Sivek, G., Suresh, A.T.: Agnostic federated learning. In: International Conference on Machine Learning, pp. 4615–4625 (2019)
14. Namkoong, H., Duchi, J.C.: Stochastic gradient methods for distributionally robust optimization with f-divergences. In: Advances in Neural Information Processing Systems, vol. 29 (2016)
15. Nouiehed, M., Sanjabi, M., Huang, T., Lee, J.D., Razaviyayn, M.: Solving a class of non-convex min-max games using iterative first order methods. In: Advances in Neural Information Processing Systems, vol. 32, pp. 14934–14942 (2019)
16. Panda, A., Mahloujifar, S., Bhagoji, A.N., Chakraborty, S., Mittal, P.: Sparsefed: mitigating model poisoning attacks in federated learning with sparsification. In: International Conference on Artificial Intelligence and Statistics, pp. 7587–7624 (2022)
17. Pillutla, K., Kakade, S.M., Harchaoui, Z.: Robust aggregation for federated learning. IEEE Trans. Signal Process. **70**, 1142–1154 (2022)
18. Rasouli, M., Sun, T., Rajagopal, R.: Fedgan: federated generative adversarial networks for distributed data. arXiv preprint arXiv:2006.07228 (2020)
19. Reisizadeh, A., Farnia, F., Pedarsani, R., Jadbabaie, A.: Robust federated learning: the case of affine distribution shifts. In: Advances in Neural Information Processing Systems, vol. 33, pp. 21554–21565 (2020)
20. Shen, S., Tople, S., Saxena, P.: Auror: defending against poisoning attacks in collaborative deep learning systems. In: Proceedings of the 32nd Annual Conference on Computer Security Applications, pp. 508–519 (2016)
21. Sun, J., Li, A., DiValentin, L., Hassanzadeh, A., Chen, Y., Li, H.: FL-WBC: enhancing robustness against model poisoning attacks in federated learning from a client perspective. Adv. Neural. Inf. Process. Syst. **34**, 12613–12624 (2021)
22. Tsaknakis, I., Hong, M., Liu, S.: Decentralized min-max optimization: formulations, algorithms and applications in network poisoning attack. In: ICASSP 2020-2020 IEEE International Conference on Acoustics, Speech and Signal Processing, pp. 5755–5759 (2020)
23. Wu, C., Yang, X., Zhu, S., Mitra, P.: Mitigating backdoor attacks in federated learning. arXiv preprint arXiv:2011.01767 (2020)
24. Zhu, C., Roos, S., Chen, L.Y.: Leadfl: client self-defense against model poisoning in federated learning. In: International Conference on Machine Learning, pp. 43158–43180 (2023)

Parameterized Algorithms

Parameterized Complexity of Influence Maximization

Panfeng Liu[ID] and Biaoshuai Tao$^{(\boxtimes)}$[ID]

Shanghai Jiao Tong University, Shanghai, China
{liupf22,bstao}@sjtu.edu.cn

Abstract. The influence maximization problem studies how to select a set of nodes as initial seeds in a social network to maximize their influence. A dual problem of the influence maximization problem is the target set selection problem that asks for a minimum-cardinality seed set that can influence all users in the social network. In this paper, we consider the decision problem of deciding if we can select at most k seeds to influence at least t users in expectation, which formulates both the influence maximization problem and the target set selection problem. We study the parameterized complexity of this decision problem and consider the two most studied diffusion models: the independent cascade model (IC) and the linear threshold model (LT). We show that the problem is W[1]-hard under both models even if both k and t are given as parameters. For the special case with t being the number of the vertices in the network (which coincides with the target set selection problem), we show that the problem under the IC model is polynomial-time solvable, and the problem under the LT model, known to be NP-hard, is fixed-parameter tractable parameterized by k.

Keywords: Influence Maximization · Parameterized Complexity

1 Introduction

The social network plays a vital role in the diffusion of information, opinions, rumors, advertisements, and so on. Accordingly, the study of information diffusion in social networks, such as Facebook, Twitter, and WeChat, has garnered significant attention in the fields of communication media and advertisement marketing. Among these studies, a natural and well-studied problem is the *influence maximization* (INFMAX) problem, which is defined to *seed* a set of nodes in the social network to maximize the spreading of the resulting *cascade* [4,14,19,21,25–27], where a *cascade* is a basic social network process that captures the information diffusion in a social network, in which a number of initially infected seeds process a certain attribute of information and spread this attribute to their neighbors.

Two most well-known cascade models are the independent cascade (IC) model and the linear threshold (LT) model, both introduced in the seminal work of

F. V. Fomin and M. Xiao (Eds.): COCOON 2025, LNCS 15984, pp. 239–252, 2026.
https://doi.org/10.1007/978-981-95-0218-9_18

Kempe et al. [19]. These two models are studied almost exclusively in the past literature of this field, and we will focus on these two models in this paper. In the IC model, a newly activated node u attempts to activate each of its inactive neighbors v with a fixed probability, independently of other infections (we use the words "activation" and "infection" interchangeably throughout this work). In the LT model, if the graph is unweighted, each non-seed vertex is assigned a threshold randomly and independently from the interval $[0, 1]$. A vertex becomes active when the fraction of its infected neighbors exceeds its threshold. When the graph is edge-weighted, the above-mentioned fractions become weighted fractions. Formal definitions of the two models are available in Sect. 2.

The computational complexity and approximability of INFMAX is relatively well-understood. The problems with both the IC and LT models are APX-hard [19,29]. On the other hand, a simple greedy algorithm achieves a $(1 - 1/e)$-approximation [19]. In addition, the greedy algorithm has a slightly better approximation guarantee if the graph is undirected, for both the IC model [20] and the LT model [30].

In this paper, we investigate the *parameterized complexity* of INFMAX under both the IC and LT models. Parameterized complexity refines classical complexity theory by considering both input size and an additional parameter k that captures specific aspects of the problem. We say that a problem with input size n and parameter k is *fixed-parameter tractable* (FPT) if it can be solved in time $f(k) \cdot \text{poly}(n)$, where f is a computable function dependent only on k and $\text{poly}(\cdot)$ is a polynomial function [13,15,16,24]. The parameterized complexity hierarchy consists of the classes FPT $\subseteq$ W[1] $\subseteq$ W[2] $\subseteq \cdots \subseteq$ W[P]. We refer the readers to references [13,15] for the precise definitions of these complexity classes in the W-hierarchy. A W[1]-hard problem is unlikely to be fixed-parameter tractable. To establish W[1]-hardness, one can provide a *parameterized reduction* from a known W[1]-hard problem. A parameterized reduction maps an instance (I, k) of problem A_1 to an instance (I', k') of problem A_2 in time $g(k) \cdot \text{poly}(|I|)$, where g is a computable function. In addition, the output parameter k' should satisfy $k' \leq h(k)$ for some computable function h, guaranteeing that the reduction preserves fixed-parameter tractability by bounding the growth of the parameter.

The parameterized complexity for a closely related problem, *the target set selection problem*, has been studied with the *threshold model* [1,2,9,23]. The *target set selection problem* [6] can be viewed as a "dual problem" of the influence maximization problem: it asks for a seed set with a *minimum* number of seeds such that *all* vertices in the network will be infected at the end of the cascade. In the *threshold model* [17], each vertex v is assigned a threshold $\text{thr}(v)$, and the vertex v becomes active when $\text{thr}(v)$ of its neighbors are active. The existing work on parameterized complexity of the target set selection problem with the threshold model is reviewed in Sect. 1.2.

However, the threshold model is fundamentally different from the IC and LT models. Influence maximization in the threshold model is *nonsubmodular*, whereas both the IC and LT models satisfy *submodularity*. In fact, in contrast to that INFMAX with submodular diffusion models where the greedy algorithm achieves a $(1 - 1/e)$-approximation [19,22], INFMAX with nonsubmodular diffusion models requires fundamentally different seeding strategies. A successful seeding strategy for a nonsubmodular diffusion model should consider putting seeds close to each other to create the synergy effect and carefully decide the "seed groups" [31]. This is quite challenging in general. INFMAX for nonsubmodular diffusion models typically admits strong inapproximability factors even for very restrictive simple cases [19,28]. In particular, INFMAX with the threshold model is NP-hardness to approximate to within a factor of $n^{1-\varepsilon}$ [19]. As a result, those parameterized complexity results on the threshold model cannot be extended to influence maximization under the IC and LT models, and, to the best of our knowledge, no previous work has studied the parameterized complexity of INFMAX under these two most well-known models.

In this paper, we consider the parameterized complexity of INFMAX with the classical IC and LT models, and we formulate the INFMAX problem as a *decision problem*: given k and t as inputs, decide if we can select at most k seeds to infect at least t nodes (in expectation). This captures both the influence maximization problem (where k is the input and t is the objective to be maximized) and the target set selection problem (where t is fixed to be the number of the nodes in the graph and k is the objective to be minimized).

1.1 Our Results

We study the parameterized complexity of the decision version of INFMAX, which we denoted by DIM, based on both the IC and LT models on both directed and undirected graphs. DIM takes as inputs a weighted social network $G = (V, E, P)$ (which can be directed or undirected), a diffusion model (IC or LT in this paper), an integer k, and a real number t. A DIM instance is a yes instance if and only if we can find a set of at most k seeds in G such that the expected number of infected vertices by the end of the cascade is at least t.

We show that DIM is W[1]-hard under both the IC and LT models even if both k and t are given as parameters. In addition, DIM is W[2]-hard under the IC model for directed graphs, parameterized by k. We also study the special case of DIM with $t = |V|$ which coincides with the target set selection problem (i.e., deciding if we can choose a set of k seeds to make every vertex infected with probability 1), for which we denote by E-DIM (short for "exact decision influence maximization"). We show that E-DIM under the IC model is polynomial-time solvable, and the problem under the LT model is fixed-parameter tractable parameterized by k. Our results are summarized in Table 1.

1.2 Related Work

Previous work has studied the target set selection problem under the threshold model [2,10–12,18]. With respect to the parameterized complexity, the target set selection problem with the threshold model is proven to be W[2]-hard parameterized by the seed size, even on bipartite graphs with a diameter of four and majority thresholds or thresholds of at most two [23]. Additionally, it is

Table 1. Summary of the parameterized complexity of DIM and E-DIM.

Model	Problem	Parameter(s)	Result
IC	DIM	k, t	W[1]-hard (Theorem 1)
	DIM	k	W[2]-hard (Theorem 3)
	E-DIM	-	Poly-time (Theorem 4)
LT	DIM	k, t	W[1]-hard (Theorem 2)
	E-DIM	k	FPT (Theorem 5)

W[1]-hard parameterized by "treewidth", "cluster" "vertex deletion number", and "pathwidth" [1,2,9]. On the positive side, the problem becomes fixed-parameter tractable when parameterized by "vertex cover number", "feedback edge set size", or "bandwidth" [1,23]. If the input graph is complete or has bounded treewidth and bounded thresholds, then the problem is polynomial-time solvable [9,23]. However, as we mentioned before, the threshold model is nonsubmodular, which is fundamentally different from the IC and LT models studied in this paper.

2   Preliminaries

Throughout this paper, a weighted social network is represented by $G = (V, E, P)$ that may or may not be directed, where V is the node (i.e., user) set with $|V| = n$, E is the edge (i.e., the connections between users) set with $|E| = m$ and $P = \{p_e\}_{e \in E}$ is the edge weights, where it is assumed that $p_e \in (0, 1]$ for each edge $e = (u, v) \in E$. For a directed edge $(u, v) \in E$, we say u (resp. v) is the incoming (resp. outgoing) neighbor of v (resp. u), and (u, v) is the outgoing (resp. incoming) edge of u (resp. v). We use $\deg(v)$ to denote the degree of vertex v when G is undirected and the *in-degree* of vertex v when G is directed. We use $N(v)$ to denote the set of neighbors (in-neighbors) for undirected (directed) graphs. Accordingly, given $S \subseteq V$, we use $N(S) = \cup_{s \in S} N(s)$ to represent the set of neighbors (in-neighbors) of S for undirected (directed) graphs.

Diffusion Models. A *diffusion model* Γ is a (possibly random) function that maps from a vertex set S (the seeds that are initially infected) to a vertex set $\Gamma(S)$ (the set of influenced vertices at the end of the spreading). Below, we define the two most studied diffusion models: *the independent cascade model* (IC) and *the linear threshold model* (LT).

Definition 1 (Independent Cascade Model (IC) [19]). *Given a social network (directed graph) $G = (V, E, P)$, the IC model assigns the state of each node either active or inactive. On the input seed set $S \subseteq V$, IC outputs the set $\Gamma_{IC}(S)$ as follows:*

1. At timestamp 0, only nodes in S are active.

2. At each timestamp $t = 1, 2, \ldots$, each newly activated node u from the previous timestamp gets one chance to activate its inactive outgoing neighbor v with probability $p_{(u,v)}$. The attempts to activate neighbors are independent of each other. If multiple incoming neighbors of an inactive node attempt to activate it, each attempt is considered separately with its own probability.

3. The diffusion process terminates when no additional activation occurs, and IC outputs $\Gamma_{\mathrm{IC}}(S)$ as the set of active nodes.

The IC model can be equivalently described using the *live-edge interpretation*. Let $\widehat{G} = (V, \widehat{E})$ be a *live-edge graph* of G where each edge $(u, v) \in E$ is included in $\widehat{E}$ with probability $p_{(u,v)}$, and let $\widehat{\Gamma}_{\mathrm{IC}}(S)$ be the set of nodes that are reachable from S. Then $\widehat{\Gamma}_{\mathrm{IC}}(S)$ and $\Gamma_{\mathrm{IC}}(S)$ have the same distribution [19]. Moreover, given a seed set S, letting A_t be the set of nodes that become active at timestamp t in the diffusion process and B_t be the set of nodes in $\widehat{G}$ each of which is at distance t from S, the proof in Kempe et al. [19] also implies A_t and B_t have the same distribution.

Definition 2 (Linear Threshold Model (LT) [19]). *Given a social network (directed graph)* $G = (V, E, P)$ *with the edge weights* P *satisfying* $\sum_{u \in N(v)} p_{(u,v)} \leq 1$ *for each* $v \in V$, *the LT model assigns the state of each node either active or inactive. Given the input seed set* $S \subseteq V$, *LT outputs the set* $\Gamma_{\mathrm{LT}}(S)$ *as follows:*

1. At the beginning, for each vertex v, a threshold $\theta_v \in [0, 1]$ is sampled uniformly at random independently.
2. At timestamp 0, only nodes in S are active.
3. At each timestamp $t = 1, 2, \ldots$, a node v is active if the sum of the weights of edges from v's active in-neighbors to v exceeds the threshold θ_v:

$$\sum_{u : u \in N(v) \text{ and } u \text{ is active}} p_{(u,v)} \geq \theta_v$$

4. The diffusion process terminates when no additional activation occurs, and LT outputs $\Gamma_{\mathrm{LT}}(S)$ as the set of active nodes.

The linear threshold model also has a live-edge interpretation, though it is less intuitive than that of the independent cascade model.

Let $\widehat{G} = (V, \widehat{E})$ be a live-edge graph of G. For a node $v \in V$ with its in-neighbors indexed by $N(v) = \{u_1, \ldots, u_{|N(v)|}\}$, a real number γ is sampled from $[0, 1]$ uniformly at random. The i-th incoming edge (u_i, v) is included in $\widehat{E}$ if and only if $\gamma \in \left[\sum_{t=1}^{i-1} p_{(u_t,v)}, \sum_{t=1}^{i} p_{(u_t,v)} \right)$, and no edge is included in $\widehat{E}$ if $\gamma \geq \sum_{t=1}^{|N(v)|} p_{(u_t,v)}$. Notice that each node v has at most one in-neighbor in $\widehat{G}$, and it has exactly one in-neighbor if we further have $\sum_{t=1}^{|N(v)|} p_{(u_t,v)} = 1$. Let $\widehat{\sigma}_{\mathrm{LT}}(S)$ be the set of nodes that are reachable from S, then $\widehat{\Gamma}_{\mathrm{LT}}(S)$ and $\Gamma_{\mathrm{LT}}(S)$ have the same distribution [19]. Moreover, given a seed set S, letting A_t be the set of nodes that become active at timestamp t in the diffusion process and B_t

be the set of nodes in $\widehat{G}$ each of which is at distance t from S, the proof in Kempe et al. [19] also implies A_t and B_t have the same distribution.

Given a diffusion model Γ, let $\sigma_\Gamma(S) = \mathbb{E}(|\Gamma(S)|)$ be the expected number of infected vertices. The goal is to find S that maximizes $\sigma_\Gamma(S)$. It is known that computing $\sigma_\Gamma(\cdot)$ is #P-hard for both the IC and LT models [7,8]. On the other hand, by simple Monte-Carlo samplings, we can easily obtain a fully polynomial-time randomized approximation scheme (FPRAS) for computing $\sigma_\Gamma(\cdot)$. In some papers, it is assumed that $\sigma_\Gamma(\cdot)$ can be accessed by an oracle. This becomes problematic if we are dealing with computational complexity, as an oracle to the function $\sigma_\Gamma(\cdot)$ for Γ being the IC or LT model is equivalent to a #P oracle, which is too powerful. Thus, in our paper, we will not make such an assumption. We remark that all our hardness proofs in Sect. 3 do not exploit the hardness of evaluating $\sigma_\Gamma(\cdot)$. This is elaborated in Remark 2 following Theorem 1. We also remark that all our algorithms for E-DIM presented in Sect. 4 do not rely on oracle accesses to $\sigma_\Gamma(\cdot)$. In particular, the feature of E-DIM that all vertices must be infected with probability 1 makes the problem special, and, as we will see in Sect. 4, deciding if $\sigma_\Gamma(S) = |V|$ is actually easy.

In this paper, we omit the subscript Γ of σ_Γ when it is clear from the context.

Remark 1. In many parts of our paper, we consider the *undirected* graphs $G = (V, E, P)$ where P assigns a weight p_e to each *undirected edge* $e = (u, v)$. The diffusion processes under both models are defined by regarding each undirected edge (u, v) as two anti-parallel directed edges (u, v) and (v, u) with the same weight $p_{(u,v)}$. Therefore, for each diffusion model, the setting with undirected graphs is a special case of that with directed graphs. All the hardness results for the case with undirected graphs extend to the case with directed graphs, and all the algorithms for the case with directed graphs extend to the case with undirected graphs.

Reverse Reachable Set. The influence spread of the aforementioned models can be characterized by the notion of the *reverse reachable set* [3] under the live-edge graph formulation.

Definition 3 (Reverse Reachable Set). *Given a directed graph $G = (V, E, P)$ with a randomly sampled live-edge graph $\widehat{G}$ (generated based on the specific diffusion model), the reverse reachable set of a node $v \in V$, denoted by $\mathrm{RR}_{\widehat{G}}(v)$, is the set of all nodes in $\widehat{G}$ that can reach v.*

We remark that $\mathrm{RR}_{\widehat{G}}(v)$ is a random set since $\widehat{G}$ is sampled randomly. Clearly, a seed set S can infect a vertex v under the live-edge graph $\widehat{G}$ if and only if $S \cap \mathrm{RR}_{\widehat{G}}(v) \neq \emptyset$. Thus, the probability that S infects v equals the probability that S overlaps with $\mathrm{RR}_{\widehat{G}}(v)$.

The Decision Influence Maximization Problem. Traditionally, the influence maximization problem is defined as an optimization problem: given a graph $G = (V, E, P)$, a positive integer k and a diffusion model Γ, find a seed set

$S \subseteq V$ with $|S| \leq k$ that maximizes the expected spread $\sigma_\Gamma(S)$ [14,18,19,26]. In this paper, we consider the decision version of IM, defined as follows.

Definition 4 (Decision Influence Maximization (DIM)). *A decision influence maximization instance is written as $(G = (V, E, P), \Gamma, k, t)$, where $G = (V, E, P)$ represents a weighted social network, Γ denotes a diffusion model, $k \in \mathbb{Z}^+$, and $t \in \mathbb{R}^+$. It is a yes instance if there exists $S \subseteq V$ with $|S| \leq k$ such that $\sigma_\Gamma(S) \geq t$. Otherwise, it is a no instance.*

Considering the special case of DIM that the spreading of the seed is exactly all the nodes, we also define the *Exact Decision Influence maximization* (E-DIM) problem.

Definition 5 (Exact Decision Influence Maximization (E-DIM)). *The exact decision influence maximization problem is a special case of the decision influence maximization problem with $t = n = |V|$.*

We consider DIM and E-DIM in both directed and undirected graphs under both the IC and LT models. We study the parameterized complexity for DIM and E-DIM, and we consider both settings parameterized by k and t respectively.

3   Parameterized Complexity of the DIM

In this section, we show that DIM is W[1]-hard for both IC and LT models parameterized by k and t, even for undirected graphs. In addition, for DIM with the IC model and directed graphs, we show that the reduction of Kempe et al. [19] can easily imply W[2]-hardness parameterized by k.

Our reduction uses the independent set problem on regular graphs. Given an undirected graph $G = (V, E)$, we say that $D \subseteq V$ is an *independent set* if there exists no edge between any pair of vertices in D.

Definition 6 (REGULARINDSET Problem). *Given an undirected regular graph $G = (V, E)$ and an integer k, decide if G contains an independent set of k vertices.*

It is proved in the textbook [13] (Theorem 13.4 and Theorem 13.18) that REGULARINDSET is W[1]-complete parameterized by k.

Theorem 1. DIM *with the* IC *model is W[1]-hard parameterized by k and t even for undirected regular graphs.*

Proof. We prove Theorem 1 by a parameterized reduction from REGULARIND-SET. Given a REGULARINDSET instance $(G_r^d = (V_r^d, E_r^d), k_r^d)$, where G_r^d denotes an undirected d-regular graph $(\deg(v) = d$ for each $v \in V_r^d)$ with $|V_r^d| = n$ and $|E_r^d| = m$. We construct an DIM instance $(G = (V, E, P), \Gamma, k, t)$ as follows. Let p be a positive value such that $kd(1 - (1 - p)^{d-1}) + pd^2 n < 1$. Notice that this is always possible, as the limits of $kd(1 - (1 - p)^{d-1})$ and $pd^2 n$ are both 0 for $p \to 0$. It is also easy to see that such a p can be found in polynomial time. Let

$V = V_r^d$, $E = E_r^d$, $k = k_r^d$, P be such that $p_e = p$ for each $e \in E$, and Γ be given by the IC model. Finally, let $t = k + kdp(1-p)^{d-1}$. The construction can clearly be done in polynomial time. Since $k = k_r^d$ and $t = k + kdp(1-p)^{d-1} < k + 1$ (to see this, $kd(1 - (1-p)^{d-1}) + pd^2 n < 1$ implies $pd^2 n < 1$, which implies $p < \frac{1}{d^2 n}$; thus, $kdp(1-p)^{d-1} < kdp < \frac{k}{dn} < 1$ by noticing that we can assume $k = k_r^d < n$ without loss of generality, for otherwise the REGULARINDSET instance is trivial), this is a parameterized reduction for parameters k and t for DIM. We will show that, if (G_r^d, k_r^d) is a yes instance, then $(G = (V, E, P), \Gamma, k, t)$ is a YES instance, and vice versa. Before we show this, we show the following claim.

($*$) Given a seed set S, the probability that a vertex $v \notin S$ is infected in the first round falls into the closed interval $[h_v p(1-p)^{d-1}, h_v p]$, where $h_v = |N(v) \cap S|$.

To prove ($*$), first notice that the claim holds trivially for $h_v = 0$ (i.e., S contains none of v's in-neighbors). We suppose $h_v \geq 1$ from now on. The probability that v is infected in the first round is given by $1 - (1-p)^{h_v}$ (notice that $(1-p)^{h_v}$ is the probability that all the h_v in-neighbors of v fail to infect v). On the other hand, by the formula of geometric progression, we have

$$1 - (1-p)^{h_v} = p \cdot \frac{1 - (1-p)^{h_v}}{1 - (1-p)} = p \cdot \sum_{i=0}^{h_v - 1} (1-p)^i.$$

Notice that each term in the summation falls into the interval $[(1-p)^{d-1}, 1]$, we have proved ($*$).

Then we continue to prove Theorem 1. If the REGULARINDSET instance (G_r^d, k_r^d) is a yes instance such that there exists an independent set D with size k_r^d, we show that the DIM instance is a yes instance by proving $\sigma_{IC}(S) \geq t$ for $S = D$. By only considering the vertices infected in the first round, We have the following lower bound of $\sigma_{IC}(S)$:

$$\sigma_{IC}(S) \geq |S| + [\text{expected number of activated nodes in the first round}]$$
$$\geq k + \sum_{v \notin S} h_v p(1-p)^{d-1} \qquad \text{(by ($*$))}$$
$$= k + kdp(1-p)^{d-1},$$

where the last equality is due to $\sum_{v \notin S} h_v = kd$ (notice that S is an independent set and there are exactly kd edges between S and $V \setminus S$). This proves $\sigma_{IC}(S) \geq t$ since we have set $t = k + kdp(1-p)^{d-1}$.

Now, suppose (G_r^d, k_r^d) is a NO REGULARINDSET instance. For any S with $|S| = k$, the number of edges between S and $V \setminus S$ is at most $kd - 1$. We aim to find an upper bound for $\sigma_{IC}(S)$. We will do this by bounding the expected number of infected vertices in each round.

-   For round 0, the expected number of infected vertices is exactly k.
-   For round 1, by ($*$), the expected number of infected vertices is at most $\sum_{v \notin S} h_v p \leq (kd - 1)p$.

– For each of the future rounds, we upper bound to the probability that an arbitrary vertex v is infected. For a vertex v to be infected after the first round, the reverse reachable set of v must contain at least one vertex u such that $(u, v) \notin E$. This means the reverse reachable set contains a path of length at least 2 that ends at v. The probability that a single length-2 path exists is p^2, and there are d^2 such paths. Therefore, by a union bound, d^2p^2 is an upper bound to the probability that a vertex v is infected after round 1. Thus, the expected number of infected vertices after round 1 is at most d^2p^2n.

Putting together, we have $\sigma_{\mathrm{IC}}(S) \leq k + (kd - 1)p + d^2p^2n$. Since we have set p such that $kd(1 - (1 - p)^{d-1}) + pd^2n < 1$, we have $k + (kd - 1)p + d^2p^2n < t$. $\square$

Theorem 2. DIM *in the* LT *model is* W[1]-*hard parameterized by* k *and* t *even for undirected regular graphs.*

Proof. This proof is mostly identical to the proof of Theorem 1. Specifically, the construction is exactly the same. We need to prove that the claim (∗) also holds for the LT model. To show (∗), consider a seed set S and a vertex $v \notin S$, and suppose $h_v = |N(v) \cap S|$. The sum of the weights of the edges connected from S to v is h_vp. By the definition of the LT model, the probability that v is infected in the first round is the probability that $\theta_v \geq h_vp$, which is exactly h_vp since θ_v is sampled uniformly at random from $[0, 1]$. The claim (∗) holds as $h_vp \in [h_vp(1 - p)^{d-1}, h_vp]$. The remaining part of the proof is exactly the same as it is for Theorem 1. $\square$

Remark 2. We have mentioned before that computing the function $\sigma(\cdot)$ is #P-complete for both the IC and LT models. On the other hand, the reduction used in the proof of Theorem 1 and 2 does not exploit the hardness of evaluating $\sigma(\cdot)$. Intuitively, the problem remains W[1]-hard even if $\sigma(\cdot)$ can be computed in polynomial time. However, it becomes tricky to formally define what this means. As we mentioned in Sect. 2, assuming an oracle that outputs $\sigma(S)$ given *any* input $G = (V, E, P)$ is equivalent to assuming a #P oracle, which is too strong. A more reasonable way is to assume an oracle that outputs $\sigma(S)$ given any input $S \subseteq V$ for *the given* INFMAX instance (i.e., for the *particular* edge-weighted graph $G = (V, E, P)$). However, it then becomes unclear if REGULARINDSET remains W[1]-complete given this oracle (although having an oracle that computes the seeds' expected influence seems irrelevant to the independent set problem, it is hard to exclude the possibility that the expected influence for certain carefully chosen seed set reveals some structural information about the independent set problem). Thus, the proof above cannot show the W[1]-completeness of our problem. Due to these subtle issues, we will stick to the setting where no such oracle is assumed.

We also remark that our proof reveals an inapproximability gap that is larger than the gap provided by the #P-hardnesss of evaluating $\sigma_{\mathrm{IC}}(\cdot)$. As we mentioned in Sect. 2, simple Monte-Carlo sampling method gives an FPRAS: we can get an $(1 \pm \varepsilon)$-approximation to $\sigma_{\mathrm{IC}}(S)$ with a running time polynomial in $1/\varepsilon$ and the input length. On the other hand, our proof of Theorem 1 shows an $(1 - \varepsilon)$ inapproximability result for some $\varepsilon > \frac{1}{16n^4}$. This gives a larger gap as $1/\varepsilon = 16n^4$ is still of polynomial scale. The number $\frac{1}{16n^4}$ comes from some detailed calculations presented as follows.

Set $p = \frac{1}{4d^2n}$ in the proof. We have shown that a yes REGULARINDSET instance implies an influence of at least $t = k + kdp(1-p)^{d-1}$, and a no instance implies an influence of at most $t' = k + (kd-1)p + d^2p^2n$. We need to show that $\varepsilon := \frac{t-t'}{t} > \frac{1}{16n^4}$. Notice that $1 > (1-p)^{d-1} > (1 - \frac{1}{4d^2n})^d \geq 1 - \frac{1}{4dn}$, where the last inequality is due to that 1) $f(x) = (1 - \frac{a}{x})^x$ is decreasing on $[1, \infty)$ for $a < 1$, 2) $a := \frac{1}{4dn} < 1$, and 3) $d \geq 1$. With the above upper and lower bounds to $(1 - p)^{d-1}$, it is easy to bound ε: we have $t - t' = p\left(1 - kd\left(1 - (1-p)^{d-1}\right) - d^2pn\right) > p\left(1 - \frac{k}{4n} - \frac{1}{4}\right) > \frac{1}{2}p = \frac{1}{8d^2n} > \frac{1}{8n^3}$ and $t = k + kdp(1-p)^{d-1} < n + ndp = n + \frac{1}{4d} < 2n$, which implies $\varepsilon > \frac{1}{16n^4}$.

As a last remark, our hardness result is parameterized by both k and t, which is incomparable to the corresponding hardness results obtained by exploiting the #P-hardness of evaluating $\sigma(\cdot)$.

Lastly, for directed graphs based on the IC model, we have the following.

Theorem 3. DIM *with the* IC *model is* W[2]-*hard parameterized by* k *in directed graphs.*

A reduction from the set cover problem is presented in Kempe et al. [19], and the same reduction can be used to prove Theorem 3. We include it for completeness and for our remark thereafter.

Proof. We prove the Theorem 3 by a parameterized reduction from the SET-COVER problem, a well-known W[2]-complete problem. A SETCOVER instance is given by $(U, \mathcal{C}, k)$, where $U = \{u_1, u_2, \ldots\}$ is a ground set, $\mathcal{C} = \{C_1, C_2, \ldots\}$ is a collection of the subset of U, and k is a positive integer. The SETCOVER problem asks if there exists a sub-collection $\mathcal{B} \subseteq \mathcal{C}$ with $|\mathcal{B}| \leq k$ such that $\cup_{C \in \mathcal{B}} C = U$. Given a SETCOVER instance $(U, \mathcal{C}, k)$, the DIM instance is constructed as follows. We have $|U|$ vertices corresponding to elements in U, and $|\mathcal{C}|$ vertices corresponding to subsets in $\mathcal{C}$. We build a directed edge from the vertex representing $C \in \mathcal{C}$ to the vertex representing $u \in U$ if $u \in C$ in the SETCOVER instance. The probability in the IC model for each edge is set to 1. The parameter k in both the SETCOVER instance and the INFMAX instance are kept the same. It is easy to verify that we can always infect $k + |U|$ vertices deterministically if the SETCOVER instance is a yes instance, and $\sigma(S) < k + |U|$ for any seed set S if the SETCOVER instance is a no instance.

Remark 3. In contrast to Remark 2, the W[2]-hardness result continues to hold if we have an oracle access to $\sigma_{\text{IC}}(\cdot)$ for any input seed set in the *given* instance of INFMAX. This is because the INFMAX instance we constructed in the reduction is *deterministic*, with $p_e = 1$ for all edges. The computation of $\sigma_{\text{IC}}(\cdot)$ for such instances can be done in polynomial time. An oracle access to $\sigma_{\text{IC}}(\cdot)$ does not add any additional computational power.

4   Fixed-Parameter Tractability of E-DIM

We prove the E-DIM problem is fixed-parameter tractable under the LT model parameterized by k, for both undirected and directed graphs. We also prove E-DIM is polynomial-time solvable under the IC model for both directed and undirected graphs. In the remaining part of this section, all graphs are assumed to be directed. As we have remarked in Remark 1, our problem with undirected graphs can be viewed as a special case.

4.1   A Polynomial Time Algorithm for E-DIM Under IC Model

E-DIM under the IC model is polynomial-time solvable.

Theorem 4. E-DIM *in the* IC *model can be solved in polynomial time.*

Firstly, since we require each node to be infected with probability 1 in E-DIM, all edges e with $p(e) < 1$ are useless (because the live-edge graph containing none of these edges is sampled with a nonzero probability, and we need to make sure every vertex is still reachable from the seed set S under this live-edge graph sample), and we can simply remove them from the graph. Then, the problem becomes deciding if it is possible to select a set S of k nodes in an *unweighted* graph $G = (V, E)$ such that all vertices are reachable from S. This problem can be solved in polynomial time: we first identify all strongly connected components of G and the connections between these components; these connections form a directed acyclic graph H where vertices of H are strongly connected components of G (notice that H can be constructed by the standard depth-first search algorithm); then it is easy to see that we need to place a seed at each component corresponding to a source vertex (i.e., a vertex with in-degree 0) of H, and the problem is simplified to deciding if the number of the source vertices in H is no more than k. We leave the details for proving Theorem 4 to the readers.

4.2   Fixed-Parameter Tractability of E-DIM Under LT Model

Firstly, we remark that E-DIM for the LT model is NP-complete even for undirected graphs, which motivates the study of parameterized complexity. Kempe et al. [19] proved that the influence maximization problem for the LT model is NP-hard. The reduction used by Kempe et al. is from the VERTEX COVER problem. Moreover, the reduction satisfies that the VERTEX COVER instance is a yes instance if and only if we can select k seeds such that all vertices are infected with probability 1. This directly shows the NP-hardness of E-DIM. To show the containment in NP, notice that we can just set $\theta_v = 1$ for each vertex v since we require each vertex to be infected with probability 1. Then the diffusion process is deterministic and it is easy to compute the set of infected vertices given a seed set S.

Then we demonstrate that E-DIM is fixed-parameter tractable parameterized by k based on the LT model. We will deal with directed graphs as mentioned before.

Theorem 5. E-DIM *is fixed-parameter tractable parameterized by* k *under the* LT *model.*

To show this, we present a parameterized reduction from E-DIM to the *feedback vertex set* problem in directed graphs, which is a known fixed-parameter tractable problem. Given a directed graph $G = (V, E)$, we say $F \subseteq V$ is a *feedback vertex set* if every directed cycle in G contains at least one vertex from F. Equivalently, removing F from G results in a directed acyclic graph.

Definition 7 (DIRECTEDFEEDBACKVERTEXSET Problem). *Given a directed graph* $G = (V, E)$ *and an integer* k*, decide if* G *contains a feedback vertex set* F *of at most* k *nodes.*

Chen et al. [5] present an algorithm for this problem with running time $4^k \cdot k! \cdot n^{O(1)}$, indicating fixed-parameter tractability of the problem parameterized by k. To prove Theorem 5, it then remains to present a parameterized reduction from E-DIM to the DIRECTEDFEEDBACKVERTEXSET problem. Before this, we prove the following lemma first.

In the lemma below and the remaining part of this section, we denote $L = \{v \mid \sum_{u \in N(v)} p_{(u,v)} < 1\}$. Given a directed graph $G = (V, E, P)$ and a vertex set $S \subseteq V$, we use $G - S$ to denote the graph by deleting S and all the incident edges (incoming edges and outgoing edges) from G.

Lemma 1. *An* E-DIM *instance* $(G = (V, E, P), \text{LT}, k)$ *is a yes instance if and only if there exists* $S \subseteq V$ *with* $|S| \leq k$ *such that*

1. $L \subseteq S$, *and*
2. $G - S$ *contains no directed cycle.*

Proof. To see the if-direction, suppose S satisfies 1 and 2. We show that all vertices will be infected with probability 1 if S is the seed set. Consider an arbitrary vertex $v \in V$, and consider an arbitrary live edge graph $\widehat{G}$ with a non-zero probability that defines a reverse reachable set $\text{RR}_{\widehat{G}}(v)$ of v. By our discussion after Definition 2, each vertex u in $\widehat{G}$ has at most one in-neighbor, and it has exactly one in-neighbor if $u \notin L$. Therefore, if we reverse the edges in $\text{RR}_{\widehat{G}}(v)$, the reverse reachable set is a *directed path* that starts from v and ends at either a vertex in L or a vertex that is already visited. In the former case, $\text{RR}_{\widehat{G}}(v)$ contains a seed since $L \subseteq S$; in the latter case, $\text{RR}_{\widehat{G}}(v)$ contains a directed cycle, and it contains a seed since $G - S$ contains no cycle. Therefore, we have proved that v is reachable from S for any live edge graph $\widehat{G}$ with a non-zero probability. Thus, v is infected with probability 1. Since v is arbitrary, we have proved the if-direction.

To prove the only-if-direction, consider an arbitrary seed set S with $|S| \leq k$, and suppose either 1 or 2 fails. We show that there is a vertex v whose infection probability is strictly less than 1. If there exists $\ell \in L$ such that $\ell \notin S$, then ℓ will not be infected with probability 1: when the threshold θ_ℓ is between $\sum_{u \in N(\ell)} p_{(u,\ell)}$ and 1, we know that ℓ will never be infected even if all of ℓ's in-neighbors are infected; this happens with probability $1 - \sum_{u \in N(\ell)} p_{(u,\ell)} > 0$ since $\ell \in L$. If $G - S$ contains a directed cycle, let C be a directed cycle in $G - S$. With a positive probability, C is a part of the live edge graph and C is isolated (i.e., form a weakly connected component). In this case, no vertex in C can be infected. $\qquad\square$

With Lemma 1, we know that a seed set S can infect all vertices with probability 1 if and only if $L \subseteq S$ and $S \setminus L$ is a feedback vertex set of the graph $G - L$. The parameterized reduction from E-DIM with the LT model to the DIRECTEDFEEDBACKVERTEXSET problem becomes straightforward: given an E-DIM instance $(G = (V, E, P), \mathrm{LT}, k)$, we construct a DIRECTEDFEEDBACKVERTEXSET instance $(G^d = (V^d, E^d), k^d)$ with $G^d = G - L$ and $k^d = k - |L|$.

References

1. Bazgan, C., Chopin, M., Nichterlein, A., Sikora, F.: Parameterized approximability of maximizing the spread of influence in networks. J. Discrete Algorithms **27**, 54–65 (2014)
2. Ben-Zwi, O., Hermelin, D., Lokshtanov, D., Newman, I.: Treewidth governs the complexity of target set selection. Discret. Optim. **8**(1), 87–96 (2011)
3. Borgs, C., Brautbar, M., Chayes, J., Lucier, B.: Maximizing social influence in nearly optimal time. In: Proceedings of the Twenty-Fifth Annual ACM-SIAM Symposium on Discrete Algorithms, pp. 946–957. SIAM (2014)
4. Brown, J.J., Reingen, P.H.: Social ties and word-of-mouth referral behavior. J. Consum. Res. **14**(3), 350–362 (1987)
5. Chen, J., Liu, Y., Lu, S., O'sullivan, B., Razgon, I.: A fixed-parameter algorithm for the directed feedback vertex set problem. In: Proceedings of the Fortieth Annual ACM Symposium on Theory of Computing, pp. 177–186 (2008)
6. Chen, N.: On the approximability of influence in social networks. SIAM J. Discret. Math. **23**(3), 1400–1415 (2009)
7. Chen, W., Wang, C., Wang, Y.: Scalable influence maximization for prevalent viral marketing in large-scale social networks. In: Proceedings of the 16th ACM SIGKDD International Conference on Knowledge Discovery and Data Mining, pp. 1029–1038 (2010)
8. Chen, W., Yuan, Y., Zhang, L.: Scalable influence maximization in social networks under the linear threshold model. In: 2010 IEEE International Conference on Data Mining, pp. 88–97. IEEE (2010)
9. Chopin, M., Nichterlein, A., Niedermeier, R., Weller, M.: Constant thresholds can make target set selection tractable. Theory Comput. Syst. **55**, 61–83 (2014)
10. Cordasco, G., Gargano, L., Mecchia, M., Rescigno, A.A., Vaccaro, U.: A fast and effective heuristic for discovering small target sets in social networks. In: Combinatorial Optimization and Applications: 9th International Conference, 2015, Houston, TX, USA, pp. 193–208. Springer (2015)
11. Cordasco, G., Gargano, L., Rescigno, A.A.: On finding small sets that influence large networks. Soc. Netw. Anal. Min. **6**(1), 1–20 (2016). https://doi.org/10.1007/s13278-016-0408-z
12. Cordasco, G., Gargano, L., Rescigno, A.A.: Active influence spreading in social networks. Theoret. Comput. Sci. **764**, 15–29 (2019)

13. Cygan, M., et al.: Parameterized Algorithms, vol. 5. Springer (2015)
14. Domingos, P., Richardson, M.: Mining the network value of customers. In: Proceedings of the Seventh ACM SIGKDD International Conference on Knowledge Discovery and Data Mining, KDD 2001, pp. 57–66. ACM, New York (2001)
15. Downey, R.G., Fellows, M.R., et al.: Fundamentals of Parameterized Complexity, vol. 4. Springer (2013)
16. Downey, R.G., Fellows, M.R.: Parameterized Complexity. Springer (2012)
17. Granovetter, M.: Threshold models of collective behavior. Am. J. Sociol. **83**(6), 1420–1443 (1978)
18. Kempe, D., Kleinberg, J., Tardos, É.: Influential nodes in a diffusion model for social networks. In: Caires, L., Italiano, G.F., Monteiro, L., Palamidessi, C., Yung, M. (eds.) ICALP 2005. LNCS, vol. 3580, pp. 1127–1138. Springer, Heidelberg (2005). https://doi.org/10.1007/11523468_91
19. Kempe, D., Kleinberg, J.M., Tardos, É.: Maximizing the spread of influence through a social network. Theory Comput. **11**, 105–147 (2015)
20. Khanna, S., Lucier, B.: Influence maximization in undirected networks. In: Proceedings of the Twenty-Fifth Annual ACM-SIAM Symposium on Discrete Algorithms, pp. 1482–1496. SIAM (2014)
21. Lerman, K., Ghosh, R.: Information contagion: an empirical study of the spread of news on Digg and Twitter social networks. In: Proceedings of the International AAAI Conference on Web and Social Media, vol. 4, no. 1, pp. 90–97 (2010)
22. Mossel, E., Roch, S.: Submodularity of influence in social networks: from local to global. SIAM J. Comput. **39**(6), 2176–2188 (2010)
23. Nichterlein, A., Niedermeier, R., Uhlmann, J., Weller, M.: On tractable cases of target set selection. Soc. Netw. Anal. Min. **3**, 233–256 (2013)
24. Niedermeier, R.: Invitation to Fixed-Parameter Algorithms, vol. 31. OUP Oxford (2006)
25. Pastor-Satorras, R., Vespignani, A.: Epidemic dynamics and endemic states in complex networks. Phys. Rev. E **63**(6), 066117 (2001)
26. Richardson, M., Domingos, P.: Mining knowledge-sharing sites for viral marketing. In: Proceedings of the Eighth ACM SIGKDD International Conference on Knowledge Discovery and Data Mining, pp. 61–70 (2002)
27. Rigobon, R.: Contagion: how to measure it? In: Preventing Currency Crises in Emerging Markets, pp. 269–334. NBER Chapters, National Bureau of Economic Research, Inc (2002)
28. Schoenebeck, G., Tao, B.: Beyond worst-case (in) approximability of nonsubmodular influence maximization. ACM Trans. Comput. Theory (TOCT) **11**(3), 1–56 (2019)
29. Schoenebeck, G., Tao, B.: Influence maximization on undirected graphs: toward closing the $(1 - 1/e)$ gap. ACM Trans. Econ. Comput. (TEAC) **8**(4), 1–36 (2020)
30. Schoenebeck, G., Tao, B., Yu, F.Y.: Limitations of greed: influence maximization in undirected networks re-visited. In: Proceedings of the 19th International Conference on Autonomous Agents and MultiAgent Systems, pp. 1224–1232 (2020)
31. Schoenebeck, G., Tao, B., Yu, F.Y.: Think globally, act locally: on the optimal seeding for nonsubmodular influence maximization. Inf. Comput. **285**, 104919 (2022)

Improved Parameterized Algorithms for Scheduling with Precedence Constraints and Time Windows

Feng Shi[1(✉)] [ID], Yicong Zhu[1], Guangwei Wu[2], Jingyi Liu[1], and Jianxin Wang[1]

[1] School of Computer Science and Engineering, Central South University,
Changsha 410083, People's Republic of China
fengshi@csu.edu.cn
[2] College of Computer and Information Engineering, Central South University
of Forestry and Technology, Changsha 410004, People's Republic of China

Abstract. Within the paper, we study several variants of the decision problem, Scheduling with precedence constraints and time windows, denoted by $P \mid prec, r_i, d_i \mid \star$, and present improved fixed-parameter algorithms parameterized by the maximum processing time $p_{\max}$ and the maximum number μ of overlapping time windows, defined as $\mu = \max_{t \in \mathbb{N}} |\{i \in S \mid r_i \leq t < d_i\}|$. Firstly, we propose an algorithm for $P \mid prec, r_i, d_i \mid \star$ with time complexity $O((p_{\max} + 2)^{\mu} p_{\max} n^3)$, where n is the number of tasks. This significantly improves the previously best-known algorithm with time complexity $O(p_{\max}^{2\mu} \cdot 16^{\mu} \sqrt{\mu} \cdot n^3)$. Then for the unit processing time case $P \mid prec, p_i = 1, r_i, d_i \mid \star$, we further develop an algorithm with time complexity $O(2^{\mu} \mu m n^3)$, where m is the number of machines, improving the previously best-known algorithm with time complexity $O(16^{\mu} n^4)$. Finally, we extend the two algorithms to the typed machine setting, solving $P \mid \mathcal{M}_j(type), prec, r_i, d_i \mid \star$ and $P \mid \mathcal{M}_j(type), prec, p_i = 1, r_i, d_i \mid \star$, with time complexities $O((p_{\max} + 2)^{\mu} p_{\max} n^3)$ and $O(2^{\mu} \mu m^k n^3)$, respectively, where k is the number of machine types.

Keywords: Scheduling · Precedence constraints · Time windows · Fixed-parameter algorithm · Pathwidth

1 Introduction

Scheduling with resource limitations and precedence constraints is a well-known research area that has garnered widespread attention since the last century [2,3,12,17]. Within the paper, we study a fundamental scheduling problem with a set S of n tasks (or jobs) and m identical machines as input. Each task

This work is supported in part by the National Natural Science Foundation of China under Grants 62472449 and 62072476, the Open Project of Xiangjiang Laboratory (No. 22XJ03005), and the Hunan Provincial Natural Science Foundation of China under Grant 2025JJ50395.

F. V. Fomin and M. Xiao (Eds.): COCOON 2025, LNCS 15984, pp. 253–267, 2026.
https://doi.org/10.1007/978-981-95-0218-9_19

$i \in S$ is characterized by a processing time p_i, a release time r_i, and a deadline d_i. Furthermore, the tasks are subject to precedence constraints, represented by a partial order $\preceq$. Specifically, if tasks $i \prec j$, then task i must be completed before task j starts execution. The aim of the problem is to look for a *feasible* schedule for the tasks in S that satisfies the precedence constraints and time windows, and optimizes a given objective function, or return NULL if no feasible schedule exists. The problem is denoted by $P \mid prec, r_i, d_i \mid \gamma$ using the standard notation [8], where γ is the objective function, and can be $C_{\max}$ (makespan), $L_{\max}$ (maximum lateness), $\sum C_i$ (sum of completion times), or $\star$ (without objective function). Our main focus is on the fixed-parameter algorithms for the variants of $P \mid prec, r_i, d_i \mid \star$, which just ask for the existence of a feasible schedule. Note that the algorithm for $P \mid prec, r_i, d_i \mid \star$ combined with a binary search [16], can solve the optimization problems $P \mid prec, r_i, d_i \mid C_{\max}$ and $P \mid prec, r_i, d_i \mid L_{\max}$. In the following, we review the results on several simplified variants of $P \mid prec, r_i, d_i \mid \star$.

The problem $P \parallel C_{\max}$ without precedence constraints and time windows is the Minimum Makespan Scheduling problem, which is well-known to be NP-hard [10] even for $m = 2$ (i.e., $P2 \parallel C_{\max}$). When introducing a time window for each task, the NP-hardness of $1 \mid r_i, d_i \mid C_{\max}$ still holds, even for $m = 1$ [10]. Garey and Johnson [7] showed that $P2 \mid prec, p_i = 1, r_i, d_i \mid C_{\max}$ with unit processing times is solvable in $O(n^3)$ runtime. However, when the number of machines is given as a part of the input, $P \mid prec, p_i = 1 \mid C_{\max}$ becomes NP-hard [20]. Note that the computational complexity of $Pm \mid prec, p_i = 1 \mid C_{\max}$ for $m > 2$ remains unknown [13].

For many scheduling problems, their NP-hardness implies that no efficient exact deterministic algorithms exist under the assumption P $\neq$ NP. Parameterized computation has been introduced as a promising algorithmic framework to address this challenge [4]. A parameterized problem is *fixed-parameter tractable* (abbr. FPT) [4] if it can be solved in runtime $f(k) \cdot n^{O(1)}$, where n is the input size and f is a computable function depending solely on the parameter k. Beyond FPT, parameterized complexity theory provides a refined hierarchy of intractability classes, such as *W-hardness* and *para-NP-hardness*. A problem is *para-NP-hard* if it is NP-hard even for constant parameter values. Previous work has explored scheduling problems w.r.t. various parameters such as the number of machines m, the height h and width w of the partial order $\preceq$. Lenstra and Rinnooy Kan [11] proved that the problem $P \mid prec, p_i = 1, h \leq 3 \mid C_{\max} \leq 3$ is NP-complete, implying that the problem is para-NP-complete when parameterized by h. Dolev and Warmuth [6] developed an $O(n^{h(m-1)+1})$-algorithm for the problem $P \mid prec, p_i = 1 \mid C_{\max}$. Later, Möhring [15] proposed an $O(n^w)$-algorithm for the same problem. Bodlaender and Fellows [1] showed that $P \mid prec, p_i = 1 \mid C_{\max}$ is W[2]-hard when parameterized by both m and w. Van Bevern et al. [21] proved that the problem $P \mid prec, p_i \in \{1, 2\} \mid C_{\max}$ is W[2]-hard when parameterized by w. Fixed-parameter algorithms have been developed w.r.t. various time-related parameters. Mnich and Wiese [14] showed $P \parallel C_{\max}$ is FPT w.r.t. the parameter $p_{\max}$, where $p_{\max}$ is the maximum processing time of the considered tasks. In another line of work, Van Bevern et al. [21] proposed a fixed-parameter algorithm for the Resource Constrained Project Scheduling Problem (abbr. RCPSP) parameterized by w and λ, where λ is the maximum allowed lag of a task i from the earliest possible start time s_i with precedence constraints (specifically, $s_i = \max_{j \prec i} s_j + p_j$). The RCPSP is closely related to $P \mid prec \mid C_{\max}$, but with a key distinction: $P \mid prec \mid C_{\max}$ considers a single type of resource, but the RCPSP considers k different types of resources.

Recently, Munier Kordon [16] proposed a fixed-parameter algorithm with complexity $O(2^{4 \cdot \mathrm{pw}(\mathcal{I})} \cdot n^4)$ for $P \mid prec, p_i = 1, r_i, d_i \mid \star$, parameterized by the pathwidth $\mathrm{pw}(\mathcal{I})$ of the interval graph constructed from task time windows (r_i, d_i). Later Hanen and Munier Kordon [9] extended this to $P \mid \mathcal{M}_j(type), prec, r_i, d_i \mid \star$, where tasks are restricted to specific machine classes. Their algorithm runs in $O(f(\mu, \rho)^2 \cdot (n^4 + n \cdot \mu^{\mu+2} \cdot \mu!))$, where $f(\mu, \rho) = \rho^{\mu^2}(\mu + 1)^{\mu^2} 2^\mu$, $\mu = \max_{t \in \mathbb{N}} |\{i \in S \mid r_i \leq t < d_i\}|$, $\rho = \min\{p_{\max}, sl_{\max}\}$, and $sl_{\max} = \max_{i \in S}(d_i - r_i - p_i)$. Notably, $\mu = \mathrm{pw}(\mathcal{I}) + 1$. Further advancing this research, Tarhan et al. [19] developed a fixed-parameter algorithm for $P \mid prec, r_i, d_i \mid \star$, parameterized by μ and $p_{\max}$, with complexity $O(p_{\max}^{2\mu} \cdot 16^\mu \sqrt{\mu} \cdot n^3)$. Their approach builds on the Demeulemeester and Herroelen algorithm (ALG-DH) [5], which uses a branch-and-bound strategy for the RCPSP. Each node u in the search tree maintains a time point t and a temporary feasible partial schedule σ, where tasks in σ start by time t. The schedule σ is temporary, meaning that the tasks in progress at time point t may be delayed in future steps. To generate a child node, the algorithm computes the earliest completion time t' of tasks in progress at t, adds eligible tasks (whose predecessors are completed by t) to σ, and delays a minimal subset of tasks if resource constraints are violated. Tarhan et al. [19] compressed the search tree into a graph, where each vertex is a quadruple (V, t, P, M). Here, V is the set of scheduled tasks, t is a time point, P is the set of tasks in progress, and M is a vector of completion times for tasks in P. By analyzing the number of states for fixed V, they bounded the number of sets P by 2^μ and vectors M by $p_{\max}^\mu$. With at most $\binom{2\mu}{\mu} \cdot n$ possible sets V, the total number of states is bounded by $\binom{2\mu}{\mu} \cdot p_{\max}^\mu \cdot 2^\mu \cdot n$. These insights yield the stated complexity.

By a thorough analysis of the algorithm due to Tarhan et al. [19], we identified two key shortcomings: (1) The process of generating new nodes in the search tree of ALG-DH is intricate, as it may require removing tasks in progress to maintain feasibility. This results in a flexible and diverse parent-child relationship, making the compression into a graph overly complex and difficult to describe; (2) To obtain the upper bound of the graph size, they analyzed the ranges of the variables V, P, and M separately. However, in reality, these three variables are closely interrelated. Motivated by these observations, we proposed targeted solutions and successfully designed an improved algorithm for the problem $P \mid prec, r_i, d_i \mid \star$.

Specifically, our algorithm for $P \mid prec, r_i, d_i \mid \star$ is based the Precedence Tree Algorithm (ALG-PT) by Sprecher [18] for the RCPSP. In ALG-PT, each node u in the search tree represents a sequence of tasks and is associated with a feasible partial schedule σ that respects the sequence's order. To generate a child u' of u, an eligible task i (whose predecessor tasks have been scheduled by σ) is added to the sequence. The start time of i in the new schedule of u' is set as the earliest precedence- and resource-feasible time which is not earlier than the start times of the tasks scheduled in σ. Obviously, the parent-child relationship in the search tree of ALG-PT is simpler than that of ALG-DH, as each child node adds exactly one more task to the sequence. Based on this structure, we define an "equivalence relation" among states in the search tree, partitioning the states into different "equivalence classes", each of which contains states that are equivalent on the equivalence relation, and corresponds to a vertex in the "state graph". To bound the size of the state graph precisely, we introduce the concept of "maximal perfect time intervals," and show that the start times of the tasks in an optimal schedule lie within these intervals. This refinement reduces the state graph size. As a result, our algorithm achieves a time complexity of $O((p_{\max} + 2)^\mu p_{\max} n^3)$, a substantial improvement over previous results.

For the special variant $P \mid prec, p_i = 1, r_i, d_i \mid \star$ with unit processing times, our algorithm achieves a time complexity of $O(3^\mu n^3)$. We further improve this to $O(2^\mu \mu m n^3)$ by introducing two key enhancements: (1) We require the associated schedule to be consistent with a given linear order of tasks, reducing the search space; (2) We refine the definition of equivalence classes to minimize the number of vertices in the state graph. In addition, we successfully adapt these two algorithms to two new variants $P \mid \mathcal{M}_j(type), prec, r_i, d_i \mid \star$ and $P \mid \mathcal{M}_j(type), prec, p_i = 1, r_i, d_i \mid \star$, where the m machines are of k different type classes, with time complexities $O((p_{\max} + 2)^\mu p_{\max} n^3)$ and $O(2^\mu \mu m^k n^3)$, respectively. Due to space limits, the proofs of the lemmas and theorems are omitted in this version but will be provided in the full version of the paper.

2   Preliminaries

Consider an instance $\mathcal{I}$ of the problem $P \mid prec, r_i, d_i \mid \star$ with a set S of n tasks and m identical machines. The tasks are subject to a partial order $\preceq$ on S, where $i \prec j$ indicates that i is a *predecessor* of j, and j is a *successor* of i. If $i \prec j$ and there is no task k such that $i \prec k \prec j$, then i is an *immediate predecessor* of j, and j is an *immediate successor* of i. For each task $i \in S$, denote by $\Gamma^-(i)$ (resp., $\Gamma^+(i)$) the set of immediate predecessors (resp., immediate successors) of task i. W.l.o.g., we assume that for any two tasks $i \prec j$, $r_i + p_i \leq r_j$. A *schedule* σ for S is an assignment of a start time $\sigma(i)$ for each task $i \in S$. The schedule σ for S is *feasible* if it satisfies the three constraints:

- **Time Window Constraint:** For each task $i \in S$, $r_i \leq \sigma(i) \leq d_i - p_i$;
- **Precedence Constraint:** For any pair of tasks $i \prec j$, then $\sigma(i) + p_i \leq \sigma(j)$ (i.e., task i must complete before task j starts);
- **Resource Constraint:** At any time point t, at most m tasks are in progress, i.e., $|\{i \in S \mid \sigma(i) \leq t < \sigma(i) + p_i\}| \leq m$.

An instance is shown in Fig. 1 and Table 1. W.l.o.g., we assume that the processing times, release times and deadlines of the tasks considered are positive integers. Thus we also assume that the start time $\sigma(i)$ for each $i \in S$ in any schedule considered in the paper is a positive integer. A *lower-set* of S is a subset $S' \subseteq S$ such that if $x \in S'$ then $y \in S'$ for any $y \prec x$. A *partial schedule* for S is a schedule for a lower-set of S. Given two (partial) feasible schedules σ and σ' for a lower-set $S' \subset S$, σ *dominates* σ' if $\sigma(i) \leq \sigma'(i)$ for all tasks $i \in S'$ and $\sigma(j) < \sigma'(j)$ for at least one task $j \in S'$. A (partial) schedule is *minimal* among a set $\mathcal{S}$ of feasible schedules for the same task set, if it is not dominated by any other schedule in $\mathcal{S}$.

By the time windows (r_i, d_i) of the tasks in $\mathcal{I}$, an interval graph can be constructed, whose *pathwidth* is denoted by $pw(\mathcal{I})$. Let $\mu = pw(\mathcal{I}) + 1$. It can be shown that μ corresponds to the maximum number of overlapping time windows, i.e., $\mu = \max_{t \in \mathbb{N}} |\{i \in S \mid r_i \leq t < d_i\}|$. Thus for any feasible schedule for S, at most μ tasks are in progress at any time point. If $\mu \leq m$, we can schedule each task as early as possible without resource conflict, and the instance $\mathcal{I}$ can be solved by a greedy algorithm. Thus we assume that $\mu > m$ in this paper.

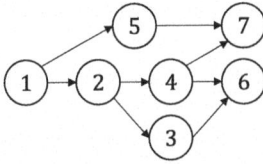

Fig. 1. The hasse diagram of an partial order on 7 tasks.

Table 1. Processing times, release times and deadlines of the 7 tasks given in Fig. 1.

i	1	2	3	4	5	6	7
p_i	1	1	2	1	3	1	2
r_i	0	1	2	3	1	4	4
d_i	3	4	6	5	4	7	6

	0	1	2	3	4	6	7
Machine 1	1		5		7		
Machine 2		2		4	3	6	

Fig. 2. A feasible schedule for the instance specified by Fig. 1 and Table 1.

3   Fixed-Parameter Algorithms for $P \mid prec, r_i, d_i \mid \star$

As mentioned earlier, our algorithms are based on the algorithm (ALG-PT), originally proposed by Sprecher [18] for the Resource-Constrained Project Scheduling Problem (abbr. RCPSP). In this subsection, we introduce the main idea of ALG-PT using our own terminology, for solving the problem $P \mid prec \mid C_{\max}$.

3.1   ALG-PT

For ease of description, we define two notions *state* and *associated schedule*.

Definition 1 (State). *A state $u = \langle u_1, u_2, \ldots, u_l \rangle$ is an ordered sequence of the tasks in a lower-set of S that satisfies the Precedence Constraint, i.e., for each task u_i ($1 \leq i \leq l$) in u, $\Gamma^-(u_i) \subseteq \{u_1, u_2, \ldots, u_{i-1}\}$.*

For a state $u = \langle u_1, u_2, \ldots, u_l \rangle$, denote by $S(u) = \{u_1, u_2, \ldots, u_l\}$ the set of scheduled tasks, and by $S_e(u) = \{i \in S \setminus S(u) \mid \Gamma^-(i) \subseteq S(u)\}$ the set of *eligible* tasks. Given two states $u = \langle u_1, u_2, \ldots, u_l \rangle$ and $u' = \langle u_1, u_2, \ldots, u_l, i \rangle$ such that $S(u') = S(u) \cup \{i\}$ and $i \in S_e(u)$, we say that u' is a *successor state* of u w.r.t. task i. A feasible schedule σ for state u is *monotone* if $\sigma(i) \leq \sigma(j)$ for any two tasks $i, j \in S(u)$ such that task i is located before task j in the sequence u.

Definition 2 (Associated Schedule). *The associated schedule σ_u of the state $u = \langle u_1, u_2, \ldots, u_l \rangle$, is a minimal one among all monotone schedules for u.*

By the above definitions, it is not hard to see that each state has a unique associated schedule. Using the relationship between states and their successor states, a (rooted) *state tree* T can be constructed, where each state is the parent of its successor states. In particular, the root ρ of T is referred to as the *initial state*, where $\rho = \langle \rangle$ and $S(\rho) = \emptyset$. Based on the tree structure of T, the associated schedule σ_u of each state u in T can be constructed iteratively. We now describe how to construct the associated schedule $\sigma_{u'}$ for a successor state u' of u w.r.t. task i, based on the associated schedule σ_u of u. First, $\sigma_{u'}$ schedules the tasks of $S(u)$ in the same way as in σ_u. For the additional task i, it is scheduled to start at the earliest precedence- and resource-feasible time, which is defined as $\max\{t(u), t_p(u, i), t_r(u)\}$, where $t(u) = \max_{j \in S(u)} \sigma_u(j)$ is the latest assigned start time in σ_u, $t_p(u, i) = \max_{j \in \Gamma^-(i)} \sigma_u(j) + p_j$ is the earliest precedence-feasible time for task i, $t_r(u) = \min\{\tau \geq t(u) \mid |S_{\mathrm{IP}}(u, \tau)| < m\}$ is the earliest resource-feasible time, and $S_{\mathrm{IP}}(u, \tau) = \{j \in S(u) \mid \sigma_u(j) \leq \tau < \sigma_u(j) + p_j\}$ is the set of tasks in progress at time point τ in σ_u.

The algorithm ALG-PT can be viewed as performing a Depth-First Search (DFS) on T, where the goal is to find a state u^* such that $S(u^*) = S$ and its associated schedule has the minimum makespan. It is not hard to see that the time complexity of ALG-PT for the $P \mid prec \mid C_{\max}$ is $O^*(n!)$. In the following two subsections, we aim to adapt ALG-PT to solve the problems $P \mid prec, r_i, d_i \mid \star$, $P \mid prec, p_i = 1, r_i, d_i \mid \star$, and the corresponding variants with typed machines.

3.2 $P \mid prec, r_i, d_i \mid \star$

The ALG-PT does not consider the Time Window Constraint. To address this, we introduce a rule called the *Time Window Rule*. Given an instance $\mathcal{I}$ of $P \mid prec, r_i, d_i \mid \star$, let $\mathcal{T}$ be the state tree for $\mathcal{I}$.

Time Window Rule (abbr. TW Rule). For a state $u = \langle u_1, u_2, \ldots, u_l \rangle$, if there exists a task $i \in S(u)$ with $d_i < \sigma_u(i) + p_i$, or a task $i \in S \setminus S(u)$ with deadline $d_i < \sigma_u(u_l) + p_i$, then the state i can be discarded.

Lemma 1. *TW Rule is correct.*

It is easy to see that ALG-PT, when augmented with the TW Rule, can solve $P \mid prec, r_i, d_i \mid \star$. In the following discussion, we assume that all states in $\mathcal{T}$, on which the TW Rule is applicable, have been discarded. Note that the number of states in $\mathcal{T}$ may still be $O(n!)$, so we aim to further reduce the number of states. To achieve this, we construct a (directed) *state graph* $\mathcal{G}$. Specifically, the state graph $\mathcal{G} = \mathcal{T}/\bowtie$ is a *quotient graph* of $\mathcal{T}$ w.r.t. an equivalence relation $\bowtie$, where the related definitions and notations are given as follows. Two states u and u' in the state tree $\mathcal{T}$ are *equivalent* w.r.t. the equivalence relation $\bowtie$ if and only if the following four conditions are satisfied:

1. **Scheduled Task Set Equality:** $S(u) = S(u')$;
2. **Latest Start Time Equality:** $\max_{i \in S(u)} \sigma_u(i) = \max_{i' \in S(u')} \sigma_{u'}(i')$ (denoted by t in the following condition);
3. **Completed Task Set Equality:** $\{i \in S(u) \mid \sigma_u(i) + p_i \le t\} = \{i' \in S(u') \mid \sigma_{u'}(i') + p_{i'} \le t\}$ (denoted by C in the following condition);
4. **Start Time Consistency for Tasks In Process:** $\sigma_u(i) = \sigma_{u'}(i)$ for each task $i \in S(u) \setminus C$.

The equivalence relation $\bowtie$ partitions the states in $\mathcal{T}$ into *equivalence classes*, which we refer to as *g-states*. Each vertex of the state graph $\mathcal{G}$ corresponds uniquely to one such equivalence class (or g-state) in this partition. In the following discussion, we say that a g-state v contains a state u if u is in the equivalence class represented by v. For each arc $\langle u, u' \rangle$ in $\mathcal{T}$ where u' is a successor state of u (note that u and v are of different equivalence classes because $S(u) \ne S(u')$), there is a corresponding arc $\langle [u]_\bowtie, [u']_\bowtie \rangle$ in $\mathcal{G}$, where $[u]_\bowtie$ (resp., $[u']_\bowtie$) denotes the equivalence class containing u (resp., u'). Here, $[u']_\bowtie$ is called a *successor* *g-state* of $[u]_\bowtie$. In particular, the equivalence class $[\rho]_\bowtie$ containing the initial state ρ of $\mathcal{T}$ is referred to as the *initial g-state* of $\mathcal{G}$. The state graph can be viewed as a graph constructed by merging several states in the state tree into a g-state. This merging process transforms $\mathcal{G}$ into a directed acyclic graph.

By the definition of equivalence, we introduce the following notations for g-states in $\mathcal{G}$. Given a g-state $v \in \mathcal{G}$ and a state $u \in \mathcal{T}$ with $v = [u]_\bowtie$,

- $S(v) = S(u)$: The set of scheduled tasks for v;
- $S_e(v) = S_e(u)$: The set of eligible tasks for v;
- $t(v) = \max_{i \in S(u)} \sigma_u(i)$: The latest assigned start time in σ_u;
- $S_c(v) = \{i \in S(u) \mid \sigma_u(i) + p_i \le t(v)\}$: The set of tasks completed before time point $t(v)$ in σ_u;
- $r(v, i) = \max\{\sigma_u(i) + p_i - t(v), 0\}$: The remaining processing time of task i at time point $t(v)$ in σ_u. In particular, let $r(v, i) = +\infty$ if $i \notin S(v)$.

Observe that $S(v)$, $S_c(v)$ and $S_e(v)$ can be expressed as $S(v) = \{i \in S \mid r(v, i) \in \mathbb{N}\}$, $S_c(v) = \{i \in S \mid r(v, i) = 0\}$, and $S_e(v) = \{i \in S \setminus S(v) \mid \Gamma^-(i) \subseteq S(v)\}$. Thus the pair $(t(v), r(v, \ast))$ fully specifies a g-state v. In particular, for the initial g-state $[\rho]_\bowtie$, $t([\rho]_\bowtie) = 0$ and $r([\rho]_\bowtie, i) = +\infty$ for each task $i \in S$.

Given a g-state $v = [u]_\bowtie$ and a successor g-states $v' = [u']_\bowtie$, where u' is a successor state of u w.r.t. a task i, we can calculate the pair $(t(v'), r(v', *))$ of v' based on the pair $(t(v), r(v, *))$ of v. As discussed earlier for associated schedules, the start time of task i in $\sigma_{u'}$ is $t(v') = \sigma_{u'}(i) = \max\{r_i, t(v), t_p(v, i), t_r(v)\}$, where $\max\{t(v), t_p(v, i)\} = \max_{j \in \Gamma^-(i)} t(v) + r(v, j)$ $(t_p(v, i) = \max_{j \in \Gamma^-(i)} \sigma_u(j) + p_j$ is the earliest precedence-feasible time), and $t_r(v) = \min\{\tau \geq t(v) \mid |S_{IP}(v, \tau)| < m\}$ is the earliest resource-feasible time. Here $S_{IP}(v, \tau) = \{j \in S(v) \mid r(v, j) + t(v) > \tau\}$ is the set of tasks in progress at time point $\tau \geq t(v)$. Thus the pair $(t(v'), r(v', *))$ of the successor g-state v' w.r.t. task i can be calculated as follows:

$$t(v') = \max\{r_i, t(v), t_p(v, i), t_r(v)\},$$
$$r(v', j) = \begin{cases} 0, & \text{if } r(v, j) \in [0, t(v') - t(v)] \\ r(v, j) - (t' - t), & \text{if } r(v, j) \in (t(v') - t(v), +\infty) \\ +\infty, & \text{if } j \neq i \wedge r(v, j) = +\infty \\ p_i, & \text{if } j = i \end{cases} \quad (1)$$

Thus we can construct the state graph $\mathcal{G}$ instead of the state tree $\mathcal{T}$, where each g-state v in $\mathcal{G}$ is associated with the pair $(t(v), r(v, *))$. Note that the TW Rule is not applicable to the state graph $\mathcal{G}$, thus we redefine the TW Rule in terms of the pair $(t(v), r(v, *))$, whose correctness is obvious.

General Time Window Rule (abbr. GTW Rule). For a g-state v, if there exists a task $i \in S(v) \setminus S_c(v)$ such that $d_i < t(v) + r(v, i)$, or a task $i \in S \setminus S(v)$ such that the deadline $d_i < t(v) + p_i$, then the g-state v can be discarded.

For the instance specified in Fig. 1 and Table 1, the corresponding state tree and state graph are shown in Fig. 3 and Fig. 4, respectively. For each vertex (i.e., g-state) v in Fig. 4, the first number represents $t(v)$, and the second set represents the function $r(v, *)$. Specifically, each pair $(i, r(v, i))$ in the second set represents the remaining processing time of task i at time point $t(v)$, and the tasks with remaining processing time 0 or $+\infty$ are omitted.

This compact representation of the state tree $\mathcal{T}$ allows us to solve the instance $\mathcal{I}$ by identifying a g-state v^* in $\mathcal{G}$ such that $S(v^*) = S$. The corresponding feasible schedule σ^* can be constructed along a path connecting the initial g-state $[\rho]_\bowtie$ with v^* in $\mathcal{G}$. Specifically, let $P = \langle v_0 = [\rho]_\bowtie, v_1, v_2, \ldots, v_n = v^* \rangle$ be an arbitrary path connecting the initial g-state with v^* that contains exactly $n+1$ g-states. For each $0 \leq i \leq n - 1$, the start time $\sigma^*(j)$ of the unique task j in $S(v_{i+1}) \setminus S(v_i)$ is set to $t(v_{i+1})$. It is easy to verify the feasibility of the schedule σ^* by the definition of $t(*)$. The above discussion is formulated in the following lemma.

Lemma 2. *Given an instance $\mathcal{I}$ of the problem $P \mid prec, r_i, d_i \mid \star$, it has a feasible schedule if and only if there exists a g-state v^* in $\mathcal{G}$ with $S(v^*) = S$.*

Our remaining job is to analyze the size of $\mathcal{G}$. Firstly, we analyze the number of g-states w.r.t. a given time point under the assumption that the g-states on which the GTW Rule is applicable have been removed from $\mathcal{G}$.

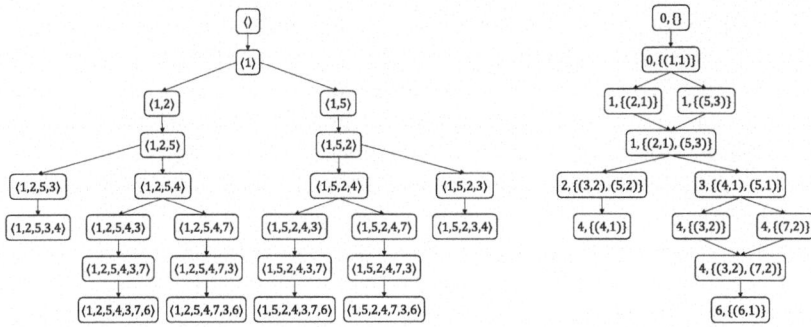

Fig. 3. State tree.   **Fig. 4.** State graph.

Lemma 3. *Given a time point t, there are at most $(p_{\max} + 2)^{\mu}$ g-states v with $t(v) = t$ in the state graph $\mathcal{G}$, where $\mu = \max_{t \in \mathbb{N}} |\{i \in S \mid r_i \leq t < d_i\}|$.*

Lemma 3 bounds the number of g-states for a fixed time point t. However, the number of time points depends on the time windows $[r_i, d_i]$ of these tasks, which can not be bounded by n and $p_{\max}$. To further bound the number of time points, we focus on minimal schedules among all feasible schedules. Note that the associated schedules of states may be not minimal among all feasible schedules. We will handle this later. First we introduce the following definitions about time intervals, where only left-closed and right-open time intervals are considered:

- *dense time interval:* A time interval $[t_1, t_2)$ is *dense* if $(2p_{\max} - 1) \cdot |S_r(t_1, t_2)| \geq t_2 - t_1$, where $S_r(t_1, t_2) = \{i \in S \mid t_1 \leq r_i < t_2\}$.
- *perfect time interval:* A time interval $[t_1, t_2)$ is *perfect* if $(2p_{\max} - 1) \cdot |S_r(t_1, t_2)| = t_2 - t_1$ and the interval $[t_1, t)$ is dense for each $t_1 \leq t < t_2$.
- *maximal perfect time interval:* A perfect time interval $[t_1, t_2)$ is *maximal* if no other perfect time interval $[t_1', t_2')$ properly contains $[t_1, t_2)$, i.e., $t_1' < t_1, t_2 \leq t_2'$ or $t_1' \leq t_1, t_2 < t_2'$.

Lemma 4. *Maximal perfect time intervals do not intersect.*

Lemma 5. *For each task $i \in S$, there exists exactly one maximal perfect time interval $[t_1, t_2)$ such that $i \in S_r(t_1, t_2)$.*

By Lemma 5, each task $i \in S$ is associated with exactly one maximal perfect time interval $[t_1, t_2)$. We say that task i belongs to the maximal perfect time interval $[t_1, t_2)$.

Lemma 6. *There are at most $n(2p_{\max} - 1)$ (integer) time points contained in these maximal perfect time intervals.*

Lemma 7. *For any minimal schedule σ among all feasible schedules for S in $\mathcal{I}$, the execution of each task (in σ) is in its maximal perfect time interval.*

Lemmas 6 and 7 establish that the number of potential execution time points in any minimal feasible schedule can be bounded by $n(2p_{\max} - 1)$. However, the associated schedules of states may not be minimal among all feasible schedules, thus we require additional techniques to bound the number of time points in the associated schedules. To this end, we consider reducing the time windows. Note that this reduction may affect the feasibility of non-minimal schedules, but it can preserve the feasibility of all minimal schedules.

Let $\tau_1, \tau_2, \ldots, \tau_N$ be the time points contained in these maximal perfect time intervals of $\mathcal{I}$, ordered such that $\tau_1 < \tau_2 < \ldots < \tau_N$, where $N \leq n(2p_{\max} - 1)$. Let $\tau_{N+1} = \max\{\tau_N + 1, \max_{i \in S} d_i\}$, and $T_r = \{\tau_1, \tau_2, \ldots, \tau_N, \tau_{N+1}\}$. Let $c(t) = |\{\tau \in T_r \mid \tau \leq t\}|$ be a function counting the number of time points in T_r that are not greater than t. Now we construct the reduced instance $\mathcal{I}'$ based on the instance $\mathcal{I}$ as follows: $\mathcal{I}'$ shares the set S of tasks, processing times p_i, and the partial order $\preceq$ as $\mathcal{I}$. For each task $i \in S$, the release time r_i and deadline d_i are replaced with the reduced values $r_i' = c(r_i)$ and $d_i' = c(d_i - 1) + 1$. This reduction ensures that the time points in $\mathcal{I}'$ are aligned with the maximal perfect time intervals of $\mathcal{I}$, simplifying the problem while preserving its essential structure.

Since each release time r_i lies within a maximal perfect time interval, r_i' is the index of value r_i in T_r. For the deadline d_i, if it is in T_r, d_i' is the index of value d_i in T_r; otherwise d_i' is the index of the smallest time point in T_r that is not less than d_i, both cases can be computed as $d_i' = c(d_i - 1) + 1$.

Lemma 8. *Instance $\mathcal{I}'$ is* YES *if and only if instance $\mathcal{I}$ is* YES.

We now present the algorithm for solving the problem $P \mid prec, r_i, d_i \mid \star$, called ALG-SPT (see Algorithm 1). The algorithm iteratively generates all g-states in the state graph $\mathcal{G}$. To organize computation, g-states are partitioned into $n + 1$ levels $\mathcal{V}_0, \mathcal{V}_1, \ldots, \mathcal{V}_n$, where level $\mathcal{V}_l$ contains g-states v with $|S(v)| = l$.

Algorithm 1. ALG-SPT

Input: An instance $\mathcal{I}$ of $P \mid prec, r_i, d_i \mid \star$;
Output: A feasible schedule if there exists a feasible schedule; otherwise, NULL.
 1: construct the maximal perfect time intervals;
 2: construct the reduced instance $\mathcal{I}'$ using the maximal perfect time intervals;
 3: $\mathcal{V}_l \leftarrow \emptyset$ for $l \in \{1, 2, \ldots, n\}$;
 4: $\mathcal{V}_0 = \{[\rho]_{\bowtie}\}$;
 5: **for each** $l \in \{0, 1, \ldots, n-1\}$ **do**
 6:     **for each** $v \in \mathcal{V}_l$ **do**
 7:         **for each** $i \in S_e(v)$ **do**
 8:             compute the pair $(t(v'), r(v', *))$ of the successor g-state v' of g-state v w.r.t. task i by Equation (1);
 9:             **if** v' cannot be discarded by the GTW Rule **then**
10:                 $\mathcal{V}_{l+1} \leftarrow \mathcal{V}_{l+1} \cup \{v'\}$;
11: **if** $\mathcal{V}_n = \emptyset$ **then**
12:     **return** NULL;
13: select a g-state v^* from $\mathcal{V}_n$;
14: construct schedule σ^* by a path from $[\rho]_{\bowtie}$ to v^*;
15: **return** the schedule for original instance $\mathcal{I}$ constructed based on σ^*.

Theorem 1. *Given an instance $\mathcal{I}$ of the problem $P \mid prec, r_i, d_i \mid \star$, the algorithm* ALG-SPT *can solve it with time complexity $O((p_{\max} + 2)^\mu p_{\max} n^3)$, where $n = |S|$ is the number of tasks, $p_{\max} = \max_{i \in S} p_i$ is the maximum processing time, and $\mu = \max_{t \in \mathbb{N}} |\{i \in S \mid r_i \leq t < d_i\}|$.*

3.3 $P \mid prec, p_i = 1, r_i, d_i \mid \star$

In this subsection, we consider the problem $P \mid prec, p_i = 1, r_i, d_i \mid \star$, where each task $i \in S$ considered in the instance $\mathcal{I}_1$ of the problem has a unit processing time. Directly calling ALG-SPT yields a time complexity of $O(3^\mu n^3)$, which can be further improved to $O(2^\mu \mu m n^3)$. The relevant definitions and notations follow those given in the previous subsection, unless otherwise stated. First of all, let ⊲ be an arbitrary *reverse topological order* of the Hasse diagram specified by the partial order $\preceq$, i.e., $i \lhd j$ (or $j \rhd i$) if task $i \in S$ precedes task $j \in S$ in the order. The following rule is inspired by the Extended Single Enumeration Rule [18].

Reverse Topological Order Rule (abbr. RTO Rule). Let $u = \langle u_1, u_2, \ldots, u_l \rangle$ be a state of the state tree $\mathcal{T}$ for $\mathcal{I}_1$. If there are two indexes $i < j$ such that $\sigma_u(u_i) = \sigma_u(u_j)$ and $u_j \lhd u_i$, then the state u can be discarded.

Lemma 9. *The RTO Rule is correct, i.e., given a lower-set S' of S, for any minimal schedule σ among all feasible (partial) schedules for S', there exists a state u such that $S(u) = S'$, u cannot be discarded by the RTO Rule and $\sigma_u = \sigma$.*

To enforce the RTO rule, we introduce a restricted variant of the associated schedule, called the RTO-associated schedule. This rule ensures that no state is discarded due to the RTO rule w.r.t. its associated schedule.

Definition 3 (RTO-Associated Schedule). *The RTO-associated schedule σ'_u of state u, is a minimal one among all feasible schedules for u such that if task i precedes task j in u, then (1) $\sigma'_u(i) < \sigma'_u(j)$, or (2) $\sigma'_u(i) = \sigma'_u(j)$ and $i \lhd j$.*

If a state u cannot be discarded by the RTO Rule w.r.t. its associated schedule, then $\sigma'_u = \sigma_u$. Combining this with Lemma 9, we have the following corollary.

Corollary 1. *Given a lower-set S' of S, for any minimal schedule σ among all (partial) feasible schedules for S', there is a state u with $S(u) = S$ and $\sigma'_u = \sigma$.*

Remark that in the RTO-associated schedule σ'_u, tasks with the same start time are ordered according to ⊲. Denote by $\mathcal{T}_1$ the state tree for $\mathcal{I}_1$. As the algorithm for $P \mid prec, r_i, d_i \mid \star$, the algorithm for $P \mid prec, p_i = 1, r_i, d_i \mid \star$ is based on a state graph $\mathcal{G}_1$, which is derived from $\mathcal{T}_1$ w.r.t. an equivalence relation $\bowtie_1$. Given two states u and u' in $\mathcal{T}_1$, they are *equivalent* under $\bowtie_1$ if and only if

1. **Scheduled Task Set Equality:** $S(u) = S(u')$;
2. **Latest Start Time Equality:** $\max_{i \in S(u)} \sigma'_u(i) = \max_{i' \in S(u')} \sigma'_{u'}(i')$ (simply denoted by t in the following conditions);

3. **Number of Tasks Starting at Time Point t Equality:** The numbers of tasks starting at the time point t in the two RTO-associated schedules are the same, i.e., $|\{i \in S(u) \mid \sigma'_u(i) = t\}| = |\{i' \in S(u') \mid \sigma'_{u'}(i') = t\}|$;
4. **Maximum Task Starting at Time Point t Equality:** The maximum tasks w.r.t. $\lhd$ starting at time point t in $\sigma'_{u'}$ and σ'_u are the same, i.e., $\max\{i \in S(u) \mid \sigma'_u(i) = t\} = \max\{i' \in S(u') \mid \sigma'_{u'}(i') = t\}$.

We have the following notations for the characteristics of states in an equivalence class w.r.t. $\bowtie_1$. Given a g-state $v \in \mathcal{G}_1$ and a state $u \in \mathcal{T}_1$ with $v = [u]_{\bowtie_1}$,

- $\mathrm{card}_{\mathrm{IP}}(v) = |\{i \in S(u) \mid \sigma'_u(i) = t(v)\}|$: The number of tasks starting at time point $t(v)$ in σ'_u;
- $\max_{\mathrm{IP}}(v) = \max\{i \in S(u) \mid \sigma'_u(i) = t(v)\}$: The maximum task among the tasks starting at time point $t(v)$ in σ'_u, w.r.t. $\lhd$.

Now a g-state v can be specified by a quadruple $(S(v), t(v), \mathrm{card}_{\mathrm{IP}}(v), \max_{\mathrm{IP}}(v))$ (the notations $S(v)$ and $t(v)$ follow the ones given previous subsection). Given a g-state $v = [u]_{\bowtie_1}$ and one of its successor g-states $v' = [u']_{\bowtie_1}$ such that u' is a successor state of u w.r.t. task i, we calculate the quadruple of v' based on the quadruple of v. Observe that $t(v') = \sigma'_{u'}(i)$. For $\sigma'_{u'}(i)$, it depends on the start times of the tasks in $S_{\mathrm{IP}}(u)$, where $S_{\mathrm{IP}}(u) = \{i \in S(u) \mid \sigma'_u(i) = t(v)\}$ is the set of tasks starting at time point $t(v)$ in the RTO-associated schedule of u. Note that for two states u, u' in v, $S_{\mathrm{IP}}(u)$ and $S_{\mathrm{IP}}(u')$ may be different.

If $i \rhd \max_{\mathrm{IP}}(v)$ (i.e., task i is greater than all tasks in $S_{\mathrm{IP}}(u)$ w.r.t. $\lhd$), then none of the predecessors of task i can be in $S_{\mathrm{IP}}(u)$. Thus the Precedence Constraint is satisfied at time point $t(v)$ if and only if $\Gamma^-(i) \subseteq S(v)$. If $\Gamma^-(i) \subseteq S(v)$ and $\mathrm{card}_{\mathrm{IP}}(v) < m$ then $\sigma'_{u'}(i) = \max\{r_i, t(v)\}$; otherwise, $\sigma'_{u'}(i) = \max\{r_i, t(v)+1\}$.

If $i \lhd \max_{\mathrm{IP}}(v)$, then by the RTO Rule, i cannot start at time point $t(v)$. The earliest possible start time of task i is $t(v)+1$, at which the Precedence Constraint and Resource Constraint are satisfied. Thus $\sigma'_{u'}(i) = \max\{r_i, t(v) + 1\}$.

Thus the quadruple $(S(v'), t(v'), \mathrm{card}_{\mathrm{IP}}(v'), \max_{\mathrm{IP}}(v'))$ of the successor g-state v' w.r.t. task i can be calculated as follows:

$$
\begin{aligned}
S(v') &= S(v) \cup \{i\} \\
t(v') &= \begin{cases} \max\{r_i, t(v)\}, & \text{if } i \rhd \max(v) \wedge \Gamma^-(i) \subseteq S(v) \wedge \mathrm{card}_{\mathrm{IP}}(v) < m \\ \max\{r_i, t(v) + 1\}, & \text{if } i \lhd \max(v) \vee \Gamma^-(i) \nsubseteq S(v) \vee \mathrm{card}_{\mathrm{IP}}(v) = m \end{cases} \\
\mathrm{card}_{\mathrm{IP}}(v') &= \begin{cases} \mathrm{card}_{\mathrm{IP}}(v) + 1, & \text{if } t(v') = t(v) \\ 1, & \text{if } t(v') > t(v) \end{cases} \\
\max_{\mathrm{IP}}(v') &= i.
\end{aligned}
$$

$$(2)$$

Additionally, the Time Window Rule can be redefined in the quadruple form:

Unit Time Window Rule. (abbr. UTW Rule). For a g-state v, if there exists a task $i \in S \setminus S(v)$ such that $d_i \leq t(v)$, then the g-state v can be discarded.

We now show that ALG-SPT can solve $P \mid prec, p_i = 1, r_i, d_i \mid \star$ with minor modifications: replacing the Equation (1) in Step 8 by Equation (2), and replacing the GTW Rule in Step 9 by the UTW Rule.

Lemma 10. *Given a time point t, there are at most $2^{\mu} \mu m$ g-states v with $t(v) = t$ in the state graph $\mathcal{G}_1$, where $\mu = \max_{t \in \mathbb{N}} |\{i \in S \mid r_i \leq t < d_i\}|$.*

By Lemma 6 with $p_{\max} = 1$, there are at most n possible time points. Thus using the same proof in Theorem 1 we have the theorem given below.

Theorem 2. *Given an instance $\mathcal{I}_1$ of the problem $P \mid prec, p_i = 1, r_i, d_i \mid \star$, the algorithm ALG-SPT can solve it with time complexity $O(2^{\mu} \mu m n^3)$, where $n = |S|$ is the number of tasks, and $\mu = \max_{t \in \mathbb{N}} |\{i \in S \mid r_i \leq t < d_i\}|$.*

4   Typed Processors

In the section we show that ALG-SPT can be adapted to solve the problem with typed machines $P \mid \mathcal{M}_j(type), prec, r_i, d_i \mid \star$. Given an instance $\mathcal{I}_t$ of $P \mid \mathcal{M}_j(type), prec, r_i, d_i \mid \star$, the machines are partitioned into k classes, denoted by $C_1, C_2, \ldots, C_k$, where each class C_p contains m_p machines. Each task i can only be executed on a specific machine class C_{π_i}, where $1 \leq \pi_i \leq k$. Denote by S_p the set containing the tasks that are executed on machines of C_p. A schedule σ for the tasks in S on typed machines is *feasible* if it satisfies the Time Window Constraint, Precedence Constraint (the two constraints are defined in the same way as that in Preliminaries), and Typed Resource Constraint:

- **Typed Resource Constraint.** At any time point t, the number of tasks in progress on machines of class C_p cannot exceed m_p, i.e., $|\{i \in S_p \mid \sigma(i) \leq t < \sigma(i) + p_i\}| \leq m_p$ for any $1 \leq p \leq k$.

The discussion for $P \mid \mathcal{M}_j(type), prec, r_i, d_i \mid \star$ and $P \mid \mathcal{M}_j(type), p_i = 1, prec, r_i, d_i \mid \star$ runs in a similar way to that for $P \mid prec, r_i, d_i \mid \star$ and $P \mid p_i = 1, prec, r_i, d_i \mid \star$, respectively.

Theorem 3. *Given an instance $\mathcal{I}_t$ of the problem $P \mid \mathcal{M}_j(type), prec, r_i, d_i \mid \star$, the algorithm ALG-SPT can solve it with time complexity $O((p_{\max}+2)^{\mu} p_{\max} n^3)$, where $n = |S|$ is the number of tasks, $p_{\max} = \max_{i \in S} p_i$ is the maximum processing time, and $\mu = \max_{t \in \mathbb{N}} |\{i \in S \mid r_i \leq t < d_i\}|$.*

Theorem 4. *Given an instance $\mathcal{I}_{t,1}$ of the problem $P \mid \mathcal{M}_j(type), prec, p_i = 1, r_i, d_i \mid \star$, the algorithm ALG-SPT can solve it with time complexity $O(2^{\mu} \mu m^k n^3)$, where $n = |S|$ is the number of tasks, m is the number of machines, and $\mu = \max_{t \in \mathbb{N}} |\{i \in S \mid r_i \leq t < d_i\}|$.*

References

1. Bodlaender, H.L., Fellows, M.R.: W[2]-hardness of precedence constrained k-processor scheduling. Oper. Res. Lett. **18**(2), 93–97 (1995)
2. Brucker, P.: Scheduling Algorithms, 4th edn. Springer (2004)

3. Chen, B., Potts, C.N., Woeginger, G.J.: A review of machine scheduling: complexity, algorithms and approximability. In: Handbook of Combinatorial Optimization: Volume 1–3, pp. 1493–1641 (1998)

4. Cygan, M., et al.: Parameterized Algorithms, vol. 5. Springer (2015)

5. Demeulemeester, E., Herroelen, W.: A branch-and-bound procedure for the multiple resource-constrained project scheduling problem. Manage. Sci. **38**(12), 1803–1818 (1992)

6. Dolev, D., Warmuth, M.K.: Scheduling precedence graphs of bounded height. J. Algorithms **5**(1), 48–59 (1984)

7. Garey, M.R., Johnson, D.S.: Two-processor scheduling with start-times and deadlines. SIAM J. Comput. **6**(3), 416–426 (1977)

8. Graham, R.L., Lawler, E.L., Lenstra, J.K., Kan, A.R.: Optimization and approximation in deterministic sequencing and scheduling: a survey. Ann. Discrete Math. **5**, 287–326 (1979)

9. Hanen, C., Munier Kordon, A.: Fixed-parameter tractability of scheduling dependent typed tasks subject to release times and deadlines. J. Sched. **27**(2), 119–133 (2024)

10. Lenstra, J.K., Kan, A.R., Brucker, P.: Complexity of machine scheduling problems. Ann. Discrete Math. **1**, 343–362 (1977)

11. Lenstra, J.K., Rinnooy Kan, A.: Complexity of scheduling under precedence constraints. Oper. Res. **26**(1), 22–35 (1978)

12. Leung, J.Y.: Handbook of Scheduling: Algorithms, Models, and Performance Analysis. Chapman and Hall/CRC (2004)

13. Mnich, M., Van Bevern, R.: Parameterized complexity of machine scheduling: 15 open problems. Comput. Oper. Res. **100**, 254–261 (2018)

14. Mnich, M., Wiese, A.: Scheduling meets fixed-parameter tractability. arXiv preprint arXiv:1311.4021 (2013)

15. Möhring, R.H.: Computationally tractable classes of ordered sets. In: Algorithms and Order, pp. 105–193. Springer (1989)

16. Munier Kordon, A.: A fixed-parameter algorithm for scheduling unit dependent tasks on parallel machines with time windows. Discret. Appl. Math. **290**, 1–6 (2021)

17. Prot, D., Bellenguez-Morineau, O.: A survey on how the structure of precedence constraints may change the complexity class of scheduling problems. J. Sched. **21**(1), 3–16 (2018)

18. Sprecher, A.: Solving the RCPSP efficiently at modest memory requirements. Technical report, Manuskripte aus den Instituten für Betriebswirtschaftslehre der Universität Kiel (1996)

19. Tarhan, I., Carlier, J., Hanen, C., Jouglet, A., Munier Kordon, A.: Parameterized analysis of a dynamic programming algorithm for a parallel machine scheduling problem. In: Cano, J., Dikaiakos, M.D., Papadopoulos, G.A., Pericàs, M., Sakellariou, R. (eds.) Euro-Par 2023: Parallel Processing, pp. 139–153. Springer, Cham (2023)

20. Ullman, J.D.: Np-complete scheduling problems. J. Comput. Syst. Sci. **10**(3), 384–393 (1975)
21. Van Bevern, R., Bredereck, R., Bulteau, L., Komusiewicz, C., Talmon, N., Woeginger, G.J.: Precedence-constrained scheduling problems parameterized by partial order width. In: International Conference on Discrete Optimization and Operations Research, pp. 105–120. Springer (2016)

Pareto Optimal Matching with Multilayer Preferences: How Hard Can It Be?

Yinghui Wen[1], Xin Tong[2], Jiong Guo[2], and Aizhong Zhou[3](✉)

[1] Shandong Institute of Information Technology Industry Development, Jinan, China
[2] Shandong University, Qingdao, China
xtong@mail.sdu.edu.cn, jguo@sdu.edu.cn
[3] Ocean University of China, Qingdao, China
zhouaizhong@ouc.edu.cn

Abstract. We study Pareto optimal matching under multilayer preferences, where each agent has more than one preference list with each list representing a criterion based on which the agents of the opposite side are evaluated. We introduce four intuitive concepts of Pareto optimality with multilayer preferences and study parameterized complexity of them. We obtain W[1]-hardness, W[2]-hardness and para-NP-hardness results for most parameters except n, the number of men/women. Although n is FPT, we show that $O^*(n!)$ time algorithm is essentially optimal for most of the concepts. In addition, almost no concept admits polynomial kernels with respect to n. These results even hold for combined parameters. We then consider cases where preferences satisfy certain desirable properties, that is, uniformity, single-layer and master list. We show that if preferences are uniform or single-layered, all of them are simply trivial and can be determined in polynomial time. However, the problem soon becomes NP-hard even when the maximum Hamming distance of preferences is a constant. For the case of master list, we find that even when there are only three layers and the preference lists on each side are all derived from the same single master list, the four problems remain NP-hard.

Keywords: Computational social choice · Pareto optimal matching · Compuational complexity · Parameterized complexity

1 Introduction

The two-sided preference-based matching model is largely attributed to their diverse applications in real-world scenarios, such as the allocation of students to educational institutions [1,2], the pairing of kidney patients with donors [27, 30,31], and the resettlement of refugees in host nations [5,6]. Various criteria

This work was supported by the National Natural Science Foundation of China (Grants No. 61772314, 61761136017 and 62072275) and the Natural Science Foundation of Shandong (Grants No. ZR202111300228).

exist to evaluate the quality of a matching, one of which is known as Pareto optimality [3,5,8,13,14]. A matching M is considered Pareto optimal if there is no matching M' dominating M. Here, M' dominates M if in M' no agent has a worse partner than in M and at least one agent has a better partner than in M.

In many practical applications, the standard preference-based model is inadequate, as it only allows each agent to provide a single preference list. Consider a situation in which firms and candidates select one another. Candidates can evaluate a company based on a variety of factors, including reputation, workplace atmosphere, location, and compensation. Instead of providing a single criterion, applicants can provide a list of preferences for each. The objective is to find a matching that satisfies each preference list. Another example is the marriage problem. Every agent can easily give a series of preference lists based on different criteria, such as age, nationality, occupation, interests, etc. Allowing agents to independently aggregate their preferences and report their global preferences is a popular solution to similar situations, and it has been investigated for decades [11,22,25]. However, as pointed out by Farczadi et al. [19], aggregated preferences could be cyclic or even intransitive. Thus, it makes sense to examine the scenario of finding a matching within the multilayered context.

In this paper, we study two-sided one-to-one matching with multilayer preferences. With multiple preference lists for each agent, there are various intuitive approaches to extend the classic Pareto optimality concept. We introduce four intuitive concepts of Pareto optimality and provide a concise overview of them here, with the formal definitions postponed to Sect. 2.

(i) The first two models are called α-*layer global Pareto optimality* (α-*LGP*) and k-*agent global Pareto optimality* (k-*AGP*), which extend the original Pareto optimality concept in a straightforward way. α-LGP tries to find a size-α subset of layers, such that there exists a matching which is Pareto optimal in the α layers; while k-AGP seeks two size-k subsets $U' \subseteq U$ and $W' \subseteq W$, such that there exists a matching which is Pareto optimal in all layers when all agents in $U \setminus U'$ and $W \setminus W'$ are removed.

(ii) The third is called α-*dominating-free layers optimality* (α-*DFL*), which tries to find a matching M with no other matching dominating M in more than $\ell - \alpha$ layers, where ℓ is the number of layers. In other words, for each $M' \neq M$, there exist α layers such that M' cannot dominate M in each of these layers.

(iii) The fourth concept is called k-*trading-cycle-free optimality* (k-*TCF*), which determines whether there is a matching M that does not admit any size-k' trading cycle in all layers for each $k' > k$, where a trading cycle is formed by a subset of agents who can benefit by exchanging their partners among themselves. This concept is inspired by Ashlagi et al. [4]. They showed that in the kidney exchange program, large trading cycles are more worthy of being performed since they can gain the most benefit from large cycles, and only they might justify changing the status quo. As a consequence, matchings without large trading cycles in all layers can be considered to be "stable" to some extent. A similar concept has been studied by Steindl and Zehavi [34], who investigated a related but more general Pareto optimality concept of assignment problems.

See Example 1 for a concrete example of the concepts.

Example 1. Consider two sets of agents, $U = \{u_1, u_2, u_3\}$ and $W = \{w_1, w_2, w_3\}$, three layers l_1, l_2 and l_3, the preferences of the agents in the three layers are defined as follows.

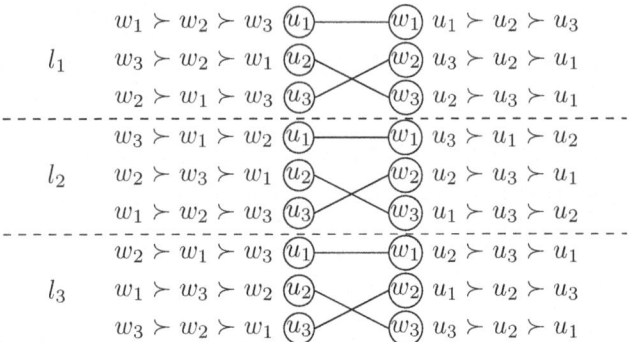

Let $M_1 = \{\{u_1, w_1\}, \{u_2, w_3\}, \{u_3, w_2\}\}$. We have the following observations.

(i) M_1 is 1-LGP, since M_1 is only Pareto optimal in l_1, and dominated by $M_2 = \{\{u_1, w_3\}, \{u_2, w_2\}, \{u_3, w_1\}\}$ in l_2 and $M_3 = \{\{u_1, w_2\}, \{u_2, w_1\}, \{u_3, w_3\}\}$ in l_3, respectively. Therefore, there is only one layer that M_1 is Pareto optimal, which implies that M_1 is 1-LGP.

(ii) M_1 is 2-DFL. There are six possible matchings in this instance. By examining all of them, we can find that matchings other than M_2 and M_3 cannot dominate M_1 in any layer. Thus, M_1 is 2-DFL since for each $M' \neq M_1$, there are at least two layers that M' cannot dominate M_1.

(iii) This instance admits two subsets $U' = \{u_1, u_2\} \subseteq U$ and $W' = \{w_1, w_3\} \subseteq W$ with $|U'| = |W'| = 2$ and a matching $M_1' = \{\{u_1, w_1\}, \{u_2, w_3\}\}$ with M_1' being 2-AGP. It is easy to check that M_1' is Pareto optimal in all layers, since by removing u_3 from the original preferences, the preferences of u_1 become $w_1 \succ w_3$, $w_3 \succ w_1$, and $w_1 \succ w_3$ in l_1, l_2, and l_3, respectively; similarly, the preferences of u_2 become $w_3 \succ w_1$, $w_3 \succ w_1$ and $w_1 \succ w_3$ in l_1, l_2 and l_3, respectively. Thus, there is no matching of U' and W' can dominate M_1' in any layer, which makes M_1' a 2-AGP matching.

(iv) No matching is k-TCF in this example, since matching other than M_1 (resp. M_2, M_3) is dominated by M_1 (resp. M_2, M_3) in l_1 (resp. l_2, l_3), and in at least one layer all agents can benefit by rematching them as M_1, M_2, or M_3. After removing u_2 and w_2 from the instance, $M = \{\{u_1, w_3\}, \{u_3, w_1\}\}$ is 1-TCF and 2-TCF, since M is Pareto optimal in all layers.

Related Work. Steindl and Zehavi [33] studied the assignment problem with multiple preferences, which is closely related to our model. Here, the assignment problem studied by them can also be seen as a two-sided Pareto optimal matching problem, where all women consider all men acceptable and treat them indifferently, and the preferences of men can be incomplete. They investigated the parameterized complexity of the problem with respect to various parameters and obtained a result similar to ours, where the problem is to determine whether there exists a Pareto optimal assignment of the given agents and items. Since completeness of preferences and whether preferences contain ties have a significant influence on the solvability of preference-based problems [20,26], it is meaningful to study Pareto optimal matching with multiple preferences when preferences are complete and do not contain ties.

Steindl and Zehavi [34] continued their own work and studied the parameterized complexity of the verification problems of multilayered assignment problems. They proposed three new optimality concepts, one of which is (k, a)-subset optimality, which asks whether for each $k' \geq k$, there is no trading cycle of each size-k' subset of agents in at least α layers. Obviously, k-TCF is a specific version of (k, a)-subset optimality by setting $\alpha = \ell$ with ℓ being the number of layers.

Wen et al. [35] also investigated Pareto optimal matching with multilayer preferences. Compared to our models, their definition of Pareto optimal matching with multilayer preferences is more layer-oriented, while ours is more classical.

Matching with multiple preferences has also been studied with other optimality notions, such as stability and popularity. Chen et al. [15] studied stable matching with multilayer preferences. They introduced three new models and focused on computational complexity aspects of the models with respect to these new scenarios. Miyazaki and Okamoto [28] studied the same scenarios as Chen et al. and found that even when the length of preference lists is bounded by 4 for both men and women sides, this problem is NP-hard. Bentert et al. [7] continued their research, who studied stable matching problems with multilayer approval preferences: instead of providing a strict ranking, agents approve some agents and treat them indifferently and disapprove all others. Csáji [16] studied popular matching with multilayer preferences and proved that finding a matching popular in all layers is NP-hard.

Our Contributions. First, we introduce four adaptions of the traditional Pareto optimality when considering multiple preferences: α-LGP, α-DFL, k-AGP, and k-TCF. Then we study the relations between the four concepts with the result summarized as Lemma 1 and Fig. 1.

Second, we examine the computational complexity and parameterized complexity of the four concepts. We show that all of them are NP-hard or coNP-hard even with (1) most of the parameters being constants, and (2) all preferences are complete or the length of each preference is no greater than a constant. Then, we study the parameterized complexity of them. Unfortunately, we obtain W[1]-hardness, W[2]-hardness and para-(co)NP-hard results for most parameters except n, where n is the number of men/women. Here, "para-(co)NP-hard" stands for (co)NP-hardness holds even with the corresponding parameter being a constant. Although n is FPT, we still cannot find an FPT algorithm other than exhaustively enumerating all possible matchings, nor can we find a polynomial kernel for n. This makes us wonder whether enumerating all matchings is the optimal algorithm for these problems. Unfortunately, our results confirm our assumption. We show that under Exponential-Time Hypothesis (defined in Sect. 2), $O^*(n!)$ time algorithm is essentially optimal for all concepts. Even worse, we show that no concept admits polynomial kernels with respect to n. These results even hold for combined parameters, such as $n+\alpha$, $n+\overline{\alpha}$ and $n+k$.

Third, we investigate computational complexity of the models when preferences satisfy some desirable property. We begin our study when preferences are *uniform* or *single-layered*. The first property requires that the preferences of all agents on each side are uniform in each layer; the second assumes that on one side, the preference lists of each agent are identical in all layers. We find that the four problems become very easy to determine under the two settings. We then examine three parameters related to similarity of agents: the maximum Hamming distance of two preferences from different layers (resp. agents) of the same agent (resp. layer) Δ_L (resp. Δ_A), and the categories of agents τ, where two agents are in the same category if they have the same preference among all layers. We find that the four problems are para-NP-hard with respect to Δ_L and Δ_A, and are W[1]-hard with respect to τ. Finally, we also investigate the case where all preferences from the same gender are derived from a *master list*. Matching problems with master preference lists have also been studied for decades [12, 21, 23, 24, 32]. We find that even when there are only three layers and the preference lists on each side are all derived from the same single master list, the four problems remain NP-hard.

Refer to Table 1 for an overview of parameterized complexity results. Here, n is the number of agents, ℓ is the number of layers. d is the maximum length of preferences. $\overline{\alpha} = \ell - \alpha$ and $\overline{k} = n - k$. Δ_L (resp. Δ_A) is the maximum Hamming distance of two preferences from different layers (resp. agents) of the same agent (resp. layer). τ represents the number of categories of agents, where two agents are considered to be in the same category if they have the same preference in each layer.

2   Preliminaries

Given a non-negative integer z, let $[z]$ denote the set $\{1, \ldots, z\}$. Let $U = \{u_1, \cdots, u_n\}$ and $W = \{w_1, \cdots, w_n\}$ be two n-elements disjoint sets of agents. We call the members in U men, and the members in W women. There are ℓ layers of preferences. For a certain layer l_r with $r \in [\ell]$, the preference of an agent $u \in U$ in l_r is a strict linear order of the members in W, denoted as $\succ_u^r$. The symbol $\succ_w^r$ is defined analogously. A matching $M \subseteq \{\{u, w\} | u \in U \wedge w \in W\}$ is a set of pairwise disjoint pairs. We say M is a perfect matching if $|M| = n$. If $\{u, w\} \in M$, we say that w is the partner of u matched by M, denoted as $M(u)$, and vice versa. Given an agent a, another agent b from the opposite gender of a, and a layer l_r, we use $P_a^r(b)$ to stand for the position of b in $\succ_a^r$.

Table 1. Complexity of the concepts studied in this paper. FPT* means the concept is FPT but does not admit polynomial kernel with respect to the parameter, and moreover, the concept already obtains essentially tight conditional lower bounds for the running times of algorithms.

	α-LGP	α-DFL	k-AGP	k-TCF
n	FPT* (Theorem 9 to 13, Corollary 1 to 2)			
ℓ	para-NP-hard (Theorem 1, Theorem 2)			para-coNP-hard (Theorem 8)
d	para-NP-hard (Theorem 2)			para-coNP-hard (Theorem 8)
α/k	para-NP-hard (Theorem 3)	coW[1]-hard (Theorem 4)	W[1]-hard (Theorem 6)	?
$\overline{\alpha}/\overline{k}$	para-NP-hard (Theorem 3)	coW[2]-hard (Theorem 5)	para-NP-hard (Theorem 7)	para-coNP-hard (Theorem 8)
combined	$n+\alpha$ — FPT* (Theorem 9, Theorem 10, Theorem 12)	$n+\alpha$ — FPT (Theorem 9)	$n+k$ — FPT* (Theorem 9, Corollary 1, Corollary 2)	$n+k$ — FPT (Theorem 9)
	$n+\overline{\alpha}$ — FPT* (Theorem 9, Theorem 11, Theorem 13)	$n+\overline{\alpha}$ — FPT* (Theorem 9, Corollary 1, Corollary 2)	$n+\overline{k}$ — FPT* (Theorem 9, Corollary 1, Corollary 2)	$n+\overline{k}$ — FPT (Theorem 9)
Δ_L	para-NP-hard (Corollary 3);		P if $\Delta_L = 0$ (single-layered, Theorem 14)	
Δ_A	para-NP-hard (Corollary 3);		P if $\Delta_A = 0$ (uniform, Theorem 14)	
τ	W[1]-hard (Theorem 15)			
master list	NP-hard (Theorem 16)			

2.1 Pareto Optimality

Given two disjoint sets U and W with $|U| = |W|$, a layer l_r, and two matchings M, M' of U and W, we say M' *dominates* M if

1. $P_a^r(M'(a)) \leq P_a^r(M(a))$ for all $a \in U \cup W$;
2. $P_a^r(M'(a)) < P_a^r(M(a))$ for at least one $a \in U \cup W$;

We say M is *Pareto optimal* in l_r if no matching can dominate M in l_r.

Now we formally define all the concepts. Let U and W be two disjoint sets with $|U| = |W|$, $L = \{l_1, \cdots, l_\ell\}$ be a set of layers, and M be a matching of U and W.

Definition 1 (α-LGP). M is α-layer global Pareto optimal (α-LGP) *if there exists a subset $L' \subseteq L$ with $|L'| = \alpha$, such that M is Pareto optimal in all layers of L',*

Definition 2 (α-DFL). M is α-dominating-free layers optimality (α-DFL) *if for each matching M' of U and W, there exists a subset $L' \subseteq L$ with $|L'| = \alpha$, such that M' cannot dominate M in each layer of L',*

Given two subsets $U' \subseteq U$ and $W' \subseteq W$ with $|U'| = |W'|$ and a layer l_r, by removing all $\overline{w} \in W \setminus W'$ (resp. $\overline{u} \in U \setminus U'$) from all men's (resp. women's) preferences, we can form a new preference layer l'_r. We say l'_r is an *induced preference layer* of l_r with respect to U' and W'.

Definition 3 (k-AGP). M is k-agent global Pareto optimality (k-AGP) *if there exist two subsets $U' \subseteq U$ and $W' \subseteq W$, such that $M \cap \{\{u, w\} | u \in U' \wedge w \in W'\}$ is Pareto optimal in each layer of L', where $L' = \{l'_1, \cdots, l'_\ell\}$ is the set of induced layers of L with respect to U' and W'.*

Given a layer l_r and a matching M, we say M admits a size-k *trading cycle* in l_r, if there exist a subset $U' \subseteq U$ and an order $u'_1, \cdots, u'_k$ of U', such that u'_i prefers $M(u'_{i+1})$ to $M(u'_i)$ and $M(u'_{i+1})$ prefers u'_i to u'_{i+1} in l_r with $i \in [k]$ and $u'_{k+1} = u'_1$.

Definition 4 (k-TCF). M is k-trading-cycle-free optimality (k-TCF) *if $M does not admit any size-k trading cycle in all layers.*

Next, we formally define the problem of finding an α-layer global Pareto optimal matching. The other three problem can be defined analogously.

α-LAYER GLOBAL PARETO OPTIMAL MATCHING PROBLEM (α-LGP)
Input: two disjoint sets U and W with $|U| = |W|$, a set of layers $L = \{l_1, \cdots, l_\ell\}$, and an integer α.
Question: Is there a α-LGP matching exists?

We then show the relations between different optimality concepts for different values of α and k with $\alpha \in [\ell]$ and $k \in [n]$.

Lemma 1 ($\star$). *Let M be a matching, the following holds.*

(i) *If M is ℓ-LGP, then M is also ℓ-DFL, n-AGP, 1-TCF, and 2-TCF, and vice versa;*
(ii) *If M is α-LGP, then M is also α-DFL with $1 < \alpha < \ell$;*
(iii) *If M is 1-LGP, then M is also 1-DFL, and vice versa.*

Refer to Fig. 1 for an overview of the relations between different optimality concepts for different values of α and k.

2.2   Parameterized Complexity

We assume basic knowledge of parameterized complexity and refer to the textbook [17,18,29] for more details.

Cross-Composition. A problem Π OR-cross-composes into a parameterized problem Π' if there exists a polynomial-time algorithm, called an OR-cross-composition, that given instances $I_1, I_2, \ldots, I_t$ of Π for some $t \in \mathbb{N}$ that are of the same size s for some $s \in \mathbb{N}$, outputs an instance (I, k) of Π' such that the following conditions are satisfied.

(1) $k \leq p(s + \log t)$ for some polynomial function p;
(2) (I, k) is a YES-instance of Π if and only if at least one of the instances $I_1, I_2, \ldots, I_t$ is a YES-instance of Π.

The notion of AND-Cross-Composition is defined similarly, with the second condition changing to that the output instance is a YES-instance if and only if all the input instances should be YES-instances. The following proposition is the key to show that a problem has no polynomial kernel with respect to a certain

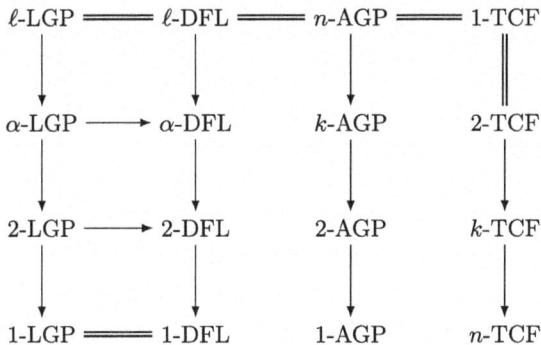

Fig. 1. Overview of the relations between different optimality concepts for different values of α and k. Herein, an arrow "$\longrightarrow$" from one concept to another implies that the first implies the second; An equals sign "$=\!\!=\!\!=$" between two concepts implies the two concepts are equivalent.

parameter: Let Π be an NP-hard problem that OR-cross-composes (or AND-cross-composes) into a parameterized problem Π'. Then, Π' does not admit a polynomial compression, unless NP $\subseteq$ coNP/poly [9,10].

Exponential-Time Hypothesis (ETH). To obtain tight lower bounds for the running times of algorithms, we rely on the well-known Exponential-Time Hypothesis (ETH) [17]. ETH asserts that 3-CNF-SAT cannot be solved in time $O(2^{o(n)})$.

3   Parameterized Complexity

Next, we present classical computational complexity and parameterized complexity results of α-LGP, α-DFL, k-AGP, and k-TCF.

3.1   NP-Hardness and W-Hardness

We first show that ℓ-LGP, ℓ-DFL, and n-AGP are NP-hard even when all preferences are complete or when the length of preference is no greater than 4. k-TCF is also very hard, which becomes NP-hard even with $k = n$ and $\ell = 1$.

Theorem 1. *α-LGP, α-DFL, and k-AGP are NP-hard even with $\alpha = \ell = 2$, $k = n$ and $d = n$.*

Proof. We show ℓ-LGP with $\ell = 2$ is NP-hard by reducing the 3-SAT problem to it. Given a variable set V and a clause set C with each clause containing exactly three literals, 3-SAT asks whether there exists a satisfying truth assignment that sets at least one literal in each clause to be true. By Lemma 1, the fact that ℓ-LGP with $\ell = 2$ is NP-hard implies that α-LGP, α-DFL, and k-AGP are NP-hard even if $\alpha = \ell = 2$, $d = n$, and $k = n$.

Let $I = (V = \{v_1, \cdots, v_n\}, C = \{c_1, \cdots, c_m\})$ be an instance of 3-SAT. For each $v_i \in V$, we create two men $u_i, \overline{u}_i$ and two women $w_i, \overline{w}_i$. For each $c_j \in C$, we create three men a_j, b_j, p_j and three women d_j, e_j, f_j. Create three layers $L = \{l_1, l_2\}$. Before setting the preferences in the other two layers, we first define some sets.

$$D_i^+ = \{d_j \mid c_j \in C \land v_i \text{ occurs as positive literal in } c_j\}$$
$$D_i^- = \{d_j \mid c_j \in C \land v_i \text{ occurs as negative literal in } c_j\}$$
$$A_i^+ = \{a_j, b_j, p_j \mid c_j \in C \land v_i \text{ occurs as positive literal in } c_j\}$$
$$A_i^- = \{a_j, b_j, p_j \mid c_j \in C \land v_i \text{ occurs as negative literal in } c_j\}$$
$$U_j^+ = \{u_i \mid v_i \in V \land v_i \text{ occurs as positive literal in } c_j\}$$
$$U_j^- = \{u_i \mid v_i \in V \land v_i \text{ occurs as negative literal in } c_j\}$$

Then we set the preferences in l_1 as follows. Here, $i \in [|V|]$ and $j \in [|C|]$. Given a $c_j \in C$, we use $v_{x_j}, v_{y_j}, v_{z_j}$ to denote the three variables that occur in c_j, and remove the agent in $\boxed{box}$ if the corresponding variable of the agent occurs as a *negative* literal in c_j. "$\cdots$" denotes an arbitrary but fixed order of the other agents. $\overleftarrow{S}$ denotes the agents in S are arranged in descending order of their indices. For instance, $\overleftarrow{\mathcal{W}}$ as defined as follows: $w_{|V|} \succ \overline{w}_{|V|} \succ w_{|V|-1} \succ \overline{w}_{|V|-1} \succ \ldots \succ w_2 \succ \overline{w}_2 \succ w_1 \succ \overline{w}_1$.

$u_i: w_i \succ \overrightarrow{D_i^+} \succ \overline{w}_i \succ \cdots \succ \overleftarrow{\mathcal{W}} \setminus \{w_i, \overline{w}_i\}$ $w_i: \overline{u}_i \succ u_i \succ \cdots \succ \overleftarrow{\mathcal{U}} \setminus \{u_i, \overline{u}_i\}$

$\overline{u}_i: w_i \succ \overline{w}_i \succ \cdots \succ \overleftarrow{\mathcal{W}} \setminus \{w_i, \overline{w}_i\}$ $\overline{w}_i: \overline{u}_i \succ \overleftarrow{A_i^+} \succ u_i \succ \cdots \succ \overleftarrow{\mathcal{U}} \setminus \{u_i, \overline{u}_i\}$

$a_j: e_j \succ f_j \succ \boxed{\overline{w}_{x_j}} \succ d_j \succ \cdots \succ \mathcal{W} \setminus \{\boxed{\overline{w}_{x_j}}\}$ $d_j: \overleftarrow{U_j^+} \succ a_j \succ b_j \succ p_j \succ \cdots \succ \overleftarrow{\mathcal{U}} \setminus U_j^+$

$b_j: e_j \succ f_j \succ \boxed{\overline{w}_{y_j}} \succ d_j \succ \cdots \succ \mathcal{W} \setminus \{\boxed{\overline{w}_{y_j}}\}$ $e_j: a_j \succ b_j \succ p_j \succ \cdots \succ \overleftarrow{\mathcal{U}}$

$p_j: e_j \succ f_j \succ \boxed{\overline{w}_{z_j}} \succ d_j \succ \cdots \succ \mathcal{W} \setminus \{\boxed{\overline{w}_{z_j}}\}$ $f_j: a_j \succ b_j \succ p_j \succ \cdots \succ \overleftarrow{\mathcal{U}}$

Then we set the preferences in l_2 as follows. Here, $i \in [|V|]$ and $j \in [|C|]$. Given a $c_j \in C$, we remove the agent in $\boxed{box}$ if the corresponding variable of the agent occurs as a *positive* literal in c_j.

$u_i: \overline{w}_i \succ \overrightarrow{D_i^-} \succ w_i \succ \cdots \succ \overleftarrow{\mathcal{W}} \setminus \{w_i, \overline{w}_i\}$ $w_i: \overline{u}_i \succ \overleftarrow{A_i^-} \succ u_i \succ \cdots \succ \overleftarrow{\mathcal{U}} \setminus \{u_i, \overline{u}_i\}$

$\overline{u}_i: \overline{w}_i \succ w_i \succ \cdots \succ \overleftarrow{\mathcal{W}} \setminus \{w_i, \overline{w}_i\}$ $\overline{w}_i: \overline{u}_i \succ u_i \succ \cdots \succ \overleftarrow{\mathcal{U}} \setminus \{u_i, \overline{u}_i\}$

$a_j: e_j \succ f_j \succ \boxed{w_{x_j}} \succ d_j \succ \cdots \succ \mathcal{W} \setminus \{\boxed{\overline{w}_{x_j}}\}$ $d_j: \overrightarrow{U_j^-} \succ a_j \succ b_j \succ p_j \succ \cdots \succ \overleftarrow{\mathcal{U}} \setminus U_j^-$

$b_j: e_j \succ f_j \succ \boxed{w_{y_j}} \succ d_j \succ \cdots \succ \mathcal{W} \setminus \{\boxed{\overline{w}_{y_j}}\}$ $e_j: a_j \succ b_j \succ p_j \succ \cdots \succ \overleftarrow{\mathcal{U}}$

$p_j: e_j \succ f_j \succ \boxed{w_{z_j}} \succ d_j \succ \cdots \succ \mathcal{W} \setminus \{\boxed{\overline{w}_{z_j}}\}$ $f_j: a_j \succ b_j \succ p_j \succ \cdots \succ \overleftarrow{\mathcal{U}}$

Then we show the correctness of the construction. Here, given a variable v_i, we say v_i is matched as "TRUE"-matching in M if $\{\{u_i, w_i\}, \{\overline{u}_i, \overline{w}_i\}\} \subseteq M$; otherwise, we say v_i is matched as "FALSE"-matching. In addition, given a clause $c_j \in C$ with v_{x_j}, v_{y_j} and v_{z_j} occurring in c_j, we say v_{x_j} (resp. v_{y_j}, v_{z_j}) is chosen to satisfy c_j in M if a_j (resp. b_j, c_j) is matched to d_j.

"$\Rightarrow$:" We create a matching M by the truth assignment of the original instance as follows. For each variable v_i that is set to true, match v_i as "TRUE"-matching in M; otherwise, match v_i as "FALSE"-matching in M. For each $c_j \in C$ with v_{x_j}, v_{y_j} and v_{z_j} occurring in c_j, we match the corresponding agents of c_j as follows. Let $\{a_j, d_j\} \in M$ and match the other four agents arbitrarily if one of the two conditions is satisfied.

- v_{x_j} occurs as a positive literal in c_j and v_{x_j} is set to true; or
- v_{x_j} occurs as a negative literal in c_j and v_{x_j} is set to false.

If v_{x_j} does not satisfy the above two conditions, check v_{y_j} and v_{z_j} and set the agents in a similar way. Since M is created according to a truth assignment that sets at least one literal in each clause to be true, at least one of v_{x_j}, v_{y_j} and v_{z_j} must satisfy the conditions. Then M must be a perfect matching.

Then we show that M is Pareto optimal in all layers. Suppose that M is not Pareto optimal in l_1, then by the preferences of l_1, we have $\{\{u_i, \overline{w}_i\}, \{\overline{u}_i, w_i\}\} \subseteq M$. Since, if $\{\{u_i, w_i\}, \{\overline{u}_i, \overline{w}_i\}\} \subseteq M$, each agent in $\{u_i, w_i, \overline{u}_i, \overline{w}_i\}$ prefers the partner matched by M to the agents of clauses, no matching can dominate M in l_1. Assuming that M' is the dominating matching of M with $M' = M \cup \{\{u_i, d_j\}, \{a_j, \overline{w}_i\}\} \setminus \{\{u_i, w_i\}, \{a_j, d_j\}\}$, v_{x_j} must occur as a negative literal, since we only match a_j to d_j if v_{x_j} occurs as a negative literal in c_j and v_{x_j} is set to false. Therefore, $a_j \notin \overrightarrow{A_i^+}$, which implies that $\overline{w}_i$ prefers u_i to a_j, a contradiction to the fact that M' dominating M in l_1. Therefore, M is Pareto optimal in l_1. The other two cases that $M' = M \cup \{\{u_i, d_j\}, \{b_j, \overline{w}_i\}\} \setminus \{\{u_i, w_i\}, \{b_j, d_j\}\}$ and $M' = M \cup \{\{u_i, d_j\}, \{c_j, \overline{w}_i\}\} \setminus \{\{u_i, w_i\}, \{c_j, d_j\}\}$ can be proved in a similar way. Therefore, M is Pareto optimal in l_1. We can proved that M is Pareto optimal in l_2 in a similar way.

"$\Leftarrow$:" Let M be a l-LGB matching of the constructed instance. We build a truth assignment as follows. For each variable $v_i \in V$, if v_i is matched as "TRUE"-matching, then let v_i be true; otherwise, let v_i be false. We only need to show that in each clause, there must be a literal set to true.

Claim 1 ($\star$). *If M is Pareto optimal in all layers, then M satisfies the following.*

(i) For each $i \in [|V|]$, $\{M(u_i), M(\overline{u}_i)\} = \{w_i, \overline{w}_i\}$;
(ii) For each $j \in [|C|]$, $\{M(a_j), M(b_j), M(p_j)\} = \{d_j, e_j, f_j\}$.

Therefore, for each $j \in [|C|]$, one agent in $\{a_j, b_j, c_j\}$ must be matched to d_j, which implies that the corresponding variable is chosen to satisfy c_j in M. Assume that in M, we have $M(a_j) = d_j$, that is, v_{x_j} is chosen to satisfy c_j.

Claim 2 ($\star$). *If M is Pareto optimal in all layers with $M(a_j) = d_j$, then M satisfies the following.*

(i) If v_{x_j} occurs as a positive literal in c_j, then $\{\{u_{x_j}, w_{x_j}\}, \{\overline{u}_{x_j}, \overline{w}_{x_j}\}\} \subseteq M$;
(ii) If v_{x_j} occurs as a negative literal in c_j, then $\{\{u_{x_j}, \overline{w}_{x_j}\}, \{\overline{u}_{x_j}, w_{x_j}\}\} \subseteq M$.

Therefore, if v_{x_j} occurs as a negative (resp. positive) literal in c_j, v_{x_j} is matched as "FALSE"-matching (resp. "TRUE"-matching) in M. The other cases where $M(b_j) = d_j$ or $M(c_j) = d_j$ can be proved in a similar way. Thus, in the assignment built by M, there must be a literal that is set to true in each clause. □

Theorem 2 ($\star$). *α-LGP, α-DFL, and k-AGP are NP-hard even with $\alpha = \ell = 4$, $k = n$ and $d \in \{4, n\}$.*

Theorem 3 ($\star$). *α-LGP is para-NP-hard with respect to α and $\overline{\alpha}$ for $\alpha < \ell$.*

Theorem 4 ($\star$). *α-DFL is coW[1]-hard with respect to α.*

Theorem 5 ($\star$). *α-DFL is coW[2]-hard with respect to $\overline{\alpha}$.*

Theorem 6 ($\star$). *k-AGP is W[1]-hard with respect to k.*

Theorem 7 ($\star$). *k-AGP is para-NP-hard with respect to $\overline{k}$.*

Theorem 8 ($\star$). *k-TCF is coNP-hard even with $k = n$, $\ell = 1$, and $d \in \{3, n\}$.*

3.2 FPT and ETH-Based Lower Bounds

First, we show that all problems are fixed-parameter tractability (FPT) with respect to different parameters.

Theorem 9 ($\star$). *α-LGP, α-DFL, k-AGP and k-TCF are FPT with respect to n. Furthermore, α-LGP and α-DFL are FPT with respect to $n + \alpha$ and $n + \overline{\alpha}$; k-AGP and k-TCF are FPT with respect to $n + k$ and $n + \overline{k}$; and k-AGP is XP with respect to k.*

Then, we show that the running times of the FPT algorithms are already obtain (essentially) tight conditional lower bounds under Exponential-Time Hypothesis (ETH). The following proofs are based on the proposition shown in [17]: Suppose that there is a polynomial-time parameterized reduction from problem A to problem B such that if the parameter of an instance of A is k, then the parameter of the constructed instance of B is at most $g(k)$ for some non-decreasing function g. Then an $O^*(2^{o(f(k))})$-time algorithm for B for some non-decreasing function f implies an $O^*(2^{o(f(g(k)))})$-time algorithm for A.

Theorem 10 ($\star$). *Unless ETH fails, there does not exist an algorithm for α-LGP with running time $O^*(2^{o(\kappa \log \kappa)})$ where $\kappa = n + \alpha$.*

Theorem 11 ($\star$). *Unless ETH fails, there does not exist an algorithm for α-LGP with running time $O^*(2^{o(\kappa \log \kappa)})$ where $\kappa = n + \overline{\alpha}$.*

The corollary follows from Lemma 1 and Theorem 11.

Corollary 1. *Unless ETH fails, there does not exist an algorithm with running time $O^*(2^{o(\kappa \log \kappa)})$ for (1) k-AGP with $\kappa = n + k$ or $\kappa = n + \overline{k}$, (2) α-DFL with $\kappa = n + \overline{\alpha}$, and (3) k-TCF with $\kappa = n$.*

3.3 Non-existence of Polynomial Kernels

Theorem 12 ($\star$). *There does not exist a polynomial kernel for α-LGP with respect to the parameter $\kappa = n + \alpha$, unless $NP \subseteq coNP/poly$.*

Theorem 13 ($\star$). *There does not exist a polynomial kernel for α-LGP with respect to the parameter $\kappa = n + \overline{\alpha}$, unless $NP \subseteq coNP/poly$.*

The corollary follows from Lemma 1 and Theorem 13.

Corollary 2. *Unless $NP \subseteq coNP/poly$, there does not exist a polynomial kernel for (1) k-AGP with $\kappa = n + k$ or $\kappa = n + \overline{k}$, (2) α-DFL with $\kappa = n + \overline{\alpha}$, and (3) k-TCF with $\kappa = n$.*

4   Special Cases

In this section, we investigate computational complexity of the models when preferences satisfy some desirable property, namely *uniform*, *single-layered*, and *master list*.

4.1   Similarity of Agents

We first consider the case that preferences are *uniform* or *single-layered*. The first property requires that the preferences of all agents on each side are uniform in each layer. The second assumes that on one side the preference list of each agent remains the same in all layers. We find that all concepts become easy to determine in the two settings.

Theorem 14 ($\star$). *If the preferences of the agents on one side are single-layered or uniform in all layers, α-LGP, α-DFL, k-AGP, and k-TCF can be solved in $O(n^2)$ time*

Obviously, similarity of preferences must play an important role in the solvability of the concepts. Then, we examine three parameters related to similarity: the maximum Hamming distance of two preferences from different layers (resp. agents) of the same agent (resp. layer) Δ_L (resp. Δ_A), and the categories of agents τ, where two agents are in the same category if they have the same preferences among all layers. We find that all concepts are para-NP-hard with respect to Δ_L and Δ_A, and are W[1]-hard with respect to τ.

The corollary follows from Theorem 2, in the proof of which both Δ_A and Δ_L are constants.

Corollary 3. *α-LGP, α-DFL, k-AGP, and k-TCF are NP-hard even with Δ_A and Δ_L being constants.*

Theorem 15 ($\star$). *α-LGP, α-DFL, k-AGP, and k-TCF are W[1]-hard with respect to τ.*

4.2 Master List

Next, we focus on a special case where the preference lists of each agent are derived from a *master list*. Here, a master list of men consists of a single list that contains all men, and each woman's preference list contains her acceptable partners ranked precisely following the master list. The master list of women is defined analogously. We find that even when there are only three layers and the preference lists on each side are all derived from the same single master list, the four problems remain NP-hard. Note that master lists of the same gender across different layers are the same.

Theorem 16 (⋆). *α-LGP, α-DFL, k-AGP, and k-TCF are NP-hard even when $\ell = 3$, and the preference lists on each side are all derived from the same single master list.*

5 Conclusion

In this paper, we present four adapted Pareto optimal matching concepts with multilayer preferences. We study their parameterized complexity for various parameters. We also consider cases where preferences exhibit desirable properties and get both tractable and intractable results.

For future work, it might be interesting to study the concepts of popular matching with multilayer preferences. Although Csáji [16] has already shown that determining whether there exists a matching that is popular in at least α layers is NP-hard, there are still many ways to adapt the concept of popular matching. For instance, one can check whether there is a matching M such that no M' is more popular than M in at least $\alpha + 1$ layers. In addition, studying the four concepts with multilayer approval preferences might be another interesting topic. Bentert et al. [7] have shown that stable matching with multilayer approval preferences can be harder than that with multilayer strict preferences.

Disclosure of Interests. The authors have no competing interests to declare that are relevant to the content of this article.

References

1. Abdulkadiroğlu, A., Pathak, P.A., Roth, A.E.: The New York city high school match. Am. Econ. Rev. **95**(2), 364–367 (2005)
2. Abdulkadiroğlu, A., Pathak, P.A., Roth, A.E., Sönmez, T.: The Boston public school match. Am. Econ. Rev. **95**(2), 368–371 (2005)
3. Abraham, D.J., Cechlárová, K., Manlove, D.F., Mehlhorn, K.: Pareto optimality in house allocation problems. In: Proceedings of ISAAC-2015, pp. 3–15 (2004)
4. Ashlagi, I., Gamarnik, D., Rees, M.A., Roth, A.E.: The need for (long) chains in kidney exchange. Technical report, National Bureau of Economic Research (2012)
5. Aziz, H., Chen, J., Gaspers, S., Sun, Z.: Stability and pareto optimality in refugee allocation matchings. In: Proceedings of AAMAS-2018, pp. 964–972 (2018)

6. Aziz, H., Gaspers, S., Sun, Z., Walsh, T.: From matching with diversity constraints to matching with regional quotas. In: Proceedings of AAMAS-2019, pp. 377–385 (2019)
7. Bentert, M., Boehmer, N., Heeger, K., Koana, T.: Stable matching with multilayer approval preferences: approvals can be harder than strict preferences. Games Econ. Behav. **142**, 508–526 (2023)
8. Biró, P., Gudmundsson, J.: Complexity of finding pareto-efficient allocations of highest welfare. Eur. J. Oper. Res. **291**(2), 614–628 (2021)
9. Bodlaender, H.L., Downey, R.G., Fellows, M.R., Hermelin, D.: On problems without polynomial kernels. J. Comput. Syst. Sci. **75**(8), 423–434 (2009)
10. Bodlaender, H.L., Jansen, B., Kratsch, S.: Kernelization lower bounds by cross-composition. SIAM J. Discret. Math. **28**(1), 277–305 (2014)
11. Bottero, M., Ferretti, V., Figueira, J.R., Greco, S., Roy, B.: On the choquet multiple criteria preference aggregation model: theoretical and practical insights from a real-world application. Eur. J. Oper. Res. **271**(1), 120–140 (2018)
12. Bredereck, R., Heeger, K., Knop, D., Niedermeier, R.: Multidimensional stable roommates with master list. In: Proceedings of WINE-2020, vol. 12495, pp. 59–73 (2020)
13. Cechlárová, K., Eirinakis, P., Fleiner, T., Magos, D., Mourtos, I., Potpinková, E.: Pareto optimality in many-to-many matching problems. Discret. Optim. **14**, 160–169 (2014)
14. Cechlárová, K., Fleiner, T.: Pareto optimal matchings with lower quotas. Math. Soc. Sci. **88**, 3–10 (2017)
15. Chen, J., Niedermeier, R., Skowron, P.: Stable marriage with multi-modal preferences. In: Proceedings of EC-2018, pp. 269–286 (2018)
16. Csáji, G.: Popular and dominant matchings with uncertain and multimodal preferences. In: Proceedings of IJCAI-2024, pp. 2740–2747 (2024)
17. Cygan, M., et al.: Parameterized Algorithms. Springer (2015)
18. Downey, R., Fellows, M.: Parameterized Complexity. Springer (2012)
19. Farczadi, L., Georgiou, K., Könemann, J.: Stable marriage with general preferences. Theory Comput. Syst. **59**(4), 683–699 (2016)
20. Gale, D., Shapley, L.S.: College admissions and the stability of marriage. Am. Math. Mon. **69**(1), 9–15 (1962)
21. Irving, R.W., Manlove, D.F., Scott, S.: The stable marriage problem with master preference lists. Discret. Appl. Math. **156**(15), 2959–2977 (2008)
22. Jin, L., Mesiar, R., Yager, R.R.: Parameterized preference aggregation operators with improved adjustability. Int. J. Gen. Syst. **49**(8), 843–855 (2020)
23. Kamiyama, N.: Stable matchings with ties, master preference lists, and matroid constraints. In: Proceedings of SAGT-2015, vol. 9347, pp. 3–14 (2015)
24. Kamiyama, N.: Many-to-many stable matchings with ties, master preference lists, and matroid constraints. In: Proceedings of AAMAS-2019, pp. 583–591 (2019)
25. Lukasiewicz, T., Malizia, E.: Complexity results for preference aggregation over (m)CP-nets: Pareto and majority voting. Artif. Intell. **272**, 101–142 (2019)
26. Manlove, D.F., Irving, R.W., Iwama, K., Miyazaki, S., Morita, Y.: Hard variants of stable marriage. Theor. Comput. Sci. **276**(1–2), 261–279 (2002)
27. Manlove, D.F., O'Malley, G.: Paired and altruistic kidney donation in the UK: algorithms and experimentation. ACM J. Exp. Algorithmics **19**(1) (2014)
28. Miyazaki, S., Okamoto, K.: Jointly stable matchings. J. Comb. Optim. **38**(2), 646–665 (2019)
29. Niedermeier, R.: Invitation to Fixed-Parameter Algorithms. Oxford University Press (2006)

30. Roth, A.E., Sönmez, T., Ünver, M.U.: Pairwise kidney exchange. J. Econ. Theory **125**(2), 151–188 (2005)
31. Roth, A.E., Sönmez, T., Ünver, M.U.: Efficient kidney exchange: coincidence of wants in markets with compatibility-based preferences. Am. Econ. Rev. **97**(3), 828–851 (2007)
32. Schlotter, I.: Recognizing when a preference system is close to admitting a master list. Theor. Comput. Sci. **994**, 114445 (2024)
33. Steindl, B., Zehavi, M.: Parameterized analysis of assignment under multiple preferences. In: Proceedings of EUMAS-2021, pp. 160–177 (2021)
34. Steindl, B., Zehavi, M.: Verification of multi-layered assignment problems. Auton. Agents Multi Agent Syst. **36**(1), 15 (2022)
35. Wen, Y., Zhou, A., Guo, J.: Position-based matching with multi-modal preferences. In: Proceedings of AAMAS-2022, pp. 1373–1381 (2022)

An FPT Factor-11 Approximation Algorithm for TSP

Jianqi Zhou, Zhongyi Zhang, and Jiong Guo[✉]

School of Computer Science and Technology, Shandong University, Qingdao, China
jqzhou@mail.sdu.edu.cn, zhangzhongyi@mail.du.edu.cn, jguo@sdu.edu.cn

Abstract. In this paper, we study the Traveling Salesman Problem (TSP), where given a weighted undirected complete graph, the goal is to find a minimum-weight cycle visiting every vertex exactly once. Motivated by different polynomial-time approximability of TSP, that is, TSP on general graphs cannot be approximated within any factor, while on metric graphs, where the edge weights satisfy the triangle inequality, admits a $(1.5 - \varepsilon)$-approximation, Zhou et al. [ISAAC '22] introduced a parameter β to measure the distance from a given instance to a metric graph and proposed a $(6\beta + 9)$-approximation algorithm for TSP running in $\beta^{O(\beta)} \cdot n^3$ time, where β is the number of vertices, whose removal from a given graph results in a metric graph. Bampis et al. posed it as an open question, whether there exists an FPT constant approximation algorithm for TSP, parameterized by β. In this paper, we answer this open question affirmatively by providing an 11-approximation algorithm for TSP running in $\beta^{O(\beta)} \cdot n^3$ time, which greatly improves the result by Zhou et al.

Keywords: Traveling Salesman Problem · Fixed-parameter tractable · Parameterized approximation · Non-metric graphs

1 Introduction

The Traveling Salesman Problem (TSP) is one of the most prominent graph and combinatorial optimization problems with wide applications. In TSP, we are given an undirected complete graph $G = (V, E)$ with n vertices and an edge weight function $w : E \mapsto R_{\geqslant 0}$, the goal is to find a minimum-weight cycle visiting every vertex exactly once. TSP is NP-hard [8,12] and can be exactly solved by the dynamic programming algorithm in $O(n^2 \cdot 2^n)$ time, independently proposed by Bellman [4] and Held and Karp [9].

With respect to approximation algorithms, TSP on general graphs cannot be approximated in polynomial time within $\rho(n)$ factor for any computable function ρ, unless P = NP [16]. There is a special class of graphs called metric graphs, where the edge weights satisfy the triangle inequality, that is

The work was supported by the National Natural Science Foundation of China (No. 61772314, 61761136017, and 62072275).

$w(v_1, v_2) \leq w(v_1, v_3) + w(v_3, v_2)$ for every $v_1, v_2, v_3 \in V$. On metric graphs, there are classical double-tree algorithm and Christofides-Serdyukov algorithm [7,17] for TSP, which achieve the approximation factors of 2 and 1.5, respectively. The ratio of 1.5 has been the best approximation ratio for metric TSP for more than 40 years, until recently it was slightly improved to $1.5 - \varepsilon$ [10,11]. On the other hand, it is NP-hard to approximate metric TSP within $\frac{123}{122}$ [13].

As a generalization from metric TSP to non-metic TSP, the parameterized triangle inequality is a relaxation of the triangle inequality, which allows $w(v_1, v_2) \leq \tau \cdot (w(v_1, v_3) + w(v_3, v_2))$ for every $v_1, v_2, v_3 \in V$ and some $\tau > 1$. Andreae and Bandelt [2] gave a $(3\tau^2 + \tau)/2$-approximation algorithm for TSP satisfying the parameterized triangle inequality. For $\tau > 7/3$, Bender and Chekuri [5] improved this approximation factor to 4τ and showed that it is impossible to get a ratio sublinear in τ, unless $P = NP$. There are some further research results for some small values of τ [1,6,15]. Based on different input instances, the value of τ is not bounded and might even be much larger than the size n of input.

Motivated by different polynomial-time approximability of general TSP and metric TSP, Zhou et al. [19] proposed the concept of "distance from approximability", introduced two parameters α and β to measure the distance from a general graph to a metric graph, and designed approximation algorithms with the exponential running time restricted to the parameters. Hereby, α is the number of triangles violating the triangle inequality and β is the minimum number of vertices, whose removal transforms a given graph to a metric graph. It always holds that $\beta \leq \alpha$. With α as parameter, Zhou et al. [19] proposed a 3-approximation algorithm for TSP running in $O(\alpha! \, 2^\alpha \cdot n^2 + n^3)$ time. Recently, Bampis et al. [3] noted that for TSP, the number α' of vertices appearing in at least one triangle violating the triangle inequality is a more intuitive parameter and might be much smaller than α. Obviously, $\alpha' \leq 3\alpha$ and $\beta \leq \alpha'$. Using α' as parameter, Bampis et al. [3] improved the approximation factor from 3 to 2.5 in FPT time. With β as parameter, Zhou et al. [19] gave a $(6\beta + 9)$-approximation algorithm for TSP running in $\beta^{O(\beta)} \cdot n^3$ time. There remains an open question whether there exists an FPT constant approximation algorithm for TSP parameterized by β.

Our Contributions. In this paper, we focus on designing an FPT constant approximation algorithm for TSP parameterized by the second parameter β introduced in [19]. More specifically, we focus on the number β of vertices, whose removal results in a metric graph, and propose an 11-approximation algorithm for TSP on general graphs with running time of $\beta^{O(\beta)} \cdot n^3$, which greatly improves the approximation ratio of $6\beta + 9$ in [19]. This first constant-approximation result also gives a positive answer to the open question, proposed in [3].

We follow the basic idea of Zhou et al. [19]. Roughly speaking, they first compute the β special vertices, whose removal results in a metric graph, and guess their orders and gaps in an optimal solution to get some chains consisting of special vertices. Then the "nearest" vertices are assigned to both ends of the chains to "wrap" all special vertices, so that the remaining triangles satisfy the triangle inequality. This step forms the most important part of this approach and determines the approximation ratio. The difficulty behind this step is that we cannot guess the end-vertices of the chains in an optimal solution in FPT time. Zhou et al. completed this step based on the local optimal strategy, achieving a $(6\beta + 9)$-approximation. In this paper, we adopt a completely different strategy to find the assignment of the vertices to the chains. More specifically, we compute the shortest path between each two consecutive chains to ensure the global optimum and thus achieve a constant-approximation. Moreover, to prove the approximation ratio of our algorithm, we derive a new lower bound of the weight of the optimal solution, and based on it, construct a Hamilton cycle, whose weight is at most 5 times the optimal solution. The construction of the Hamilton cycle might be of independent interest and provide a new tool for solving other path and cycle problems.

2   Preliminaries

We introduce some definitions, following the notations in [19]. For a positive integer i, set $[i] = \{1, \cdots, i\}$. Throughout the paper, we consider a simple undirected complete graph $G = (V, E)$ with a non-negative edge weight function $w : E \mapsto R_{\geqslant 0}$. The graph G is metric if $w(v_1, v_2) \leq w(v_1, v_3) + w(v_3, v_2)$ for every $v_1, v_2, v_3 \in V$. A set of vertices is called a *violating vertex set*, if its removal transforms a non-metric graph into a metric graph. For an edge set $E' \subseteq E$, the weight of E' is the sum of the weights of the edges in E' and is denoted by $w(E')$. And we use $V(E')$ to denote the set of the end-vertices of the edges in E'. For a vertex set $V' \subseteq V$, $E(V')$ is the set of the edges that connect two vertices in V', and $G[V'] = (V', E(V'))$ is the subgraph of G induced by V'. A t-forest is an acyclic graph consisting of t connected components. When a spanning subgraph of G is a t-forest, it is called a t-spanning forest of G. A t-spanning forest with the minimum weight is called a t-minimum spanning forest (t-MSF) of G.

3   An FPT 11-Approximation for TSP Parameterized by β

In this section, we consider an instance of general TSP with the minimum size β of violating vertex sets. A violating vertex set of at most β vertices can be computed in $O(3^\beta \cdot n^3)$ time by a simple search tree. The basic idea is that at least one vertex of each triangle violating the triangle inequality has to appear in a violating vertex set. The vertices in such a minimum violating vertex set are called *bad* and the remaining vertices are called *good*. So we partition all vertices in V into the set V^b of bad vertices and the set V^g of good vertices. Obviously, we have $|V^b| = \beta$. Note that a triangle consisting of three good vertices must satisfy the triangle inequality and a triangle containing at least one bad vertex might violate the triangle inequality.

First, we give a high-level description of our algorithm.

The first step follows the same approach as in [19], that is, we "guess" the relative positions of bad vertices in an optimal TSP-solution opt, which mean the occurrence order of bad vertices and the "gaps" between bad vertices where good vertices are inserted. This can be done by enumerating all possible permutations of bad vertices and for each permutation, enumerating all possible partitions respecting the order of the corresponding permutation. Here, Zhou et al. [19] defined the concept of "bad chains" to describe the bad vertices in one subset of each partition case. A *bad chain* is one bad vertex or a path consisting of distinct bad vertices. It can be denoted by $q = (b_1, b_2, \ldots, b_l)$ with $l \geq 1$, where $b_i \in V^b$ for $i \in [l]$ and there is an edge connecting b_i and b_{i+1} for each $i \in [l-1]$. We use $b_s(q)$, $b_e(q)$ to denote the starting and ending vertices, respectively, and use $w(q)$ to denote the total weight of the edges in q. If a bad chain q consists of only one bad vertex b_1, then $b_s(q) = b_1$, $b_e(q) = b_1$ and $w(q) = 0$. We assume that in opt, the bad vertices in a bad chain occur together and obey the order in the chain, the bad chains occur according to the permutation order, and there is at least one good vertex between two consecutive bad chains.

At the second step, we assign good vertices to both ends of each bad chain, so that all bad vertices are "wrapped" around good vertices. In other words, we fix the direct neighbors of bad chains in a TSP-solution. Here, Zhou et al. [19] defined the concept of "alternating chains" to extend the concept of bad chains. An *alternating chain* is a path consisting alternatively of distinct good vertices and bad chains. It can be denoted by $h = (g_1, q_1, g_2, \ldots, g_l, q_l, g_{l+1})$ with $l \geq 1$, where $g_1, \ldots, g_{l+1} \in V^g$, and $q_1, \ldots, q_l$ are vertex-disjoint bad chains. Moreover, g_{i+1} for $i \in [l-1]$ is connected by edges to $b_e(q_i)$ and $b_s(q_{i+1})$, g_1 is adjacent to $b_s(q_1)$, and g_{l+1} is adjacent to $b_e(q_l)$. We use $g_s(h)$, $g_e(h)$ to denote the starting and ending good vertices, respectively, and use $w(h)$ to denote the total weight of the edges in h, i.e., $w(h) = \sum_{i=1}^{l} w(q_i) + \sum_{i=1}^{l} w(g_i, b_s(q_i)) + \sum_{i=1}^{l} w(b_e(q_i), g_{i+1})$. If two consecutive bad chains share the same good vertex, then we merge them together into one alternating chain. Since the number of bad vertices is at most β, we can enumerate all possible cases of bad chains in FPT time. Among these cases, there must exist one in which we obtain the same bad chains as in opt. However, we cannot "guess" the same alternating chains as in opt in FPT time, since the number of possible good end-vertices is not bounded by a function of β. This is the key difficulty for the parameterization of β, which we solve by a completely different approach than in [19], resulting in a better approximation factor. Briefly speaking, in Procedure EnumAC (will be given in Sect. 3.1), we enumerate some possible cases of alternating chains, and for each of these cases, we perform the operation of the following third step. Here, the number of possible cases can be bounded by a function of β.

At the third step, we apply Procedure FindPaths (will be given in Sect. 3.1) to find paths consisting of all remaining good vertices to connect consecutive alternating chains in the same way as in [19], resulting in a TSP-solution.

3.1 Algorithm and Time Complexity

We use Algorithm β-TSP to denote the overall framework of our algorithm and introduce the details of each step later.

Algorithm β-TSP: An FPT 11-approximation for TSP parameterized by β

1. Compute a violating vertex set V^b of β vertices by a simple search tree and the corresponding set V^g of good vertices.
2. Enumerate all possible permutations of bad vertices. For each permutation, enumerate all possible β_1-partitions of bad vertices for each $\beta_1 \in [\beta]$, respecting the corresponding permutation order. For each β_1-partition, do:
   (a) Connect the bad vertices in each subset of the partition in the order of the permutation, resulting in a collection of β_1 bad chains $Q = (q_1, \ldots, q_{\beta_1})$, ordered according to the permutation.
   (b) Apply Procedure EnumAC to enumerate some cases of alternating chains.
   (c) For each collection of alternating chains $H = (h_1, \cdots, h_{\beta_2})$ with $\beta_2 \leq \beta_1$ enumerated in Procedure EnumAC, apply Procedure FindPaths to obtain a TSP-solution $\mathcal{A}$ for this enumeration case.
3. Return the solution $\mathcal{A}^{\min}$ with the minimum weight among all enumeration cases.

Next, we show how to assign good vertices to both ends of each bad chains to get alternating chains in EnumAC, which differs from the approach in [19]. Given β_1 bad chains, we divide all good vertices into β_1 connected components with the minimum edge weight by computing a β_1-minimum spanning forest. Zhou et al. [19] just choose some good vertices in each connected components which are closest to the starting and ending bad vertices of each bad chain, and enumerate all possible cases where each chosen good vertex may be the end-vertex of the bad chain. Namely, they choose nearest good vertices as the end-vertices of each bad chain based on a locally optimal strategy, resulting in corresponding alternating chains.

Hereby, we construct alternating chains based on the global optimum. For each two consecutive bad chains, we enumerate some paths with only one or two internal good vertices to form a set of paths connecting the two bad chains. These paths should satisfy that on the one hand these good internal vertices are from all possible β_1 connected components, and on the other hand, the weights of these paths are as low as possible. The number of enumerated paths can be bounded by a function of β_1 with $\beta_1 \leq \beta$. For each enumerated path, we fix the internal good vertices adjacent to two corresponding bad chains as the end-vertices of these two bad chains, resulting in the alternating chains for this enumeration case. More details are given in Procedure EnumAC below.

The intuition behind our algorithm is that we can shortcut all good vertices not adjacent to bad chains in opt to get a cycle consisting of alternating chains in opt. All triangles involved in the shortcut process contain no bad vertex and thus, satisfy the triangle inequality. Hence, the weight of the resulting cycle is a lower bound of the weight of opt. Our algorithm aims to find a similar cycle with the minimum weight to get the corresponding alternating chains.

Note that for each two consecutive bad chains, EnumAC will enumerate more than β cases to find vertex-disjoint alternating chains. The bounds $5\beta_1$, $8\beta_1$ and $4\beta_1$ in EnumAC are determined by the requirement of the proof of Lemma 4, and these constant-factors do not affect the FPT running time.

Procedure EnumAC: Get β_2 alternating chains based on β_1 bad chains

---

1. Apply the algorithm in [14] to get a β_1-minimum spanning forest of $G[V^g]$, denoted by $\mathcal{F} = (\mathcal{T}_1, \ldots, \mathcal{T}_{\beta_1})$. Let $V_j = V(\mathcal{T}_j)$ for each $j \in [\beta_1]$.
2. For each $i \in [\beta_1]$ and $j \in [\beta_1]$, let $P(q_i, V_j, q_{i+1})$ be the set of $\min\{5\beta_1, |V_j|\}$ minimum-weight paths of the form $(b_e(q_i), g^j, b_s(q_{i+1}))$, where $g^j \in V_j$.
3. For each $i \in [\beta_1]$ and $j, j' \in [\beta_1]$, initially set $P(q_i, V_j, V_{j'}, q_{i+1}) = \emptyset$ and from $l = 1$ to $l = \min\{8\beta_1, |V_j|, |V_{j'}|\}$, do the following:
   (a) Let $(b_e(q_i), g_l^j, g_l^{j'}, b_s(q_{i+1}))$ denote a minimum-weight path of the form $(b_e(q_i), g^j, g^{j'}, b_s(q_{i+1}))$ and add it to $P(q_i, V_j, V_{j'}, q_{i+1})$, where $g^j \in V_j \setminus \{g_1^j, \cdots, g_{l-1}^j\}$, $g^{j'} \in V_{j'} \setminus \{g_1^{j'}, \cdots, g_{l-1}^{j'}\}$ and $g^j \neq g^{j'}$.
   (b) Add $\min\{4\beta_1, |V_{j'}|\}$ minimum-weight paths of the form $(b_e(q_i), g_l^j, g^{j'}, b_s(q_{i+1}))$ to $P(q_i, V_j, V_{j'}, q_{i+1})$, where g_l^j is defined in Step 3(a), $g^{j'} \in V_{j'}$ and $g^{j'} \neq g_l^j$.
   (c) Add $\min\{4\beta_1, |V_j|\}$ minimum-weight paths of the form $(b_e(q_i), g^j, g_l^{j'}, b_s(q_{i+1}))$ to $P(q_i, V_j, V_{j'}, q_{i+1})$, where $g_l^{j'}$ is defined in Step 3(a), $g^j \in V_j$ and $g^j \neq g_l^{j'}$.
4. Set $P(q_i, q_{i+1}) = (\cup_{j \in [\beta_1]} P(q_i, V_j, q_{i+1})) \bigcup (\cup_{j,j' \in [\beta_1]} P(q_i, V_j, V_{j'}, q_{i+1}))$, for each $i \in [\beta_1]$.
5. For each possible β_1 paths $p(q_1, q_2), p(q_2, q_3), \cdots, p(q_{\beta_1}, q_1)$ satisfying Conditions (1) $p(q_i, q_{i+1}) \in P(q_i, q_{i+1})$ for each $i \in [\beta_1]$ and (2) $V(p(q_i, q_{i+1})) \cap V(p(q_{i'}, q_{i'+1})) \cap V^g = \emptyset$ for each $i, i' \in [\beta_1]$ and $i \neq i'$, where $V(p(q_i, q_{i+1}))$ and $V(p(q_{i'}, q_{i'+1}))$ denote the sets of vertices in $p(q_i, q_{i+1})$ and $p(q_{i'}, q_{i'+1})$, do the following: (Note that $q_{\beta_1+1} = q_1$.)
   (a) Connect β_1 bad chains by β_1 paths $p(q_1, q_2), p(q_2, q_3), \cdots, p(q_{\beta_1}, q_1)$ to get a cycle $\hat{A}$.
   (b) If there is no edge connecting two good vertices in $\hat{A}$, then go to the next iteration; otherwise, delete all the edges connecting two good vertices in $\hat{A}$, resulting in a collection of alternating chains $H = (h_1, \ldots, h_{\beta_2})$ with $\beta_2 \leq \beta_1$ for this case.

For each collection of β_2 alternating chains $H = (h_1, \cdots, h_{\beta_2})$ with $\beta_2 \leq \beta_1$ enumerated in EnumAC, FindPaths needs to find paths consisting of all remaining good vertices to connect consecutive alternating chains, resulting in a TSP-solution $\mathcal{A}$ for this case. Hereby, FindPaths is the same as in [19]. In the following, given a tour $(\cdots, g_1, g_2, g_3, \cdots)$ with $g_1, g_2, g_3 \in V^g$, we can obtain a new tour $(\cdots, g_1, g_3, \cdots)$ by skipping the good vertex g_2. This operation is called *shortcutting* the vertex g_2.

Procedure FindPaths: Find β_2 paths consisting of all remaining good vertices to connect β_2 alternating chains to get a TSP-solution

1. Compute a β_2-minimum spanning forest (β_2-MSF) $\mathbb{F} = (\mathbb{T}_1, \ldots, \mathbb{T}_{\beta_2})$ of $G[V^g \setminus \cup_{j=1}^{\beta_2}(V(h_j) \setminus \{g_e(h_j)\})]$ rooted at $\{g_e(h_j) \mid j \in [\beta_2]\}$.
2. For each $j \in [\beta_2]$, double the edges of $\mathbb{T}_j$, compute an Euler tour of this double tree and shortcut repeated vertices in the Euler tour, resulting in O_j.
3. For each $j \in [\beta_2]$, do: if $O_j = (g_e(h_j))$, then set $p_j = (g_e(h_j), g_s(h_{j+1}))$; if $O_j = (g_e(h_j), g_1, \cdots, g_{l_j}, g_e(h_j))$ with $l_j \geq 1$, then set $p_j = (g_e(h_j), g_1, \cdots, g_{l_j}, g_s(h_{j+1}))$. Hereby, $h_{\beta_2+1} = h_1$.
4. Connect $g_e(h_j)$ and $g_s(h_{j+1})$ by p_j for each $j \in [\beta_2]$, and obtain a TSP-solution $\mathcal{A} = (h_1, p_1, h_2, p_2, \ldots, h_{\beta_2}, p_{\beta_2})$.

Next, we give the running time of the whole algorithm.

Lemma 1. *Algorithm β-TSP runs in $\beta^{O(\beta)} \cdot n^3$ time.*

Proof. In Algorithm β-TSP, Step 1 computes a violating vertex set of β vertices in $O(3^\beta \cdot n^3)$ time by a simple search tree. The number of β_1-partitions of permutations enumerated in Step 2 is bounded by $\beta! \, 2^\beta$. For each collection of β_1 bad chains in Step 2(a) of Algorithm β-TSP, Step 2(b) applies EnumAC. In EnumAC, Step 1 takes $O(n^2 \log n)$ time [14]. The number of β_1 paths enumerated in Step 5 of EnumAC is bounded by $O((\beta_1 \cdot 5\beta_1 + \beta_1^2 \cdot 8\beta_1 \cdot 8\beta_1)^{\beta_1}) = O(64^\beta \cdot \beta^{4\beta})$. For each collection of β_2 alternating chains in EnumAC, FindPaths needs $O(n^2)$ time [19]. Hence, Algorithm β-TSP runs in $O(3^\beta \cdot n^3 + \beta! \, 2^\beta \cdot 64^\beta \cdot \beta^{4\beta} \cdot n^2 \log n) = \beta^{O(\beta)} \cdot n^3$ time. □

3.2 Analysis of Approximation Ratio

In Step 2 of Algorithm β-TSP, we enumerate all possible permutations and partitions of bad vertices. Thus, in some partition case, we get the same ordered bad chains as in an optimal solution opt, denoted by $Q^{\text{opt}} = (q_1^{\text{opt}}, \ldots, q_{\beta_1}^{\text{opt}})$. In Step 1 of EnumAC, we compute a β_1-MSF $\mathcal{F} = (\mathcal{T}_1, \ldots, \mathcal{T}_{\beta_1})$ of the graph $G[V^g]$, and set $V_j = V(\mathcal{T}_j)$ for $j \in [\beta_1]$. The following lemma is obvious.

Lemma 2 (Lemma 11 in [19]). $w(\text{opt}) \geq w(\mathcal{F})$.

We first give an overview of the remaining analysis, which completely differs from the one in [19]. Note that the main difficulty is that the alternating chains in opt might not be enumerated in EnumAC. Thus, we will modify opt to construct another TSP-solution $\tilde{\mathcal{A}}$, which satisfies that on the one hand, the ordered alternating chains in $\tilde{\mathcal{A}}$ occur in some case enumerated by Step 5(b) of EnumAC; on the other hand, the weight of $\tilde{\mathcal{A}}$ is bounded, that is, $w(\tilde{\mathcal{A}}) = O(1) \cdot w(\text{opt})$. Finally, we will show that FindPaths returns a solution $\mathcal{A}$ with $w(\mathcal{A}) \leq 2 \cdot w(\tilde{\mathcal{A}}) + w(\text{opt})$.

We give a high-level description of the construction of $\tilde{\mathcal{A}}$. First, shortcut all the good vertices in opt, which are not directly adjacent to bad chains to obtain a cycle opt*. Second, for each two consecutive bad chains q_i^{opt} and q_{i+1}^{opt} in opt*, if the path in opt* connecting them does not occur in $P(q_i^{\text{opt}}, q_{i+1}^{\text{opt}})$ defined in Step 4 of EnumAC, then choose some less-weight path in $P(q_i^{\text{opt}}, q_{i+1}^{\text{opt}})$ to replace it to obtain a new cycle $\hat{\mathcal{A}}^*$. Here, a less-weight path always exists, since EnumAC adds a sufficiently large number of minimum-weight paths to $P(q_i^{\text{opt}}, q_{i+1}^{\text{opt}})$. Third, add all remaining good vertices according to their orders in the β_1-MSF $\mathcal{F}$ of $G[V^g]$ to obtain a TSP-solution $\tilde{\mathcal{A}}$.

Construction of opt* from opt. The first step shortcuts all good vertices in opt, which are not directly adjacent to bad chains, resulting in a cycle opt*.

Lemma 3. $w(\text{opt}^*) \leq w(\text{opt})$.

Proof. The triangles involved in the shortcut process contain no bad vertex and thus, satisfy the triangle inequality. Hence, we have $w(\text{opt}^*) \leq w(\text{opt})$. □

Construction of $\hat{\mathcal{A}}^*$ from opt*. Next we construct a cycle $\hat{\mathcal{A}}^*$ from opt* by replacing some paths in opt*, which do not occur in the enumeration cases of EnumAC. Since opt* does not shortcut bad vertices and the good vertices adjacent to bad vertices in opt, the bad vertices $b_s(q_i^{\text{opt}})$ and $b_e(q_i^{\text{opt}})$ for each $i \in [\beta_1]$ are adjacent in opt* to the same good vertices as in opt, denoted by $g_s(q_i^{\text{opt}})$ and $g_e(q_i^{\text{opt}})$. Then, the path in opt* connecting q_i^{opt} and q_{i+1}^{opt} can be denoted by $p^*(q_i^{\text{opt}}, q_{i+1}^{\text{opt}}) = (b_e(q_i^{\text{opt}}), g_e(q_i^{\text{opt}}), b_s(q_{i+1}^{\text{opt}}))$ with $g_e(q_i^{\text{opt}}) = g_s(q_{i+1}^{\text{opt}})$ or $p^*(q_i^{\text{opt}}, q_{i+1}^{\text{opt}}) = (b_e(q_i^{\text{opt}}), g_e(q_i^{\text{opt}}), g_s(q_{i+1}^{\text{opt}}), b_s(q_{i+1}^{\text{opt}}))$.

In $P(q_i^{\text{opt}}, q_{i+1}^{\text{opt}})$ defined in Steps 2–4 of EnumAC, there are many paths of the similar form to $p^*(q_i^{\text{opt}}, q_{i+1}^{\text{opt}})$ with less weight, which can replace $p^*(q_i^{\text{opt}}, q_{i+1}^{\text{opt}})$ without increasing total weight. But there are two conditions that should be considered. First, new replacing paths should be vertex-disjoint with old replacing paths. We use the set R^* to contain the good vertices that have been used in old replacing paths. Second, for some i, if $p^*(q_i^{\text{opt}}, q_{i+1}^{\text{opt}}) = (b_e(q_i^{\text{opt}}), g_e(q_i^{\text{opt}}), b_s(q_{i+1}^{\text{opt}}))$ with $g_e(q_i^{\text{opt}}) \in V_j$, then there are some good vertices in V_j that are forbidden to be used to replace $g_e(q_i^{\text{opt}})$. In opt, $g_e(q_i^{\text{opt}})$ is adjacent to two bad vertices, while some good vertices in V_j are adjacent to some good vertices in $V_{j'}$ with $j' \neq j$. The latter can provide the locations for adding the vertices in $V_{j'}$ in the next step of the whole construction, and thus, cannot be placed between two bad vertices. Hereby, for each $j \in [\beta_1]$, we use Z_j^* to denote the set of forbidden vertices in V_j. Initially, set $Z_j^* = \emptyset$. For each $j' \in [\beta_1] \setminus \{j\}$, if there exists an edge $\{g^j, g^{j'}\}$ in opt with $g^j \in V_j$ and $g^{j'} \in V_{j'}$, then add g^j to Z_j^* and go to the next iteration. Thus, we have $|Z_j^*| \leq \beta_1 - 1$.

We apply the following procedure to opt*. Initially, set $R^* = \emptyset$.

Procedure ReplacePaths: Replace the paths in opt* not occurring in Enu-mAC

From $i = 1$ to $i = \beta_1$, do the following operations to opt*:

1. If $p^*(q_i^{\mathrm{opt}}, q_{i+1}^{\mathrm{opt}}) = (b_e(q_i^{\mathrm{opt}}), g_e(q_i^{\mathrm{opt}}), b_s(q_{i+1}^{\mathrm{opt}}))$, then consider V_j satisfying $g_e(q_i^{\mathrm{opt}}) \in V_j$.
   If $p^*(q_i^{\mathrm{opt}}, q_{i+1}^{\mathrm{opt}}) \notin P(q_i^{\mathrm{opt}}, V_j, q_{i+1}^{\mathrm{opt}})$, then
   (a) Pick a minimum-weight path denoted by $(b_e(q_i^{\mathrm{opt}}), g_e^*(q_i^{\mathrm{opt}}), b_s(q_{i+1}^{\mathrm{opt}}))$ from $P(q_i^{\mathrm{opt}}, V_j, q_{i+1}^{\mathrm{opt}})$ satisfying $g_e^*(q_i^{\mathrm{opt}}) \cap (R^* \cup V(\mathrm{opt}^*) \cup Z_j^*) = \emptyset$.
   (b) Replace $p^*(q_i^{\mathrm{opt}}, q_{i+1}^{\mathrm{opt}})$ by $(b_e(q_i^{\mathrm{opt}}), g_e^*(q_i^{\mathrm{opt}}), b_s(q_{i+1}^{\mathrm{opt}}))$.
   (c) Add $g_e^*(q_i^{\mathrm{opt}})$ to R^*.
2. Otherwise, we have $p^*(q_i^{\mathrm{opt}}, q_{i+1}^{\mathrm{opt}}) = (b_e(q_i^{\mathrm{opt}}), g_e(q_i^{\mathrm{opt}}), g_s(q_{i+1}^{\mathrm{opt}}), b_s(q_{i+1}^{\mathrm{opt}}))$.
   Then consider V_j and $V_{j'}$ with $g_e(q_i^{\mathrm{opt}}) \in V_j$ and $g_s(q_{i+1}^{\mathrm{opt}}) \in V_{j'}$.
   If $p^*(q_i^{\mathrm{opt}}, q_{i+1}^{\mathrm{opt}}) \notin P(q_i^{\mathrm{opt}}, V_j, V_{j'}, q_{i+1}^{\mathrm{opt}})$, then
   (a) Pick a minimum-weight path denoted by $(b_e(q_i^{\mathrm{opt}}), g_e^*(q_i^{\mathrm{opt}}), g_s^*(q_{i+1}^{\mathrm{opt}}), b_s(q_{i+1}^{\mathrm{opt}}))$ from $P(q_i^{\mathrm{opt}}, V_j, V_{j'}, q_{i+1}^{\mathrm{opt}})$ satisfying $\{g_e^*(q_i^{\mathrm{opt}}), g_s^*(q_{i+1}^{\mathrm{opt}})\} \cap (R^* \cup (V(\mathrm{opt}^*) \setminus \{g_e(q_i^{\mathrm{opt}}), g_s(q_{i+1}^{\mathrm{opt}})\})) = \emptyset$.
   (b) Replace $p^*(q_i^{\mathrm{opt}}, q_{i+1}^{\mathrm{opt}})$ by $(b_e(q_i^{\mathrm{opt}}), g_e^*(q_i^{\mathrm{opt}}), g_s^*(q_{i+1}^{\mathrm{opt}}), b_s(q_{i+1}^{\mathrm{opt}}))$.
   (c) Add $g_e^*(q_i^{\mathrm{opt}})$ and $g_s^*(q_{i+1}^{\mathrm{opt}})$ to R^*.

Let $\hat{A}^*$ denote the result of ReplacePaths. The following lemma ensures the correctness of ReplacePaths and gives an upper bound on $w(\hat{A}^*)$.

Lemma 4. $\hat{A}^*$ *is a cycle and* $w(\hat{A}^*) \leq w(\mathrm{opt}^*)$.

Proof. For the case $p^*(q_i^{\mathrm{opt}}, q_{i+1}^{\mathrm{opt}}) = (b_e(q_i^{\mathrm{opt}}), g_e(q_i^{\mathrm{opt}}), b_s(q_{i+1}^{\mathrm{opt}}))$ shown in Step 1 of ReplacePaths, if $g_e(q_i^{\mathrm{opt}}) \in V_j$ and $p^*(q_i^{\mathrm{opt}}, q_{i+1}^{\mathrm{opt}}) \notin P(q_i^{\mathrm{opt}}, V_j, q_{i+1}^{\mathrm{opt}})$, which mean $|V_j| > 5\beta_1$, then we have $|P(q_i^{\mathrm{opt}}, V_j, q_{i+1}^{\mathrm{opt}})| = 5\beta_1$ and the weight of each path in $P(q_i^{\mathrm{opt}}, V_j, q_{i+1}^{\mathrm{opt}})$ is at most $w(p^*(q_i^{\mathrm{opt}}, q_{i+1}^{\mathrm{opt}}))$ by the definition of $P(q_i^{\mathrm{opt}}, V_j, q_{i+1}^{\mathrm{opt}})$ in Step 2 of EnumAC. Since $|R^*| \leq 2\beta_1 - 2$, $|V(\mathrm{opt}^*) \cap V^g| \leq 2\beta_1$ and $|Z_j^*| \leq \beta_1$, we have $|P(q_i^{\mathrm{opt}}, V_j, q_{i+1}^{\mathrm{opt}})| - |R^*| - |V(\mathrm{opt}^*) \cap V^g| - |Z_j^*| \geq 2$. Thus, we can find a path $(b_e(q_i^{\mathrm{opt}}), g_e^*(q_i^{\mathrm{opt}}), b_s(q_{i+1}^{\mathrm{opt}}))$ from $P(q_i^{\mathrm{opt}}, V_j, q_{i+1}^{\mathrm{opt}})$, which satisfies $g_e^*(q_i^{\mathrm{opt}}) \cap (R^* \cup V(\mathrm{opt}^*) \cup Z_j^*) = \emptyset$ and $w(b_e(q_i^{\mathrm{opt}}), g_e^*(q_i^{\mathrm{opt}}), b_s(q_{i+1}^{\mathrm{opt}})) \leq w(p^*(q_i^{\mathrm{opt}}, q_{i+1}^{\mathrm{opt}}))$.

For the case $p^*(q_i^{\mathrm{opt}}, q_{i+1}^{\mathrm{opt}}) = (b_e(q_i^{\mathrm{opt}}), g_e(q_i^{\mathrm{opt}}), g_s(q_{i+1}^{\mathrm{opt}}), b_s(q_{i+1}^{\mathrm{opt}}))$ shown in Step 2 of ReplacePaths, if $g_e(q_i^{\mathrm{opt}}) \in V_j$, $g_s(q_{i+1}^{\mathrm{opt}}) \in V_{j'}$ and $p^*(q_i^{\mathrm{opt}}, q_{i+1}^{\mathrm{opt}}) \notin P(q_i^{\mathrm{opt}}, V_j, V_{j'}, q_{i+1}^{\mathrm{opt}})$, then there are three cases to discuss. In the following, we use the set $P'(q_i^{\text{opt}}, V_j, V_{j'}, q_{i+1}^{\text{opt}})$ to contain the paths $(b_e(q_i^{\text{opt}}), g_l^j, g_l^{j'}, b_s(q_{i+1}^{\text{opt}}))$'s for all $l \in [\min\{8\beta_1, |V_j|, |V_{j'}|\}]$, which are defined in Step 3(a) of EnumAC. Obviously, $P'(q_i^{\text{opt}}, V_j, V_{j'}, q_{i+1}^{\text{opt}}) \subseteq P(q_i^{\text{opt}}, V_j, V_{j'}, q_{i+1}^{\text{opt}})$.

Case 1: If $(b_e(q_i^{\text{opt}}), g_e(q_i^{\text{opt}}), g_s(q_{i+1}^{\text{opt}}), b_s(q_{i+1}^{\text{opt}})) \notin P(q_i^{\text{opt}}, V_j, V_{j'}, q_{i+1}^{\text{opt}})$ and there exists a path $(b_e(q_i^{\text{opt}}), g_l^j, g_l^{j'}, b_s(q_{i+1}^{\text{opt}}))$ in $P'(q_i^{\text{opt}}, V_j, V_{j'}, q_{i+1}^{\text{opt}})$ for some l satisfying $g_l^j = g_e(q_i^{\text{opt}})$ and $g_l^{j'} \neq g_s(q_{i+1}^{\text{opt}})$, then there are $4\beta_1$ paths in $P(q_i^{\text{opt}}, V_j, V_{j'}, q_{i+1}^{\text{opt}})$ of the form $(b_e(q_i^{\text{opt}}), g_l^j, g^{j'}, b_s(q_{i+1}^{\text{opt}}))$ where $g^{j'} \in V_{j'}$ and the weight of each of these $4\beta_1$ paths is at most $w(p^*(q_i^{\text{opt}}, q_{i+1}^{\text{opt}}))$, according to Step 3(b) of EnumAC. By the fact that $|R^*| \leq 2\beta_1 - 2$ and $|V(\text{opt}^*) \cap V^g| \leq 2\beta_1$, we can find a path $(b_e(q_i^{\text{opt}}), g_l^j, g_s^*(q_{i+1}^{\text{opt}}), b_s(q_{i+1}^{\text{opt}}))$ from $P(q_i^{\text{opt}}, V_j, V_{j'}, q_{i+1}^{\text{opt}})$, which satisfies $\{g_l^j, g_s^*(q_{i+1}^{\text{opt}})\} \cap (R^* \cup (V(\text{opt}^*) \setminus \{g_e(q_i^{\text{opt}}), g_s(q_{i+1}^{\text{opt}})\})) = \emptyset$ and $w(b_e(q_i^{\text{opt}}), g_l^j, g_s^*(q_{i+1}^{\text{opt}}), b_s(q_{i+1}^{\text{opt}})) \leq w(p^*(q_i^{\text{opt}}, q_{i+1}^{\text{opt}}))$.

Case 2: If $(b_e(q_i^{\text{opt}}), g_e(q_i^{\text{opt}}), g_s(q_{i+1}^{\text{opt}}), b_s(q_{i+1}^{\text{opt}})) \notin P(q_i^{\text{opt}}, V_j, V_{j'}, q_{i+1}^{\text{opt}})$ and there exists a path $(b_e(q_i^{\text{opt}}), g_l^j, g_l^{j'}, b_s(q_{i+1}^{\text{opt}}))$ in $P'(q_i^{\text{opt}}, V_j, V_{j'}, q_{i+1}^{\text{opt}})$ for some l satisfying $g_l^j \neq g_e(q_i^{\text{opt}})$ and $g_l^{j'} = g_s(q_{i+1}^{\text{opt}})$, then the case is similar to Case 1.

Case 3: If $(b_e(q_i^{\text{opt}}), g_e(q_i^{\text{opt}}), g_s(q_{i+1}^{\text{opt}}), b_s(q_{i+1}^{\text{opt}})) \notin P(q_i^{\text{opt}}, V_j, V_{j'}, q_{i+1}^{\text{opt}})$ and for every l, the path $(b_e(q_i^{\text{opt}}), g_l^j, g_l^{j'}, b_s(q_{i+1}^{\text{opt}}))$ in $P'(q_i^{\text{opt}}, V_j, V_{j'}, q_{i+1}^{\text{opt}})$ satisfies $g_l^j \neq g_e(q_i^{\text{opt}})$ and $g_l^{j'} \neq g_s(q_{i+1}^{\text{opt}})$, then we have $|P'(q_i^{\text{opt}}, V_j, V_{j'}, q_{i+1}^{\text{opt}})| = 8\beta_1$ and the weight of each path in $P'(q_i^{\text{opt}}, V_j, V_{j'}, q_{i+1}^{\text{opt}})$ is at most $w(p^*(q_i^{\text{opt}}, q_{i+1}^{\text{opt}}))$. Since $|R^*| \leq 2\beta_1 - 2$ and $|V(\text{opt}^*) \cap V^g| \leq 2\beta_1$, we have $|P'(q_i^{\text{opt}}, V_j, V_{j'}, q_{i+1}^{\text{opt}})| - 2 \cdot (|R^*| + |V(\text{opt}^*) \cap V^g|) \geq 4$. Thus, we can find a path $(b_e(q_i^{\text{opt}}), g_e^*(q_i^{\text{opt}}), g_s^*(q_{i+1}^{\text{opt}}), b_s(q_{i+1}^{\text{opt}}))$ from $P'(q_i^{\text{opt}}, V_j, V_{j'}, q_{i+1}^{\text{opt}})$, which satisfies $\{g_e^*(q_i^{\text{opt}}), g_s^*(q_{i+1}^{\text{opt}})\} \cap (R^* \cup (V(\text{opt}^*) \setminus \{g_e(q_i^{\text{opt}}), g_s(q_{i+1}^{\text{opt}})\})) = \emptyset$ and $w(b_e(q_i^{\text{opt}}), g_e^*(q_i^{\text{opt}}), g_s^*(q_{i+1}^{\text{opt}}), b_s(q_{i+1}^{\text{opt}})) \leq w(p^*(q_i^{\text{opt}}, q_{i+1}^{\text{opt}}))$.

Based on all above discussion, $\hat{A}^*$ is a cycle and $w(\hat{A}^*) \leq w(\text{opt}^*)$. $\qquad \square$

Combining the construction of $\hat{A}^*$ from opt^* with the proof of Lemma 4, we conclude that in some case in Step 5(a) of EnumAC, we get a cycle which is the same as $\hat{A}^*$. Then Step 5(b) deletes all the edges in $\hat{A}^*$ connecting two good vertices, resulting in a collection of alternating chains $H^* = (h_1^*, \cdots, h_{\beta_2^*}^*)$.

Construction of $\tilde{A}$ from $\hat{A}^*$. Finally, we add all remaining vertices to $\hat{A}^*$ to get a TSP-solution $\tilde{A}$. For each $j \in [\beta_1]$, we double the edges in $\mathcal{T}_j$, compute an Euler tour of this double tree and shortcut the repeated vertices in the Euler tour, resulting in a cycle $\mathcal{O}_j$, where $V(\mathcal{O}_j) = V_j$ and $w(\mathcal{O}_j) \leq 2 \cdot w(\mathcal{T}_j)$. We apply the following procedure to $\hat{A}^*$ to add the remaining vertices in $V_j \setminus V(\hat{A}^*)$ to $\hat{A}^*$ according to their orders in $\mathcal{O}_j$. Note that we update $\hat{A}^*$ by adding a cycle or a walk in each iteration of while-loops.

Procedure AddCycles: Add all remaining good vertices

1. While there exists a good vertex g_s^j in $\hat{A}^*$ satisfying the following three conditions: (i) in $\hat{A}^*$, g_s^j is adjacent to at least one good vertex, (ii) $g_s^j \in V_j$ and (iii) $V_j \setminus V(\hat{A}^*) \neq \emptyset$, do the following:
   (a) Traverse $\mathcal{O}_j$ starting from g_s^j and denote it by $\mathcal{O}_j = (g_s^j, \cdots, g_e^j, g_s^j)$.
   (b) Add the cycle $\mathcal{O}_j$ to $\hat{A}^*$.
2. While there exists an edge $\{g, g_s^j\}$ in opt satisfying the following four conditions: (i) $g \in V(\hat{A}^*)$, (ii) in $\hat{A}^*$, g is adjacent to at least one good vertex, (iii) $g_s^j \in V_j$ and (iv) $V_j \setminus V(\hat{A}^*) \neq \emptyset$, do the following:
   (a) Traverse $\mathcal{O}_j$ starting from g_s^j and denote it by $\mathcal{O}_j = (g_s^j, \ldots, g_e^j, g_s^j)$.
   (b) Add the walk $\mathcal{O}_j' = (g, g_s^j, \cdots, g_e^j, g_s^j, g)$ to $\hat{A}^*$.
3. Compute the Euler tour of $\hat{A}^*$ and shortcut repeated good vertices in the Euler tour which are not adjacent to bad chains to get a cycle $\tilde{A}$.

After AddCycles, we obtain a Hamilton cycle $\tilde{A}$. The following lemma proves the correctness of AddCycles and gives an upper bound on $w(\tilde{A})$.

Lemma 5. $\tilde{A}$ is a TSP-solution and $w(\tilde{A}) \leq 5 \cdot w(\text{opt})$.

Proof. First, we show $\tilde{A}$ is a TSP-solution. After Step 1 of AddCycles, either $V_j \subseteq V(\hat{A}^*)$ or every good vertex in $V_j \cap V(\hat{A}^*)$ is adjacent to two bad vertices for each $j \in [\beta_1]$. Here, we set $\mathcal{V}_1 = (\cup_{j=1}^{\beta_1} V_j) \cap V(\hat{A}^*)$, $\mathcal{V}_2 = (\cup_{j=1}^{\beta_1} V_j) \setminus V(\hat{A}^*)$ and $\mathcal{Z}^* = \cup_{j=1}^{\beta_1} Z_j^*$, where Z_j^* is defined before Procedure ReplacePaths. The sets $\mathcal{V}_1$ and $\mathcal{V}_2$ update according to the update of $\hat{A}^*$. In each iteration of while-loop in Step 2 of AddCycles, there always exists an edge in opt connecting two vertices between $\mathcal{V}_1 \cap \mathcal{Z}^*$ and $\mathcal{V}_2$ until $\mathcal{V}_2 = \emptyset$ based on the definition of Z_j^*'s, since, otherwise, opt would not be a TSP-solution. Every good vertex in $\mathcal{V}_1 \cap \mathcal{Z}^*$ is not placed between two bad vertices and thus, is always adjacent to in $\hat{A}^*$ at least one good vertex. Hence, such an edge in opt can satisfy the four conditions in Step 2 of AddCycles and then Step 2(b) adds all remaining vertices in some V_j into $\hat{A}^*$. After Step 2, we have $V(\hat{A}^*) = V^b \cup V^g$. Then Step 3 of AddCycles can compute the Euler tour of $\hat{A}^*$, since each time when Step 1 or 2 adds a cycle or a walk to a good vertex in $\hat{A}^*$, the degree of this good vertex increases by 2. Finally, since every repeated good vertex is adjacent to bad vertices at most once in the Euler tour, Step 3 can shortcut repeated good vertices which are not adjacent to bad chains to get a Hamilton cycle $\tilde{A}$ without increasing weight.

Next we show that there is only a small weight increase caused by the addition of all remaining vertices. We use Δw_j to denote the weight increase caused by adding the vertices in $V_j \setminus V(\hat{A}^*)$ and give an upper bound on Δw_j. If the vertices in $V_j \setminus V(\hat{A}^*)$ are added in Step 1 of AddCycles, we add the cycle $\mathcal{O}_j = (g_s^j, \cdots, g_e^j, g_s^j)$ to $\hat{A}^*$. Thus, $\Delta w_j = w(\mathcal{O}_j) \leq 2 \cdot w(\mathcal{T}_j)$. Otherwise, the vertices in $V_j \setminus V(\hat{A}^*)$ are added in Step 2 of AddCycles, that is, we add the walk $\mathcal{O}_j' = (g, g_s^j, \cdots, g_e^j, g_s^j, g)$ to $\hat{A}^*$, where $\{g, g_s^j\}$ is an edge in opt. Thus,

$\Delta w_j = w(\mathcal{O}_j) + 2 \cdot w(g, g_s^j) \leq 2 \cdot w(\mathcal{T}_j) + 2 \cdot w(g, g_s^j)$. Since Step 3 of AddCycles shortcuts repeated good vertices which are not adjacent to bad chains, there is no weight increase caused by Step 3. By Lemmas 2–4, we conclude $w(\tilde{\mathcal{A}}) = w(\mathcal{A}^*) + \sum_{j=1}^{\beta_1} \Delta w_j \leq w(\mathcal{A}^*) + 2 \cdot w(\mathcal{F}) + 2 \cdot w(\text{opt}) \leq 5 \cdot w(\text{opt})$.   □

Since Step 3 of AddCycles does not shortcut good vertices which are adjacent to bad chains, the order of all alternating chains in $\tilde{\mathcal{A}}$ is the same as H^*, which is also the order of all alternating chains in $\hat{\mathcal{A}}^*$. In other words, we get the same ordered alternating chains as in $\tilde{\mathcal{A}}$ in some case in Step 5(b) of EnumAC. Next, we give a precise analysis of the result $w(\mathcal{A})$ returned by FindPaths in this case.

Lemma 6. $w(\mathcal{A}) \leq 11 \cdot w(\text{opt})$.

Proof. The alternating chains in $\tilde{\mathcal{A}}$ are denoted by $H^* = (h_1^*, \cdots, h_{\beta_2^*}^*)$. Let opt^{H^*} denote a minimum-weight TSP-solution, where the vertices in $\cup_{j=1}^{\beta_2^*} V(h_j^*)$ occur in the same alternating chains with the same order as specified in H^*. We have $w(\text{opt}^{H^*}) \leq w(\tilde{\mathcal{A}})$. We use p_j^* and p_j to denote the paths in opt^{H^*} and $\mathcal{A}$, respectively, which connect $g_e(h_j^*)$ and $g_s(h_{j+1}^*)$ and consist solely of good vertices. Here, $h_{\beta_2^*+1}^* = h_1^*$. Then $w(\text{opt}^H) = \sum_{j=1}^{\beta_2^*} w(h_j^*) + \sum_{j=1}^{\beta_2^*} w(p_j^*)$ and $w(\mathcal{A}) = \sum_{j=1}^{\beta_2^*} w(h_j^*) + \sum_{j=1}^{\beta_2^*} w(p_j)$.

In Step 1 of FindPaths, we compute a β_2^*-minimum spanning forest $\mathbb{F} = (\mathbb{T}_1, \ldots, \mathbb{T}_{\beta_2^*})$ of $G[V^g \setminus \cup_{j=1}^{\beta_2^*}(V(h_j^*) \setminus \{g_e(h_j^*)\})]$ rooted at $\{g_e(h_j^*) \mid j \in [\beta_2^*]\}$. If we remove from opt^{H^*} all alternating chains except their ending vertices, then we can obtain a β_2^*-spanning forest rooted at these ending vertices. Thus, we have $w(\text{opt}^{H^*}) \geq w(\mathbb{F})$. Based on Steps 2 and 3 of FindPaths, it is easy to verify $w(p_j) = w(\mathcal{O}_j) - w(g_{l_j}, g_e(h_j^*)) + w(g_{l_j}, g_s(h_{j+1}^*)) \leq 2 \cdot w(\mathbb{T}_j) + w(g_e(h_j^*), g_s(h_{j+1}^*))$. Note that $(g_e(h_j^*), g_s(h_{j+1}^*))$ is also the path in $\hat{\mathcal{A}}^*$ connecting h_j^* and h_{j+1}^*. Finally, combining all above discussion with Lemmas 3–5, we conclude that $w(\mathcal{A}) = \sum_{j=1}^{\beta_2^*} w(h_j^*) + \sum_{j=1}^{\beta_2^*} w(p_j) \leq \sum_{j=1}^{\beta_2^*} w(h_j^*) + \sum_{j=1}^{\beta_2^*} w(g_e(h_j^*), g_s(h_{j+1}^*)) + 2\sum_{j=1}^{\beta_2^*} w(\mathbb{T}_j) = w(\hat{\mathcal{A}}^*) + 2 \cdot w(\mathbb{F}) \leq w(\text{opt}) + 2 \cdot w(\tilde{\mathcal{A}}) \leq 11 \cdot w(\text{opt})$.   □

Based on Lemmas 1 and 6, we arrive at the main result of this section.

Theorem 1. *Algorithm β-TSP is an 11-approximation algorithm for TSP on general graphs running in $\beta^{O(\beta)} \cdot n^3$ time, where β is the minimum number of vertices whose removal results in a metric graph.*

4 Conclusion

In [3,19], three parameters α, α' and β have been introduced to measure the distance from a general graph to a metric graph. It always holds that $\beta \leq \alpha$ and $\beta \leq \alpha'$. Using α or α' as parameter, Bampis et al. [3] proposed an FPT 2.5-approximation algorithm for TSP. With β as parameter, we achieve an FPT 11-approximation for TSP. An interesting research direction is to study whether there exists a gap between the lower bounds of the FPT approximation ratios for TSP parameterized by α' (or α) and β.

4.1 Subsequent Work

After the submission of this paper, some better FPT approximations have been developed for TSP. Zhao et al. [18] followed the works [3,19] to study FPT approximation algorithms for TSP with respect to these two parameters α' and β. For the parameterization by α', they proposed an FPT 1.5-approximation algorithm, improving the result of Bampis et al. [3]. For the parameterization by β, they gave an FPT 3-approximation algorithm, improving our approximation result in this paper.

Acknowledgements. The authors are grateful to the reviewers of COCOON 2025 for their valuable comments and constructive suggestions.

Disclosure of Interests. The authors have no competing interests to declare that are relevant to the content of this article.

References

1. Andreae, T.: On the traveling salesman problem restricted to inputs satisfying a relaxed triangle inequality. Networks **38**(2), 59–67 (2001)
2. Andreae, T., Bandelt, H.: Performance guarantees for approximation algorithms depending on parametrized triangle inequalities. SIAM J. Discret. Math. **8**(1), 1–16 (1995)
3. Bampis, E., Escoffier, B., Xefteris, M.: Improved FPT approximation for non-metric TSP. CoRR **abs/2407.08392** (2024)
4. Bellman, R.: Dynamic programming treatment of the travelling salesman problem. J. ACM **9**(1), 61–63 (1962)
5. Bender, M.A., Chekuri, C.: Performance guarantees for the TSP with a parameterized triangle inequality. Inf. Process. Lett. **73**(1–2), 17–21 (2000)
6. Böckenhauer, H.J., Hromkovič, J., Klasing, R., Seibert, S., Unger, W.: Towards the notion of stability of approximation for hard optimization tasks and the traveling salesman problem. Theor. Comput. Sci. **285**(1), 3–24 (2002)
7. Christofides, N.: Worst-case analysis of a new heuristic for the travelling salesman problem. Technical report 388, Carnegie-Mellon University (1976)
8. Garey, M.R., Johnson, D.S.: Computers and Intractability: A Guide to the Theory of NP-Completeness. W. H Freeman (1979)
9. Held, M., Karp, R.M.: A dynamic programming approach to sequencing problems. J. Soc. Indust. Appl. Math. **10**(1), 196–210 (1962)
10. Karlin, A.R., Klein, N., Gharan, S.O.: A (slightly) improved approximation algorithm for metric TSP. In: Proceedings of STOC 2021, pp. 32–45 (2021)
11. Karlin, A.R., Klein, N., Gharan, S.O.: A deterministic better-than-3/2 approximation algorithm for metric TSP. In: Proceedings of IPCO 2023, pp. 261–274 (2023)
12. Karp, R.M.: Reducibility among combinatorial problems. In: Proceedings of Complexity of Computer Computations, pp. 85–103 (1972)
13. Karpinski, M., Lampis, M., Schmied, R.: New inapproximability bounds for TSP. J. Comput. Syst. Sci. **81**(8), 1665–1677 (2015)
14. Khachay, M., Neznakhina, K.: Approximability of the minimum-weight k-size cycle cover problem. J. Glob. Optim. **66**(1), 65–82 (2016)

15. Mömke, T.: An improved approximation algorithm for the traveling salesman problem with relaxed triangle inequality. Inf. Process. Lett. **115**(11), 866–871 (2015)
16. Sahni, S., Gonzalez, T.: P-complete approximation problems. J. ACM **23**(3), 555–565 (1976)
17. Serdyukov, A.I.: O nekotorykh ekstremal'nykh obkhodakh v grafakh. Upravliaemie systemy **17**, 76–79 (1978)
18. Zhao, J., Sheng, Z., Xiao, M.: Improved FPT approximation algorithms for TSP. CoRR **abs/2503.03642** (2025)
19. Zhou, J., Li, P., Guo, J.: Parameterized approximation algorithms for TSP. In: Proceedings of ISAAC 2022, pp. 50:1–50:16 (2022)

From Metric to General Graphs: FPT Constant-Factor Approximation Algorithms for Three Location Problems

Jianqi Zhou[1], Zhongyi Zhang[1], Yinghui Wen[2], and Jiong Guo[1(✉)]

[1] School of Computer Science and Technology, Shandong University,
Qingdao, China
jguo@sdu.edu.cn
[2] Shandong Institute of Information Technology Industry Development,
Jinan, China

Abstract. Location problems form an important class of combinatorial optimization problems, where we are given a set of n vertices representing locations and the goal is to pick some locations to satisfy some optimization objective. In this paper, we study three prominent problems in the location theory: k-center, k-supplier and k-dispersion problems, where k-supplier is a generalization of k-center. All three problems admit tight polynomial-time constant-approximations on metric graphs, where the distances between vertices satisfy the triangle inequality, while on general graphs, none can be approximated in polynomial time within a factor of $\rho(n)$ for any computable function ρ.

Motivated by different approximability for metric and non-metric cases, we focus on two parameters α and β measuring the distance from a general graph to a metric graph, where α is the number of vertices that appear in the triangles, in which distances violate the triangle inequality, and β is the minimum number of vertices, whose removal from a given graph results in a metric graph. It always holds that $\beta \leq \alpha$. We obtain the following results: (1) A tight 3-approximation in $O(2^\alpha \cdot n^5)$ time and a 13-approximation in $\beta^{O(\beta)} \cdot \mathrm{poly}(n)$ time for k-supplier; (2) A tight $1/2$-approximation in $O(3^\beta \cdot n^4)$ time for k-dispersion. Since k-center is a special case of k-supplier, our results imply two FPT approximation algorithms for k-center, parameterized by α or β, which achieve approximation factors of 3 and 13, respectively. To our best knowledge, these algorithms are the first FPT constant-approximation algorithms for all three problems on general graphs parameterized by α or β.

Keywords: k-Center · k-Supplier · p-Dispersion · Non-metric · Fixed-parameter tractable · Parameterized approximation

1 Introduction

Location problems form an important class of combinatorial optimization problems, where given a set of n vertices representing locations and the goal is to

The work was supported by the National Natural Science Foundation of China (No. 61772314, 61761136017, and 62072275).

pick some locations to satisfy some optimization objective. In the location theory, many problems model the placement of "desirable" facilities, such as warehouses and hospitals. Here, we study the k-supplier problem on general graphs. The input consists of an undirected complete graph $G = (C \cup F, E)$, a distance function d mapping each vertex pair to a non-negative real number and a positive integer k, where C is the set of clients, F is the set of facilities and $|C \cup F| = n$. The goal is to find a set $S \subseteq F$ with $|S| = k$ so as to minimize $\max_{v \in C} d(v, S)$, where $d(v, S) = \min_{v' \in S} d(v, v')$. The k-center problem is a special case of k-supplier with $C = F$.

In general, k-center is NP-hard and cannot be approximated in polynomial time within a factor of $\rho(n)$ for any computable function ρ, unless P = NP [12]. Following the reduction in [12] from k-Dominating Set (DS) to k-center, it is trivial to get the FPT inapproximability of k-center parameterized by k as follows. Suppose that there is a ρ-approximation algorithm in $\varphi(k) \cdot n^{O(1)}$ time for k-center, for any computable functions ρ and φ. We construct a k-center instance from an arbitrary DS-instance: for every two vertices u and v, if there is an edge between u and v in the DS-instance, then set $d(u, v) = 1$ in the instance of k-center; otherwise, set $d(u, v) = \rho + 1$. If there exists a k-DS, then there exist k centers "covering" all vertices within radius 1; otherwise, the radius is at least $\rho + 1$. If $d(v, v') \leq r$, then we say v covers v' within a radius r. The gap between the radii shows that a ρ-approximation for k-center can exactly solve k-DS in $\varphi(k) \cdot n^{O(1)}$ time, which contradicts the W[2]-hardness of k-DS parameterized by the solution size k [3,8]. This FPT inapproximability result also holds for general k-supplier. There is a special class of graphs called metric graphs, where the distances between vertices satisfy the triangle inequality, that is, $d(v_1, v_2) \leq d(v_1, v_3) + d(v_3, v_2)$ for every three vertices v_1, v_2 and v_3. On metric graphs, there are a 3-approximation for k-supplier [7] and a 2-approximation for k-center [5,6] in polynomial time. These two results are tight, unless P = NP [5–7].

There is another kind of location problems, called dispersion models, which require that the distances between the facilities are as large as possible. The k-dispersion problem has as input an undirected complete graph $G = (V, E)$ with $|V| = n$, a non-negative distance function d defined on vertex pairs and a positive integer k, and the goal is to find a set $S \subseteq V$ with $|S| = k$ so as to maximize $\min_{v, v' \in S} d(v, v')$. The k-dispersion problem is equivalent to the dual problem of $(k - 1)$-center on a tree network [2,10]. Erkut [4] proved that k-dispersion is NP-complete even if the distance function satisfies the triangle inequality. Ravi et al. [9] showed that k-dispersion, in general, cannot be approximated in polynomial time within a factor of $\rho(n)$ for any computable function ρ, unless P = NP. Following their reduction [9] from k-Clique to k-dispersion, where k-Clique is W[1]-complete parameterized by the solution size k [3,8], we conclude that k-dispersion cannot be approximated in $\varphi(k) \cdot n^{O(1)}$ time within a factor of $\rho(n)$ for any computable function φ and ρ, unless FPT = W[1]. On metric graphs, where the distance function satisfies the triangle inequality, there is a $1/2$-approximation in polynomial time for k-dispersion [9,11], and Ravi et al. [9] showed that the approximation ratio is tight, unless P = NP.

Our Contributions. Motivated by the polynomial-time inapproximability in general and constant approximations on metric graphs for k-center, k-supplier and k-dispersion, we apply the "distance from approximability" scheme proposed in [13] to these problems, that is, using a parameter to measure the distance between the inapproximable and approximable cases and designing approximation algorithms with the exponential running time restricted to the parameter. Here, we use general graphs to denote the inapproximable case of the problems and use metric graphs to denote the approximable case, and focus on two parameters α and β to measure the distance between them. Specifically, α is the number of vertices that appear in the triangles, in which distances violate the triangle inequality, and β is the minimum number of vertices, whose removal transforms the input general graph to a metric graph. Given an input instance, the value of α can be easily computed in $O(n^3)$ time and the value of β can be determined in $O(3^\beta \cdot n^3)$ time by using a simple search tree [3,8]. It always holds that $\beta \le \alpha$.

For k-supplier on general graphs, using α or β as parameter, we obtain the following two FPT constant-approximation algorithms: (1) A 3-approximation algorithm running in $O(2^\alpha \cdot n^5)$ time; (2) A 13-approximation algorithm running in $\beta^{O(\beta)} \cdot \text{poly}(n)$ time. The 3-approximation result is tight, due to the inapproximability bound for k-supplier on metric graphs [7]. Moreover, since k-center is a special case of k-supplier, our results imply two FPT approximation algorithms for k-center parameterized by α or β, which achieve approximation factors of 3 and 13, respectively. Here, the approximation ratio of 3 is not necessarily tight for k-center, for which the inapproximability bound is 2 [5,6].

For k-dispersion on general graphs, using β as parameter, we propose a $1/2$-approximation algorithm running in $O(3^\beta \cdot n^4)$ time. The approximation ratio is also tight, due to the lower bound of the approximation ratio on metric graphs [9]. Due to $\beta \le \alpha$, this also implies a $1/2$-approximation for k-dispersion with α as parameter. To our best knowledge, the above algorithms are the first non-trivial FPT constant-factor approximation algorithms for k-center, k-supplier and k-dispersion on general graphs parameterized by α or β. Due to lack of space, most proofs are omitted.

Related Work. The concept "distance from approximability" was first proposed by Zhou et al. [13]. They applied this concept to design two FPT approximation algorithms for TSP on general graphs parameterized by α or β, which achieve the approximation ratios of 3 and $6\beta + 9$, respectively. Then the approximation factor of 3 was improved to 2.5 by Bampis et al. [1] in FPT time parameterized by α. This concept has also been applied to Correlation Clustering, introducing a parameter to measure the distance from a general graph to a complete graph and designing an FPT constant approximation algorithm [14].

2   Preliminaries

We introduce some basic definitions and notations. For a positive integer i, set $[i] = \{1, \cdots, i\}$. We consider an undirected complete graph $G = (V, E)$ with a non-negative distance function d defined on vertex pairs. The distance $d(v, v')$

between two vertices $v, v' \in V$ is also called the cost of the edge $\{v, v'\}$. For a set $V' \subseteq V$, $E(V')$ is the set of the edges that connect two vertices in V', and $G[V'] = (V', E(V'))$ is the subgraph of G induced by V'. For a set $E' \subseteq E$, $V(E')$ is the set of the end-vertices of the edges in E' and $G - E'$ denote the graph $G' = (V, E \setminus E')$. A set of vertices whose removal transforms a non-metric graph into a metric graph, is called *a violating vertex set*. For a vertex $v \in V$ and a set $V' \subseteq V$, set $d(v, V') = \min_{v' \in V'} d(v, v')$, where $d(v, v) = 0$ and $d(v, \emptyset) = +\infty$.

In k-supplier, we call the vertices in C the *clients* and the vertices in F the *facilities*. For $V' \subseteq (C \cup F)$ and $F' \subseteq F$ with $F' \neq \emptyset$, we define $d(V', F') = \max_{v \in V'} d(v, F')$, where $d(\emptyset, F') = 0$. If $d(V', F') \leq r$, then the facilities in F' are said to *cover* the vertices in V' within a *radius* r. Sometimes, we may omit the radius r and just say that F' covers V', since the radius can be computed, given F' and V'. We use S^* to denote an optimal solution such that $d(C, S^*) \leq d(C, S)$ for every $S \subseteq F$ with $|S| = k$, and use *centers* to refer to the facilities in S^*. Obviously, the centers in S^* cover all clients in C within *the optimal radius* $d(C, S^*)$. To avoid ambiguity, for each vertex $v \in C \cup F$, *the center in S^* covering v* or *the center in S^* of v* refers to an arbitrary but fixed center in $\{s \mid s \in S^*, d(v, s) = d(v, S^*)\}$.

3   An FPT 3-Approximation for k-Supplier Parameterized by α

In this section, we consider an instance of k-supplier with α vertices that appear in the triangles violating the triangle inequality. The α vertices can be computed in $O(n^3)$ time by checking all triangles, that is, size-3 vertex subsets, in the input and are called *bad*. The remaining vertices are called *good*. Then, we partition the set C of clients into the set C^b of bad clients and the set C^g of good clients. Similarly, the set F of facilities can be partitioned into the set F^b of bad facilities and the set F^g of good facilities. Obviously, we have $|C^b \cup F^b| = \alpha$. The following observation about bad and good vertices is crucial for the algorithm: a triangle consisting of three bad vertices might violate the triangle inequality, while a triangle containing at least one good vertex must satisfy the triangle inequality.

Our algorithm consists mainly of three steps. The first step "guesses" the optimal radius r^* by trying all distinct distances between the vertices in C and F. The second step "guesses" the bad centers in the optimal solution S^*, that is, the centers in $S^b = S^* \cap F^b$. This can be done by enumerating all possible 2-partitions of F^b in $O(2^\alpha)$ time. The bad facilities in $F^b \setminus S^b$ are irrelevant. Then we partition the set C into C_1 and $C \setminus C_1$, where $C_1 = \{c \mid c \in C, d(c, S^b) > r^*\}$. The clients in $C \setminus C_1$ can be covered by the bad centers in S^b within r^* and thus, are irrelevant. The third step picks $k - |S^b|$ facilities from F^g to cover the clients in C_1 by applying the polynomial-time 3-approximation [7] to a metric auxiliary graph H. The key for the construction of H is that after partitioning C_1 into $C_1^b = C_1 \cap C^b$ and $C_1^g = C_1 \cap C^g$, only the distances between the bad clients in C_1^b might violate the triangle inequality and should be reset. For more details, see Step 2(c) of Algorithm α-kS as follows.

Algorithm α-kS: An FPT 3-approximation for k-supplier parameterized by α

1. Compute all triangles violating the triangle inequality and C^b, C^g, F^b, F^g being the set of bad clients, good clients, bad facilities, good facilities, respectively.
2. For each two distinct vertices c and f with $c \in C$ and $f \in F$, do:
   For each 2-partition of F^b into S^b and $F^b \setminus S^b$ satisfying $|S^b| \leq k$, do:
   (a) Set $r^* = d(c, f)$.
   (b) Set $C_1 = \{c \mid c \in C, d(c, S^b) > r^*\}$, $C_1^b = C_1 \cap C^b$ and $C_1^g = C_1 \cap C^g$.
   (c) Construct an auxiliary graph H from G as follows: delete all the vertices not in $C_1 \cup F^g$ and all the edges incident to them; for each two distinct bad clients b, b' in C_1^b, set $d(b, b')$ to be equal to the length of a shortest path with the form (b, g, b') where $g \in C_1^g \cup F^g$.
   (d) Apply the 3-approximation algorithm [7] to H to get a set $S^g \subseteq F^g$ with $|S^g| = k - |S^b|$, which covers the clients in C_1.
   (e) The solution of this iteration is set to $S = S^b \cup S^g$, and compute the corresponding radius $r = d(C, S)$.
3. Return the solution $S^{\min}$ with the minimum radius $r^{\min}$ among all iterations.

Theorem 1. *Algorithm α-kS is a 3-approximation algorithm for general k-supplier running in $O(2^\alpha \cdot n^5)$ time, where α is the number of vertices that appear in the triangles violating the triangle inequality.*

4   An FPT 13-Approximation for k-Supplier Parameterized by β

In this section, we consider an instance of k-supplier, parameterized by the minimum size β of a violating vertex set, that is, the minimum number of vertices, whose removal transforms the given instance into a metric graph. A violating vertex set of at most β vertices can be computed in $O(3^\beta \cdot n^3)$ time by a simple search tree [3,8]. The basic idea is that at least one vertex of every triangle, which violates the triangle inequality, has to be added to the violating vertex set, which also implies $\beta \leq \alpha$. Here, the vertices in the violating vertex set are called *bad* and the remaining vertices are called *good*. Similarly, we define C^b, C^g, F^b and F^g as the set of bad clients, good clients, bad facilities and good facilities, respectively. In contrast to Sect. 3, a triangle containing at least one bad vertex might violate the triangle inequality; only the triangle consisting of three good vertices must satisfy the triangle inequality, which means that the idea of constructing an auxiliary metric graph does not apply to the parameterization of β.

We first describe the high-level idea and intuition of our algorithm and introduce the details of each step later. See Fig. 1 for an illustration.

The first step, as in Algorithm α-kS, "guesses" the optimal radius r^* and the set S^b of bad centers in S^*. Then, we partition C into three subsets $\hat{C}$, C_1^g and

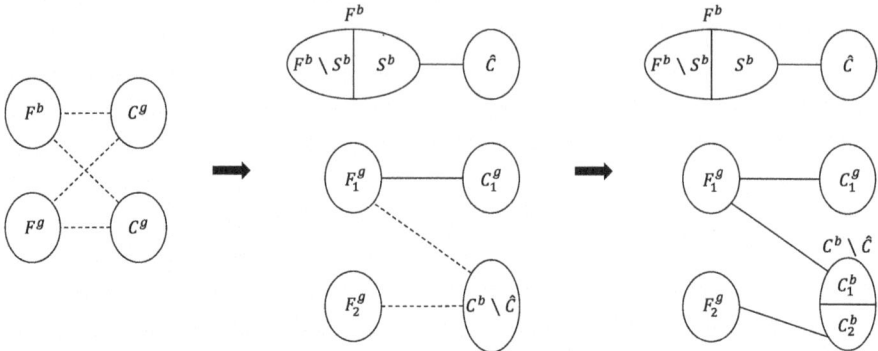

Fig. 1. An illustration of Algorithm β-kS. A dashed line between two sets means that there exist some edges between them, whose costs are at most r^*. A solid line between two sets means that in the optimal solution S^*, there exist some facilities in the left set, which can cover all clients in the right set within a radius r^*.

$C^b \setminus \hat{C}$, where $\hat{C} = \{c \mid c \in C, d(c, S^b) \le r^*\}$, $C_1^g = \{c \mid c \in C^g, d(c, S^b) > r^*\}$ and $C^b \setminus \hat{C} = \{c \mid c \in C^b, d(c, S^b) > r^*\}$. The clients in $\hat{C}$ can be covered by the bad centers in S^b within r^* and thus, are irrelevant. Next, we need to pick $k - |S^b|$ facilities in F^g to cover the clients in C_1^g and $C^b \setminus \hat{C}$. We partition F^g into F_1^g and F_2^g according to the distances between the facilities in F^g and the good clients in C_1^g, where $F_1^g = \{f \mid f \in F^g, d(f, C_1^g) \le r^*\}$ and $F_2^g = \{f \mid f \in F^g, d(f, C_1^g) > r^*\}$. The idea behind the partition is that the facilities in F_2^g can only cover some bad clients in $C^b \setminus \hat{C}$, while the facilities in F_1^g can cover good clients in C_1^g and other bad clients in $C^b \setminus \hat{C}$. See the middle figure in Fig. 1.

The second step "guesses" a partition of $C^b \setminus \hat{C}$ into C_1^b and C_2^b to satisfy that the centers in S^* covering the bad clients in C_1^b are in F_1^g and the centers in S^* covering the bad clients in C_2^b are in F_2^g. This can be done by enumerating all possible 2-partitions of $C^b \setminus \hat{C}$. In some partition case, the clients in $C_1^b \cup C_1^g$ can be covered by the centers in F_1^g within r^* and the clients in C_2^b can be covered by the centers in F_2^g within r^*. See the right figure in Fig. 1.

The third step "guesses" the number β_2 of the centers in S^* covering the bad clients in C_2^b with $\beta_2 \le |C_2^b| \le \beta$, and computes a set $S_2^g \subseteq F_2^g$ with $|S_2^g| = \beta_2$ to cover C_2^b by Algorithm `far_facilities` and a set $S_1^g \subseteq F_1^g$ with $|S_1^g| \le k'$ to cover $C_1^b \cup C_1^g$ by Algorithm `near_facilities`, where $k' = k - |S^b| - \beta_2 \le k$. Finally, merge the sets S^b, S_1^g and S_2^g together to a solution S of k facilities.

In the following, we use Algorithm β-kS to denote the overall framework of our algorithm for k-supplier parameterized by β. Recall that for $C' \subseteq C$ and $F' \subseteq F$, we define $d(C', F') = \max_{c \in C'} d(c, F')$ and $d(\emptyset, F') = 0$.

Algorithm β-kS: An FPT 13-approximation for k-supplier parameterized by β

1. Compute a violating vertex set of β vertices by a simple search tree and C^b, C^g, F^b, F^g being the set of bad clients, good clients, bad facilities, good facilities, respectively.
2. For each two distinct vertices c and f with $c \in C$ and $f \in F$, do:
   For each 2-partition of F^b into S^b and $F^b \setminus S^b$, satisfying $|S^b| \leq k$, do:
   (a) Set $r^* = d(c, f)$.
   (b) Set $\hat{C} = \{c \mid c \in C, d(c, S^b) \leq r^*\}$, $C_1^g = \{c \mid c \in C^g, d(c, S^b) > r^*\}$, $F_1^g = \{f \mid f \in F^g, d(f, C_1^g) \leq r^*\}$ and $F_2^g = \{f \mid f \in F^g, d(f, C_1^g) > r^*\}$.
   (c) For each 2-partition of $C^b \setminus \hat{C}$ into C_1^b and C_2^b, satisfying $d(C_1^b, F_1^g) \leq r^*$ and $d(C_2^b, F_2^g) \leq r^*$, do:
       For each $\beta_2 \in [\min\{k - |S^b|, |C_2^b|, |F_2^g|\}] \cup \{0\}$, do:
       i. Apply `far_facilities`$(G[C_2^b \cup F_2^g], d, \beta_2, r^*)$. If the output is "NO", then go to the next iteration; otherwise, let S_2^g denote the output.
       ii. Set $k' = k - |S^b| - \beta_2$ and apply `near_facilities`$(G[C_1^g \cup C_1^b \cup F_1^g], d, k', r^*)$ to get a set S_1^g of k' facilities.
       iii. The solution of this iteration is set to $S = S^b \cup S_1^g \cup S_2^g$, and compute the corresponding radius $r = d(C, S)$.

3. Return the solution $S^{\min}$ with the minimum radius $r^{\min}$ among all iterations.

4.1   Find Far Facilities to Cover Bad Clients

As indicated in Step 2(c)i of Algorithm β-kS, `far_facilities` needs to compute a set $S_2^g \subseteq F_2^g$ with $|S_2^g| = \beta_2$, which covers the bad clients in C_2^b, where $\beta_2 \leq \beta$. Hereby, we "guess" a partition of C_2^b into β_2 non-empty subsets to satisfy that the bad clients in each subset are covered by one distinct center in $S^* \cap F_2^g$ within r^* and thus, can be regarded as one bad client. This can be done by trying all possible β_2-partitions of C_2^b. Then, we "contract" the bad clients in each subset to one client, delete the edges with costs more than r^* and compute a matching between β_2 contracted clients and the facilities in F_2^g, which is perfect with respect to the former. Obviously, the β_2 facilities in the matching can cover all bad clients in C_2^b within r^*. Note that $C_2^b \cap F_2^g = \emptyset$, which means that contracting the bad clients in C_2^b does not affect the good facilities in F_2^g.

Algorithm `far_facilities`: Find facilities in F_2^g to cover bad clients in C_2^b.
   Input: a complete graph $G[C_2^b \cup F_2^g]$ with a set C_2^b of bad clients and a set F_2^g of good facilities, a distance function d, an integer β_2 and a distance r^*.
   Output: a set $S_2^g \subseteq F_2^g$ with $|S_2^g| = \beta_2$, which can cover C_2^b within r^* or "NO".

1. If $C_2^b = \emptyset$, then return the set of β_2 arbitrary facilities in F_2^g as output and end `far_facilities`.
2. If $C_2^b \neq \emptyset$ and $\beta_2 = 0$, then return "NO" as output and end `far_facilities`.

3. If $C_2^b \neq \emptyset$ and $\beta_2 \geq 1$, then for each possible partition of C_2^b into β_2 non-empty subsets, do:
   (a) Construct an auxiliary graph H from $G[C_2^b \cup F_2^g]$ as follows.
       For each $i \in [\beta_2]$, contract the clients in the i-th subset to a client x_i as follows: add a client x_i; for each $f \in F_2^g$, connect x_i and f by an edge and set $d(x_i, f)$ to the maximum distance between the clients in this subset and f; delete the clients in this subset and the edges incident to them. Delete the edges with costs more than r^*, resulting in $H = G[X \cup F_2^g]$, where $X = \{x_1, x_2, \cdots, x_{\beta_2}\}$.
   (b) If there exists a matching E' in H between X and F_2^g, which is perfect with respect to X, then return $S_2^g = V(E') \cap F_2^g$ as output and end `far_facilities`; otherwise, go to the next iteration.
4. Return "NO" as output.

4.2   Find Near Facilities to Cover Bad and Good Clients

As indicated in Step 2(c)ii of Algorithm β-kS, `near_facilities` needs to compute a set $S_1^g \subseteq F_1^g$ with $|S_1^g| = k'$ to cover the clients in $C_1^g \cup C_1^b$, where $k' = k - |S^b| - \beta_2 \leq k$.

Here, we describe the high-level idea of our algorithm. See Fig. 2 for an illustration. First, we compute a subset $\hat{S} = \{\hat{s}_1, \hat{s}_2, \cdots, \hat{s}_{k'}\}$ of F_1^g to cover C_1^g by applying the 3-approximation algorithm [7] to the metric graph $G[C_1^g \cup F_1^g]$. In Fig. 2, $k' = 6$. Then, we compute k' "clusters" $N(\hat{s}_1), \cdots, N(\hat{s}_{k'})$ containing all vertices in $C_1^g \cup F_1^g$, which satisfy that $\hat{s}_i$ is the facility in $\hat{S}$, which is closest to the vertices in $N(\hat{s}_i)$ for each $i \in [k']$. Since we also need to cover the bad clients in C_1^b, some facilities in $\hat{S}$ should be replaced, which means that some clusters should be "dissolved" and the clients in these clusters can be covered by remaining facilities in $\hat{S}$. Thus, we apply Procedure `dissolve_clusters` to dissolve some clusters, such as $N(\hat{s}_4)$ in Fig. 2. Finally, we pick some facilities in the remaining clusters to cover the bad clients in C_1^b. In Fig. 2, we pick three facilities to cover bad clients. The three facilities can replace $\hat{s}_5$ and $\hat{s}_6$ to cover the clients in $N(\hat{s}_5)$ and $N(\hat{s}_6)$. For more details, see Algorithm `near_facilities`.

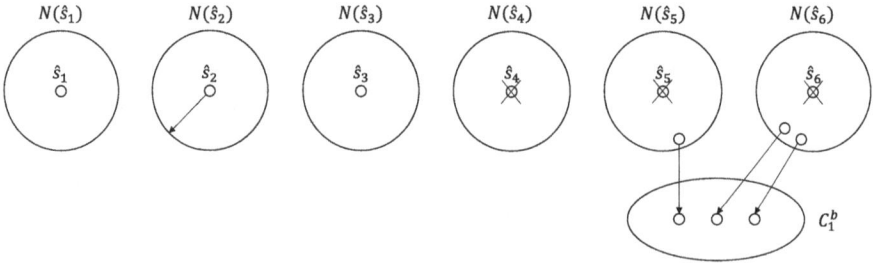

Fig. 2. An illustration of Algorithm `near_facilities`

Next, we give some explanations for some steps in our algorithm. In Step 2 of `near_facilities`, we order the facilities in $\hat{S}$ according to their minimal distances to other facilities in $\hat{S}$ in a non-ascending order. The idea behind the ordering is that the facilities ordered last should be replaced, since the good clients that are initially covered by these facilities can be covered by other facilities without increasing the approximation ratio too much.

Another point that needs to be explained is how to "guess" the number of facilities in $\hat{S}$, which need to be replaced. In Steps 4-4(a), as in `far_facilities`, we "guess" a partition of C_1^b into β_1 non-empty subsets and contract the bad clients in C_1^b into β_1 bad clients $X = \{x_1, x_2, \cdots, x_{\beta_1}\}$, where $\beta_1 \leq \beta$. Here, the bad clients in the same subset are regarded as one bad client in X, which means that they are covered by one center in S^*. Two different clients in X are assumed to be covered by two different centers in S^*. Then in Step 4(b), we "guess" a partition of X into β_1' subsets $X_1, X_2, \cdots, X_{\beta_1'}$, with $\beta_1' \leq \beta_1$, to satisfy that the centers in S^* covering the clients in each subset are contained in "one distinct cluster" among $N(\hat{s}_1), N(\hat{s}_2), \cdots, N(\hat{s}_{k'})$. Here and in the following, with *one distinct cluster*, we mean that for every $i, j \in [\beta_1']$ and $i \neq j$, the $|X_i|$ centers in S^* covering the clients in X_i are contained in one cluster, while the $|X_j|$ centers in S^* covering the clients in X_j are contained in another different cluster.

For example, in Fig. 2, $\beta_1 = 3$ and $\beta_1' = 2$. It is assumed that there are three good centers in the optimal solution S^* covering all bad clients in C_1^b. We contract all bad clients into three clients x_1, x_2, x_3, each of which is covered by one good center in S^*. The three good centers are assumed to be contained in two clusters among $N(\hat{s}_1), N(\hat{s}_2), \cdots, N(\hat{s}_{k'})$. In other words, to cover all bad clients, S^* has to choose three good centers from two clusters. Thus, one cluster $N(\hat{s}_4)$ contains no center in S^*. Otherwise, there would be more than k' good centers in S^* covering the clients in $C_1^g \cup C_1^b$, which contradicts our assumption. We conclude that there are $\beta_1 - \beta_1'$ clusters, which contain no center in S^* and can be "dissolved". Hereby, dissolving some cluster $N(\hat{s}_j)$ means to delete $\hat{s}_j$ from $\hat{S}$. The clients in the dissolved clusters can be covered by other remaining facilities in $\hat{S}$.

Algorithm `near_facilities`: Find facilities in F_1^g to cover clients in $C_1^g \cup C_1^b$.

Input: a complete graph $G[C_1^g \cup C_1^b \cup F_1^g]$ with a set C_1^g of good clients, a set C_1^b of bad clients and a set F_1^g of good facilities, a distance function d, an integer k' and a distance r^*.

Output: a set $S_1^g \subseteq F_1^g$ with $|S_1^g| = k'$, which covers the clients in $C_1^g \cup C_1^b$.

1. Apply the 3-approximation algorithm [7] to $G[C_1^g \cup F_1^g]$ to get a set $\hat{S} \subseteq F_1^g$ with $|\hat{S}| = k'$, which covers the good clients in C_1^g.
2. Order the facilities in $\hat{S}$ as follows. Initially set $\hat{S}' = \hat{S}$ and $\hat{S} = \emptyset$. Pick an arbitrary facility from $\hat{S}'$, add it to $\hat{S}$ as the first facility denoted by $\hat{s}_1$ and delete it from $\hat{S}'$. From $l = 2$ to $l = k'$, do: pick an arbitrary facility $s \in \hat{S}'$, which is furthest from $\hat{S}$, that is $d(s, \hat{S}) \geq d(s', \hat{S})$ for every $s' \in \hat{S}'$, add s to $\hat{S}$ as the l-th facility denoted by $\hat{s}_l$ and delete it from $\hat{S}'$. After above process, we get k' ordered facilities $\hat{S} = \{\hat{s}_1, \hat{s}_2, \cdots, \hat{s}_{k'}\}$.
3. For each $i \in [k']$, initially set $N(\hat{s}_i) = \{\hat{s}_i\}$. For each $v \in (C_1^g \cup F_1^g) \setminus \hat{S}$, pick an arbitrary facility $\hat{s}$ in $\hat{S}$ satisfying $d(v, \hat{s}) \leq d(v, \hat{S})$, and add v to $N(\hat{s})$.

4. For each $\beta_1 \in [\min\{k', |C_1^b|, |F_1^g|\}] \cup \{0\}$, do:

   if $C_1^b = \emptyset$, then the solution of this iteration is set to $\hat{S}$; if $C_1^b \neq \emptyset$ and $\beta_1 = 0$, then go to the next iteration; if $C_1^b \neq \emptyset$ and $\beta_1 \geq 1$, then for each possible partition of C_1^b into β_1 non-empty subsets, do:

   (a) Construct an auxiliary graph H_1 from $G[C_1^b \cup F_1^g]$ as follows.

   For each $i \in [\beta_1]$, contract the clients in the i-th subset to a client x_i as follows: add a client x_i; for each $f \in F_1^g$, connect x_i and f by an edge and set $d(x_i, f)$ to the maximum distance between the clients in this subset and f; delete the clients in this subset and the edges incident to them.

   Delete the edges with costs more than r^*, resulting in $H_1 = G[X \cup F_1^g]$, where $X = \{x_1, x_2, \cdots, x_{\beta_1}\}$.

   (b) For each $\beta_1' \in [\beta_1]$, enumerate all possible partitions of X into β_1' non-empty subsets $X_1, X_2, \cdots, X_{\beta_1'}$.

   For each β_1'-partition, apply Procedure `dissolve_clusters`.

   For each case enumerated in Procedure `dissolve_clusters`, let $\tilde{S} = \{\tilde{s}_1, \tilde{s}_2, \cdots, \tilde{s}_{k''}\}$ denote the resulting set after deleting $\beta_1 - \beta_1'$ facilities from $\hat{S}$ with $k'' = k' - (\beta_1 - \beta_1') \leq k'$, and do:

   i. Construct an auxiliary bipartite graph H_2 as follows.

   Add β_1' vertices $y_1, y_2, \cdots, y_{\beta_1'}$ and k'' vertices $z_1, z_2, \cdots, z_{k''}$ to H_2. For each $i \in [\beta_1']$ and each $j \in [k'']$, if there exists a matching in H_1 between X_i and $N(\tilde{s}_j) \cap F_1^g$, which is perfect with respect to X_i, then we add an edge between y_i and z_j.

   ii. If there is no matching between Y and Z in $H_2 = G[Y \cup Z]$, which is perfect with respect to Y, where $Y = \{y_1, y_2, \cdots, y_{\beta_1'}\}$ and $Z = \{z_1, z_2, \cdots, z_{k''}\}$, then go to the next iteration.

   iii. Otherwise, compute such a perfect matching E_2' between Y and Z in H_2. Initially, set $S_1^g = \tilde{S}$.

   For each edge $\{y_i, z_j\}$ in E_2', do: compute a matching $E_{i,j}'$ in H_1 between X_i and $N(\tilde{s}_j) \cap F_1^g$, which is perfect with respect to X_i; set $S_1^g = S_1^g \setminus \{\tilde{s}_j\}$ and $S_1^g = S_1^g \cup (V(E_{i,j}') \cap N(\tilde{s}_j) \cap F_1^g)$.

   iv. The solution of this iteration is set to S_1^g, and compute the corresponding radius $d(C_1^g \cup C_1^b, S_1^g)$.

5. Return the solution with the minimum radius among all iterations.

In `dissolve_clusters`, we need to dissolve $\beta_1 - \beta_1'$ clusters, which means to delete $\beta_1 - \beta_1'$ facilities from $\hat{S}$, under the assumption that there are at least $\beta_1 - \beta_1'$ clusters containing no center in S^*. We will require that each dissolved cluster $N(\hat{s}_i)$ satisfies the following two conditions:

(1) $N(\hat{s}_i)$ contains no center in S^* of any bad client in C_1^b;
(2) there exists some undissolved cluster $N(\hat{s}_j)$ with $j \neq i$ satisfying $d(\hat{s}_i, \hat{s}_j) \leq 6r^*$.

The basic idea is that if Condition (2) is satisfies, then $\hat{s}_j$ can cover all clients in $N(\hat{s}_i)$ within $9r^*$. Moreover, it is easy to verify that if $N(\hat{s}_i)$ satisfies the following condition:

(2′) the center in S^* of $\hat{s}_i$ is contained in some undissolved cluster $N(\hat{s}_j)$,

then $d(\hat{s}_i, \hat{s}_j) \leq 6r^*$. Thus, if a cluster satisfies Conditions (1) and (2) (or (1) and (2′)), then we can dissolve it without increasing the approximation factor too much.

Next, we give details of `dissolve_clusters`. If $|\hat{S}| \leq 3\beta_1$, then we "guess" a partition of $\hat{S}$ into $\hat{S}_1$ and $\hat{S}_2$ to satisfy that (1) for every $\hat{s} \in \hat{S}_1$, $N(\hat{s})$ contains no center in S^*; (2) for every $\hat{s}' \in \hat{S}_2$, $N(\hat{s}')$ contains at least one center in S^*. Thus, we delete $\beta_1 - \beta_1'$ arbitrary facilities in $\hat{S}_1$ from $\hat{S}$.

Otherwise, we have $|\hat{S}| > 3\beta_1$. Then let T denote the set of the last $2\beta_1$ facilities in $\hat{S}$. We "guess" a partition of T into three subsets T_1, T_2 and T_3 to satisfy that (1) for every $t \in T_1$, $N(t)$ contains no center in S^* ; (2) for every $t' \in T_2$, $N(t')$ contains at least one center in S^*, which covers at least one bad client in C_1^b or at least one facility in T_1; (3) for every $t'' \in T_3$, $N(t'')$ contains at least one center in S^*, while contains no center in S^* covering any bad client in C_1^b or any facility in T_1. Obviously, we can delete arbitrary facilities in T_1 and cannot delete the facilities in T_2. There are three cases to consider.

Case 1: If $|T_1| \geq \beta_1 - \beta_1'$, then we delete $\beta_1 - \beta_1'$ arbitrary facilities in T_1.

Case 2: If $|T_1| < \beta_1 - \beta_1'$ and there exist two facilities $\hat{s}, \hat{s}'$ in $\hat{S} \setminus T$ satisfying $d(\hat{s}, \hat{s}') \leq 6r^*$, then for every facility $t \in T_3$, there exists a facility $\hat{s}''$ in $\hat{S} \setminus T$ satisfying $d(t, \hat{s}'') \leq 6r^*$, according to the ordering of the facilities in $\hat{S}$ in Step 2 of `near_facilities`. Thus, we delete all facilities in T_1 and $\beta_1 - \beta_1' - |T_1|$ arbitrary facilities in T_3.

Case 3: If $|T_1| < \beta_1 - \beta_1'$ and $d(s, s') > 6r^*$ for every two facilities $\hat{s}, \hat{s}'$ in $\hat{S} \setminus T$, then for every $\hat{s} \in \hat{S} \setminus T$, the center in S^* covering $\hat{s}$ can only be in $N(\hat{s})$ or $N(t)$ for some $t \in T_2 \cup T_3$, since $\hat{s}$ is too far from $N(\hat{s}')$ for all $\hat{s}' \in \hat{S} \setminus (T \cup \{\hat{s}\})$. For this case, there exist at least $\beta_1 - \beta_1' - |T_1|$ facilities in $\hat{S} \setminus T$, such that for each $\hat{s}$ of these facilities, $N(\hat{s})$ contains no center in S^*, and the center in S^* covering $\hat{s}$ is contained in $N(t)$ for some $t \in T_2 \cup T_3$. Hereby, we use U to denote the set of β_1 facilities in $\hat{S} \setminus T$, which are "closest" to $T_2 \cup T_3$, that is, $d(\hat{s}, T_2 \cup T_3) \leq d(\hat{s}', T_2 \cup T_3)$ for every $\hat{s} \in U$ and $\hat{s}' \in \hat{S} \setminus (T \cup U)$. Then, we "guess" a subset U_1 of U with $|U_1| = \beta_1 - \beta_1' - |T_1|$ to satisfy that the facilities in U_1 are closest to $T_2 \cup T_3$ on the premise that for every $u \in U_1$, (1) $N(u)$ is not one of at most β_1' clusters containing the centers in S^* of the bad clients in C_1^b; (2) $N(u)$ is not one of at most $|T_1|$ clusters containing the centers in S^* of the facilities in T_1. Finally, we delete all facilities in $T_1 \cup U_1$.

Procedure `dissolve_clusters`: Delete $\beta_1 - \beta_1'$ facilities from $\hat{S}$.

1. If $|\hat{S}| \leq 3\beta_1$, then for each 2-partition of $\hat{S}$ into $\hat{S}_1$ and $\hat{S}_2$ satisfying $|\hat{S}_1| \geq \beta_1 - \beta_1'$, do: set $W = \emptyset$, add $\beta_1 - \beta_1'$ arbitrary facilities in $\hat{S}_1$ to W, and set $\tilde{S} = \hat{S} \setminus W$.
2. Otherwise, we have $|\hat{S}| > 3\beta_1$, then let T denote the set of the last $2\beta_1$ facilities $\hat{s}_{k'-2\beta_1+1}, \cdots, \hat{s}_{k'}$ in $\hat{S}$ and for each 3-partition of T into T_1, T_2 and T_3 satisfying $|T_2| \leq |T_1| + \beta_1'$, do:
   (a) Case 1: If $|T_1| \geq \beta_1 - \beta_1'$, then set $W = \emptyset$, add $\beta_1 - \beta_1'$ arbitrary facilities in T_1 to W, and set $\tilde{S} = \hat{S} \setminus W$.

(b) Case 2: If $|T_1| < \beta_1 - \beta_1'$ and there exist two facilities $\hat{s}, \hat{s}'$ in $\hat{S} \setminus T$ satisfying $d(\hat{s}, \hat{s}') \leq 6r^*$, then set $W = \emptyset$, add all facilities in T_1 and $\beta_1 - \beta_1' - |T_1|$ arbitrary facilities in T_3 to W, and set $\tilde{S} = \hat{S} \setminus W$.

(c) Case 3: If $|T_1| < \beta_1 - \beta_1'$ and $d(\hat{s}, \hat{s}') > 6r^*$ for every two facilities $\hat{s}, \hat{s}'$ in $\hat{S} \setminus T$, then do:

  i. Let U denote the set of β_1 facilities in $\hat{S} \setminus T$, which are closest to the facilities in $T_2 \cup T_3$, that is, $d(\hat{s}, T_2 \cup T_3) \leq d(\hat{s}', T_2 \cup T_3)$ for every $\hat{s} \in U$ and $\hat{s}' \in \hat{S} \setminus (T \cup U)$.

  ii. For each subset U_1 of U satisfying $|U_1| = \beta_1 - \beta_1' - |T_1|$, do: set $W = T_1 \cup U_1$ and $\tilde{S} = \hat{S} \setminus W$.

Here, we briefly introduce the idea of proving the approximation ratio of `near_facilities`. By a similar operation as in Step 3 of `near_facilities`, we define k' clusters $M(\hat{s}_1), M(\hat{s}_2), \cdots, M(\hat{s}_{k'})$ containing the the clients in C_1^g, which satisfy that $\hat{s}_i$ is the facility in $\hat{S}$ closest to the clients in $M(\hat{s}_i)$ for each $i \in [k']$. It is easy to verify that the clients in $M(\hat{s}_i)$ can be covered by $\hat{s}_i$ within $3r^*$ and the vertices in $N(\hat{s}_i)$ can be covered by $\hat{s}_i$ within $4r^*$ for every $i \in [k']$.

We consider the worst case in `near_facilities`, as illustrated in Fig. 3, where

1) the cluster $N(\hat{s}_4)$ is dissolved;
2) $N(\hat{s}_5)$ is not dissolved and $\hat{s}_5$ is the remaining facility in $\tilde{S}$, which is closest to $\hat{s}_4$;
3) Step 4(b)iii picks the facility f^* in $N(\hat{s}_5)$ to cover some bad client in C_1^b and replace $\hat{s}_5$.

For this case, `dissolve_clusters` guarantees $d(\hat{s}_4, \hat{s}_5) \leq 6r^*$. Thus, we conclude that $f^* \in S_1^g$ can cover the bad client in C_1^b within r^*, cover the good clients in $M(\hat{s}_5)$ within $7r^*$ and cover the good clients in $M(\hat{s}_4)$ within $13r^*$.

Theorem 2. *Algorithm β-kS is a 13-approximation algorithm for general k-supplier running in $\beta^{O(\beta)} \cdot \mathrm{poly}(n)$ time, where β is the minimum number of vertices, whose removal results in a metric graph.*

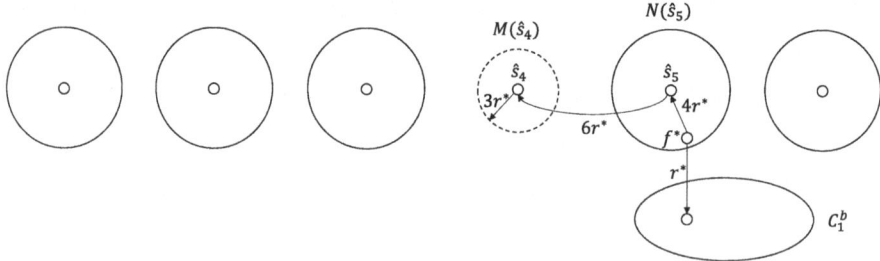

Fig. 3. An illustration for the analysis of the approximation ratio of `near_facilities`

5 An FPT 1/2-Approximation for k-Dispersion Parameterized by β

In this section, we consider k-dispersion parameterized by β. Similarly, let V^b denote the violating vertex set, whose vertices are called *bad*. The remaining vertices are called *good*. In this section, we use S^* to denote an optimal solution such that $\min_{v,v' \in S^*} d(v, v') \geq \min_{v,v' \in S} d(v, v')$ for every feasible solution S. For a set $V' \subseteq V$, we use *the internal distance of* V' to refer to $\min_{v,v' \in V'} d(v, v')$.

Our algorithm consists mainly of two steps. The first step "guesses" the optimal internal distance r^* of S^* and the bad vertices in $S^b = S^* \cap V^b$. The second step applies the polynomial-time 1/2-approximation algorithm [9] to compute a set S^g of $k - |S^b|$ good vertices in the metric graph induced by the good vertices whose minimal distances to all bad vertices in S^b are at least r^*. Then we merge S^b and S^g together to get a solution. Recall that for $v \in V$ and $V' \subseteq V$, we set $d(v, V') = \min_{v' \in V'} d(v, v')$, where $d(v, \emptyset) = +\infty$. The algorithm is as follows.

Algorithm β-kD: An FPT 1/2-approximation for k-dispersion parameterized by β

1. Compute a violating vertex set V^b of β vertices by a simple search tree and the corresponding set V^g of good vertices.
2. For each two distinct vertices v and v' with $v, v' \in V$, do:
   For each 2-partition of V^b into S^b and $V^b \setminus S^b$ satisfying $|S^b| \leq k$, do:
   (a) Set $r^* = d(v, v')$.
   (b) Set $V_1^g = \{v \mid v \in V^g, d(v, S^b) \geq r^*\}$ and apply the 1/2-approximation algorithm [9] to the graph $G[V_1^g]$ to compute a set S^g of $k - |S^b|$ vertices.
   (c) The solution of this iteration is set to $S = S^b \cup S^g$, and compute the internal distance $r = \min_{v,v' \in S} d(v, v')$ of S.
3. Return the solution $S^{\max}$ with the maximum internal distance $r^{\max}$ among all iterations.

Theorem 3. *Algorithm β-kD is a 1/2-approximation algorithm for general k-dispersion running in $O(3^\beta \cdot n^4)$ time, where β is the minimum number of vertices, whose removal results in a metric graph.*

6 Future Work

From metric to general graphs, we use the parameters α or β to measure the distance between inapproximable and approximable cases of k-center, k-supplier and k-dispersion, and design FPT constant-approximation algorithms for these three problems. The similar idea was also applied to the Traveling Salesman Problem, and approximation results have been successfully obtained in FPT time parameterized by α and β [13]. Guessing the performance of bad vertices in an optimal solution and for each possibility, applying constant approximation algorithms to the remaining metric graphs and combining the guess and the approximative result seems to be a general approach for all four problems.

An interesting research direction is to identify a class of polynomial-time inapproximable problems, which admit constant approximations on metric graphs, and establish a general framework for designing FPT constant-approximation algorithms for these problems. We believe the "distance from approximability" scheme might lead to theoretical and practical impulse for more problems.

Acknowledgements. The authors are grateful to the reviewers of COCOON 2025 for their valuable comments and constructive suggestions.

Disclosure of Interests. The authors have no competing interests to declare that are relevant to the content of this article.

References

1. Bampis, E., Escoffier, B., Xefteris, M.: Improved FPT approximation for non-metric TSP. CoRR **abs/2407.08392** (2024)
2. Chandrasekaran, R., Tamir, A.: Polynomially bounded algorithms for locating p-centers on a tree. Math. Program. **22**(1), 304–315 (1982)
3. Downey, R.G., Fellows, M.R.: Fundamentals of Parameterized Complexity. Springer (2013)
4. Erkut, E.: The discrete p-dispersion problem. Eur. J. Oper. Res. **46**(1), 48–60 (1990)
5. Gonzalez, T.F.: Clustering to minimize the maximum intercluster distance. Theor. Comput. Sci. **38**, 293–306 (1985)
6. Hochbaum, D.S., Shmoys, D.B.: A best possible heuristic for the k-center problem. Math. Oper. Res. **10**(2), 180–184 (1985)
7. Hochbaum, D.S., Shmoys, D.B.: A unified approach to approximation algorithms for bottleneck problems. J. ACM **33**(3), 533–550 (1986)
8. Niedermeier, R.: Invitation to Fixed-Parameter Algorithms. Oxford University Press (2006)
9. Ravi, S.S., Rosenkrantz, D.J., Tayi, G.K.: Heuristic and special case algorithms for dispersion problems. Oper. Res. **42**(2), 299–310 (1994)
10. Shier, D.R.: A min-max theorem for p-center problems on a tree. Transp. Sci. **11**(3), 243–252 (1977)
11. Tamir, A.: Obnoxious facility location on graphs. SIAM J. Discret. Math. **4**(4), 550–567 (1991)
12. Vazirani, V.V.: Approximation Algorithms. Springer (2001)
13. Zhou, J., Li, P., Guo, J.: Parameterized approximation algorithms for TSP. In: Proceedings of ISAAC 2022, pp. 50:1–50:16 (2022)
14. Zhou, J., Zhang, Z., Guo, J.: An FPT constant-factor approximation algorithm for correlation clustering. In: Proceedings of COCOON 2024, pp. 518–530 (2024)

String Algorithms and Discrete Structures

Revisit the Partial Coloring Method: Prefix Spencer and Sampling

Dongrun Cai[1], Xue Chen[1,2]($\boxtimes$) (iD), Wenxuan Shu[1], Haoyu Wang[1], and Guangyi Zou[1]

[1] College of Computer Science, University of Science and Technology of China, Hefei 230026, China
{cdr,wxshu}@mail.ustc.edu.cn, xuechen1989@ustc.edu.cn
[2] Hefei National Laboratory, University of Science and Technology of China, Hefei 230088, China

Abstract. As the most powerful tool in discrepancy theory, the partial coloring method has wide applications in many problems including the Beck-Fiala problem [12] and Spencer's celebrated result [4,28,37,41]. Currently, there are two major algorithmic approaches for the partial coloring method: the first approach uses linear algebraic tools to update the partial coloring for many rounds [4,6,7,25,28]; and the second one, called Gaussian measure algorithm [36,37], projects a random Gaussian vector to the feasible region that satisfies all discrepancy constraints in $[-1,1]^n$.

In this work, we explore the advantages of these two approaches and show the following results for them separately.

Spencer [42] conjectured that the prefix discrepancy of any $\mathbf{A} \in \{0,1\}^{m \times n}$ is $O(\sqrt{m})$, i.e., $\exists \mathbf{x} \in \{\pm 1\}^n$ such that $\max_{t \le n} \| \sum_{i \le t} \mathbf{A}(\cdot, i) \cdot \mathbf{x}(i) \|_\infty = O(\sqrt{m})$ where $\mathbf{A}(\cdot, i)$ denotes column i of $\mathbf{A}$. Combining small deviations bounds of Gaussian processes and the Gaussian measure algorithm [37], we show how to find a partial coloring with prefix discrepancy $O(\sqrt{m})$ and $\Omega(n)$ entries in $\{\pm 1\}$ efficiently. While this bound was folklore before [34,43], our argument is simpler and more direct than previous proofs. Moreover, our conceptual contribution here is a connection between Gaussian processes and the Gaussian measure algorithms.

Our second result extends the linear algebraic approach to a sampling algorithm in Spencer's classical setting. On the first hand, besides the six deviation bound [41], Spencer also proved that there are 1.99^m good colorings with discrepancy $O(\sqrt{m})$ for any $\mathbf{A} \in \{0,1\}^{m \times n}$. Hence a natural question is to design efficient random sampling algorithms in Spencer's setting. On the second hand, some applications of discrepancy theory, such as experimental design, prefer a random solution instead of a fixed one [20,22,44]. Our second result is an efficient sampling algorithm whose random output has min-entropy $\Omega(n)$ and discrepancy $O(\sqrt{m})$.

Supported by Innovation Program for Quantum Science and Technology 2021ZD0302901, NSFC 62372424, and CCF-HuaweiLK2023006.

F. V. Fomin and M. Xiao (Eds.): COCOON 2025, LNCS 15984, pp. 313–326, 2026.
https://doi.org/10.1007/978-981-95-0218-9_23

1   Introduction

The partial coloring method, as one of the most powerful techniques in discrepancy theory, has been applied to obtain the best known bounds for a variety of problems. This method was first introduced by Beck [11]. Later on, Spencer [41] successfully applied it to prove the discrepancy of any set system of m subsets is $\leq 6\sqrt{m}$. Formally, given $\mathbf{A} \in \mathbb{R}^{m \times n}$, we call $\min_{\mathbf{x} \in \{\pm 1\}^n} \|\mathbf{A}\mathbf{x}\|_\infty$ the discrepancy of $\mathbf{A}$ where $\mathbf{x} \in \{\pm 1\}^n$ is a bi-coloring on the columns of A. In this work, we use $\mathbf{A}(i, \cdot)$ to denote row i of $\mathbf{A}$ and $\mathbf{A}(\cdot, j)$ to denote its column j. For convenience, we consider each row $\mathbf{A}(i, \cdot)$ as a constraint on $\mathbf{x}$ and call $\langle \mathbf{A}(i, \cdot), \mathbf{x} \rangle$ the discrepancy of this row given $\mathbf{x}$. A partial coloring is a relaxation of $\mathbf{x}$ from the discrete Boolean cube to $[-1, 1]^n \subset \mathbb{R}^n$ with $\Omega(n)$ entries in $\{\pm 1\}$.

In particular, Spencer's classical result [41] shows that for any $\mathbf{A} \in \{0, 1\}^{m \times m}$, there exists a partial coloring $\mathbf{x} \in \{-1, 0, 1\}^m$ with 99% entries in $\{\pm 1\}$ such that $\|\mathbf{A}\mathbf{x}\|_\infty = O(\sqrt{m})$. Via a reduction from n variables to m variables [11,12], the discrepancy of any $\mathbf{A} \in \{0, 1\}^{m \times n}$ is $O(\sqrt{m})$ by recursively applying Spencer's partial coloring. A celebrated line of research provides efficient algorithms to find such a partial coloring, to name a few [4,9,25,28,33,37]. Recently, the partial coloring method has been successfully applied to many other problems, such as the *prefix* Spencer conjecture [7,42] and the vector balancing problem [5,6,13].

There are two major algorithmic approaches for the partial coloring method. The first one keeps updating the partial coloring in $\mathbb{R}^n$ via linear-algebraic tools, called linear-algebraic framework in this work. In [25,28,33], the algorithms find a safe subspace H_t and pick a (random) vector $\mathbf{v}_t \in H_t$ to update the coloring. Other works [4–7] have used semi-definite programs to find the subspace and update the partial coloring. The second approach measures the feasible region by the (standard) Gaussian distribution, called Gaussian measure algorithm [36,37]. Here the feasible region is the convex body $K \subset \mathbb{R}^n$ of all x satisfying $\|\mathbf{A}\mathbf{x}\|_\infty \leq L$ for some discrepancy bound L. Then a partial coloring is a point $\mathbf{x} \in K \cap [-1, 1]^n$ with $\Omega(n)$ entries in $\{\pm 1\}$. The key idea in the 2nd approach is to use the (standard) Gaussian measure to lower bound the size of K [21]. Then Rothvoss [37] proved that the projection of a random Gaussian vector to $K \cap [-1, 1]^n$ would be a good partial coloring (w.h.p.).

While both approaches obtain the same algorithmic results for Spencer's bound, they are quite different. On the first hand, the linear-algebraic framework is more flexible such that one could impose extra linear constraints and pick any vector in the safe subspace. On the second hand, the Gaussian measure approach could comply many more discrepancy constraints than the linear-algebraic framework, since it only needs a lower bound on the Gaussian measure of the feasible region K. In this work, we continue the study of the partial coloring method and explore the advantages of these two methods separately.

1.1   Our Results

For ease of exposition, we focus on Spencer's setting in this section, whose input $\mathbf{A} \in \{0,1\}^{m \times n}$ satisfies $n \geq m$. Our first result is an application of the Gaussian measure algorithm [37] to the prefix Spencer conjecture. Instead of bounding $\|\mathbf{Ax}\|_\infty$, prefix discrepancy problems bound $\max_{t \leq n} \|\sum_{i=1}^{t} \mathbf{A}(\cdot, i) \cdot \mathbf{x}(i)\|_\infty$ where $\mathbf{A}(\cdot, i)$ denotes column i of $\mathbf{A}$ such that $\sum_{i=1}^{t} \mathbf{A}(\cdot, i) \cdot \mathbf{x}(i)$ is a prefix summation of vectors in $\mathbf{A} \cdot \mathbf{x}$. While Spencer [42] conjectured the prefix discrepancy is $O(\sqrt{m})$ the same as the discrepancy of the original problem [41], the best known bounds were $2m$ [14] and $O(\sqrt{m \log n})$ [3,7] for *arbitrarily large* n.

For the prefix Spencer conjecture, we show how to find a partial coloring $\mathbf{x}$ such that (1) it has $\Omega(n)$ entries in $\{\pm 1\}$ and (2) its prefix discrepancy is $O(\sqrt{m})$. We remark that this upper bound on the prefix discrepancy is *optimal* up to constants [41,42]. However, different from the classical Spencer problem, there is no reduction from n variables to m variables in prefix discrepancy theory. Thus repeating this partial coloring leads to a *full* coloring of prefix discrepancy $\sqrt{m} \cdot \log \frac{O(n)}{m}$ instead of $O(\sqrt{m})$ in Spencer's classical setting.

Theorem 1. *Given any $\mathbf{A} \in \{0,1\}^{m \times n}$ with $n \geq m$, the Gaussian measure algorithm finds a partial coloring $\mathbf{x} \in [-1, 1]^n$ such that*

1. *its prefix discrepancy $\max_{t \leq n} \|\sum_{i=1}^{t} \mathbf{A}(\cdot, i) \cdot \mathbf{x}(i)\|_\infty$ is $O(\sqrt{m})$;*
2. *$\mathbf{x}$ has $\Omega(n)$ entries in $\{\pm 1\}$.*

Moreover, there exists an efficient algorithm to find a full coloring $\mathbf{x} \in \{\pm 1\}^n$ whose prefix discrepancy is $\sqrt{m} \cdot \log \frac{O(n)}{m}$.

We remark that although this partial coloring bound was folklore [34,43], previous methods are more complicated and involved which design extra discrepancy constraints delicately and rely on advanced partial coloring bounds. Our algorithm is simpler: it applies the Gaussian measure algorithm to projecting a random Gaussian vector to the feasible region directly. Also, our argument is more direct: we explore a connection between small ball probabilities of Gaussian processes and Gaussian measures of convex bodies. Hence, we believe our argument of Theorem 1 sheds more insights on prefix discrepancy problems and this connection could find more applications in discrepancy theory.

Specifically, Theorem 1 uses the crucial fact that to apply the Gaussian measure algorithm, we only need to lower bound the Gaussian measure of feasible $\mathbf{x} \in \mathbb{R}^n$. While the number of prefix constraints is nm, greater than the number of variables n, these constraints are highly correlated. The key step in Theorem 1 is to apply a small-deviation bound for (sub-)Gaussian processes to bound correlated prefix constraints: for some small constant c,

$$\Pr_{\mathbf{x} \sim N(0,1)^n} \left[\forall t \leq n, \left| \sum_{i=1}^{t} \mathbf{x}(i) \right| = O(\sqrt{m}) \right] \geq 2^{-c \cdot n/m}. \tag{1}$$

While inequality (1) holds for $x \sim \{\pm 1\}^n$ also, it is unclear how to satisfy m constraints simultaneously for $x \sim \{\pm 1\}^n$. When $x \sim N(0,1)^n$, we apply the classical Šidák-Khatri inequality [23,38,40] to lower bound the probability of satisfying all prefix constraints in $\mathbf{A}$. This allows us to apply the classical Gaussian measure algorithm by Rothvoss [37].

Our second result studies the sampling question in Spencer's setting via the linear-algebraic framework. In his original paper [41], Spencer had shown that there are exponentially many colorings satisfying $\|\mathbf{Ax}\|_\infty \leq 10\sqrt{m}$ for any $\mathbf{A} \in \{0,1\}^{m \times m}$. Hence a natural question is to sample from these good colorings $\mathbf{x}$ (satisfying $\|\mathbf{Ax}\|_\infty \leq 10\sqrt{m}$). Moreover, some applications of discrepancy theory, such as experimental design [19,20], seek a random balanced coloring rather than a fixed one. Specifically, experimental design requires the coloring to be both balanced and robust. While there are several ways to define robustness, randomized solutions are considered as the most reliable way [44]. For example, Harshaw et al. [22] considered how to generate a random coloring whose covariance matrix is balanced. In this work, we use the *min-entropy* of the coloring as an alternative way to measure its randomness and robustness. Equivalently, we bound the probability that the sampling algorithm outputs any fixed string.

Theorem 2. *There exists a sampling algorithm such that given any* $\mathbf{A} \in \{0,1\}^{m \times n}$ *with* $n \geq m$,

1. *its output* $\mathbf{x} \in \{\pm 1\}^n$ *always satisfies* $\|\mathbf{Ax}\|_\infty \leq O(\sqrt{m})$;
2. *for any good coloring* $\epsilon \in \{\pm 1\}^n$, *the probability that the output equals* ϵ *is* $O(1.9^{-0.9n})$.

Our algorithm extends the linear-algebraic framework by incorporating leverage scores of randomized matrix algorithms. Although previous algorithms [4,28,37] for Spencer's problem use a large amount of randomness to generate a partial coloring, it is unclear how to measure the *min-entropy* of their outputs. Our main idea is to use leverage scores of the safe subspace to find an entry k_t in every update t such that the partial coloring could either fix $\mathbf{x}(k_t) = 1$ or $\mathbf{x}(k_t) = -1$ arbitrarily. This guarantees that different partial colorings lead to different *full* colorings. In fact, in every update, our algorithm could fix 99% uncolored entries in this way. Compared to previous algorithms in this framework [4,7,9,25,28,33], this provides an alternate way to control the partial coloring; and our algorithm can be viewed as an application of this idea to sampling.

Specifically, we consider the leverage scores on coordinates of a subspace H. In particular, the leverage score $\tau_i(H)$ of coordinate i indicates the closeness of the indicator vector e_i to H. While indicator vectors $e_1, \ldots, e_n$ are not in the safe subspace H, we use leverage scores of H to find an approximation $\mathbf{v} \in H$ for indicator vectors such that updating along $\mathbf{v}$ or $-\mathbf{v}$ could set an entry to be 1 or -1 arbitrarily.

Finally, we extend the sampling algorithm to the Beck-Fiala problem [12] and the vector-balancing problem [2]. Analogue to Theorem 2 in the Spencer setting, our algorithm outputs a good coloring $\mathbf{x}$ satisfying the Beck-Fiala bound with a sufficiently large min-entropy. For the vector-balancing problem, we extend the online algorithm by Liu et al. [27] which uses a random walk with a Gaussian stationary distribution. Our contributions here are several new properties about this Gaussian stationary random walk. Due to the space constraint, we describe the details of these two algorithms in the full version.

1.2 Related Works

Discrepancy Algorithms in the Linear-Algebraic Framework. Spencer's discrepancy bound $O(\sqrt{m})$ is tight for the Hadamard matrix. In a breakthrough, Bansal [4] used a SDP-based random walk to find a partial coloring efficiently and provided the first algorithmic method of Spencer's bound. Lovett and Meka [28] simplified Bansal's algorithm via walking on edges of convex bodies. Very recently, Pesenti and Vladu [33] used a regularized random walk to improve the big-O constant in Spencer's bound.

Several works have extended the linear algebraic framework to obtain efficient algorithms for Banaszczyk's bound [2] on the vector balancing problem. The vector balancing problem assumes each column $\mathbf{A}(\cdot, j)$ of $\mathbf{A}$ is a unit vector such that the discrepancy problem becomes balancing $\| \sum_{j=1}^{n} \mathbf{x}(j) \cdot \mathbf{A}(\cdot, j) \|_{\infty}$. The seminal works by Bansal et al. [5,6] showed efficient algorithms with discrepancy $O(\sqrt{\log n})$ via guaranteeing that $\mathbf{A}\mathbf{x}$ is a random sub-Gaussian vector. Subsequent works [9,25] provided unified approaches to obtain both Banszczyk's result and Spencer's result. In particular, Levy et al. [25] gave the first deterministic algorithms by derandomizing these random walk algorithms via a carefully selected multiplicative weight update method.

Recent work by Harrow et al. [22] improved the sub-Gaussian constant of [6] to 1. Moreover, they consider balancing the covariance matrix of $\mathbf{x}$ as a way to measure its randomness since randomization is the major method to guarantee the robustness in experimental design [44].

Prefix Discrepancy Theory. Bárány and Grinberg [14] proved that the prefix discrepancy of any $\mathbf{A} \in \{0,1\}^{m \times n}$ is at most $2m$ independent with n. Later on, Spencer [42] conjectured the prefix discrepancy is $O(\sqrt{m})$ and proved this for the special case $n = m$. Banaszczyk [3] used techniques from convex geometry to prove the existence of $\mathbf{x}$ with prefix discrepancy $O(\sqrt{m \log n})$ for Spencer's prefix conjecture.

In [7], Bansal and Garg provided an elegant prefix discrepancy algorithm in the linear-algebraic framework that matches Banaszczyk's bound. Technically, they showed how to find an update vector by semi-definite programs such that $\sum_{j \leq t} \mathbf{A}(\cdot, j)\mathbf{x}(j)$ satisfies certain sub-Gaussian properties. While our work is *inspired* by their elegant algorithm, our results and techniques are incomparable. First of all, while Bansal and Garg's algorithm produces a full coloring with discrepancy $O(\sqrt{m \log n})$, it leaves a gap $O(\sqrt{\log n})$ to Spencer's conjecture. On the other hand, our prefix algorithm finds a *partial* coloring whose prefix discrepancy matches the bound in Spencer's conjecture. Secondly, the technique of our sampling algorithm is to find a vector in subspace H_t that is very close to indicator vectors $e_1, \ldots, e_n$. But Bansal and Garg's algorithm looks for a "well-spread" random vector with certain sub-Gaussian properties.

At the same time, previous arguments [34,43] implied a partial coloring with prefix discrepancy $O(\sqrt{m})$ for the prefix Spencer conjecture. However, all previous approaches are more involved than our argument: their algorithms incorporate extra constraints and their proofs rely on refined versions of partial colorings. On the other hand, our algorithm simply projects a Gaussian vector to the feasible region without any extra constraint; and its analysis based on the small-ball probability is more intuitive.

Recent work by Bansal et al. [10] conjectured that the prefix discrepancy of the Beck-Fiala problem (where each column of $\mathbf{A}$ has $O(1)$ ones) is $O(1)$ and provided an application to approximation algorithms in scheduling theory.

The Gaussian Measure Algorithm. Gluskin [21] used the Gaussian measure to show the existence of a good partial coloring. Rothvoss [37] designed an elegant algorithm to find a partial coloring directly based on the Gaussian measure of convex bodies. While this framework is less flexible than the first linear-algebraic framework, it could comply more discrepancy constraints. For example, this partial coloring method has been used in matrix sparsification [35] and matrix Spencer's conjecture [8]. Furthermore, Reis and Rothvoss [36] has shown the Gaussian measure could be exponentially small for any constant base. Our technical contribution here is a connection between Gaussian processes and the Gaussian measure algorithm.

Finally, there are many other settings and applications of discrepancy theory including online discrepancy [1,24,27], hereditary discrepancy [31], discrepancy of permutations [32,43]. We refer to textbooks [15,30] for an overview.

1.3   Discussion

We explore the advantages of the two algorithmic frameworks for the partial coloring method and apply them to two problems in Spencer's setting separately.

For the prefix Spencer conjecture, our algorithm finds a partial coloring whose prefix discrepancy is almost-optimal (up to constants). However, this leads to a full coloring of discrepancy $\sqrt{m} \cdot \log \frac{O(n)}{m}$. While it is well known how to reduce n in classical discrepancy problems, much less is known about prefix problems. It would be interesting to investigate the reduction of variables for prefix discrepancy problems.

Technically, we apply the small deviation bounds of Gaussian processes to the Gaussian measure algorithm. Inspired by the chaining argument in Gaussian processes, we use the correlation of prefix constraints to reduce the loss of applying the Šidák-Khatri inequality directly. Similar ideas have been studied in other works about matrix sparsification [18,35]. It would be interesting to apply this idea to other problems in discrepancy theory including the Kadison-Singer problem [29] and ℓ_1 subspace embedding (a.k.a. sparsification of zonotopes) [13,45].

There are many more open questions in prefix discrepancy theory. While the prefix discrepancy of the Beck-Fiala problem is conjectured to be $O(1)$ [10], the best known upper bound is $O(\sqrt{\log n})$ from Banaszczyk's existence proof [3]. In fact, there is no *efficient* algorithm to output a coloring with prefix discrepancy

$O(\sqrt{\log n})$ in this setting yet [24]. This is in contrast with the prefix Spencer conjecture where Bansal and Garg [7] provided an efficient algorithm matching Banaszczyk's bound $O(\sqrt{m \log n})$. Furthermore, an intriguing direction is to design *efficient* online algorithms for the vector balancing problem with a prefix discrepancy $O(\sqrt{\log n})$.

For the linear-algebraic framework, our algorithm admits the partial coloring to fix an entry $\mathbf{x}(k_t)$ to 1 or -1 arbitrarily in every update t. While we state it as a sampling algorithm, one could apply it to control those $\{\pm 1\}$-entries in partial colorings. For example, in the Beck-Fiala setting, given the current partial coloring $\mathbf{x}_t$ at step t, let $\mathcal{F}_t$ be the set of fixed $\{\pm 1\}$-entries in $\mathbf{x}_t$. For unfixed entries in $\mathbf{x}_t([n] \setminus \mathcal{F}_t)$, our algorithm guarantees that $0.99 \cdot (n - |\mathcal{F}_t|)$ entries could be chosen to get fixed (to ± 1) in this step. In another word, at each step, it could set *almost* any unfixed entry to either 1 or -1. This provides a way to strengthen the standard partial coloring method. One open question would be to use stronger partial coloring methods to obtain a bound $o(d)$ for the Beck-Fiala problem of degree-d.

Organization. The rest of this paper is organized as follows. In Sect. 2, we provide basic notations and definitions. In Sect. 3, we prove Theorem 1 about the prefix discrepancy conjecture. In Sect. 4, we provide a random sampling algorithm in Spencer's setting and prove Theorem 2. Due to the space constraint, we leave some proof and statement of algorithms to the full version of this paper in arXiv.

2   Preliminaries

Notations. We use bold letters like $\mathbf{A}$ and $\mathbf{v}$ to denote matrices and vectors. For a vector $\mathbf{v} \in \mathbb{R}^n$, let $\mathsf{supp}(\mathbf{v})$ denote the set of non-zero entries in $\mathbf{v}$ as the support of $\mathbf{v}$ and $\mathbf{v}(S)$ for a subset $S \subseteq [n]$ denote the punctured-vector in $\mathbb{R}^S$ such that $\mathbf{v}(i)$ denotes its entry i. Moreover, we use $\mathbf{e}_i$ to denote the indicator vector for coordinate i.

For any matrix $\mathbf{A}$ whose dimension is $m \times n$, we use $\mathbf{A}(S, T)$ to denote the submatrix on $S \times T$ for any $S \subseteq [n]$ and $T \subseteq [m]$. In this work, $\mathbf{A}(\cdot, i)$ denotes column i of $\mathbf{A}$ and $\mathbf{A}(j, \cdot)$ denotes its row j. Moreover, $\mathbf{A}(\cdot, [t])$ will denote the submatrix of the first t columns.

Discrepancy Theory. Through this work, we always use $\mathbf{A} \in \mathbf{R}^{m \times n}$ to denote the input matrix for various settings. So discrepancy theory looks for an $\mathbf{x} \in \{\pm 1\}^n$ minimizing $\|\mathbf{Ax}\|_\infty$. Since $\|\mathbf{Ax}\|_\infty = \max_{i \in [m]} |\langle \mathbf{A}(i, \cdot), \mathbf{x} \rangle|$, we consider each row $\mathbf{A}(i, \cdot)$ as a constraint in the discrepancy problem. Moreover, we call $\mathbf{x} \in [-1, 1]^n$ a partial coloring and $\mathbf{x} \in \{\pm 1\}^n$ a *full* coloring.

Prefix Discrepancy. In the prefix problem, the goal is to bound the prefix summation of any ℓ terms instead of the total summation $\mathbf{A} \cdot \mathbf{x} = \sum_{i \le n} \mathbf{A}(\cdot, i) \cdot \mathbf{x}(i)$:

$$\min_{x \in \{\pm 1\}^n} \max_{\ell \le n} \left\| \sum_{i=1}^{\ell} \mathbf{A}(\cdot, i) \cdot \mathbf{x}(i) \right\|_\infty. \tag{2}$$

Since $\mathbf{A}(\cdot, [\ell])$ denotes the first ℓ columns and $\mathbf{x}([\ell])$ denotes the first ℓ entries, we simplify the prefix summation as $\sum_{i \le \ell} \mathbf{A}(\cdot, i) \cdot \mathbf{x}(i) = \mathbf{A}(\cdot, [\ell]) \cdot \mathbf{x}([\ell])$. Also, the prefix discrepancy in (2) provides an upper bound on the discrepancy of any consecutive subset of indices (up to factor 2) [10].

Spencer [42] conjectured that for any $\mathbf{A} \in [-1, 1]^{m \times n}$, there always exists $\mathbf{x} \in \{-1, 1\}^n$ such that the prefix discrepancy defined in (2) is $O(\sqrt{m})$.

3   Spencer's Prefix Conjecture

We show how to find an optimal partial coloring for the prefix Spencer conjecture in this section. In this section, ϵ is a small constant to be fixed later (in Lemma 1).

Theorem 3. *There exists $\epsilon > 0$ such that for any $\mathbf{A} \in [-1, 1]^{m \times n}$ with $n \ge m$, Function PREFIXPARTIALCOLORING in Algorithm 1 takes input $\mathbf{A}$ and returns a partial coloring $\mathbf{x}$ such that with high probability,*

1. *$\mathbf{x}$ has εn entries in $\{\pm 1\}$;*
2. *its prefix discrepancy is $O(\sqrt{m})$, i.e. $\max_{\ell \le n} \|\mathbf{A}(\cdot, [\ell]) \cdot (\mathbf{x}([\ell]))\|_\infty = O(\sqrt{m})$.*

In the rest of this section, we finish the proof of Theorem 3. The key of Function PREFIXPARTIALCOLORING is the Gaussian measure algorithm [36,37], which is reformulated as Function GAUSSIANSAMPLING. One remark is that while Function GAUSSIANSAMPLING is formulated for the general case of finding a partial coloring in $\mathbb{R}^S$, one could assume $S = [n]$ in this section. In the full version, we apply the general form to obtaining a full coloring with prefix discrepancy $\sqrt{m} \cdot \log \frac{O(n)}{m}$.

The main property of Function GAUSSIANSAMPLING is that when Q is large enough, its output $\mathbf{x}^*$ has $\left| \left\{ i : \mathbf{x}^*(i) \in \{\pm 1\} \right\} \right| = \Omega(|S|)$. In particular, we use the Gaussian measure $\gamma(\cdot)$ to measure the size of Q, whose definition is

$$\gamma_S(K) := \Pr_{\mathbf{g} \sim N(0,1)^S}[\mathbf{g} \in K] \quad \text{for any measurable set } K \subset \mathbb{R}^S.$$

$N(0, 1)^S$ denotes a random vector in $\mathbb{R}^S$ whose entries are sampled from $N(0, 1)$ independently. We state Theorem 7 by Rothvoss [37] for this property.

Lemma 1. *Let $\epsilon := 10^{-4}$ and $\delta := \frac{9}{5000}$. If $K \subset \mathbb{R}^S$ is a symmetric convex body with $\gamma_S(K) \ge e^{-\delta n}$, then the output x^* of function GAUSSIANSAMPLING(K, S) has at least $\epsilon \cdot |S|$ many coordinates in $\{-1, 1\}$ with probability $1 - e^{-\Omega(|S|)}$.*

In this section, we fix ϵ and δ as defined in Lemma 1. In order to apply Lemma 1, the analysis of Theorem 3 needs several properties about Gaussian random variables.

Algorithm 1. partial coloring with optimal prefix discrepancy

---

1: **function** GAUSSIANSAMPLING(Q and S s.t. $Q \subset \mathbb{R}^S$ is a symmetric convex body)
2:     Sample a Gaussian vector $\boldsymbol{y}^* \sim N(0,1)^S$.
3:     Compute $\mathbf{x}^* = \arg\min_{z \in Q \cap [-1,1]^S}\{\|\boldsymbol{y}^* - \boldsymbol{z}\|_2\}$.
4:     Return $\mathbf{x}^*$.
5: **end function**
6: **function** PREFIXPARTIALCOLORING($\mathbf{x} \in [-1,1]^n, \mathbf{A} \in \mathbb{R}^{m \times n}$).
7:     $Q \leftarrow \left\{ \boldsymbol{g} \in \mathbb{R}^n : \max_{j \in [m]} \max_{1 \leq k \leq n} \left| \sum_{i=1}^k \boldsymbol{g}(i) \cdot \mathbf{A}(j,i) \right| \leq 112\sqrt{m} \right\}$.
8:     $\mathbf{x} \leftarrow$ GAUSSIANSAMPLING$(Q, [n])$
9:     **return** $\mathbf{x}$
10: **end function**

Gaussian Random Variables. We discuss several useful facts about the Gaussian random variables. The starting point of the prefix discrepancy algorithm is Lévy's inequality [26] which bounds the deviation of the prefix of symmetric random variables like $X_i \sim \{\pm 1\}$.

Lemma 2 (Lévy's inequality). *Let $X_1, X_2, \ldots, X_n$ be independent symmetric random variables and $\mathbf{S}_k = \sum_{i=1}^k X_i$ be their partial summations of $k \in [n]$. For any $\theta > 0$ we have* $\Pr\left[\max_{k \in [n]}\{|\mathbf{S}_k|\} \geq \theta\right] \leq 2 \cdot \Pr\left[|\mathbf{S}_n| \geq \theta\right]$.

While Lévy's inequality has been used for the special case $n = m$ [42], this is not strong enough to bound the prefix discrepancy by $O(\sqrt{m})$ for arbitrarily large n. Even for a single constraint $\mathbf{A}(j, \cdot)$, $\sum_{i=1}^n \mathbf{A}(j,i)\mathbf{x}(i)$ could have variance $\Theta(n)$; and $\Pr\left[|\sum_{i=1}^n \mathbf{A}(j,i)\mathbf{x}(i)| = \omega(\sqrt{m})\right]$ is close to 1 when $n = \omega(m)$.

A new ingredient is a *lower bound* on the probability of a sequence of Gaussian random variables (a.k.a. a Gaussian process) with small deviations. We remark this estimation is tight up to constants in the exponent [17,39].

Lemma 3. *Let $\{X_i\}_{i \in [n]}$ be i.i.d Gaussian random variables, with $\mathbb{E}[X_i] = 0, \mathbb{E}[X_i^2] \leq 1$. Let $S_n = \sum_{i=1}^n X_i$ and $S_n^* = \max_{1 \leq i \leq n} |S_i|$. Then for any ℓ, we have* $\Pr[S_n^* < 4\sqrt{\ell}] \geq 4^{-\frac{n}{\ell}}$.

We use the deviation bound of Lemma 3 for $\ell = \Theta(\sqrt{m})$. For completeness, we provide an elementary proof in the full version.

The last one is the Šidák-Khatri inequality about the correlation of two bodies in Gaussian measures [38,40].

Lemma 4. *Given any two symmetric convex bodies $T_1 \subseteq \mathbb{R}^n$ and $T_2 \subseteq \mathbb{R}^n$, $\gamma_n(T_1 \cap T_2) \geq \gamma_n(T_1) \cdot \gamma_n(T_2)$.*

Proof of Theorem 3. Define the convex body satisfying the prefix constraint of row j as $Q_j := \left\{ \boldsymbol{g} \in \mathbb{R}^n : \max_{1 \leq k \leq n} \left| \sum_{i=1}^k \boldsymbol{g}(i) \cdot \mathbf{A}(j,i) \right| \leq 112\sqrt{m} \right\}$. Then by Lemma 3, $\gamma_n(Q_j) = \Pr_{\boldsymbol{g} \sim N(0,1)^n}\left[\max_{1 \leq k \leq n} \left| \sum_{i=1}^k \boldsymbol{g}(i) \cdot \mathbf{A}(j,i) \right| \leq 112\sqrt{m} \right] \geq e^{-\frac{n}{28^2 m}} \geq e^{-\frac{9}{5000} \cdot n/m}$.

Since Q_j is symmetric and $Q = \cap_j Q_j$, Lemma 4 implies their intersection $\gamma_n(Q) \geq \prod_{j=1}^m \gamma_n(Q_j) = e^{-\frac{9}{5000}n}$. Then by Lemma 1, GAUSSIANSAMPLING can find a partial coloring $\mathbf{x} \in Q \cap [-1,1]^n$ so that $\mathbf{x}$ has at least $\epsilon \cdot n$ many entries in $\{\pm 1\}$ with probability $1 - e^{-\Omega(n)}$.   □

4   Random Sampling in Spencer's Setting

We will prove Theorem 2 in this section, which provides a random sampling algorithm in Spencer's setting. It extends the classical linear algebraic framework by incorporating leverage scores such that the min-entropy of the output is close to n. In fact, we will show an extended version of Theorem 2 as follows.

Theorem 4. *For any $n \geq m$, there exist a constant C and an efficient randomized algorithm such that for any input $\mathbf{A} \in [-1,1]^{m \times n}$,*

1. *its output $\mathbf{x}$ always satisfies $\|\mathbf{A}\mathbf{x}\|_\infty \leq C\sqrt{m}$.*
2. *for any $\epsilon \in \{\pm 1\}^n$, the probability of $\mathbf{x} = \epsilon$ is $O(1.9^{-0.9n})$.*

We summarize the classical bounds by Spencer [41] and its algorithmic results by [4,25,28,37]. We state the following version whose starting point $\mathbf{x}_0$ could be arbitrary.

Theorem 5. *Given any $\mathbf{A} \in [-1,1]^{m \times n}$ with $n \geq m$, there exist exponentially many $\mathbf{x} \in \{\pm 1\}^n$ such that $\|\mathbf{A}\mathbf{x}\|_\infty = O(\sqrt{m})$. Moreover, there exist efficient algorithms that given any starting point $\mathbf{x}_0$, they find $\mathbf{x} \in \{\pm 1\}^n$ with $\mathbf{x}(i) = \mathbf{x}_0(i)$ for each $\mathbf{x}_0(i) \in \{\pm 1\}$ satisfying $\|\mathbf{A}(\mathbf{x} - \mathbf{x}_0)\|_\infty = O(\sqrt{m})$.*

Before describing our sampling algorithm, we need to introduce the *leverage score* at first. The standard leverage score is defined for every row of a full-rank matrix $\mathbf{G} \in \mathbb{R}^{n \times d}$ with $n \geq d$:

$$\tau_i = \mathbf{G}(i, \cdot)^\top (\mathbf{G}^\top \mathbf{G})^{-1} \mathbf{G}(i, \cdot) \text{ for row } i \text{ in } \mathbf{G}. \tag{3}$$

One could generalize it for every coordinate of a subspace by basic properties of leverage scores [16]. We summarize two properties of τ_i here and provide a self-contained proof and the corresponding algorithm in the full version.

Proposition 1. *Given a linear subspace $H \subseteq \mathbb{R}^n$, for each coordinate $i \in n$, let $\tau_i(H) := \max_{\mathbf{u} \in H \setminus \{\mathbf{0}\}} \frac{|\mathbf{u}(i)|^2}{\|\mathbf{u}\|_2^2}$. Then $\sum_{i \in [n]} \tau_i(H) = dim(H)$.*

Moreover, given H and i, there exists an algorithm that can compute the corresponding vector $\mathbf{u}$ such that $\frac{\mathbf{u}(i)^2}{\|\mathbf{u}\|_2^2} = \tau_i(H)$ and $\mathbf{u}(i) = 1$.

Our algorithm has two parts. When $n \geq C^* \cdot m$ for a fixed constant $C^* \geq 10^6$, we apply the Function COLUMNREDUCTIONSAMPLING (defined in Algorithm 2) to get a random partial coloring. Then we apply it to Theorem 5 as the starting point $\mathbf{x}_0$ to get the full coloring. Otherwise for $n \leq C^* \cdot m$ we employ the Function SAMPLINGSPENCER (defined in the full version).

We state the correctness of Function COLUMNREDUCTIONSAMPLING and Function SAMPLINGSPENCER in the following two lemmas respectively.

Lemma 5. *Given* $\mathbf{A} \in [-1,1]^{m \times n}$ *where* $n \geq C^* \cdot m$, *with probability* 0.99 *over* $\mathbf{r}$, *Function* COLUMNREDUCTIONSAMPLING *in Algorithm 2 returns* $\mathbf{x}_T \in [-1,1]^n$ *with* $|\{i \in [n] : |\mathbf{x}(i)| = 1\}| = 0.9n$ *and* $\|\mathbf{A}\mathbf{x}_T\|_\infty = 0$.

Moreover,
for different seeds $\mathbf{r}$ *and* $\mathbf{r}'$, $\mathbf{x}_T \leftarrow$ COLUMNREDUCTIONSAMPLING($\mathbf{r}$) *and* $\mathbf{x}'_T \leftarrow$ COLUMNREDUCTIONSAMPLING($\mathbf{r}'$) *have a different* $\{\pm 1\}$-*entry.*

Algorithm 2. Reduction Sampling in Spencer's Setting when $n \geq C^* \cdot m$

Input: $\mathbf{A} \in [-1,1]^{m \times n}$, $\eta = 0.1$, $T = 0.9n$, $\gamma = 0.0001$, $\delta = 0.05$.
Random seed $\mathbf{r} \in \{\pm 1\}^T$ will be generated on the fly.
Output: $\mathbf{x}_T \in [-1,1]^n$.

1: **function** COLUMNREDUCTIONSAMPLING($\mathbf{r}$)
2:     $\mathbf{x}_0 = \mathbf{0}$.
3:     **for** $0 \leq t \leq T - 1$ **do**
4:         $\mathcal{F}_t \leftarrow \{i \in [n] : |\mathbf{x}_t(i)| = 1\}$.
5:         $\mathcal{D}_t \leftarrow \{i \in [n] : |\mathbf{x}_t(i)| \geq 1 - \eta\}$.
6:         $H_t :=$ the subspace in $\mathbb{R}^{[n] \backslash \mathcal{D}_t}$ orthogonal to $\mathbf{x}_t([n] \backslash \mathcal{D}_t)$ and $\{\mathbf{A}(j, [n] \backslash \mathcal{D}_t) : j \in [m]\}$.
7:         Compute the leverage scores of $\tau_i(H_t)$ for each $i \in [n] \backslash \mathcal{D}_t$.
8:         Find the first coordinate $k_t \notin \mathcal{D}_t$ such that $\tau_{k_t}(H_t) \geq 1 - \gamma$ and $|\mathbf{x}_t(k_t)| \leq \delta$
— return $\perp$ if such a coordinate does not exist.
9:         Let $\mathbf{u}_t$ be the vector $\mathbf{u}_t(k_t) = 1$ and $\|\mathbf{u}_t\|_2^2 = 1/\tau_i$ (By Proposition 1).
10:         For $\delta_t := \mathbf{x}_t(k_t)$, set $\mathbf{r}(t+1) = 1$ w.p. $\frac{1+\delta_t}{2}$; otherwise $\mathbf{r}(t+1) = -1$.
11:         Choose $\alpha_t \in \mathbb{R}$ such that $\mathbf{x}_{t+1} \leftarrow \mathbf{x}_t + \alpha_t \cdot \mathbf{u}_t$ satisfies $\mathbf{x}_{t+1}(k_t) = \mathbf{r}(t+1)$.
12:     **end for**
13:     **return** $\mathbf{x}_T$
14: **end function**

Lemma 6. *Given* $\mathbf{A} \in [-1,1]^{m \times n}$ *where* $n \leq C \cdot m$ *for any constant* $C = O(1)$, *with probability* 0.99 *over* $\mathbf{r}$, *Function* SAMPLINGSPENCER *(defined in the full version) returns* $\mathbf{x}_T \in [-1,1]^n$ *with* $|\{i \in [n] : |\mathbf{x}(i)| = 1\}| = 0.9n$ *and* $\langle \mathbf{A}(j, \cdot), \mathbf{x}_T \rangle = O(\sqrt{m})$ *for any* $j \in [m]$.

Moreover, for different seeds $\mathbf{r}$ *and* $\mathbf{r}'$, $\mathbf{x}_T \leftarrow$ SAMPLINGSPENCER($\mathbf{A}, \mathbf{r}$) *and* $\mathbf{x}'_T \leftarrow$ SAMPLINGSPENCER($\mathbf{A}, \mathbf{r}'$) *have a different* $\{\pm 1\}$-*entry.*

Finally we prove Theorem 4 as follows.

Proof of Theorem 4. While both functions could fail and output $\perp$, one could repeat Function COLUMNREDUCTIONSAMPLING (when $n \geq C^* \cdot m$) or Function SAMPLINGSPENCER (when $n \leq C^* \cdot m$) n times. Thus with probability at least $1 - 0.01^n$ we could get a partial coloring $\mathbf{x}_T \in [-1,1]^n$. Otherwise, if all the n repeats fail and return $\perp$, we apply Theorem 5 directly to output $\mathbf{x} \in \{\pm 1\}^n$ with $\|\mathbf{A}\mathbf{x}\|_\infty \leq O(\sqrt{m})$.

Then we bound the discrepancy for the two cases $n < C^* \cdot m$ and $n \geq C^* \cdot m$ separately. When $n \leq C^* \cdot m$, note that we run Function COLUMNREDUCTION-SAMPLING on $\mathbf{A}' := \left(\begin{smallmatrix} \mathbf{A} \\ -\mathbf{A} \end{smallmatrix}\right)$ instead of $\mathbf{A}$. By Lemma 6, $\max\limits_{j \in [2m]} \langle \mathbf{A}'(j, \cdot), \mathbf{x}_T \rangle = O(\sqrt{m})$. Then we apply Theorem 5 on $\mathbf{x}_T$ and $\mathbf{A}'$ to get the full coloring $\mathbf{x} \in \{\pm 1\}^n$ with $\|\mathbf{A}'(\mathbf{x} - \mathbf{x}_T)\|_\infty \leq O(\sqrt{m})$. So the discrepancy is

$$\|\mathbf{A}\mathbf{x}\|_\infty = \max_{j \in [2m]} \langle \mathbf{A}'(j, \cdot), \mathbf{x} \rangle \leq \max_{j \in [2m]} \langle \mathbf{A}'(j, \cdot), \mathbf{x}_T \rangle + \|\mathbf{A}'(\mathbf{x} - \mathbf{x}_T)\|_\infty \leq O(\sqrt{m}).$$

When $n \geq C^* \cdot m$, we apply Theorem 5 on $\mathbf{x}_T$ and $\mathbf{A}$ to get full coloring $\mathbf{x}$ as well. By Lemma 5, it's clear that $\|\mathbf{A}\mathbf{x}\|_\infty \leq \|\mathbf{A}\mathbf{x}_T\|_\infty + \|\mathbf{A}(\mathbf{x} - \mathbf{x}_T)\|_\infty \leq O(\sqrt{m})$.

Next, we discuss the probability $\Pr[\mathbf{x} = \boldsymbol{\epsilon}]$. Let us consider the case $n \geq C^* \cdot m$. Lemma 5 guarantees that for any two different random seeds $\boldsymbol{r}$ and $\boldsymbol{r}'$, $\mathbf{x}_T \neq \mathbf{x}'_T$. Let $\mathbf{x}$ and $\mathbf{x}'$ be the full colorings obtained by applying Theorem 5 on $\mathbf{x}_T$ and $\mathbf{x}'_T$ respectively. By the guarantee of Theorem 5, $\mathbf{x} \neq \mathbf{x}'$ for different seeds $\boldsymbol{r}$ and $\boldsymbol{r}'$. Hence we bound the probability of generating any $\boldsymbol{r} \in \{\pm 1\}^T$. Recall that for any $0 \leq t \leq T - 1$, $\delta_t := \mathbf{x}_t(k_t)$ is $\leq \delta := 0.05$. From Line 10 of Algorithm 2,

$$\boldsymbol{r}(t+1) = \begin{cases} 1 & \text{with probability } \frac{1+\delta_t}{2}, \\ -1 & \text{with probability } \frac{1-\delta_t}{2}. \end{cases}$$

So for any $\boldsymbol{\epsilon} \in \{\pm 1\}^n$, $\Pr[\mathbf{x} = \boldsymbol{\epsilon}] \leq (\frac{1+\delta}{2})^T \leq 0.525^{0.9n}$. When $n \geq C^* \cdot m$,

$$\Pr[\mathbf{x} = \boldsymbol{\epsilon}] = \Pr\left[\mathbf{x} = \boldsymbol{\epsilon} \wedge \exists \text{ one call of COLUMNREDUCTIONSAMPLING succeeds}\right]$$
$$+ \Pr[\mathbf{x} = \boldsymbol{\epsilon} \wedge \text{All calls of COLUMNREDUCTIONSAMPLING} = \bot]$$
$$\leq 0.525^{0.9n}/(1 - 0.01) + 0.01^n \leq O(1.9^{-0.9n}).$$

When $n \leq C^* \cdot m$, we run Function SAMPLINGSPENCER. By the same argument above, $\Pr[\mathbf{x} = \boldsymbol{\epsilon}] \leq O(1.9^{-0.9n})$ holds as well. $\qquad\square$

Acknowledgement. Thank the anonymous referee for the helpful comments on the previous version.

References

1. Alweiss, R., Liu, Y.P., Sawhney, M.: Discrepancy minimization via a self-balancing walk. In: STOC 2021: 53rd Annual ACM SIGACT Symposium on Theory of Computing, pp. 14–20. ACM (2021)
2. Banaszczyk, W.: Balancing vectors and gaussian measures of n-dimensional convex bodies. Random Struct. Algorithms **12**(4), 351–360 (1998)
3. Banaszczyk, W.: On series of signed vectors and their rearrangements. Random Struct. Algorithms **40**(3), 301–316 (2012)
4. Bansal, N.: Constructive algorithms for discrepancy minimization. In: 51th Annual IEEE Symposium on Foundations of Computer Science, FOCS 2010, pp. 3–10. IEEE Computer Society (2010)

5. Bansal, N., Dadush, D., Garg, S.: An algorithm for Komlós conjecture matching Banaszczyk's bound. SIAM J. Comput. **48**(2), 534–553 (2019)
6. Bansal, N., Dadush, D., Garg, S., Lovett, S.: The gram-Schmidt walk: a cure for the Banaszczyk blues. Theory Comput. **15**, 1–27 (2019)
7. Bansal, N., Garg, S.: Algorithmic discrepancy beyond partial coloring. In: Proceedings of the 49th Annual ACM SIGACT Symposium on Theory of Computing, STOC 2017, pp. 914–926. ACM (2017)
8. Bansal, N., Jiang, H., Meka, R.: Resolving matrix spencer conjecture up to polylogarithmic rank. In: Proceedings of the 55th Annual ACM Symposium on Theory of Computing, STOC 2023, pp. 1814–1819. ACM (2023)
9. Bansal, N., Laddha, A., Vempala, S.S.: A unified approach to discrepancy minimization. In: APPROX/RANDOM 2022. LIPIcs, vol. 245, pp. 1:1–1:22. Schloss Dagstuhl - Leibniz-Zentrum für Informatik (2022)
10. Bansal, N., Rohwedder, L., Svensson, O.: Flow time scheduling and prefix Beck-Fiala. In: Proceedings of the 54th Annual ACM SIGACT Symposium on Theory of Computing, pp. 331–342 (2022)
11. Beck, J.: Roth's estimate of the discrepancy of integer sequences is nearly sharp. Combinatorica **1**, 319–325 (1981)
12. Beck, J., Fiala, T.: "integer-making" theorems. Discrete Appl. Math. **3**(1), 1–8 (1981)
13. Bozzai, R., Reis, V., Rothvoss, T.: The vector balancing constant for zonotopes. In: 64th IEEE Annual Symposium on Foundations of Computer Science, FOCS 2023, pp. 1292–1300. IEEE (2023)
14. Bárány, I., Grinberg, V.: On some combinatorial questions in finite-dimensional spaces. Linear Algebra Appl. **41**, 1–9 (1981)
15. Chazelle, B.: The Discrepancy Method (Randomness and Complexity). Cambridge University Press (2002)
16. Chen, X., Price, E.: Active regression via linear-sample sparsification. In: Conference on Learning Theory, COLT 2019, 25–28 June 2019. Proceedings of Machine Learning Research, vol. 99, pp. 663–695. PMLR (2019)
17. Chung, K.L.: On the maximum partial sums of sequences of independent random variables. Trans. Am. Math. Soc. **64**(2), 205–233 (1948)
18. Dadush, D., Jiang, H., Reis, V.: A new framework for matrix discrepancy: partial coloring bounds via mirror descent. In: STOC 2022: 54th Annual ACM SIGACT Symposium on Theory of Computing, pp. 649–658. ACM (2022)
19. Fisher, R.A.: The arrangement of field experiments. J. Minist. Agric. **33**, 503–515 (1926)
20. Fisher, R.: Statistical Methods for Research Workers. Biological monographs and manuals, Oliver and Boyd (1925)
21. Gluskin, E.D.: Extremal properties of orthogonal parallelepipeds and their applications to the geometry of banach spaces. Sbornik Math. **64**(1), 85–96 (1989)
22. Harshaw, C., Sävje, F., Spielman, D.A., Zhang, P.: Balancing covariates in randomized experiments with the gram–schmidt walk design. J. Am. Stat. Assoc. **0**(0), 1–13 (2024)
23. Khatri, C.G.: On certain inequalities for normal distributions and their applications to simultaneous confidence bounds. Ann. Math. Stat., 1853–1867 (1967)
24. Kulkarni, J., Reis, V., Rothvoss, T.: Optimal online discrepancy minimization. In: Proceedings of the 56th Annual ACM Symposium on Theory of Computing, STOC 2024, pp. 1832–1840. ACM (2024)

25. Levy, A., Ramadas, H., Rothvoss, T.: Deterministic discrepancy minimization via the multiplicative weight update method. In: IPCO 2017, Proceedings, vol. 10328, pp. 380–391. Springer (2017)
26. Lévy, P.: Théorie de l'addition des variables aléatoires. Gauthier-Villars, Paris (1938)
27. Liu, Y.P., Sah, A., Sawhney, M.: A gaussian fixed point random walk. In: 13th Innovations in Theoretical Computer Science Conference, ITCS 2022. LIPIcs, vol. 215, pp. 101:1–101:10 (2022)
28. Lovett, S., Meka, R.: Constructive discrepancy minimization by walking on the edges. In: 53rd Annual Symposium on Foundations of Computer Science, FOCS 2012, pp. 61–67. IEEE (2012)
29. Marcus, A.W., Spielman, D.A., Srivastava, N.: Interlacing families II: mixed characteristic polynomials and the Kadison–Singer problem. Ann. Math. **182**(1), 327–350 (2015)
30. Matousek, J.: Geometric Discrepancy: An Illustrated Guide. Springer Science (2009)
31. Matousek, J., Nikolov, A., Talwar, K.: Factorization norms and hereditary discrepancy. Int. Math. Res. Not. **2020**(10), 751–780 (2020)
32. Newman, A., Neiman, O., Nikolov, A.: Beck's three permutations conjecture: a counterexample and some consequences. In: 53rd Annual IEEE Symposium on Foundations of Computer Science, FOCS 2012, pp. 253–262. IEEE (2012)
33. Pesenti, L., Vladu, A.: Discrepancy minimization via regularization. In: Bansal, N., Nagarajan, V. (eds.) Proceedings of the 2023 ACM-SIAM Symposium on Discrete Algorithms, SODA 2023, pp. 1734–1758. SIAM (2023)
34. Reis, V.: Vector Balancing and Integer Programming. University of Washington (2023)
35. Reis, V., Rothvoss, T.: Linear size sparsifier and the geometry of the operator norm ball. In: Proceedings of the 2020 ACM-SIAM Symposium on Discrete Algorithms, SODA 2020, pp. 2337–2348. SIAM (2020)
36. Reis, V., Rothvoss, T.: Vector balancing in lebesgue spaces. Random Struct. Algorithms **62**(3), 667–688 (2023)
37. Rothvoss, T.: Constructive discrepancy minimization for convex sets. SIAM J. Comput. **46**(1), 224–234 (2017)
38. Royen, T.: A simple proof of the gaussian correlation conjecture extended to multivariate gamma distributions. Far East J. Theoret. Stat. **48**(2), 139–145 (2014)
39. Shao, Q.M.: A note on small ball probability of a gaussian process with stationary increments. J. Theor. Probab. **6**, 595–602 (1993)
40. Šidák, Z.: Rectangular confidence regions for the means of multivariate normal distributions. J. Am. Stat. Assoc. **62**(318), 626–633 (1967)
41. Spencer, J.: Six standard deviations suffice. Trans. Am. Math. Soc. **289**(2), 679–706 (1985)
42. Spencer, J.: Balancing vectors in the max norm. Combinatorica **6**, 55–65 (1986)
43. Spencer, J.H., Srinivasan, A., Tetali, P.: The discrepancy of permutation families. Unpublished manuscript (2001)
44. Student: Comparison between balanced and random arrangements of field plots. Biometrika **29**(3/4), 363–378 (1938)
45. Talagrand, M.: Embedding subspaces of l1 into ln 1. Proc. Am. Math. Soc. **108**(2), 363–369 (1990)

A Sparse Dynamic Programming Algorithm for Solving the Coding Sequence Design Problem

Long-Shang Cho, Kai-Wei Chang, and Chin Lung Lu[✉]

Department of Computer Science, National Tsing Hua University, Hsinchu, Taiwan
cllu@cs.nthu.edu.tw

Abstract. In this work, we study the coding sequence design problem, which involves designing a coding sequence to encode a given amino acid sequence by optimizing both its secondary structure stability and codon usage. The structural stability and codon usage are quantified by minimum free energy and codon adaptation index, respectively. The coding sequence design problem is important since it has significant potential for the development of mRNA-based vaccines. Previously, we proposed an $\mathcal{O}(L^3)$ time and $\mathcal{O}(L^2)$ space dynamic programming algorithm to solve the coding sequencing design problem, where L is the length of the coding sequence to be designed. In this study, we utilize the sparsification technique to further reduce the time complexity of this dynamic programming algorithm from $\mathcal{O}(L^3)$ to $\mathcal{O}(L^2 + ZP)$ for the problem under the base pair-based energy model, where Z and P are two sparsity parameters satisfying $Z \leq L(6 + P)$ and $P \leq 36L$.

Keywords: computational biology · sparsification · dynamic programming · coding sequence design

1 Introduction

The rapid advancement of messenger RNA (mRNA) vaccines during the COVID-19 pandemic has demonstrated the significant potential of mRNA-based therapeutics due to their safety and efficacy. However, an mRNA molecule is inherently unstable and prone to degradation, resulting in insufficient protein expression that diminishes the vaccine's ability to stimulate strong immune responses. The degradation of mRNA can be mitigated by forming a more stable secondary structure. However, in this situation, it is generally assumed that the cellular translation machinery faces greater challenges in processing secondary structures in mRNA, leading to reduced protein production. Conversely, several recent studies have indicated that the issue introduced by secondary structures may not be problematic, as increasing the secondary structure in the coding sequence (CDS)

This work was supported in part by National Science and Technology Council of Taiwan under grant NSTC113-2221-E-007-109.

of mRNA can significantly enhance its protein expression [7,8,12,14]. In addition, codon usage preference has long been recognized as a significant factor in protein expression, as it varies among different host genomes and typically correlates positively with protein expression levels [4,10]. It has been reported that optimizing the codon usage of a CDS has little effect on improving its secondary structure stability, and the reverse is also true [14]. Therefore, it is essential and desirable to develop efficient algorithms that can design the CDS of a protein sequence by optimizing both its secondary structure stability and codon usage simultaneously.

In fact, the number of feasible candidates for designing the CDS of a protein sequence is exponential due to the redundancies in the genetic code. In essence, there are 64 codons for 20 amino acids, meaning that each amino acid can be encoded by multiple codons, with some amino acids having up to 6 codons. Currently, several algorithms have been proposed to design the CDS of a protein sequence by optimizing its secondary structure stability [2,11], or by optimizing both its secondary structure stability and codon usage [3,6,14]. By extending the Zuker algorithm [15] under the constraint of an amino acid sequence, Cohen and Skiena [2] and Terai et al.. [11] independently proposed a dynamic programming algorithm to design a CDS with the minimum free energy (MFE) secondary structure among all possible candidates encoding a given protein sequence. Moreover, Terai et al. have implemented their algorithm as a very useful tool called CDSfold. Both algorithms mentioned above require $\mathcal{O}(L^3)$ time and $\mathcal{O}(L^2)$ space, where L is the length of the CDS to be designed. However, they only optimize the secondary structure stability of the designed CDS without taking codon usage into account. Recently, Zhang et al. [14] have utilized the lattice parsing technique from computational linguistics to design a CDS by simultaneously optimizing its secondary structure stability and codon adaptation index (CAI), where CAI is a common measure of codon usage in the host [10]. The time and space complexities of the algorithm developed by Zhang et al. are $\mathcal{O}(L^3)$ and $\mathcal{O}(L^2)$, respectively. In addition, Zhang et al. have utilized the beam search technique [5] to significantly reduce the search space of their algorithm without sacrificing much of its accuracy. Zhang et al. have also implemented their algorithm as a tool, called LinearDesign, which can obtain a high-quality approximate CDS in $\mathcal{O}(L)$ time. However, the beam search function in the current standalone version of LinearDesign is not yet accessible to users.

In their study [14], Zhang et al. pointed out that the extended nucleotides used in CDSfold [11] make it challenging, if not impossible, to modify its dynamic programming algorithm to optimize both the MFE and CAI of a CDS design. The so-called *extended* nucleotides introduced by Terai et al. [11] actually are different notations to represent the same nucleotide at the second position of some amino acids, such as arginine and leucine. They were used by CDSfold to deal with the nucleotide dependencies between different codon positions when executing its dynamic programming algorithm. However, in our previous study [6], we have successfully modified the dynamic programming algorithm of CDSfold to perform the same function as LinearDesign by optimizing both the MFE and

CAI of the CDS to be designed. The key to this success lies in our introduction of *new* extended nucleotides (see Appendix for details), which allows us to handle nucleotide dependencies while integrating CAI into the representation of these new extended nucleotides without creating any irreconcilable results, as noted by Zhang *et al.* in their study [14]. The time and space complexities of our dynamic programming algorithm [6] are $\mathcal{O}(L^3)$ and $\mathcal{O}(L^2)$, respectively. In addition, by further incorporating the beam search heuristic [5], our dynamic programming algorithm can obtain an approximate CDS design with high quality in $\mathcal{O}(L)$ time. We have also implemented our dynamic programming algorithm with the beam search function into a tool named LinearCDSfold [6].

Later, we learned that Gu *et al.* [3] have proposed another dynamic programming algorithm, called DERNA, to solve the CDS design problem by optimizing both the MFE and CAI of the designed CDS. However, the dynamic programming algorithm of DERNA, which runs in $\mathcal{O}(L^3)$ time and $\mathcal{O}(L^2)$ space, remains distinct from the one we used in LinearCDSfold. To address the challenges introduced by the extended nucleotides used in CDSfold, DERNA simultaneously considers all three nucleotides within a codon during the stepwise process of its dynamic programming algorithm, in contrast to our LinearCDSfold, which focuses on individual nucleotides. The benefit to DERNA is that it does not have to handle the nucleotide dependencies of different positions in the codons of arginine and leucine. However, the approach adopted by DERNA results in a significantly higher computational time compared to our LinearCDSfold in practical usage, even though both tools share the same time and space complexities. For instance, according to our experiments on 10 test protein sequences with an average length of 1,008 amino acids, our LinearCDSfold took an average of 30.9 min, whereas DERNA required 212.7 min [6].

In this study, we utilize the sparsification technique to further reduce the time complexity of our dynamic programming algorithm [6] from $\mathcal{O}(L^3)$ to $\mathcal{O}(L^2 + ZP)$ for exactly solving the CDS design problem under the base pair-based energy model [9], where Z and P are two sparsity parameters satisfying $Z \leq L(6+P)$ and $P \leq 36L$. The sparsification technique has previously been used to significantly improve the time complexities of dynamic programming algorithms for the RNA secondary structure prediction problem, which aims to identify the secondary structure with minimum free energy for an RNA sequence [1,13].

2   Preliminaries

RNA is typically a single-stranded molecule composed of a sequence of nucleotides (or bases) that can fold into a structural conformation. An RNA sequence consists of four types of bases: A, G, C, and U. Its structural conformation is formed by hydrogen bonds between these bases, and usually, each base in the sequence can form a bond with at most one other base. Typically, G pairs with C, A pairs with U, and a weaker pairing can also occur between G and U. Two bases that can pair with each other are called *complementary* bases, and the bond formed between them is called a *base pair*. The set of formed base pairs is known as the *secondary structure* and represents the *folding* of an RNA sequence. Paired bases in an RNA folding always occur in a *nested* fashion. This means that if arcs are drawn to connect base pairs on an RNA sequence, none of the arcs will cross each other. If non-nested base pairs occur in an RNA folding, they are called *pseudoknots*. In this study, only pseudoknot-free foldings are considered.

Given an amino acid sequence $A = a_1 a_2 \ldots a_l$, an RNA sequence $R = r_1 r_2 \ldots r_L$ is called a *coding sequence* (CDS) for A if three consecutive nucleotide $r_{3i-2} r_{3i-1} r_{3i}$ in R form a *codon* encoding the ith amino acid a_i in A, where $1 \leq i \leq l$ and $L = 3l$. Let $R_{i,j}$ denote the *subsequence* of R from index i to index j. However, if $i > j$, then $R_{i,j}$ is defined as an *empty* subsequence. In this study, a folding F of a subsequence $R_{i,j}$ is denoted by a set of index pairs satisfying the following conditions: (i) For every index pair (g, h) in F, it corresponds to a base pair between two complementary bases r_g and r_h, and moreover $h - g > t$, where $i \leq g < h \leq j$ and typically $t = 3$ (i.e., an RNA molecule does not fold too sharply on itself). (ii) For any two index pairs (g, h) and (g', h') in F, $g = g'$ if and only if $h = h'$ (i.e., each index can only be paired once). (iii) There are no two index pairs (g, h) and (g', h') in F such that $g < g' < h < h'$ (i.e., F is pseudoknot-free).

Given a folding F of a subsequence $R_{i,j}$, a index q with $i < q \leq j$ is called a *split-point* of F if, for every index pair $(g, h) \in F$, either $h < q$ or $q \leq g$. This means that F can be partitioned at position q into two independent foldings: one for the non-empty prefix subsequence $R_{i,q-1}$ and the other for the non-empty suffix subsequence $R_{q,j}$. In addition, q is called *paired* in F if q appears in a base pair in F. Let $Q_{i,j} = \{q : i < q \leq j\}$ be the set of all possible split-points with respect to $R_{i,j}$. A folding F of $R_{i,j}$ is called *partitionable* if it has at least one split-point in $Q_{i,j}$; otherwise, F is called *co-terminus* (or *non-partitionable*). For a subsequence $R_{i,j}$ with $j = i$ (i.e., $Q_{i,j} = \varnothing$), the only possible folding F of $R_{i,j}$ is the empty folding (i.e., $F = \varnothing$), which is classified as a co-terminus folding since $R_{i,j}$ cannot be partitioned into two non-empty subsequences in this case. For a subsequence $R_{i,j}$ with $i < j$ (i.e., $Q_{i,j} \neq \varnothing$), a folding of $R_{i,j}$ is co-terminus if it contains the index pair (i, j).

In this study, two objectives are optimized when designing a CDS for a given amino acid sequence: structural stability and codon usage. The objective of structural stability is to find a CDS with the lowest minimum-free-energy folding among all possible CDSs encoding the given amino acid sequence. Suppose that $R = r_1 r_2 \ldots r_L$ is a CDS. Let $D(R)$ denote the set of all possible foldings of R. Given a folding $F \in D(R)$, let $\Delta G(R, F)$ denote the free energy of F based on an energy model. The *minimum free energy* (MFE) of R is defined as $\text{MFE}(R) = \min_{F \in D(R)} \Delta G(R, F)$. Conversely, another objective of codon usage is to identify a CDS with the highest codon adaptation index (CAI). The CAI of R is defined by Sharp and Li [10] as the geometric mean of the relative adaptiveness values of all codons in R. More specifically, let $C = c_1 c_2 \ldots c_l$ denote the *codon sequence* of R such that $c_i = r_{3i-2} r_{3i-1} r_{3i}$ is the ith codon in C, where $1 \leq i \leq l$. In addition, let $w(c_i)$ be the *relative adaptiveness* of the codon

c_i, which is the frequency of c_i divided by the frequency of its most frequent synonymous codon. Essentially, $0 \leq w(c_i) \leq 1$. The CAI of R is defined as $\mathrm{CAI}(R) = (\prod_{1 \leq i \leq l} w(c_i))^{\frac{1}{l}}$. By definition, $\mathrm{CAI}(R)$ is a positive value between 0 and 1, while $\mathrm{MFE}(R)$ usually is a negative value proportional to the sequence length of R. To reconcile such a difference, Zhang *et al.* [14] defined a joint objective $\mathrm{MFECAI}_\lambda(R)$ for R that incorporates both $\mathrm{MFE}(R)$ and $\mathrm{CAI}(R)$ as follows.

$$\mathrm{MFECAI}_\lambda(R) = \mathrm{MFE}(R) - \lambda l \log \mathrm{CAI}(R) \tag{1}$$

In Eq. (1), λ is a scaling parameter used to balance the contributions of $\mathrm{MFE}(R)$ and $\mathrm{CAI}(R)$ to $\mathrm{MFECAI}_\lambda(R)$. By expanding $\mathrm{CAI}(R)$ in Eq. (1), the joint objective $\mathrm{MFECAI}_\lambda(R)$ can be expressed as $\mathrm{MFE}(R)$ plus the λ-scaled sum of the negative logarithm of the relative adaptiveness of each codon, as shown in the following equation.

$$\mathrm{MFECAI}_\lambda(R) = \mathrm{MFE}(R) + \lambda \sum_{1 \leq i \leq l} -\log w(c_i) \tag{2}$$

Using Eq. (2), the CDS design problem we study in this paper is formulated as follows. Given a target amino acid sequence $A = a_1 a_2 \ldots a_l$ and a non-negative constant λ, the *CDS design problem* aims to design a CDS that has the lowest value of MFECAI_λ among all possible CDS candidates encoding A. Recall that the relative adaptiveness function w is defined on codons rather than individual nucleotides. In our previous study [6], we introduced a new function θ on nucleotides to replace the function w on codons, which enabled us to design a dynamic programming algorithm to solve the CDS problem using nucleotides instead of codons. Given a codon c_i, the function θ is defined on its three nucleotides r_{3i-2}, r_{3i-1} and r_{3i} by setting $\theta(r_{3i-2}) = \theta(r_{3i-1}) = 0$ and $\theta(r_{3i}) = -\log w(c_i)$, where $1 \leq i \leq l$. For convenience, we called $\theta(r_i)$ as a *CAI score* for nucleotide r_i. With the help of the function θ, the value of $\mathrm{MFECAI}_\lambda(R)$ for a CDS $R = r_1 r_2 \ldots r_L$ can be further expressed as follows.

$$\mathrm{MFECAI}_\lambda(R) = \mathrm{MFE}(R) + \lambda \sum_{1 \leq i \leq L} \theta(r_i) \tag{3}$$

For each $1 \leq i \leq L$, let N_i denote the set of possible nucleotides at position i of the CDS $R = r_1 r_2 \ldots r_L$ to be designed (i.e., $r_i \in N_i$). Furthermore, let $N_i|n$ represent the set of possible nucleotides allowed at position i when the nucleotide at position $i-1$ is n, and $N_i \wedge n$ denote the set of allowable nucleotides at position i when the nucleotide at position $i+1$ is n.

3 Algorithm

To simplify the design of our algorithm, we use the base pair-based energy model [9] to measure the free energy of an RNA folding in this study. In this model, the free energy of an RNA folding is determined by the sum of the scores of all base pairs in the folding. For simplicity, all G-C, A-U and G-U base pairs are assigned a score of -1 in this work. Moreover, for any two nucleotides x and y, $\delta(x,y)$ is defined as -1 if x and y are complementary; otherwise, it is defined as ∞.

Let $S_{i,j}^{n_i,n_j}$ be the set of subsequences of all CDSs encoding the target amino acid sequence in which each subsequence starts at position i with nucleotide n_i and ends at position j with nucleotide n_j. Let $\gamma_{i,j}^{n_i,n_j} = \min\{\mathrm{MFECAI}_\lambda(R') : R' \in S_{i,j}^{n_i,n_j}\}$ be the minimum MFECAI$_\lambda$ value among all the subsequences in $S_{i,j}^{n_i,n_j}$. Suppose that $R' = r_i r_{i+1} \dots r_j$ is a subsequence in $S_{i,j}^{n_i,n_j}$ and F is a folding in $D(R')$. Let $\mathrm{FECAI}_\lambda(R',F) = \Delta G(R',F) + \lambda \sum_{i \leq k \leq j} \theta(r_k)$. Then, $\gamma_{i,j}^{n_i,n_j}$ can be rewritten as the equation: $\gamma_{i,j}^{n_i,n_j} = \min\{\mathrm{FECAI}_\lambda(R',F) : R' \in S_{i,j}^{n_i,n_j}$ and $F \in D(R')\}$. In addition, a subsequence $R' \in S_{i,j}^{n_i,n_j}$ and a folding $F \in D(R')$ are called an *optimal CDS design* and an *optimal CDS folding*, respectively, with respect to $S_{i,j}^{n_i,n_j}$ if $\mathrm{FECAI}_\lambda(R',F) = \gamma_{i,j}^{n_i,n_j}$. Let $\Omega_{i,j}^{n_i,n_j} = \min\{\mathrm{FECAI}_\lambda(R',F) : R' \in S_{i,j}^{n_i,n_j}, F \in D(R')$ and F is co-terminus$\}$ and let $\Psi_{i,j}^{n_i,n_j} = \min\{\mathrm{FECAI}_\lambda(R',F) : R' \in S_{i,j}^{n_i,n_j}, F \in D(R')$ and F is partitionable$\}$. In other words, $\Omega_{i,j}^{n_i,n_j}$ (respectively, $\Psi_{i,j}^{n_i,n_j}$) is the minimum FECAI$_\lambda$ value among all the subsequences in $S_{i,j}^{n_i,n_j}$ with co-terminus (respectively, partitionable) foldings. Based on the above definitions, we can further rewrite $\gamma_{i,j}^{n_i,n_j}$ as the following equation:

$$\gamma_{i,j}^{n_i,n_j} = \min\{\Omega_{i,j}^{n_i,n_j}, \Psi_{i,j}^{n_i,n_j}\} \tag{4}$$

A set $S_{i,j}^{n_i,n_j}$ is called *trivial* if $j = i$. For any subsequence R' in a trivial set $S_{i,j}^{n_i,n_j}$, it has no partitionable folding, since $|R'| = 1$ and thus R' cannot be partitioned into two non-empty subsequences. Hence, we define $\Psi_{i,i}^{n_i,n_i} = \infty$. Note that the only folding of R' mentioned above is the empty folding, which is considered co-terminus by definition. Hence, we have $\Omega_{i,i}^{n_i,n_i} = \lambda\theta(n_i)$ and as a result, $\gamma_{i,i}^{n_i,n_i} = \Omega_{i,i}^{n_i,n_i} = \lambda\theta(n_i)$. For any non-trivial set $S_{i,j}^{n_i,n_j}$ with $i < j$, we can derive the recursive equations of $\Omega_{i,j}^{n_i,n_j}$ and $\Psi_{i,j}^{n_i,n_j}$ as follows:

$$\Omega_{i,j}^{n_i,n_j} = \min_{\substack{j-i>t \\ \delta(n_i,n_j)=-1 \\ n_{i+1} \in N_{i+1}|n_i \\ n_{j-1} \in N_{j-1} \wedge n_j}} \{\gamma_{i+1,j-1}^{n_{i+1},n_{j-1}} + \delta(n_i,n_j) + \lambda(\theta(n_i) + \theta(n_j))\} \tag{5}$$

$$\Psi_{i,j}^{n_i,n_j} = \min_{\substack{q \in Q_{i,j} \\ n_q \in N_q \\ n_{q-1} \in N_{q-1} \wedge n_q}} \{\gamma_{i,q-1}^{n_i,n_{q-1}} + \gamma_{q,j}^{n_q,n_j}\} \tag{6}$$

Based on Eqs. (4), (5) and (6), we can design a dynamic programming algorithm with a time complexity of $\mathcal{O}(\mu^4 L^3)$ and a space complexity of $\mathcal{O}(\mu^2 L^2)$ to solve the CDS design problem, as done in our previous study [6], where

$\mu = \max_{1 \le i \le L} |N_i| = 6$. Below, we utilize the sparsification technique to further reduce the time complexity of this dynamic programming algorithm from $\mathcal{O}(\mu^4 L^3)$ to $\mathcal{O}(\mu^4 L^2 + ZP)$, where Z and P are two sparsity parameters satisfying $Z \le L(\mu + P)$ and $P \le \mu^2 L$.

A set $S_{i,j}^{n_i, n_j}$ is defined as *optimally co-terminus* (OCT) if every optimal CDS folding with respect to $S_{i,j}^{n_i, n_j}$ is co-terminus (i.e., $\gamma_{i,j}^{n_i, n_j} = \Omega_{i,j}^{n_i, n_j} < \Psi_{i,j}^{n_i, n_j}$). By this definition, any trivial set $S_{i,j}^{n_i, n_j}$ with $i = j$ is an OCT, since any subsequence in $S_{i,j}^{n_i, n_j}$ has no partitionable folding. For a non-empty, non-trivial set $S_{i,j}^{n_i, n_j}$ with $i < j$, an index q with $i < q \le j$ is called an *optimal split-point* with respect to $S_{i,j}^{n_i, n_j}$ if there are $n_q \in N_q$ and $n_{q-1} \in N_{q-1} \wedge n_q$ such that $\Psi_{i,j}^{n_i, n_j} = \gamma_{i,q-1}^{n_i, n_{q-1}} + \gamma_{q,j}^{n_q, n_j}$.

Lemma 1. *For a non-empty set $S_{i,j}^{n_i, n_j}$ with $i < j$, $n_i \in N_i$ and $n_j \in N_j$, there is an optimal split-point q with respect to $S_{i,j}^{n_i, n_j}$ such that $S_{q,j}^{n_q, n_j}$ is an OCT.*

Proof. Let q be an optimal split-point with respect to $S_{i,j}^{n_i, n_j}$ such that $j - q$ is minimal. If $q = j$, then $S_{q,j}^{n_q, n_j}$ is an OCT and hence the lemma holds. Otherwise (i.e., $q < j$), suppose that p be an index with $q < p \le j$. Then we have $\gamma_{i,q-1}^{n_i, n_{q-1}} + \gamma_{q,j}^{n_q, n_j} < \gamma_{i,p-1}^{n_i, n_{p-1}} + \gamma_{p,j}^{n_p, n_j}$ for any $n_p \in N_p$ and $n_{p-1} \in N_{p-1} \wedge n_p$. By definition, we have $\gamma_{i,p-1}^{n_i, n_{p-1}} \le \Psi_{i,p-1}^{n_i, n_{p-1}} \le \gamma_{i,q-1}^{n_i, n_{q-1}} + \gamma_{q,p-1}^{n_q, n_{p-1}}$. Therefore, we have $\gamma_{i,q-1}^{n_i, n_{q-1}} + \gamma_{q,j}^{n_q, n_j} < \gamma_{i,p-1}^{n_i, n_{p-1}} + \gamma_{p,j}^{n_p, n_j} \le \gamma_{i,q-1}^{n_i, n_{q-1}} + \gamma_{q,p-1}^{n_q, n_{p-1}} + \gamma_{p,j}^{n_p, n_j}$. As a result, $\gamma_{q,j}^{n_q, n_j} < \gamma_{q,p-1}^{n_q, n_{p-1}} + \gamma_{p,j}^{n_p, n_j}$ for any $q < p \le j$. In other words, we have $\gamma_{q,j}^{n_q, n_j} < \Psi_{q,j}^{n_q, n_j}$. Therefore, $S_{q,j}^{n_q, n_j}$ is an OCT and the lemma holds.

Based on Lemma 1, we can rewrite the recursive formula of $\Psi_{i,j}^{n_i, n_j}$ as follows:

$$\Psi_{i,j}^{n_i, n_j} = \min_{\substack{q \in Q_{i,j} \\ n_q \in N_q \\ n_{q-1} \in N_{q-1} \wedge n_q \\ S_{q,j}^{n_q, n_j} \text{ is an OCT}}} \{ \gamma_{i,q-1}^{n_i, n_{q-1}} + \gamma_{q,j}^{n_q, n_j} \} \tag{7}$$

Let $Z = |\{ S_{i,j}^{n_i, n_j} : 1 \le i \le j \le L, n_i \in N_i, n_j \in N_j \text{ and } S_{i,j}^{n_i, n_j} \text{ is an OCT} \}|$ be the number of all the OCT sets. By definition, we have $L \le Z \le \mu^2 L^2 = 36 L^2$, where the inequality $L \le Z$ holds since any trivial set $S_{i,i}^{n_i, n_i}$ is an OCT for $1 \le i \le L$. In the following, we further restrict the set of split-points to be examined in the computation of $\Psi_{i,j}^{n_i, n_j}$. Given a non-empty set $S_{i,j}^{n_i, n_j}$ with $i < j$, $n_i \in N_i$ and $n_j \in N_j$, it is called a *STEP* if $\gamma_{i,j}^{n_i, n_j} < \gamma_{i,i}^{n_i, n_i} + \gamma_{i+1,j}^{n_{i+1}, n_j}$ for all $n_{i+1} \in N_{i+1} | n_i$. It means that none of the optimal CDS foldings with respect to $S_{i,j}^{n_i, n_j}$ have index $i + 1$ as the split-point. In other words, if $S_{i,j}^{n_i, n_j}$ is a STEP, then base i is paired in any optimal CDS folding with respect to $S_{i,j}^{n_i, n_j}$. Note that if $S_{i,j}^{n_i, n_j}$ itself is an OCT and $i < j$, then $S_{i,j}^{n_i, n_j}$ must also be a STEP.

Lemma 2. *For any non-empty set $S_{i,j}^{n_i, n_j}$ with $i < j$, $n_i \in N_i$ and $n_j \in N_j$, there is an optimal split-point q with respect to $S_{i,j}^{n_i, n_j}$ such that either $q = i + 1$, or $S_{i,q-1}^{n_i, n_{q-1}}$ is a STEP and $S_{q,j}^{n_q, n_j}$ is an OCT.*

Proof. Based on Lemma 1, there is an optimal split-point q with respect to $S_{i,j}^{n_i,n_j}$ such that $\Psi_{i,j}^{n_i,n_j} = \gamma_{i,q-1}^{n_i,n_{q-1}} + \gamma_{q,j}^{n_q,n_j}$, where $n_q \in N_q$, $n_{q-1} \in N_{q-1} \wedge n_q$ and the suffix set $S_{q,j}^{n_q,n_j}$ is an OCT. If the prefix set $S_{i,q-1}^{n_i,n_{q-1}}$ is a STEP, then the lemma holds. Suppose that $S_{i,q-1}^{n_i,n_{q-1}}$ is not a STEP. Then there is an $n_{i+1} \in N_{i+1}|n_i$ such that $\gamma_{i,q-1}^{n_i,n_{q-1}} = \gamma_{i,i}^{n_i,n_i} + \gamma_{i+1,q-1}^{n_{i+1},n_{q-1}}$ (i.e., there is at least an optimal CDS folding with respect to $S_{i,q-1}^{n_i,n_{q-1}}$ that has the split-point $i+1$). Hence, we have $\Psi_{i,j}^{n_i,n_j} = \gamma_{i,q-1}^{n_i,n_{q-1}} + \gamma_{q,j}^{n_q,n_j} = \gamma_{i,i}^{n_i,n_i} + \gamma_{i+1,q-1}^{n_{i+1},n_{q-1}} + \gamma_{q,j}^{n_q,n_j} \geq \gamma_{i,i}^{n_i,n_i} + \gamma_{i+1,j}^{n_{i+1},n_j} \geq \Psi_{i,j}^{n_i,n_j}$. In other words, $\Psi_{i,j}^{n_i,n_j} = \gamma_{i,i}^{n_i,n_i} + \gamma_{i+1,j}^{n_{i+1},n_j}$ and hence index $i+1$ is an optimal split-point with respect to $S_{i,j}^{n_i,n_j}$.

Based on Lemma 2, we can further restrict the split-points that need to be examined in the computation of $\Psi_{i,j}^{n_i,n_j}$:

$$\Psi_{i,j}^{n_i,n_j} = \min \begin{cases} \min_{n_{i+1} \in N_{i+1}|n_i} \{\gamma_{i,i}^{n_i,n_i} + \gamma_{i+1,j}^{n_{i+1},n_j}\} \\ \min_{\substack{q \in Q_{i,j} \\ n_q \in N_q \\ n_{q-1} \in N_{q-1} \wedge n_q \\ S_{i,q-1}^{n_i,n_{q-1}} \text{ is a STEP} \\ S_{q,j}^{n_q,n_j} \text{ is an OCT}}} \{\gamma_{i,q-1}^{n_i,n_{q-1}} + \gamma_{q,j}^{n_q,n_j}\} \end{cases} \tag{8}$$

Let $P = \max_{1 \leq j \leq L} |\{(i,n_i,n_j) : 1 \leq i < j, n_i \in N_i, n_j \in N_j$ and $S_{i,j}^{n_i,n_j}$ is a STEP$\}|$. In other words, P represents the maximum number of STEP sets ending at index j over all $1 \leq j \leq L$ and by definition, we have $P \leq \mu^2 L$. As a result, the total number of STEP sets is less than or equal to LP. Since any OCT set $S_{i,j}^{n_i,n_j}$ with $i < j$ is also a STEP and there are at most μL OCT sets $S_{i,j}^{n_i,n_j}$ with $i = j$, we have $Z \leq \mu L + LP = L(\mu + P)$. Since $\mu = 6$, we have $Z \leq L(6 + P)$ and $P \leq 36L$.

Based on Eqs. (4), (5) and (8), we design a sparse dynamic programming algorithm as shown in Algorithm 1 to solve the CDS design problem. The procedure of Backtracking called by Algorithm 1 aims to recover the nucleotide sequence of an optimal CDS and its corresponding folding. Algorithm 1 computes the values of $\gamma_{i,j}^{n_i,n_j}$ by increasing the column index j and decreasing the row index i, with each indexed position at (i,j) composed of $|N_i| \times |N_j|$ entries that are indexed by n_i and n_j. In the computation process, *forward* dynamic programming technique is applied to maintain the following invariant: upon reaching $\gamma_{i,j}^{n_i,n_j}$, this entry already contains the value of $\Psi_{i,j}^{n_i,n_j}$. This invariant property is assured in the following manner. Let $STEP\text{-}OCT(i,j,n_i,n_j) = |\{(q,n_q) : q \in Q_{i,j}, n_q \in N_q, n_{q-1} \in N_{q-1} \wedge n_q, S_{i,q-1}^{n_i,n_{q-1}}$ is a STEP and $S_{q,j}^{n_q,n_j}$ is an OCT$\}|$. When reaching the entry $\gamma_{i,j}^{n_i,n_j}$ in column j, where $2 \leq j \leq L$, its value is updated to $\gamma_{i,i}^{n_i,n_i} + \gamma_{i+1,j}^{n_{i+1},n_j}$, as specified at lines 11–13 of Algorithm 1, if the current value of $\gamma_{i,j}^{n_i,n_j}$ is greater than $\gamma_{i,i}^{n_i,n_i} + \gamma_{i+1,j}^{n_{i+1},n_j}$. The purpose of this update is to examine the split-point $q = i + 1$ in the computation of $\Psi_{i,j}^{n_i,n_j}$ according to Eq. (8). At this point, if $i = j - 1$, then the invariant is preserved since $STEP\text{-}OCT(j-1,j,n_{j-1},n_j) = 0$, indicating that the first entry $\gamma_{j-1,j}^{n_{j-1},n_j}$

Algorithm 1: Sparse dynamic programming algorithm

1: **for** $j = 1$ to L **do**   // Initialization
2:   | **for** $i = j$ down to 1 **do**
3:   | | **for all** $n_i \in N_i$ and $n_j \in N_j$ **do**
4:   | | | **if** $i = j$ and $n_i = n_j$ **then**
5:   | | | | $\gamma_{i,i}^{n_i,n_i} \leftarrow \lambda\theta(n_i)$ and mark $S_{i,i}^{n_i,n_i}$ as an OCT;
6:   | | | **else**
7:   | | | | $\gamma_{i,j}^{n_i,n_j} \leftarrow \infty$;

8: **for** $j = 2$ to L **do**   // Compute $\Omega_{i,j}^{n_i,n_j}$ and $\Psi_{i,j}^{n_i,n_j}$
9:   | **for** $i = j - 1$ down to 1 **do**
10:   | | **for all** $n_i \in N_i$ and $n_j \in N_j$ **do**
11:   | | | **for** $n_{i+1} \in N_{i+1}|n_i$ **do**
12:   | | | | **if** $\gamma_{i,j}^{n_i,n_j} > \gamma_{i,i}^{n_i,n_i} + \gamma_{i+1,j}^{n_{i+1},n_j}$ **then**
13:   | | | | | $\gamma_{i,j}^{n_i,n_j} \leftarrow \gamma_{i,i}^{n_i,n_i} + \gamma_{i+1,j}^{n_{i+1},n_j}$ and remove the STEP mark from $S_{i,j}^{n_i,n_j}$;
14:   | | | $\Omega_{i,j}^{n_i,n_j} \leftarrow \displaystyle\min_{\substack{n_{i+1} \in N_{i+1}|n_i \\ n_{j-1} \in N_{j-1} \wedge n_j}} \{\gamma_{i+1,j-1}^{n_{i+1},n_{j-1}} + \delta(n_i, n_j) + \lambda(\theta(n_i) + \theta(n_j))\}$;
15:   | | | **if** $\Omega_{i,j}^{n_i,n_j} < \gamma_{i,j}^{n_i,n_j}$ **then**
16:   | | | | $\gamma_{i,j}^{n_i,n_j} \leftarrow \Omega_{i,j}^{n_i,n_j}$ and mark $S_{i,j}^{n_i,n_j}$ as both an OCT and a STEP;
17:   | | | **if** $S_{i,j}^{n_i,n_j}$ is an OCT **then**
18:   | | | | **for each** $(k, i-1, n_k, n_{i-1})$ such that $n_k \in N_k$, $n_{i-1} \in N_{i-1} \wedge n_i$ and $S_{k,i-1}^{n_k,n_{i-1}}$ is a STEP **do**
19:   | | | | | **if** $\gamma_{k,j}^{n_k,n_j} > \gamma_{k,i-1}^{n_k,n_{i-1}} + \gamma_{i,j}^{n_i,n_j}$ **then**
20:   | | | | | | $\gamma_{k,j}^{n_k,n_j} \leftarrow \gamma_{k,i-1}^{n_k,n_{i-1}} + \gamma_{i,j}^{n_i,n_j}$ and mark $S_{k,j}^{n_k,n_j}$ as a STEP;

21: find $\widehat{n_i}$ and $\widehat{n_j}$ such that $\gamma_{1,L}^{\widehat{n_1},\widehat{n_L}} = \min\{\gamma_{1,L}^{n_1,n_L} : n_1 \in N_1 \text{ and } n_L \in N_L\}$;
22: **call** Backtracking;   // Output an optimal CDS and its folding

---

in column j already contains the value of $\Psi_{i,j}^{n_i,n_j}$. Based on this invariant, the value of $\gamma_{i,j}^{n_i,n_j}$ is then updated to the minimum between the current entry value (i.e., $\Psi_{i,j}^{n_i,n_j}$) and the value of $\Omega_{i,j}^{n_i,n_j}$ that is computed at line 14 of Algorithm 1 based on Eq. (5). If $\Omega_{i,j}^{n_i,n_j} < \Psi_{i,j}^{n_i,n_j}$, then the set $S_{i,j}^{n_i,n_j}$ corresponding to the current entry is classified as both an OCT and a STEP. In addition, if $S_{i,j}^{n_i,n_j}$ is an OCT, then the split-point i is taken into account and its influence is carried forward into the computation of $\Psi_{k,j}^{n_k,n_j}$ for all $k < i$ when $S_{k,i-1}^{n_k,n_{i-1}}$ is a STEP. This forward computation is to update the value of $\gamma_{k,j}^{n_k,n_j}$ to the minimum between its current value and $\gamma_{k,i-1}^{n_k,n_{i-1}} + \gamma_{i,j}^{n_i,n_j}$. In this manner, the minimum value of Eq. (8) is accumulated, eventually ensuring that the invariant is consistently maintained.

Procedure: Backtracking

1: $cds[1] \leftarrow \widehat{n_1}$ and $cds[L] \leftarrow \widehat{n_L}$;
2: push$(1, L, \widehat{n_1}, \widehat{n_L})$ on stack;
3: loop_start:
4: **while** stack is not empty **do**
5:     pop(i, j, n_i, n_j);
6:     **if** $i = j$ **then**
7:        goto loop_start;

8:     **for all** $n_{i+1} \in N_{i+1} | n_i$ and $n_{j-1} \in N_{j-1} \wedge n_j$ **do**   // Case 1
9:        **if** $\gamma_{i,j}^{n_i,n_j} = \gamma_{i+1,j-1}^{n_{i+1},n_{j-1}} + \delta(n_i, n_j) + \lambda(\theta(n_i) + \theta(n_j))$ **then**
10:           $folding[i] \leftarrow$ "(" and $folding[j] \leftarrow$ ")";
11:           $cds[i+1] \leftarrow n_{i+1}$ and $cds[j-1] \leftarrow n_{j-1}$;
12:           push$(i+1, j-1, n_{i+1}, n_{j-1})$;
13:           goto loop_start;

14:     **for all** q such that $i < q \leq j$ **do**   // Case 2
15:        **for all** $n_q \in N_q$ and $n_{q-1} \in N_{q-1} \wedge n_q$ **do**
16:           **if** $\gamma_{i,j}^{n_i,n_j} = \gamma_{i,q-1}^{n_i,n_{q-1}} + \gamma_{q,j}^{n_q,n_j}$ **then**
17:              $cds[q-1] \leftarrow n_{q-1}$ and $cds[q] \leftarrow n_q$;
18:              push$(i, q-1, n_i, n_{q-1})$ and push(q, j, n_q, n_j);
19:              goto loop_start;

20: **return** cds and $folding$;

Below, we analyze the time and space complexities of Algorithm 1. The initialization at lines 1–7 takes $\mathcal{O}(\mu^2 L^2)$ time. In the algorithm, a split-point q is examined for computing $\Psi_{i,j}^{n_i,n_j}$ only when $q = i+1$ or the prefix set $S_{i,q-1}^{n_i,n_q-1}$ is a STEP and the suffix set $S_{q,j}^{n_q,n_j}$ is an OCT. In the former case (i.e., $q = i+1$), the computation performed at line 13 costs $\mathcal{O}(\mu^3 L^2)$ time in total. In the latter case, there are Z OCT sets that can serve as suffix OCT sets, with each having at most P STEP sets to be used as the corresponding prefix STEP sets. Hence, the total computation time executed at line 20 for the latter case is $\mathcal{O}(ZP)$. As a result, the time for computing all $\Psi_{i,j}^{n_i,n_j}$ is $\mathcal{O}(\mu^3 L^2 + PZ)$. Moreover, the computation time of all $\Omega_{i,j}^{n_i,n_j}$ performed at line 14 is $\mathcal{O}(\mu^4 L^2)$, and the computation of the minimum MFECAI$_\lambda$ value (i.e., $\gamma_{1,L}^{\widehat{n_1},\widehat{n_L}}$) at line 21 takes $\mathcal{O}(\mu^2)$ time. Finally, an optimal CDS and its corresponding folding returned by the backtracking procedure at line 22 require $\mathcal{O}(Z)$ time. The reason for this is described as follows. Let $OCT(i, j, n_j) = |\{(q, n_q) : q \in Q_{i,j}, n_q \in N_q$ and $S_{q,j}^{n_q,n_j}$ is an OCT$\}|$. That is, $OCT(i, j, n_j)$ corresponds to the number of all the OCT sets that begin at index q with $i < q$ and end at index j with nucleotide n_j. It can be verified that $\sum_{j=1}^{L} |N_j| \cdot OCT(1, j, n_j) = Z$. For the **while** loop at lines 3–19 in the backtracking procedure, the number of its iterations is equal to the number of push operations to be performed, since each iteration executes a single pop operation. During each iteration, if $i = j$, then no operation is performed. For $i < j$, if $S_{i,j}^{n_i,n_j}$

is an OCT (i.e., Case 1), then the backtracking procedure takes $\mathcal{O}(\mu^2)$ time to identify a tuple $(i+1, j-1, n_{i+1}, n_{j-1})$ and push it onto the stack. On the other hand, if $S_{i,j}^{n_i,n_j}$ is not an OCT (i.e., Case 2), then it costs $\mathcal{O}(OCT(i, j, n_j))$ time to find two tuples $(i, q-1, n_i, n_{q-1})$ and (q, j, n_q, n_j) and push them onto the stack. However, the set $S_{q,j}^{n_q,n_j}$ corresponding to the tuple (q, j, n_q, n_j) is an OCT, which *derives* a push operation being executed in Case 1 of the next iteration. Since the set $S_{i,q-1}^{n_i,n_{q-1}}$ corresponding to the tuple $(i, q-1, n_i, n_{q-1})$ may be not an OCT, two push operations may be derived and performed in Case 2 of the next iteration. Note that if a push operation $push(i, j, n_i, n_j)$ is derived from $push(h, k, n_h, n_k)$ in the backtracking process, then $j - i < k - h$. According to the above discussion, the instructions in Case 1 or 2 of the **while** loop will be performed at most L times. As a result, the instructions in Case 1 take $\mathcal{O}(\mu^2 L)$ time in total, while those in Case 2 require at most $\sum_{j=1}^{L} OCT(1, j, n_j) = O(Z)$ time. Therefore, the running time of the backtracking procedure is $\mathcal{O}(Z)$. Consequently, the time complexity of Algorithm 1 is $\mathcal{O}(\mu^4 L^2 + ZP)$. On the other hand, it can be verified that the space complexity of Algorithm 1 is $\mathcal{O}(\mu^2 L^2)$. Based on the above discussion and the fact that $\mu = 6$, we have the following theorem.

Theorem 1. *The sparse dynamic programming in Algorithm 1 solves the CDS design problem in $\mathcal{O}(L^2 + ZP)$ time and $\mathcal{O}(L^2)$ space, where L is the length of the CDS to be designed, and Z and P are two sparsity parameters satisfying $Z \leq L(6 + P)$ and $P \leq 36L$.*

4   Conclusion

In this work, we studied the CDS design problem, which has significant potential for the development of mRNA-based vaccines to combat diseases. Previously, we proposed a dynamic programming algorithm to exactly solve this problem in $\mathcal{O}(L^3)$ time and $\mathcal{O}(L^2)$ space, where L is the length of the CDS to be designed. In this study, we applied the sparsification technique to further improve the time complexity of this dynamic programming algorithm for the CDS problem under the base pair-based energy model by reducing it from $\mathcal{O}(L^3)$ to $\mathcal{O}(L^2 + ZP)$, where Z and P are two sparsity parameters satisfying $Z \leq L(6+P)$ and $P \leq 36L$. Our experimental results on real datasets have revealed that the sparse version of the dynamic programming algorithm achieves a speedup of 13 to 43 times compared to its non-sparse version.

Appendix

According to the genetic code for amino acids, there are nucleotide dependencies between different codon positions for some amino acids. For instance, the amino acid serine is encoded by six codons: AGC, AGU, UCA, UCG, UCC and UCU. If A appears in the first codon position of serine, then G must appear in the second codon position. Conversely, if U is in the first codon position of serine, then C must be in the second codon position. Moreover, the amino acids leucine and arginine both exhibit nucleotide dependencies between their first and third codon positions. To address the dependency between any two consecutive nucleotides, CDSfold [11] introduced two notations $N_i|n$ and $N_i \wedge n$ for correctly designing the CDS. In addition, CDSfold introduced the concept of extended nucleotides to address the nucleotide dependencies between the first and third codon positions for both arginine and leucine. For instance, arginine is encoded by six codons: AGA, AGG, CGA, CGG, CGC and CGU. Therefore, two kinds of Gs, denoted by G^{AG} and G^{CU}, are created by CDSfold for the second codon position of arginine (see Fig. 1, left panel), depending on the nucleotide at the third codon position. It means that the nucleotide at the second codon position of arginine is G^{AG} (respectively, G^{CU}) if its nucleotide at the third codon position is A or G (respectively, C or U). Similarly, two kinds of Us, denoted by U^{AG} and U^{CU}, are also used to denote the nucleotide at the second codon position of leucine (see Fig. 1, right panel). Using these extended nucleotides, the dependencies between non-adjacent nucleotides for arginine and leucine can be transformed into dependencies between adjacent nucleotides. For instance, if the nucleotide at the first codon position of arginine is A, then the nucleotide at the second codon position must be G^{AG}, which further ensures that the nucleotide at the third codon position must be A or G. Conversely, if the nucleotide at the first codon position of arginine is C, then the nucleotide at the second codon position can be both G^{AG} and G^{CU}, further implying that the nucleotide at the third codon position can be A, G, C or U.

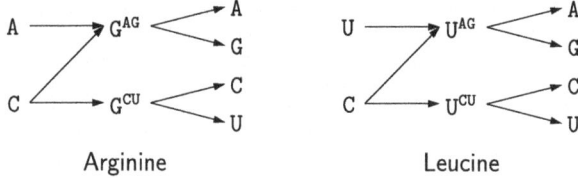

Fig. 1. Representations of extended nucleotides for the second codon position of arginine (left) and leucine (right), where the dependencies of two adjacent nucleotides are indicated by arrows.

Recall that the joint objective to be optimized in Eq. (2) factors the negative logarithm of CAI of a CDS candidate into the negative logarithm of the relative adaptiveness (NLRA for short) of each individual codon. As mentioned in the study by Zhang et al. [14], it is difficult to integrate such NLRA values of all six codons for arginine and leucine into the representation of their extended nucleotides in Fig. 1. To do this, the NLRA values of the codons must be integrated into the vertices, rather than the edges, in the extended nucleotide representation. As demonstrated in the study by Zhang et al. [14], the relative adaptiveness values for the six codons of leucine (UUA, UUG, CUA, CUG, CUC and CUU) in human genomes are 0.2, 0.3, 0.2, 1.0, 0.5, and 0.3, respectively. Thus, their corresponding NLRA values are 0.7, 0.5, 0.7, 0, 0.3, and 0.5, respectively.

As shown in Fig. 1 (right panel), the codons $UU^{AG}A$ and $CU^{AG}A$ share the same nucleotide A at the third codon position, while $UU^{AG}G$ and $CU^{AG}G$ share the same G at the third codon position. Since both $UU^{AG}A$ and $CU^{AG}A$ have the same NLRA value of 0.7, their NLRA values can be integrated into the extended nucleotide representation of leucine by assigning the nucleotide A at the third codon position a value of 0.7 and the other nucleotides at the first and second codon positions a value of 0. In fact, the values assigned to the nucleotides at the three codon positions are equivalent to the CAI scores of the three nucleotides (i.e., $\theta(r_1), \theta(r_2)$ and $\theta(r_3)$ if the three nucleotides are denoted by r_1, r_2 and r_3). However, the NLRA values of $UU^{AG}G$ and $CU^{AG}G$ are different (0.5 for the former and 0 for the latter) and therefore they cannot be simultaneously integrated into the extended nucleotide representation of leucine, as only a single NLRA value (either 0.5 or 0) can be assigned to the nucleotide G at the third codon position.

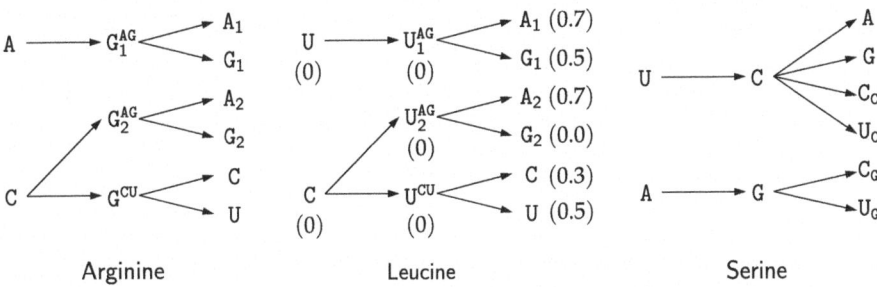

Fig. 2. Representations of new extended nucleotides for arginine (left), leucine (middle) and serine (right), where the values in parentheses in the representation of leucine are the CAI scores of nucleotides.

To address the issue mentioned above, we introduced new extended nucleotides at the second and third codon positions of both arginine and leucine in our previous study [6]. As illustrated in Fig. 2 (left and middle panels), the extended nucleotide G^{AG} at the second codon position of arginine is further replaced with two kinds of *new* extended nucleotides G_1^{AG} and G_2^{AG}, while the extended nucleotide U^{AG} at the second codon position of leucine is replaced with U_1^{AG} and U_2^{AG}. Furthermore, two kinds of As, denoted by A_1 and A_2, and two kinds of Gs, denoted by G_1 and G_2, are created to replace the original ones at the third codon position of both arginine and leucine. The purpose of these new extended nucleotides is to ensure that if G_1^{AG} and U_1^{AG} (or G_2^{AG} and U_2^{AG}, respectively) appears at the second codon position of arginine and leucine, respectively, then the nucleotide at their third codon positions must be A_1 or G_1 (or A_2 or G_2, respectively). Therefore, the NLRA values of leucine's six codons mentioned above can be easily integrated into the new extended nucleotide representation, as shown in Fig. 2 (middle panel), by assigning each vertex the CAI score of its corresponding nucleotide (i.e., $\theta(r_i)$ if the nucleotide is r_i).

In fact, as mention previously, serine has six codons: AGC, AGU, UCA, UCG, UCC and UCU. To properly assign the CAI scores to all the nucleotides in the six codons of serine, two kinds of Cs, denoted by C_C and C_G, and two kinds of Us, denoted by U_C and U_G, are created to replace the original ones at the third codon position as shown in Fig. 2 (right panel). It means that if C_C or U_C appears at the third codon position of serine, then the nucleotide at the second codon position must be C. On the other hand, if C_G or U_G appears at the third codon position of serine, then the nucleotide at the second codon position must be G.

References

1. Backofen, R., Tsur, D., Zakov, S., Ziv-Ukelson, M.: Sparse RNA folding: time and space efficient algorithms. J. Discrete Algorithms **9**, 12–31 (2011)
2. Cohen, B., Skiena, S.: Natural selection and algorithmic design of mRNA. J. Comput. Biol. **10**, 419–432 (2003)
3. Gu, X., Qi, Y., El-Kebir, M.: DERNA enables Pareto optimal RNA design. J. Comput. Biol. **31**, 179–196 (2024)
4. Gustafsson, C., Govindarajan, S., Minshull, J.: Codon bias and heterologous protein expression. Trends Biotechnol. **22**, 346–353 (2004)
5. Huang, L., et al.: LinearFold: linear-time approximate RNA folding by 5'-to-3' dynamic programming and beam search. Bioinformatics **35**, I295–I304 (2019)
6. Ju, Y.R., Cho, L.S., Lu, C.L.: A more efficient dynamic programming algorithm for designing a coding sequence by jointly optimizing its structural stability and codon usage. IEEE/ACM Trans. Comput. Biol. Bioinf. (2025, under revision)
7. Leppek, K., et al.: Combinatorial optimization of mRNA structure, stability, and translation for RNA-based therapeutics. Nat. Commun. **13**, 1536 (2022)
8. Mauger, D.M., et al.: mRNA structure regulates protein expression through changes in functional half-life. Proc. Natl. Acad. Sci. U.S.A. **116**, 24075–24083 (2019)
9. Nussinov, R., Jacobson, A.B.: Fast algorithm for predicting the secondary structure of single-stranded RNA. Proc. Natl. Acad. Sci. U.S.A. **77**, 6309–6313 (1980)
10. Sharp, P.M., Li, W.H.: The codon adaptation index - a measure of directional synonymous codon usage bias, and its potential applications. Nucleic Acids Res. **15**, 1281–1295 (1987)
11. Terai, G., Kamegai, S., Asai, K.: CDSfold: an algorithm for designing a protein-coding sequence with the most stable secondary structure. Bioinformatics **32**, 828–834 (2016)
12. Wayment-Steele, H.K., et al.: Theoretical basis for stabilizing messenger RNA through secondary structure design. Nucleic Acids Res. **49**, 10604–10617 (2021)
13. Wexler, Y., Zilberstein, C., Ziv-Ukelson, M.: A study of accessible motifs and RNA folding complexity. J. Comput. Biol. **14**, 856–872 (2007)
14. Zhang, H., et al.: Algorithm for optimized mRNA design improves stability and immunogenicity. Nature **621**, 396–403 (2023)
15. Zuker, M., Stiegler, P.: Optimal computer folding of large RNA sequences using thermodynamics and auxiliary information. Nucleic Acids Res. **9**, 133–148 (1981)

Improved Approximation Algorithm and Hardness Result for Sorting Unsigned Strings by Symmetric Reversals

Wenfeng Lai[1], Haitao Jiang[1(⊠)], Daming Zhu[1], and Binhai Zhu[2]

[1] School of Computer Science and Technology, Shandong University, Qingdao, China
wflai@mail.sdu.edu.cn, {htjiang,dmzhu}@sdu.edu.cn
[2] Gianforte School of Computing, Montana State University, Bozeman, MT 59717, USA
bhz@montana.edu

Abstract. It is widely recognized that structural variation represents a significant source of genetic variation. As a well-known type of structural variation, reversion is studied drastically by both biologists and computer scientists. Recently, scientists have found that repetitive sequences always appear at the ends of the segment where the reversal occurred on the chromosome, which has inspired new interests in models for sorting unsigned chromosomes by symmetric reversals (abbreviated $MUSR$). The problem of $MUSR$ asks for a minimum number of unsigned symmetric reversals to transform a chromosome S into another chromosome T, and requires symmetric reversals to be performed on the segment flanked by the same letter.

In this paper, we show that $MUSR$ is NP-hard through an intricate reduction from the MAX-$(3,B2)$-SAT problem. Moreover, we provide an innovative depiction of the optimal solution and then develop an improved approximation algorithm that ensures the approximation factor of $\frac{2\ln 3 + 7}{8}$ (approximately 1.15) and a time complexity of $O(n^2)$.

Keywords: Symmetric reversals · Approximation algorithms · NP-hard

1 Introduction

Structure variations are widely acknowledged as large-scale changes in the genome that involve the breakage and rejoining of DNA fragments, where reversion, insertion, deletion, and duplication are all typical structure variations that may occur during genome evolution. Biologists discovered as early as the 1980s that many species share remarkably similar genes. These species typically possess nearly identical gene content, the primary distinction being the sequence in which these genes are arranged. During the past thirty years, genome rearrangement problems have attracted significant attention within the field of computational biology. Sankoff *et al.* provided a formal definition for various genome operations,

F. V. Fomin and M. Xiao (Eds.): COCOON 2025, LNCS 15984, pp. 341–354, 2026.
https://doi.org/10.1007/978-981-95-0218-9_25

including reversals, transpositions, and translocations, effectively framing these genome rearrangement scenarios as classic combinatorial optimization problems [18]. Among these, the genome reversal problem is the most renowned and extensively researched. It involves determining the sequence of reversals needed to convert one chromosome into another [9,28].

The problems of sorting genomes by reversals are classified into two categories: signed and unsigned, based on how biologists infer gene orders. Gene orders can be deduced through whole genome sequencing and comparative physical mapping. Sequencing offers insight into gene orientation, allowing a genome to be depicted as a sequence of signed genes. Conversely, comparative physical mapping, although widely used in experiments and data collection, typically does not reveal gene orientation, resulting in a genome being portrayed as a sequence of unsigned genes.

Pevzner and Waterman laid the groundwork in 1995 by examining combinatorial problems inspired by genome rearrangements [16]. The difficulty involved in sorting permutations by reversals is significantly influenced by consideration of gene orientation. Caprara established that sorting unsigned permutations by reversals is NP-hard [6]. Kececioglu and Sankoff revealed the relationship between the reversal distance and the permutation breakpoints, then developed a factor 2 approximation algorithm [12]. Christie later refined this approximation to a factor of 1.5 [7], and Berman et al. proposed a factor-1.375 approximation for the problem of sorting an unsigned permutation by reversals [4]. Recently, Sun et al. slightly improved the approximation factor to 1.3748 [21]. Bergeron offered a simplified explanation of the Hannenhalli–Pevzner theory, an insightful method to solve this problem [3]. For sorting signed permutations by reversals, Hannenhalli and Pevzner introduced an exact algorithm with a time complexity of $O(n^4)$ [11]; advancements brought the time complexity down to $O(n^{1.5}\sqrt{\log n})$ by Tannier et al. [23], $O(n^{1.5})$ by Han [10], and $O(n \log^2 n / \log \log n)$ by Dudek and Gawrychowski [8].

Researchers have identified that the regions of genomes in which reversals occur could possess unique characteristics [15,17]. Various studies have indicated a frequent association of these breakpoints with repetitive elements [1,2,20,24]. A systematic investigation by Wang et al. comparing different bacterial strains showed that repeats are linked to the ends of rearrangement segments across multiple rearrangement events, including reversals [26,27]. Swenson et al. proposed a model for genome sorting using weighted double cut and join (DCJ) operations, in which a DCJ operation is assigned a weight of zero if it breaks two adjacencies of the same color, indicating the presence of a pair of identical repeats [19,22]. Li et al. conducted a systematic research on the problem of sorting strings by flanked block-interchanges [14]. Also, it is typical for the reversed segments to be flanked by repeat pairs [20]. Tong et al. recently investigated the sorting of genomes through signed symmetric reversals, developing an $O(n^2)$ algorithm for the decision problem [25]. On the other hand, Lai et al. focused on sorting chromosomes using unsigned symmetric reversals, where each repeat occurs at most twice, and introduced a 1.5-factor approximation algorithm for this case.

In this paper, we investigate the problem of sorting unsigned chromosomes by symmetric reversals, where repeats occur twice and genes occur once on the input chromosomes. By leveraging the observation that the input genomes share the same multi-set of adjacencies, we show that there exists an optimal solution that is based on a maximum cycle matching of this multi-set. Fortunately, achieving such a maximum cycle matching is feasible in $O(n^2)$ time. Based on this point, we present an approximation algorithm with a factor of $\frac{2\ln 3 + 7}{8}$ (approximately 1.15), which runs in $O(n^2)$ time. Lastly, we confirm the NP-hardness of the problem through a non-trivial reduction from the MAX-$(3,B2)$-SAT problem.

2 Preliminaries

The genome rearrangement problem typically involves an alphabet Σ, where each letter in Σ denotes a DNA sequence or syntenic block, which is called a gene hereafter. In a chromosome, genes usually appear just once, and short repetitive sequences at the termini of the reversed segment occur in pairs. Thus, when both genes and repetitive sequences are depicted by letters in Σ, a chromosome can be represented as a string in which each letter appears no more than twice. Given a chromosome $S = [S_1, S_2, \ldots, S_n]$, a reversal $\rho(i, j)$ inverts the segment $[S_i, \ldots, S_j]$, thereby converting S into another chromosome $S' = S \cdot \rho(i, j) = [S_1, \ldots, S_{i-1}, S_j, \ldots, S_i, S_{j+1}, \ldots, S_n]$. A reversal $\rho(i, j)$ is termed *symmetric* when S_i and S_j are identical symbols; such an instance is labeled a *symmetric reversal*.

In this paper, the relative order of genes on a chromosome is crucial, yet the direction of each gene is not. Therefore, despite the fact that a linear chromosome can be interpreted bidirectionally, we treat them as identical. For ease, we append a unique letter pair r_0 to both ends of a chromosome, enabling us to reverse the entire chromosome symmetrically about the r_0 pair. Two strings S and T are considered *related* when they contain the same number of each letter. We now formally present the problem that will be investigated in this paper.

Definition 1. *Sorting Unsigned Strings by the Minimum Number of Symmetric Reversals (**MUSR** for short).*

Instance: *Two related strings S and T with each letter appearing no more than twice in S (and similarly in T).*

Question: *A sequence of unsigned symmetric reversals that transforms S into T.*

Measurement: *Number of symmetric reversals required.*

The length of the shortest sequence of symmetric reversals that transform S into T is called the symmetric reversal distance between S and T.

Given two strings denoted by $S = [S_1, S_2, \ldots, S_n]$ and $T = [T_1, T_2, \ldots, T_n]$, on the alphabet Σ, we define a pair of sequential letters $\langle S_i, S_{i+1} \rangle$ (or $\langle T_i, T_{i+1} \rangle$ for T), with $1 \leq i \leq n-1$, as an *adjacency* for S (similarly for T). Each string S and T contains $n-1$ adjacencies. We represent the multi-sets of adjacencies in S

and T as $adj(S)$ and $adj(T)$, respectively. The i-th adjacency in S is depicted as $adj(S)_i = \langle S_i, S_{i+1} \rangle$, and in T as $adj(T)_i = \langle T_i, T_{i+1} \rangle$, where $1 \le i \le n-1$. Two adjacencies $\langle x_1, y_1 \rangle$ and $\langle x_2, y_2 \rangle$ are called *consistent* if $x_1 = x_2$ and $y_1 = y_2$, denoted by $\langle x_1, y_1 \rangle = \langle x_2, y_2 \rangle$. Conversely, they are called *opposite* if $x_1 = y_2$ and $y_1 = x_2$, denoted by $\langle x_1, y_1 \rangle = -\langle x_2, y_2 \rangle$. Regardless of whether the adjacencies are consistent or opposite, they are collectively referred to as *identical*. As stated in [13], two strings can be transformed into one another through symmetric reversals only when the multi-sets of their adjacencies are equivalent.

When an adjacency and its opposite adjacency appear three times within S (or similarly, T), they should be arranged as $[x, y, x, y]$ or $[y, x, y, x]$ within S (and respectively for T). By merging these sequences into $[x, y]$ and $[y, x]$, the frequency of the adjacency is reduced to once. One can verify that the symmetric reversal distance remains unchanged since symmetric reversals involving 'x' and 'y' are redundant. Throughout the rest of this paper, we assume $adj(S) = adj(T)$, and every adjacency along with its opposite adjacency appears no more than twice in S (or T).

2.1   The Breakpoint Graph

Given two related strings S and T, if an adjacency $adj(S)_i \in adj(S)$ and an adjacency of $adj(T)_j \in adj(T)$ are identical, they form a *match*, denoted as $adj(S)_i \sim adj(T)_j$. Note that $adj(S) = adj(T)$, there is always a *matching* between $adj(S)$ and $adj(T)$. We define $M = \{adj(S)_i \sim adj(T)_j \mid 1 \le i, j \le n-1,$ such that each adjacency in $adj(S)$ matches uniquely with exactly one in $adj(T)\}$.

Each adjacency is assigned an *id*, consisting of a positive integer and a sign, either '$+$' or '$-$'. For each adjacency $adj(T)_j$ where $1 \le j \le n_1$ within $adj(T)$, the *id* is $+j$. Considering each adjacency $adj(S)_i$ that matches $adj(T)_j$ in the matching M between $adj(S)$ and $adj(T)$, if $adj(S)_i$ and $adj(T)_j$ are consistent, then the *id* for $adj(S)_i$ is $+j$; otherwise, if they are opposite, the *id* for $adj(S)_i$ becomes $-j$. As a result, S can be considered as a permutation of the *id*s, say π.

Similarly to [11], we also create the *breakpoint graph* G_M as follows. For every adjacency $adj(S)_i$ with an *id* π_i in $adj(S)$, establish a pair of nodes: a left node and a right node for π_i, labeled as $l(\pi_i)$ and $r(\pi_i)$, respectively. If π_i has a positive sign, its head (denoted $h(\pi_i)$) is $l(\pi_i)$ and its tail (denoted $t(\pi_i)$) is $r(\pi_i)$. Conversely, if π_i has a negative sign, the head of π_i is $r(\pi_i)$ and the tail of π_i is $l(\pi_i)$. For every $1 \le i \le n-2$, connect $r(\pi_i)$ to $l(\pi_{i+1})$ with a black edge, and connect $t(\pi_i)$ to $h(\pi_j)$ with a grey edge if $|\pi_i| + 1 = |\pi_j|$, $(1 \le i, j \le n-1)$. Notably, since $l(\pi_1)$ and $r(\pi_{n-1})$ are isolated nodes in G_M, we ignore them hereafter. In the breakpoint graph G_M, each node is connected to precisely one black edge and one grey edge. Consequently, G_M consists of edge-disjoint cycles, where black and grey edges alternate. A cycle is termed an l-cycle if it contains l black (or equivalently, grey) edges. An example of the breakpoint graph is shown in Fig. 1.

Each π_i is associated with an adjacency in $adj(S)$. Observe that the adjacencies $adj(S)_i$ and $adj(S)_{i+1}$ both involve the letter S_{i+1}. Consequently, in G_M, the black edge connecting $r(\pi_i)$ to $l(\pi_{i+1})$ corresponds to S_{i+1} within S. In constructing G_M, a grey edge connects $t(\pi_i)$ to $h(\pi_j)$ provided $|\pi_i| + 1 = |\pi_j|$, where $(1 \leq i, j \leq n-1)$. This implies that the adjacencies of π_i and π_j match with the successive adjacencies $adj(T)_{|\pi_i|}$ and $adj(T)_{|\pi_i|+1}$ in T, which share the letter $T_{|\pi_i|+1}$. Therefore, the grey edge $(t(\pi_i), h(\pi_j))$ corresponds to $T_{|\pi_i|+1}$ in T. As concluded in Lemma 2 of [13], the black and grey edges within a cycle correspond to the identical letter. Consequently, cycles within G_M are limited to either 1-cycles or 2-cycles. A letter appearing once in S corresponds to a 1-cycle in G_M, whereas a letter appearing twice in S corresponds to two 1-cycles or a single 2-cycle in G_M. For each letter x, let $c_M(x)$ be the number of cycles that correspond to letter x in G_M. Obviously, $c_M(x) \in \{1, 2\}$.

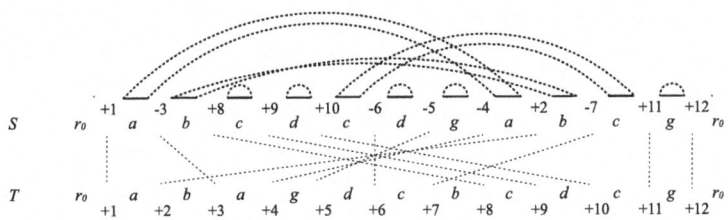

Fig. 1. An example of the breakpoint graph. Thick dashed lines represent grey edges, thin dashed lines represent the matching between $adj(S)$ and $adj(T)$. Signed numbers represent the ids of the adjacencies in permutations π and τ.

A symmetrical reversal alters the order of the adjacencies in S, thereby also modifies the ids' permutation π.

Lemma 1. *Given two strings S and T with a matching M between $adj(S)$ and $adj(T)$. Let $\pi = [\pi_1, \ldots, \pi_{n-1}]$ be the permutation of the ids of adjacencies in $adj(S)$ based on M. Suppose ρ is a symmetrical reversal applied to the segment $[S_i, \ldots, S_j]$, yielding a new string $S' = S \cdot \rho = [S_1, \ldots, S_{i-1}, S_j, \ldots, S_i, S_{j+1}, \ldots, S_n]$. We can establish a matching $M' = M \cdot \rho$ between $adj(S')$ and $adj(T)$ in $O(n)$ time, so that the permutation of adjacency ids in $adj(S')$ according to M' is $\pi' = [\pi_1, \ldots, \pi_{i-1}, -\pi_{j-1}, \ldots, -\pi_i, \pi_j, \ldots, \pi_{n-1}]$.*

Clearly, the breakpoint graph $G_{M \cdot \rho}$ can equally be derived from $\pi \cdot \rho$.

Theorem 1 ([13]). *If all cycles in G_M are 1-cycles, then $S = T$.*

Thus, it is sufficient to convert all cycles in G_M into 1-cycles by a sequence of symmetric reversals. We emphasize that any sequence of symmetric reversals converting S into T indicates a matching. This is due to the fact that these reversals can be performed on T in reverse order, subsequently converting T back to S. As stated in Lemma 1, such a matching is guaranteed to exist.

2.2   The Interleaving Graph

An edge in G_M is *oriented* if it connects either two left nodes or two right nodes; otherwise, it is *unoriented*. The *orientation* of a cycle is *oriented* when it contains at least one oriented edge; otherwise, it remains *unoriented*. A symmetric reversal may alter the cycles in G_M, which is analogous to the findings in [11].

For the input strings S and T, if $S_i = S_j$, we refer to them as a *couple*, which is represented as $\prec S_i, S_j \succ$. Two couples $\prec S_i, S_j \succ$ and $\prec S_k, S_l \succ$ *overlap* with each other if either $i < k < j < l$ or $k < i < l < j$. As each letter appears no more than twice in S, without cause of confusion, the couple of a letter x is also represented as $\prec x, x \succ$. A couple might correspond to either a 2-cycle or two 1-cycles within G_M. Two 2-cycles are termed *interleaving* if the corresponding couples overlap with each other.

We describe the overlapping graph $H_M = (V_b \cup V_w \cup V_g, E_o)$ as follows. In H_M, each couple is represented by a colored vertex. A vertex is colored black/white/green if the corresponding pair represents an oriented 2-cycle/an unoriented 2-cycle/two 1-cycles in G_M. The collection of black, white, and green vertices is denoted by V_b, V_w, and V_g, respectively. An edge is present in E_o between two vertices if and only if their corresponding couples overlap.

By removing the green vertices from the overlap graph H_M, we obtain a graph denoted by $\bar{H}_M$. In fact, in [11], $\bar{H}_M$ is referred to as the 'interleaving graph'. Each connected component found within $\bar{H}_M$ corresponds to a component from the breakpoint graph G_M. To avoid any confusion, we will not differentiate between a component in $\bar{H}_M$ and G_M. A component is *oriented* if it contains at least one oriented 2-cycle (depicted by a black vertex in H_M), and otherwise, it is *unoriented*. As stated in Lemma 13 of [11], every unoriented component in G_M contains at least two unoriented 2-cycles. Observe that symmetric reversals can only be executed on couples (vertex in H_M). As in [11], based on Lemma 1, we derive the following property.

Property 1. *Upon executing the symmetric reversal ρ at a vertex v in H_M, the resulting interleaving graph $H_{M \cdot \rho}$ can be derived from H_M through the following process.*

- *The color of its adjacent black neighbor is altered to white, while the color of its adjacent white neighbor is altered to black.*
- *For any two neighbors of vertex v in H_M, they will be connected in $H_{M \cdot \rho}$ if and only if they are disconnected in H_M.*
- *If v is black in H_M, it changes to green in $H_{M \cdot \rho}$. If v is white in H_M, it remains white in $H_{M \cdot \rho}$. If v is green in H_M, it changes to black in $H_{M \cdot \rho}$.*

Detailed investigations into oriented components can be found in [11]. These findings are encapsulated in the theorem below.

Theorem 2 ([11]). *For every oriented component composed of c 2-cycles in G_M, there exist c safe reversals (A safe reversal always transforms an oriented 2-cycle into two 1-cycles.) that transform all the c 2-cycles in this oriented component into 1-cycles.*

It is important to point out that, based on Property 1, an unoriented component in $\bar{H}_M$ can be converted into an oriented one by executing a symmetric reversal either on a white vertex within the component or on a green adjacent vertex in H_M.

We emphasize that symmetric reversals differ significantly from traditional reversals, especially when concerning unoriented components. As previously mentioned, symmetric reversals are restricted to being performed on couples, whereas traditional reversals can be applied between any two letters. Consequently, the traditional reversal problem offers greater flexibility in managing unoriented components. As outlined in [11], simply addressing the specific unoriented components known as 'hurdles' is enough to resolve the issue. Following this, we will demonstrate how to manage unoriented components using symmetric reversals.

3 Description of a Solution

A vertex in H_M *covers* a component in $\bar{H}_M$ if either it belongs to that component or is connected to a vertex within it in H_M. If a vertex in H_M covers a component in $\bar{H}_M$, it is also stated that its corresponding couple covers the corresponding component in G_M. A *cover-set* K of G_M is a set of couples such that every unoriented component in G_M is covered by at least one couple from K. Within a cover-set of G_M, a couple is termed as a *dominating* couple if it corresponds to a white or black vertex in H_M, and as a *crossing* couple if it corresponds to a green vertex in H_M. The *value* of a cover-set, containing k_1 dominating couples and k_2 crossing couples, is $k_1 + 2k_2$.

Theorem 3. *Given a matching M and a cover-set K of G_M, let k_1 and k_2 be the number of dominating couples and crossing couples in K, respectively. Let the total number of cycles in G_M be c. Then there exists a series of symmetric reversals with size at most $k_1 + 2k_2 + (n-2) - c$ which transforms S into T.*

Based on the constructive proof of Theorem 3, once a matching M and a cover-set K of G_M are identified, it is possible to convert S into T using no more than $k_1 + 2k_2 + (n-2) - c$ symmetric reversals. The detailed steps are outlined as Part-IV of Algorithm 1.

For each matching M between $adj(S)$ and $adj(T)$, $OPT(M)$ represents the shortest sequence of symmetric reversals that transform all cycles into 1 cycles in G_M, and $|OPT(M)|$ represents the number of symmetric reversals.

Theorem 4. *Given a matching M, let c^* be the number of cycles in G_M. There exists a cover-set K^* of G_M consisting of k_1^* dominating couples and k_2^* crossing couples, such that $|OPT(M)| = k_1^* + 2k_2^* + (n-2) - c^*$.*

According to Theorem 4, when a matching is established, the size of the optimal solution based on this matching is determined. Therefore, identifying a suitable matching for the global optimal solution is crucial. Despite the exponential number of possible matchings between $adj(S)$ and $adj(T)$, we successfully identify a matching M such that $OPT(M)$ serves as the global optimal solution.

4   An Approximation Algorithm with Factor 1.15

In this section, we propose an approximation algorithm for the MUSR problem. The algorithm encompasses two primary stages: identify a matching M^a comparable to that of the optimal solution, and determine a cover-set of G_{M^a}.

4.1   Find a Proper Maximum Cycle Matching

Firstly, we introduce some notation. A letter located between identical letters in S (or in T) is called an *S-inflection point* (or *T-inflection point*). Both *S-inflection point* and *T-inflection point* are called the *inflection point*. An adjacency is defined as a *variable adjacency* if it and its opposite adjacency appear totally twice in $adj(S)$ (or $adj(T)$). We denote the substring $[S_l, S_{l+1}, \ldots, S_r]$ of S by $S[l, r]$, where $1 \leq l \leq r \leq n$. A substring $S[l, r]$ is called a *variable interval* if all the adjacencies within it are variable adjacencies. A *unit variable interval* is a variable interval $S[l, r]$ in which none of the letters $[S_{l+1}, S_{l+2}, \ldots, S_{r-1}]$ are inflection points. A unit variable interval is termed a *maximal unit variable interval* (abbreviated as MUVI) if it is not a proper substring of any other unit variable interval. Within a *maximal unit variable interval*, the *fine variable interval* (abbreviated as FVI) denotes the substring that excludes any possible inflection points.

As stated in Lemma 6 of [13], there is no duplicate letter in any unit variable interval. Two strings are considered *identical* if they are exactly the same or if one is the reverse of the other. Lemma 7 in [13] indicates that for every unit variable interval in S, there is also a corresponding identical unit variable interval in S. Furthermore, we have the following lemma.

Lemma 2. *If a maximal unit variable interval contains two inflection points, then these two inflection points must be classified as an S-inflection point and a T-inflection point.*

Consequently, we refer to a maximal unit variable interval with two inflection points as an ST-MUVI. Moreover, the presence of an S-inflection point causes an ST-MUVI and its identical ST-MUVI to consistently share the same S-inflection point within the string S.

For each variable adjacency $\langle x, y \rangle$, suppose that the two identical adjacencies in S, are $\langle S_i, S_{i+1} \rangle$ and $\langle S_j, S_{j+1} \rangle$ and the two identical adjacencies in T are $\langle T_{i'}, T_{i'+1} \rangle$ and $\langle T_{j'}, T_{j'+1} \rangle$. Without loss of generality, assume that in the matching M, $\langle S_i, S_{i+1} \rangle \sim \langle T_{i'}, T_{i'+1} \rangle$ and $\langle S_j, S_{j+1} \rangle \sim \langle T_{j'}, T_{j'+1} \rangle$. Then another matching M' can be derived from M by switching the matches of $\langle x, y \rangle$: $\langle S_i, S_{i+1} \rangle \sim \langle T_{j'}, T_{j'+1} \rangle$, $\langle S_j, S_{j+1} \rangle \sim \langle T_{i'}, T_{i'+1} \rangle$, with all other matches unchanged.

Detailed information on how the number of cycles alters when the matches of a variable adjacencies are switched can be found in [13].

Lemma 3 ([13], Lemma 8, Lemma 9). *Let $\langle x, y \rangle$ be a variable adjacency, M and M' be the matching before and after switching the matches of $\langle x, y \rangle$,* *respectively. If x (resp. y) is not an inflection point, then $c_{M'}(x) - c_M(x) \in \{1, -1\}$ (resp. $c_{M'}(y) - c_M(y) \in \{1, -1\}$). If x (resp. y) is an inflection point, then $c_{M'}(x) - c_M(x) = 0$ (resp. $c_{M'}(y) - c_M(y) = 0$).*

Let M and M' be two matchings before and after switching the matches of a variable adjacency $\langle x, y \rangle$. It is possible that the switch results in different cycles in G_M and $G_{M'}$, and it potentially alters the cover-sets of G_M and $G_{M'}$. If $|OPT(M')| \leq |OPT(M)|$, we say that switching the matches of $\langle x, y \rangle$ is *safe*.

The following Lemma 4 shows how the *id*s of the variable adjacencies changes before and after switching the matches.

Lemma 4. *Given a matching M, let $\langle S_i, S_{i+1} \rangle \sim \langle T_k, T_{k+1} \rangle$ and $\langle S_j, S_{j+1} \rangle \sim \langle T_l, T_{l+1} \rangle$ be two matches in M, where $\langle S_i, S_{i+1} \rangle$ and $\langle S_j, S_{j+1} \rangle$ are identical and their ids are m_1 and m_2, respectively. After switching the two matches, the ids of $\langle S_i, S_{i+1} \rangle$ and $\langle S_j, S_{j+1} \rangle$ are changed to m_2 and m_1 if they are consistent, and $-m_2$ and $-m_1$ if they are opposite.*

The following Lemma 5, 6, 7 show types of the cycles that a letter corresponds to.

Lemma 5. *Given a matching M, let x be a letter but not an inflection point, and y be another letter. If there are two consistent adjacencies, both of which are $\langle x, y \rangle$ or $\langle y, x \rangle$, then x corresponds to two 1-cycles or an unoriented 2-cycle in G_M. If there are two opposite adjacencies, one of which is $\langle x, y \rangle$ and the other $\langle y, x \rangle$, then x corresponds to two 1-cycles or an oriented 2-cycle in G_M.*

Lemma 6. *Given a matching M. Let x be an inflection point, then x corresponds to two 1-cycles in G_M if and only if x is both an S-inflection point and a T-inflection point in S.*

Lemma 7. *Given a matching M, let x be a T-inflection point, and y is another letter. If there exist two consistent adjacencies in S, both of which are $\langle x, y \rangle$ or $\langle y, x \rangle$, then x corresponds to two 1-cycles or an oriented 2-cycle in G_M. If there are two opposite adjacencies, one of which is $\langle x, y \rangle$ and the other $\langle y, x \rangle$, then x corresponds to two 1-cycles or an unoriented 2-cycle in G_M.*

The following Lemma 8 shows how the types of the cycles alter before and after switching the matches of a variable adjacency with an inflection point involved.

Lemma 8. *Let x be an inflection point, and let y be the letter such that the substring $[y, x, y]$ appears in S or T. Let M and M' be the matching before and after switching the matches of two adjacencies in S that are identical to $\langle x, y \rangle$.*

(1) if x is both a T-inflection point and an S-inflection point, then x corresponds to two 1-cycles in both G_M and $G_{M'}$;

(2) if x is a T-inflection point but not an S-inflection point, then x corresponds to a 2-cycle in both G_M and $G_{M'}$, and they have the same orientation;

(3) if x is an S-inflection point but not a T-inflection point, then x corresponds to a 2-cycle in both G_M and $G_{M'}$, and they have different orientations.

Lemma 9 and 10 show the conditions of a safe switching.

Lemma 9. *Let M be a matching, $\langle x, y \rangle$ be a variable adjacency. If x and y are not inflection points and $c_M(x) + c_M(y) = 3$ or $c_M(x) + c_M(y) = 2$ in G_M, then switching the matches of $\langle x, y \rangle$ is safe.*

Lemma 10. *Let M be a matching, x be an inflection point, y be the letter such that the substring $[y, x, y]$ appears in S or T. If y is not an inflection point and $c_M(y) = 1$. Then switching the matches of $\langle x, y \rangle$ is safe.*

A matching M is called a *maximum cycle matching* if the number of cycles in G_M is not less than the number of cycles in the breakpoint graph of any other matching. A *maximum cycle matching* M is *proper* if the S-inflection points within the ST-MUVIs consistently corresponds to oriented 2-cycles in G_M.

Lemma 11 ([13], Lemma 10). *Given a matching M, for each variable interval I, $c_M^2(I)$ represents the total number of 2-cycles in G_M corresponding to the letters in I. M is a maximum cycle matching if and only if it satisfies: (1) for each maximal unit variable interval I without an inflection point, $c_M^2(I) \leq 1$; (2) if a maximal unit variable interval U contains at least one inflection point, then $c_M^2(I) = 0$, where I is the fine variable interval of U.*

According to [13], it is feasible to obtain a maximum cycle matching in $O(n^2)$ time, the algorithm is shown as 'Part-I' of Algorithm 1. We develop an algorithm to derive a proper maximum cycle matching by switching certain matches in a maximum cycle matching, which is shown as 'Part-II' of Algorithm 1.

Theorem 5. *Let M^* be the matching of the optimal solution of MUSR, and M^a be the matching obtained by Algorithm 1, then $|OPT(M^a)| = |OPT(M^*)|$.*

4.2 Find a Cover-Set via Solving the (1,2)-Set Cover Problem

According to Theorem 5, the optimal solution also depends on the output matching M^a of Algorithm 1. We then concentrate on identifying a cover-set of G_{M^a} in this subsection.

Let P be the set of unoriented components in G_{M^a}. For each letter x, denote P_x as the set of unoriented components covered by the couple of x, with P_x being a subset of P. Let Q represent the entire collection of all such P_x. Consequently, the pair (P, Q) forms an instance of the *Set Cover* problem. Moreover, solving this particular set cover instance yields a cover-set for G_{M^a}. According to Theorem 3, a dominating couple contributes a value of one to the solution, whereas a crossing couple contributes a value of two. Consequently, if x corresponds to two 1-cycles in G_{M^a}, P_x is given a weight of two; and if x corresponds to a 2-cycle in G_{M^a}, P_x is given a weight of one. As a result, if this (1,2)-set cover instance has a solution of size p^a, then the MUSR instance will have a solution size of $n - 2 - c + p^a$, where c represents the number of cycles in G_{M^a}. In Algorithm 1, 'Part-III' outlines the process to return a set cover for the (1–2)-Set Cover instance.

Lemma 12. *Let m be the size of the base set P. Let $\mathcal{Q}_1^a$ and $\mathcal{Q}_2^a$ be the collection of the subsets of weight 1 and 2 in $\mathcal{Q}^a$, respectively. We use q_1 (resp. q_2) to denote the size of $\mathcal{Q}_1^a$ (resp. $\mathcal{Q}_2^a$). Similarly, let $\mathcal{Q}_1^*$ and $\mathcal{Q}_2^*$ be the collection of the subsets of weight 1 and 2 in the optimal solution of the (1,2)-Set Cover instance $\mathcal{Q}^*$, respectively. Using q_1^* (resp. q_2^*) to denote the size of $\mathcal{Q}_1^*$ (resp. $\mathcal{Q}_2^*$), then, $q_1 + 2q_2 \leq 2q_2^*(\ln m - \ln q_2^*) + q_2^* + q_1^*$.*

Algorithm 1. Approximation Algorithm for MUSR

Input: Two related strings S and T such that each letter appears at most twice in S and T, and $adj(S) = adj(T)$.

Output: A series of symmetric reversals that transform S to T.

Part-I: Obtain a maximum cycle matching M_1

1: Find an arbitrary matching M, construct the breakpoint graph G_M.

2: **for each** MUVI U from left to right in S, scan U in one of the following ways, and ignore the identical MUVI of U.

    (1) if U contains no inflection point, scan U from left to right;

    (2) if U contains one inflection point, scan U from the non-inflection point to the inflection point;

    (3) if U is a ST-MUVI, scan U from the letter succeeding to the T-inflection point to the S-inflection point. **do**

3:    **for each** adjacency $\langle x, y \rangle$ **do**

4:        **if** $c_M(x) = 1$, **then** switch the matches of $\langle x, y \rangle$ and update M and G_M.

5: Let the current matching be M_1.

Part-II: Obtain a proper maximum cycle matching M^a

6: **for each** ST-MUVI U from left to right in S, scan U from the S-inflection point to the T-inflection point, and ignore the identical MUVI of U. **do**

7:    **for each** adjacency $\langle x, y \rangle$ **do**

8:        Switch the matches of $\langle x, y \rangle$ and update M and G_M.

9: Let the current matching be M^a.

Part-III: Obtain a cover-set K

10: Construct a (1-2)-Set Cover instance $(P, \mathcal{Q})$, where P represents the set of unoriented components in G_{M^a}. For each letter x, denote P_x as the set of unoriented components covered by the couple of x, the weight of P_x is 1 if x corresponds to a 2-cycle in G_{M^a}; otherwise, it is 2. $\mathcal{Q}$ represents the entire collection of all these P_x. Initiate $\mathcal{Q}^a \leftarrow \emptyset$, $Uncov \leftarrow P$, $K \leftarrow \emptyset$.

11: **while** $Uncov \neq \emptyset$ **do**

12:    Let P_y be the subset in $\mathcal{P}$ that contains the maximum number of elements in $Uncov$.

13:    **if** $|P_y \cap Uncov| \geq 2$ **then**

14:        $\mathcal{Q}^a \leftarrow \mathcal{Q}^a \cup \{P_y\}$, $Uncov \leftarrow Uncov - P_y$, $K \leftarrow K \cup \prec y, y \succ$.

15:    **else**

16:        Let P_z be a weight-1 subset and $|P_z \cap Uncov| = 1$, $\mathcal{Q}^a \leftarrow \mathcal{Q}^a \cup \{P_z\}$, $Uncov \leftarrow Uncov - P_z$, $K \leftarrow K \cup \prec z, z \succ$.

Part-IV: Obtain a series of symmetric reversals via M^a and K

17: For each cross couple $\prec S_i, S_j \succ$ in K, perform $\rho(i, j)$ and update G_{M^a}.
18: For each dominating couple $\prec S_{i'}, S_{j'} \succ$ in K (which consistently covers an unoriented component in G_{M^a}), perform $\rho(i', j')$ and update G_{M^a}.
19: Transform all 2-cycles into 1-cycles in G_{M^a} according to Lemma 2.

In conclusion, the findings of this section are summarized in the following theorem.

Theorem 6. *Algorithm 1 guarantees an approximation factor of $\frac{2\ln 3 + 7}{8}(\approx 1.15)$ and runs in $O(n^2)$ time.*

Proof. Let ALG be the set of symmetric reversals performed during the execution of Algorithm 1. Let c be the number of cycles and m be the number of the unoriented components in G_M. From Theorem 5, the optimal solution of the MUSR problem $OPT*$ is equal to the optimal solution $OPT(M)$ based on the matching M, i.e., $|OPT(M)| = |OPT*|$. Let q_1 (resp. q_2) be the number of dominate couples (resp. crossing couples) in the cover-set K. According to Theorem 3, $|ALG| = n - 2 - c + 2q_2 + q_1$. Let $K*$ be the cover-set of G_M that we argued in Theorem 4. Let q_1^* (resp. q_2^*) be the number of dominate couples (resp. crossing couples) in the cover-set K^*. According to Theorem 3, $|OPT(M)| = n - 2 - c + 2q_2^* + q_1^*$.

$$
\begin{aligned}
\frac{|ALG|}{|OPT*|} &= \frac{(n-2) - c + 2q_2 + q_1}{(n-2) - c + 2q_2^* + q_1^*} \le \frac{(n-2) - c + 2q_2^*(\ln m - \ln q_2^*) + q_2^* + q_1^*}{(n-2) - c + 2q_2^* + q_1^*} \\
&\le \frac{(n-2) - c + 2q_2^*(\ln m - \ln q_2^*) + q_2^*}{(n-2) - c + 2q_2^*} \le \frac{2m + 2q_2^*(\ln m - \ln q_2^*)}{2m + 2q_2^*} \\
&= 1 + \frac{2q_2^*\ln(\frac{m}{q_2^*}) - q_2^*}{2m + 2q_2^*} = 1 + \frac{2\ln(\frac{m}{q_2^*}) - 1}{2(\frac{m}{q_2^*}) + 2} \le 1 + \frac{2\ln 3 - 1}{8} \quad (< 1.15)
\end{aligned}
$$

The first "$\le$" holds due to Lemma 12. The third "$\le$" is based on the fact that $n - 2 - c \ge 2m$. (Let c_1 and c_2 be the number of 1-cycles and 2-cycles in G_M, then $c_1 + 2c_2 = n - 2$ and $c = c_1 + c_2$, the number of 2-cycles is $c_2 = n - 2 - c$. As stated in Lemma 13 of [11], each unoriented component contains at least two 2-cycles. Hence, we have $m \le \frac{c_2}{2} = \frac{n-2-c}{2}$). The last "$\le$" is true because if the cover-set of $OPT(M)$ contains a crossing couple that covers one or two unoriented components in G_M, we can replace it with one or two dominating couples without increasing the solution value, thus we can assume that each crossing couple in the cover-set of $OPT(M)$ covers at least three unoriented components. □

5   The Complexity of the MUSR Problem

We conduct a reduction from the *MAX-(3,B2)-SAT* problem (which is shown to be NP-hard in [5]) to the MUSR problem. In any instance of *MAX-(3,B2)-SAT*, each clause contains exactly 3 literals and each variable occurs exactly twice in its positive form and twice in its negative form.

Theorem 7. *The MUSR problem is NP-hard.*

6   Concluding Remarks

In this paper, we conduct systematic research on the problem of Sorting Unsigned Strings by Symmetric Reversals. We show the problem to be NP-hard and improve the approximation ratio from 1.5 to 1.15. It is worth to point out that we describe any solution by a matching and a so-called cover-set. Surprisingly, the matching of the optimal solution can be derived. Thus, a better cover-set implies a better approximation for the MUSR problem. Another interesting point is when a letter may occur more than twice in the input strings. Although the problem remains NP-hard, we have yet to discover any algorithm for it.

Acknowledgements. This research is supported by the National Natural Science Foundation of China (No. 62272279, 62272272, U24A20257), and NSF under project CNS-2243010.

References

1. Armengol, L., Pujana, M.A., Cheung, J., Scherer, S.W., Estivill, X.: Enrichment of segmental duplications in regions of breaks of synteny between the human and mouse genomes suggest their involvement in evolutionary rearrangements. Hum. Mol. Genet. **12**(17), 2201–2208 (2003)
2. Bailey, J.A., Baertsch, R., James Kent, W., Haussler, D., Eichler, E.E.: Hotspots of mammalian chromosomal evolution. Genome Biol. **5**, 1–7 (2004)
3. Bergeron, A.: A very elementary presentation of the Hannenhalli–Pevzner theory. Discrete Appl. Math. **146**(2), 134–145 (2005). 12th Annual Symposium on Combinatorial Pattern Matching
4. Berman, P., Hannenhalli, S., Karpinski, M.: 1.375-approximation algorithm for sorting by reversals. In: Möhring, R., Raman, R. (eds.) ESA 2002. LNCS, vol. 2461, pp. 200–210. Springer, Heidelberg (2002). https://doi.org/10.1007/3-540-45749-6_21
5. Berman, P., Karpinski, M., Scott, A.: Approximation hardness of short symmetric instances of MAX-3SAT. Technical report (2004)
6. Caprara, A.: Sorting permutations by reversals and Eulerian cycle decompositions. Siam J. Discrete Math. (1999)
7. Christie, D.A.: A 3/2-approximation algorithm for sorting by reversals. In: Proceedings of the Ninth Annual ACM-SIAM Symposium on Discrete Algorithms, 25–27 January 1998, San Francisco, California, USA, pp. 244–252. ACM/SIAM (1998)
8. Dudek, B., Gawrychowski, P., Starikovskaya, T.: Sorting signed permutations by reversals in nearly-linear time. In: 2024 Symposium on Simplicity in Algorithms (SOSA), pp. 199–214. SIAM (2024)
9. Fertin, G., Labarre, A., Rusu, I., Tannier, E., Vialette, S.: Combinatorics of genome rearrangements. In: Computational Molecular Biology. MIT Press (2009)
10. Han, Y.: Improving the efficiency of sorting by reversals. In: Proceedings 2006 International Conference on Bioinformatics & Computational Biology (BIOCOMP 2006), pp. 406–409 (2006)

11. Hannenhalli, S., Pevzner, P.A.: Transforming cabbage into turnip: polynomial algorithm for sorting signed permutations by reversals. J. ACM (JACM) **46**(1), 1–27 (1999)
12. Kececioglu, J., Sankoff, D.: Exact and approximation algorithms for sorting by reversals, with application to genome rearrangement. Algorithmica **13**(1), 180–210 (1995)
13. Lai, W., Jiang, H., Zhu, D., Zhu, B.: On sorting by unsigned symmetric reversals. In: Proceedings of the International Computing and Combinatorics Conference, pp. 421–432. Springer (2024)
14. Li, T., Jiang, H., Zhu, B., Wang, L., Zhu, D.: Flanked block-interchange distance on strings. IEEE/ACM Trans. Comput. Biol. Bioinf. (2), 21 (2024)
15. Longo, M.S., Carone, D.M.: NISC comparative sequencing program eGreen@ NHGRI. nih.gov. In: Green, E.D., O'Neill, M.J., O'Neill, R.J. (eds.) Distinct Retroelement Classes Define Evolutionary Breakpoints Demarcating Sites of Evolutionary Novelty. BMC Genomics **10**, 1–14 (2009)
16. Pevzner, P.A., Waterman, M.S.: Open combinatorial problems in computational molecular biology. In: Proceedings Third Israel Symposium on the Theory of Computing and Systems, pp. 158–173. IEEE (1995)
17. Sankoff, D.: The where and wherefore of evolutionary breakpoints. J. Biol. **8**, 1–3 (2009)
18. Sankoff, D., Leduc, G., Antoine, N., Paquin, B., Franz Lang, B., Cedergren, R.: Gene order comparisons for phylogenetic inference: evolution of the mitochondrial genome. Proc. Nat. Acad. Sci. **89**(14), 6575–6579 (1992)
19. Simonaitis, P., Swenson, K.M.: Finding local genome rearrangements. Algorithms Mol. Biol. **13**(1), 9 (2018)
20. Small, K., Iber, J., Warren, S.T.: Emerin deletion reveals a common X-chromosome inversion mediated by inverted repeats. Nat. Genet. **16**(1), 96–99 (1997)
21. Sun, C., Jiang, H., Wang, L., Zhu, D.: Can the 1.375 approximation ratio of unsigned genomes distances be improved? In: Proceedings of the International Computing and Combinatorics Conference, pp. 3–15. Springer (2024)
22. Swenson, K.M., Simonaitis, P., Blanchette, M.: Models and algorithms for genome rearrangement with positional constraints. Algorithms Mol. Biol. **11**(1), 13 (2016)
23. Tannier, E., Bergeron, A., Sagot, M.-F.: Advances on sorting by reversals. Discret. Appl. Math. **155**(6–7), 881–888 (2007)
24. Thomas, A., Varré, J.-S., Ouangraoua, A.: Genome dedoubling by DCJ and reversal. BMC Bioinf. **12**, 1–9 (2011)
25. Tong, X., et al.: Cabbage can't always be transformed into turnip: decision algorithms for sorting by symmetric reversals. In: International Computing and Combinatorics Conference, pp. 279–294. Springer (2023)
26. Wang, D., Li, S., Guo, F., Ning, K., Wang, L.: Core-genome scaffold comparison reveals the prevalence that inversion events are associated with pairs of inverted repeats. BMC Genomics **18**, 1–13 (2017)
27. Wang, D., Wang, L.: GRSR: a tool for deriving genome rearrangement scenarios from multiple unichromosomal genome sequences. BMC Bioinf. **19**, 11–19 (2018)
28. Wenger, A.M., Peluso, P., Rowell, W.J., et al.: Accurate circular consensus long-read sequencing improves variant detection and assembly of a human genome. Nat. Biotechnol. **37**(10), 1155–1162 (2019)

Longest Double-Bounded (k]-Tuple Common Substrings

Tiantian Li[1], Siqi Jiang[2], Haitao Jiang[2], and Daming Zhu[2(✉)]

[1] School of Software and Joint SDU-NTU Centre for Artificial Intelligence Research
(C-FAIR), Shandong University, Jinan, Shandong, China
[2] School of Computer Science and Technology, Shandong University, Qingdao,
Shandong, China
`201914666@mail.sdu.edu.cn`, `{htjiang,dmzhu}@sdu.edu.cn`

Abstract. A *(k]-tuple common substring* (abbr. (k]-CSS) is a sequence
of at most k common substrings of two or more strings. A longest (k]-
CSS of two strings is known retrievable in quadratic time and linear
space and even more, in subquadratic time and space if k is a constant.
Motivated by computational biology applications in need of a (k]-CSS
with designated number of consecutively matching letters, we propose
to find a longest (k]-CSS of two strings whose substrings are of length
within $[l_1, l_2]$.

We present a sliding window based dynamic programming algorithm
to find such a longest (k]-CSS of two strings whose lengths are n_1 and n_2
in $O(kn_1n_2)$ time and space, the same complexity as without the length
bounds l_1 and l_2. Through rolling array based dynamic programming
to get the longest (k]-CSS length in advance, we present a divide-and-
conquer algorithm to find such a longest (k]-CSS in $O(kn_1n_2)$ time and
$O(n_1 + kl_2n_2)$ space, which is intended to work for two much longer given
strings. We also present an algorithm to find such a longest (2]-CSS in
$O(n \log^2 n)$ time where n is the total length of input strings.

Keywords: Algorithm · Complexity · Common substring

1 Introduction

Longest Common Substring and Longest Common Subsequence (abbr. LCS) are
classical and well-studied string matching problems [7]. For two strings, Longest
Common Substring is known solvable in linear time whereas LCS in quadratic
time and space and even better, in quadratic time and linear space [8,14]. On the
other hand, an LCS of two strings of total length n cannot be found in $O(n^{2-\epsilon})$
time unless the Strong Exponential Time Hypothesis is not true [1] although it
is retrievable in $O(\frac{n^2}{\log n})$ time for finite alphabet [11].

Common substrings carry fundamental features of *conserved regions* in biological sequences [6]. To characterize such conserved regions as cDNA whose consecutive regions are in conserved order and of limited constituent, the (k]-tuple

© The Author(s), under exclusive license to Springer Nature Singapore Pte Ltd. 2026
F. V. Fomin and M. Xiao (Eds.): COCOON 2025, LNCS 15984, pp. 355–366, 2026.
https://doi.org/10.1007/978-981-95-0218-9_26

common substring (abbr. (k)-CSS), a sequence of at most k common substrings of multiple given strings, was proposed [10]. Despite that finding a longest (k)-CSS of m strings is NP-Hard [12], a longest (k)-CSS of m strings of total length n can be found in $O(mn^k)$ time and $O(kmn)$ space [10]. For arbitrary k, a longest (k)-CSS of two strings of lengths n_1 and n_2 can be found in $O(kn_1n_2)$ time and $O(n_1 + kn_2)$ space [9]. If k is a constant, a longest (k)-CSS of two strings can be found by an improved dynamic programming algorithm in $n_1 n_2^{1-\Theta(1)}$ time and space [3]. If the length l of the longest (k)-CSS of two strings is sufficiently long, a longest (k)-CSS can be found in $O(kn_2)$ time [15].

Sufficiently long consecutive string matches serve as conserved regions with higher reliability, especially in applications such as sequence alignment based structural variation prediction and pan-genome analysis [13]. Longest common subsequence with substring length constraints requiring exact or at most length t was proposed by Benson et al., who designed polynomial time algorithms for finding such subsequences between two strings [4,5]. These constrained LCS variants can be solved in $O(n^2)$ time and $O(tn)$ space using improved algorithms, where n denotes the total length of input strings [16]. In order to find common substrings carrying conserved regions in which consecutive matches are sufficiently many and moderately dispersed, we propose *double-bounded* (k)-CSS, a (k)-CSS variant whose substrings are of length within $[l_1, l_2]$ (in Sect. 2).

We present a sliding window based dynamic programming algorithm to find a longest double-bounded (k)-CSS of two strings whose lengths are n_1 and n_2 in $O(kn_1n_2)$ time and space, the same complexity as without the length bounds l_1 and l_2 (in Sect. 3.1 and 3.2). The space complexity $O(kn_1n_2)$ proves prohibitive in practice, as the algorithm frequently fails to return solutions on 16 GB memory systems when processing inputs with $n_1 + n_2 = 10,000$ and $k = 20$. To process longer input strings, we present a dynamic programming plus divide-and-conquer algorithm for finding a longest double-bounded (k)-CSS of two strings in $O(kn_1n_2)$ time and $O(n_1 + kl_2n_2)$ space (in Sect. 4). We present an algorithm to find such a longest (2)-CSS in $O(n\log^2 n)$ time at last where n is the total length of input strings (in Sect. 5).

2   Preliminaries

Let Σ denote a finite alphabet. A symbol string is referred to as *on* Σ if all symbols in the string belong to Σ. The number of symbols in a string s is referred to as the *length* of s and will be denoted by $|s|$. For a string $s = s[1]$ $... s[n]$ on Σ, we denote by $s[i, j]$ the consecutive substring $s[i] ... s[j]$ of s and $\overline{s[i,j]} = s[j] ... s[i]$ the inversion of $s[i,j]$ with $1 \le i \le j \le n$. Two strings $s_1 = s_1[1, k]$ and $s_2 = s_2[1, k]$ are referred to as *identical* if $s_1[i] = s_2[i]$ for $i \in [1,k]$. We denote it as $s_1 = s_2$ for two strings s_1 and s_2 to be identical.

Let $s = s[1, n]$ and $\lambda = \lambda[1, l]$ be two strings on Σ with $l \le n$. The substring $s[i, i+l-1]$ for $i \in [1, n-l+1]$ is referred to as an *occurrence* of λ if $\lambda = s[i, i+l-1]$. A *begin-point* (resp. *endpoint*) of λ in s is i (resp. $i+l-1$), if $s[i, i+l-1]$ is an occurrence of λ in s. Let $\lambda_1, ..., \lambda_l$ be a sequence of non-empty strings on

Σ. We refer to $s[\alpha[1], \alpha[1] + |\lambda_1| - 1]$, ..., $s[\alpha[l], \alpha[l] + |\lambda_l| - 1]$, a sequence of substring in s, as an *occurrence* of λ_1, ..., λ_l in s, if $\lambda_t = s[\alpha[t], \alpha[t] + |\lambda_t| - 1]$ for $t \in [1, l]$ in s and $\alpha[t] + |\lambda_t| \leq \alpha[t + 1]$ for $t \in [1, l - 1]$.

Let $S = \{s_1, s_2\}$ be a set of two strings on Σ. We refer to λ_1, ..., λ_l as an *l-tuple common substring* of S (abbr. *l*-CSS) and shall denote it as $\lambda_1 \oplus \lambda_2 \oplus ... \oplus \lambda_l$, if there are occurrences of λ_1, ..., λ_l in both s_1 and s_2. We refer to λ_t for $t \in [1, l]$ as the *t*-th *member* in *l*-CSS $\lambda_1 \oplus \lambda_2 \oplus ... \oplus \lambda_l$. *The length* of an *l*-CSS refers to the length sum of all its members. A 1-CSS of S is a *common substring* of S. For an arbitrary *l*-CSS $\Lambda = \lambda_1 \oplus ... \oplus \lambda_l$ of S, let $\overline{\Lambda} = \overline{\lambda_l} \oplus ... \oplus \overline{\lambda_1}$, an *l*-CSS of $\{\overline{s_1}, \overline{s_2}\}$.

Let $\Lambda = \lambda_1 \oplus \lambda_2 \oplus ... \oplus \lambda_l$ denote an *l*-CSS of S. Then λ_1 and λ_l are the *first* and the *last* members in Λ. We shall say λ_t for $t \in [1, l]$ is *with a begin-point* at $\alpha[t]$ in s_1 (resp. s_2), if $s_1[\alpha[t], \alpha[t] + |\lambda_t| - 1] = \lambda_t$ is the *t*-th member in an occurrence of Λ in s_1 (resp. s_2). If λ_t for $t \in [1, l]$ is with a begin-point at $\alpha[t]$ in s_1 (resp. s_2), then it is *with an endpoint at* $\alpha[t] + |\lambda_t| - 1$. Let Λ_1 and Λ_2 denote an l_1-CSS and an l_2-CSS of S. If the last member of Λ_1 is with an endpoint at less than a begin-point of the first member of Λ_2 in both s_1 and s_2, we shall denote by $\Lambda_1 \oplus \Lambda_2$ the $(l_1 + l_2)$-CSS of S combined from Λ_1 and Λ_2.

An *l*-CSS of S will be referred to as a $(k]$-CSS of S if $l \leq k$. A $(k]$-CSS of S is *longest*, if its length is no less than any $(k]$-CSS of S. In order to find structural common substrings carrying conserved regions in which consecutive matches are sufficiently many and moderately dispersed, members in a $(k]$-CSS should be imposed length limit from two sides. Then we encounter the following problem.

Longest double-bounded $(k]$-CSS (abbr. $LCSS_k(l_1, l_2)$):

Input: A set $S = \{s_1, s_2\}$ of two strings on Σ and three integers k, l_1 and l_2 with $1 \leq l_1 \leq l_2$.

Question: Find a longest $(k]$-CSS of S whose members are of lengths within $[l_1, l_2]$ and an occurrence of it in both s_1 and s_2.

A $(t]$-CSS for $t \in [1, k]$, if without special notation, will carry members of length within $[l_1, l_2]$.

3 An $O(kn_1n_2)$ Time Algorithm for $LCSS_k(l_1, l_2)$

This section is devoted to presenting a sliding window based dynamic algorithm for $LCSS_k(l_1, l_2)$. Assume a set $S = \{s_1[1, n_1], s_2[1, n_2]\}$ of two strings on Σ and integers k, l_1 and l_2 constitute an instance of $LCSS_k(l_1, l_2)$. A $(k]$-CSS of $\{s_1[1, n_1], s_2[1, n_2]\}$ is referred to as *regular* if its last member is with endpoints at n_1 in s_1 and n_2 in s_2.

3.1 Dynamic Programming

Let $W(i, j, t, 1)$ and $W(i, j, t, 0)$ denote the lengths of longest regular and irregular $(t]$-CSS of $\{s_1[1, i], s_2[1, j]\}$ respectively and $D(i, j, t)$ the length of a longest $(t]$-CSS of $\{s_1[1, i], s_2[1, j]\}$. Then $D(i, j, t) = \max\{W(i, j, t, 1), W(i, j, t, 0)\}$ because a $(t]$-CSS is either regular or irregular. Therefore, it suffices to obtain both lengths of a longest regular and irregular (t)-CSS of $\{s_1[1, i], s_2[1, j]\}$ for computing $D(i, j, t)$. Let us focus on the recurrence of $W(i, j, t, 0)$ and $W(i, j, t, 1)$.

Lemma 1. *For* $i \in [0, n_1]$, $j \in [0, n_2]$ *and* $t \in [0, k]$,

$$
W(i, j, t, 0) = \begin{cases} 0 & \text{if } i < l_1 \, or \, j < l_1 \, or \, t = 0; \\ \max \begin{cases} W(i-1, j, t, 0) \\ W(i-1, j, t, 1) \\ W(i, j-1, t, 0) \\ W(i, j-1, t, 1) \end{cases} & \text{otherwise.} \end{cases} \tag{1}
$$

Proof. If $i < l_1$ or $j < l_1$ or $t = 0$, then $W(i, j, t, 0) = 0$ because there is no (t)-CSS whose members are of lengths within $[l_1, l_2]$. Otherwise, $W(i, j, t, 0) = \max\{W(i-1, j, t, 1), W(i, j-1, t, 1), W(i-1, j, t, 0), W(i, j-1, t, 0)\}$ because an irregular (t)-CSS of $\{s_1[1, i], s_2[1, j]\}$ is a (t)-CSS of $\{s_1[1, i-1], s_2[1, j]\}$ or a (t)-CSS of $\{s_1[1, i], s_2[1, j-1]\}$ and vice versa. $\square$

To decide if there is a regular (t)-CSS of $\{s_1[1, i], s_2[1, j]\}$, it needs to check the longest common suffix of $s_1[1, i]$ and $s_2[1, j]$ for if it is no shorter than l_1. Let $G(i, j)$ denote the longest common suffix length of $s_1[1, i]$ and $s_2[1, j]$. Then $G(i, j) > 0$ if and only if $s_1[i] = s_2[j]$. For this reason, $G(i, j) = 0$ for $s_1[i] \neq s_2[j]$. Since $G(i, j) = \max\{r \mid s_1[i-r+1, i] = s_2[j-r+1, j]\}$ for $s_1[i] = s_2[j]$, $G(i, j)$ is computed recursively by

$$
G(i, j) = \begin{cases} G(i-1, j-1) + 1 & \text{if } s_1[i] = s_2[j]; \\ 0 & \text{otherwise.} \end{cases} \tag{2}
$$

If $G(i, j) < l_1$ or $t = 0$, then there is no regular (t)-CSS of $\{s_1[1, i], s_2[1, j]\}$, let us assign $W(i, j, t, 1) = 0$. Let $l(i, j) = \min\{G(i, j), l_2\}$.

Lemma 2. *For* $i \in [0, n_1]$, $j \in [0, n_2]$ *and* $t \in [0, k]$,

$$
W(i, j, t, 1) = \begin{cases} 0 & \text{if } G(i, j) < l_1 \, or \, t = 0; \\ \max_{l_1 \leq x \leq l(i,j)} D(i-x, j-x, t-1) + x & \text{otherwise.} \end{cases} \tag{3}
$$

Proof. Assume Λ is a longest regular (t)-CSS of $\{s_1[1, i], s_2[1, j]\}$ whose last member is of length x. Then $x \in [l_1, l(i, j)]$. Members other than the last in Λ constitute a longest ($t-1$)-CSS of $\{s_1[1, i-x], s_2[1, j-x]\}$. Otherwise, there is a longer regular (t)-CSS of $\{s_1[1, i], s_2[1, j]\}$ than Λ. It follows that $W(i, j, t, 1) = \max\{W(i-x, j-x, t-1, 0) + x, W(i-x, j-x, t-1, 1) + x \mid x \in [l_1, l(i, j)]\} = \max\{D(i-x, j-x, t-1) + x \mid x \in [l_1, l(i, j)]\}$. $\square$

A dynamic programming algorithm arises from Formulae (1), (2) and (3) (See Algorithm 1). Through the interdependent recurrence of $W(i, j, t, 1)$ and $W(i, j, t, 0)$ for $i \in [0, n_1]$, $j \in [0, n_2]$ and $t \in [0, k]$, $W(n_1, n_2, k, 1)$ and $W(n_1,$

n_2, k, 0) will be obtained. Since $D(i, j, t) = \max\{W(i, j, t, 1), W(i, j, t, 0)\}$, the endpoints of members in a longest (k)-CSS occurrence of S in s_1 and s_2 are obtainable along a traceback path from $W(n_1, n_2, k, 1)$ or $W(n_1, n_2, k, 0)$ to some $W(i, j, t, q)$ with $i \cdot j \cdot t = 0$. As long as the endpoints of members in a longest (k)-CSS occurrence in s_1 and s_2 are available, all members in a longest (k)-CSS can be identified in $O(n_1 + n_2)$ time.

Algorithm 1. $DP(S, k, l_1, l_2)$

Require: $S = \{s_1[1, n_1], s_2[1, n_2]\}$, integers k, l_1, l_2.
Ensure: The occurrences of a longest (k)-CSS of S in s_1 and s_2 whose members are of length within $[l_1, l_2]$.
 1: $W(i, j, t, 1) \leftarrow 0$, $W(i, j, t, 0) \leftarrow 0$, $G(i, j) \leftarrow 0$ for $i \cdot j \cdot t = 0$;
 2: **for** i from 1 to n_1 **do**
 3:     **for** j from 1 to n_2 **do**
 4:         Get $G(i, j)$ by Formula (2);
 5:         **for** t from 1 to k **do**
 6:             Get $W(i, j, t, 0)$ and $W(i, j, t, 1)$ by Formulas (1) and (3);
 7:         **end for**
 8:     **end for**
 9: **end for**
10: **return** a longest (k)-CSS of S by a traceback path.

It takes $O(kn_1n_2)$ cells to store $W(i, j, t, 0)$ and $W(i, j, t, 1)$ for all $i \in [0, n_1]$, $j \in [0, n_2]$ and $t \in [0, k]$. Prior to computing $W(i, j, t, 0)$ and $W(i, j, t, 1)$ for certain i, j and $t \in [1, k]$, $G(i, j)$ should be computed using Formula (2), which takes $O(1)$ time. In the execution of Algorithm 1, it takes $O(1)$ time to compute $W(i, j, t, 0)$ using Formula (1) and $O(l_2 - l_1)$ time to compute $W(i, j, t, 1)$ using Formula (3) for certain i, j and t. It follows that the time and space complexity of Algorithm 1 are $O(k(l_2 - l_1)n_1n_2)$ and $O(kn_1n_2)$.

3.2 $O(kn_1n_2)$ Time and Space

To improve the dynamic programming algorithm for $LCSS_k(l_1, l_2)$ to achieve $O(kn_1n_2)$ time complexity, we focus on achieving the following key objective: *For a maximal common substring* $s_1[p_1, p_2] = s_2[q_1, q_2]$ *with boundary conditions* $s_1[p_1 - 1] \neq s_2[q_1 - 1]$ *and* $s_1[p_2 + 1] \neq s_2[q_2 + 1]$, *compute all* $W(i, j, t, 1)$ *for* $i \in [p_1, p_2]$ *and* $j = q_1 + i - p_1$ *in* $O(p_2 - p_1)$ *time.* Unless otherwise specified, indices $i \in [p_1, p_2]$ and $j = q_1 + i - p_1$ are maintained throughout this subsection, where i, j correspond to the indices of symbols in s_1 and s_2, respectively.

If $G(i, j) < l_1$ for $i \in [p_1, p_2]$ and $j = q_1 + i - p_1$, then $W(i, j, t, 1) = 0$ by Formula (3). Assume $G(i, j) \geq l_1$ below in this subsection. To compute $W(i, j, t, 1)$, let us focus on the last member of a regular (t)-CSS of $\{s_1[1, i], s_2[1, j]\}$, which is with begin-points at $i' \in [i - l(i, j) + 1, i - l_1 + 1]$ in s_1 and $j' = j - i + i'$ in s_2. Let $P(i, j) = \{(i', j') \mid i - i' + 1 = j - j' + 1 \in [l_1, l(i, j)]\}$. Then

$(i', j') \in P(i, j)$ if and only if the last member of a regular $(t]$-CSS of $\{s_1[1, i],$ $s_2[1, j]\}$ is with begin-points at i' in s_1 and j' in s_2. By Formula (3), another recurrence of $W(i, j, t, 1)$ under the assumption $G(i, j) \geq l_1$ is formulated as

$$W(i, j, t, 1) = \max_{(i', j') \in P(i, j)} D(i' - 1, j' - 1, t - 1) + i - i' + 1. \tag{4}$$

Not all pairs in $P(i, j)$ are necessary for computing $W(i, j, t, 1)$. For (i_1, j_1), $(i_2, j_2) \in P(i, j)$ and integer t, we say (i_1, j_1) is t-dominated by (i_2, j_2) if $i_1 < i_2$ and $D(i_1 - 1, j_1 - 1, t - 1) - i_1 \leq D(i_2 - 1, j_2 - 1, t - 1) - i_2$. If (i_1, j_1) is t-dominated by (i_2, j_2), then (i_1, j_1) can be excluded from consideration in computing $W(i, j, t, 1)$. This follows because a longest regular $(t]$-CSS of $\{s_1[1, i], s_2[1, j]\}$ whose last member is with begin-points at i_1 in s_1 and j_1 in s_2 cannot exceed the length of a longest regular $(t]$-CSS of $\{s_1[1, i], s_2[1, j]\}$ whose last member is with begin-points at i_2 in s_1 and j_2 in s_2. Let $L(i, j, t) = \{(i', j') \in P(i, j) \mid (i', j')$ is not t-dominated by any element in $P(i, j)\}$ (See Fig. 1 for example).

Fig. 1. Let $s_1 = c\,1\,2\,3\,4\,5\,6\,7\,8\,9$ and $s_2 = c\,1\,2\,3\,4\,a\,1\,2\,3\,4\,5\,6\,7\,8\,9$ be two strings input to $LCSS_2(4, 6)$. It is known that $G(10, 15) = 9$ and $l(10, 15) = 6 = l_2$. Then $P(10, 15) = \{(5, 10), (6, 11), (7, 12)\}$. Since $D(4, 9, 1) + 6 = 10 = D(5, 10, 1) + 5$, the pair $(6, 11)$ t-dominates $(5, 10)$. Hence $L(10, 15, 2) = \{(6, 11), (7, 12)\}$ and $I(10, 15, 2) = 6$, then $W(10, 15, 2, 1) = D(5, 10, 1) + 5 = 10$.

We define the first index i_1 of pair $(i_1, j_1) \in L(i, j, t)$ as the *minimum* first index if $i_1 = \min\{i' \mid (i', j') \in L(i, j, t)\}$. Then $W(i, j, t, 1)$ can be computed from the minimum first index of $L(i, j, t)$ by the following lemma.

Lemma 3. *If* $(i_1, j_1) \in L(i, j, t)$ *and* $i_1 < i_2$ *for any other* $(i_2, j_2) \in L(i, j, t)$, *then* $W(i, j, t, 1) = D(i_1 - 1, j_1 - 1, t - 1) + i - i_1 + 1.$

Let $I(i, j, t)$ denote the minimum first index of $L(i, j, t)$. Then it follows from Lemma 3 and Formula (4) that $W(i, j, t, 1)$ can be expressed more explicitly as

$$W(i, j, t, 1) = D(I(i, j, t) - 1, j + I(i, j, t) - i - 1, t - 1) + i - I(i, j, t) + 1. \tag{5}$$

To compute all $W(i, j, t, 1)$ using Formula (5) in $O(p_2 - p_1)$ time, it suffices to get all $I(i, j, t)$ in $O(p_2 - p_1)$ time. For this purpose, let us revisit $P(i, j)$. For $G(i, j) > l_1$, the variation of $P(i, j)$ relative to $P(i - 1, j - 1)$ is expressed as

$$P(i, j) \setminus \{(i - l_1 + 1, j - l_1 + 1)\} = P(i - 1, j - 1) \setminus \{(i - l_2, j - l_2)\}. \tag{6}$$

Since $(i - l_1 + 1, j - l_1 + 1) \in P(i, j)$ and $i - l_1 + 1$ is the maximum first index of pairs in $P(i, j)$, $(i - l_1 + 1, j - l_1 + 1)$ is not t-dominated by any element in $P(i, j)$ thereby $(i - l_1 + 1, j - l_1 + 1) \in L(i, j, t)$. If $G(i, j) = l_1$, then $L(i, j, t) = P(i, j) = \{(i - l_1 + 1, j - l_1 + 1)\}$. Since $L(i, j, t) \subseteq P(i, j)$, it follows from Formula (6) that

$$L(i, j, t) \setminus \{(i - l_1 + 1, j - l_1 + 1)\} \subseteq P(i - 1, j - 1) \setminus \{(i - l_2, j - l_2)\}. \tag{7}$$

The following two lemmas demonstrate the derivation from $L(i - 1, j - 1, t)$ to $L(i, j, t)$.

Lemma 4. $L(i, j, t) \setminus \{(i - l_1 + 1, j - l_1 + 1)\} \subseteq L(i - 1, j - 1, t) \setminus \{(i - l_2, j - l_2)\}.$

Lemma 5. For $(i_1, j_1) \in L(i - 1, j - 1, t) \setminus \{(i - l_2, j - l_2)\}$, $(i_1, j_1) \in L(i, j, t)$ if and only if $D(i_1 - 1, j_1 - 1, t - 1) - i_1 > D(i - l_1, j - l_1, t - 1) - i + l_1 - 1.$

Note that $s_1[p_1, p_2] = s_2[q_1, q_2]$ is a maximal common substring of s_1 and s_2. If $i = p_1 + l_1 - 1$ $(G[i, j] = l_1)$, then it follows from $L(i, j, t) = \{(i - l_1 + 1, j - l_1 + 1)\}$ that $I(i, j, t) = i - l_1 + 1$. If $i \in [p_1 + l_1, p_2]$ $(G[i, j] > l_1)$ and $L = L(i - 1, j - 1, t)$ is available where the first indices of pairs are in ascending order, then by Lemma 4 and 5, L is converted into $L(i, j, t)$ by the following subroutine GetI(i, j, t).

GetI(i, j, t): (1) *Remove* $(i - l_2, j - l_2)$ *from* L *if* $(i - l_2, j - l_2) \in L$. (2) *Add* $(i - l_1 + 1, j - l_1 + 1)$ *to* L *as the last pair and remove pairs in* L *that* t-*dominated by* $(i - l_1 + 1, j - l_1 + 1)$ *one by one in descending order of their first indices.* (3) *Assign the minimum first index of* L *to* $I(i, j, t)$.

Through dynamic programming using Formula (1) to compute $W(i, j, t, 0)$ and Formula (5) to compute $W(i, j, t, 1)$ after using GetI(i, j, t) to compute $I(i, j, t)$, $LCSS_k(l_1, l_2)$ will be solved. It takes $O(kn_1 n_2)$ cells to store $W(i, j, t, 0/1)$ and $I(i, j, t)$. The index pair set $L(i, j, t)$ is used in the situation where $G(i, j) \geq l_1$ and can be organized as a linked table. Since not all of $L(i, j, t)$ but just one is maintained for $i \in [p_1, p_2]$, $j = q_1 + i - p_1$ and t, it takes $O(p_2 - p_1)$ cells to store $L(i, j, t)$. The space complexity of solving $LCSS_k(l_1, l_2)$ is $O(kn_1 n_2)$. Since there are at most $p_2 - p_1 - l_1 + 2$ index pairs added into or removed from $L(i, j, t)$ in executions of computing $L(i, j, t)$ for i from $p_1 + l_1$ to p_2, it takes $O(p_2 - p_1)$ time to compute all $I(i, j, t)$ for $i \in [p_1 + l_1 - 1, p_2]$ and $j = q_1 + i - p_1$. Follows from Formula (5) that

Lemma 6. *In the execution of solving* $LCSS_k(l_1, l_2)$, *it takes* $O(p_2 - p_1)$ *time to compute all* $W(i, j, t)$ *for* $i \in [p_1, p_2]$ *and* $j = q_1 + i - p_1$.

The values of $W(i, j, t, 0/1)$ and $I(i, j, t)$ suffice to determine a longest $(k]$-CSS of S. If $W(i, j, t, 1) \geq W(i, j, t, 0)$, then the t-th member of a longest $(k]$-CSS of S is of length $i - I(i, j, t) + 1$ and with begin-points at $I(i, j, t)$ in s_1 and $j - i + I(i, j, t)$ in s_2. The first $t - 1$ members of this longest $(k]$-CSS of S can be found by tracing back to $W(I(i, j, t) - 1, j - i + I(i, j, t) - 1, t - 1, 0/1)$. If $W(i, j, t, 0) > W(i, j, t, 1)$, the first t members of a longest $(k]$-CSS of S can be found by tracing back to $W(i, j - 1, t, 0/1)$ or $W(i - 1, j, t, 0/1)$. The pseudocode for computing a longest $(k]$-CSS of S is presented in Algorithm 2.

Algorithm 2. $DP_1(S, k, l_1, l_2)$

Require: $S = \{s_1[1, n_1], s_2[1, n_2]\}$, integers k, l_1, l_2.
Ensure: The occurrences of a longest (k)-CSS of S in s_1 and s_2 whose members are
    of length within $[l_1, l_2]$.
1: $W(i, j, t, 1) \leftarrow 0$, $W(i, j, t, 0) \leftarrow 0$, $I(i, j, t) \leftarrow 0$, $G(i, j) \leftarrow 0$, for $i \cdot j \cdot t = 0$;
2: **for** i from 1 to n_1 **do**
3:    **for** j from 1 to n_2 **do**
4:       Get $G(i, j)$ by Formula (2);
5:       **for** t from 1 to k **do**
6:          Get $W(i, j, t, 0)$ by Formula (1);
7:          **if** $G(i, j) \geq l_1$ and $t > 0$ **then**
8:             Get $I(i, j, t)$ by subroutine GetI(i, j, t);
9:             Get $W(i, j, t, 1)$ by Formula (5);
10:         **end if**
11:       **end for**
12:    **end for**
13: **end for**
14: **return** a longest (k)-CSS of S by a traceback path.

Theorem 1. *Algorithm 2 takes $O(kn_1n_2)$ time and space to return an exact solution of $LCSS_k(l_1, l_2)$, where n_1 and n_2 are lengths of the input strings.*

To compute the length of the longest (k)-CSS without explicitly constructing it, a sliding window based dynamic programming approach enhanced with a rolling array optimization, as usually used for dynamic programming [8,16], can be employed for solving $LCSS_k(l_1, l_2)$ in $O(kn_1n_2)$ time and $O(n_1 + kl_2n_2)$ space.

4 $O(kn_1n_2)$ Time and $O(n_1 + kl_2n_2)$ Space

This section is devoted to finding a longest (k)-CSS of S in $O(kn_1n_2)$ time and $O(n_1 + kl_2n_2)$ space. If $n_1 < 2l_1$ or $n_2 < 2l_1$ or $k = 1$, then a longest 1-CSS of S is a satisfactory solution that can be found in linear time. If $2l_1 \leq n_1 \leq 2l_2$, then Algorithm 2 suffices to get a longest (k)-CSS of S in $O(kn_1n_2)$ time and $O(kl_2n_2)$ space. Assume $n_1 > 2l_2$, $n_2 \geq 2l_1$ and $k > 1$ below.

For $i \in [1, n_1]$, $j \in [0, n_2]$ and $t \in [0, k]$, let Λ_1 be a longest (t)-CSS of $\{s_1[1, i], s_2[1, j]\}$ and Λ_2 a longest $(k - t)$-CSS of $\{s_1[i + 1, n_1], s_2[j + 1, n_2]\}$. Then $\Lambda_1 \oplus \Lambda_2$ is a (k)-CSS of S whose length will be denoted as $N(i, j, t)$. Let $D^*(i, j, t)$ denote the length of a longest (t)-CSS of $\{s_1[i, n_1], s_2[j, n_2]\}$. Then $N(i, j, t) = D(i, j, t) + D^*(i + 1, j + 1, k - t)$. To find i, j and t such that $N(i, j, t) = D(n_1, n_2, k)$ in a way of space-saving, we have to tighten the value range of i.

Lemma 7. *For $\hat{i} \in [l_2, n_1]$, let $\mathcal{N}(\hat{i}) = \max\{N(i, j, t) \mid i \in [\hat{i} - l_2 + 1, \hat{i}], j \in [0, n_2], t \in [0, k]\}$. Then $\mathcal{N}(\hat{i}) = D(n_1, n_2, k)$.*

Given $\hat{i} \in [l_2, n_1]$, when all $N(i, j, t)$ values are available for $i \in [\hat{i} - l_2 + 1, \hat{i}]$, $j \in [0, n_2]$ and $t \in [0, k]$, parameters $\bar{i}$, $\bar{j}$ and $\bar{t}$ satisfying $N(\bar{i}, \bar{j}, \bar{t}) =$

$D(n_1, n_2, k)$ can be identified by Lemma 7. For $\bar{i}$, $\bar{j}$ and $\bar{t}$ with $N(\bar{i}, \bar{j}, \bar{t}) = D(n_1, n_2, k)$, $LCSS_k(l_1, l_2)$ is divided into $LCSS_{\bar{t}}(l_1, l_2)$ for $\{s_1[1, \bar{i}], s_2[1, \bar{j}]\}$ and $LCSS_{k-\bar{t}}(l_1, l_2)$ for $\{s_1[\bar{i}+1, n_1], s_2[\bar{j}+1, n_2]\}$, whose solution combination constitutes a longest (k)-CSS of S.

To compute $N(i, j, t)$ for $i \in [\hat{i}-l_2+1, \hat{i}]$, $j \in [0, n_2]$ and $t \in [0, k]$, both $D(i, j, t)$ and $D^*(i+1, j, t)$ need to be precomputed first. Since these values can be obtained in $O(kl_2n_2)$ space and $O(kn_1n_2)$ time by rolling array and sliding window based dynamic programming, it takes $O(kl_2n_2)$ space and $O(kn_1n_2)$ time to identify $\bar{i} \in [0, n_1]$, $\bar{j} \in [0, n_2]$ and $\bar{t} \in [0, k]$ satisfying $N(\bar{i}, \bar{j}, \bar{t}) = D(n_1, n_2, k)$. The time complexity of dividing $LCSS_k(l_1, l_2)$ into two subproblems is $O(kn_1n_2)$.

Although any $\hat{i} \in [1, n_1]$ permits dividing s_1 and s_2 into substrings to compute the solution of $LCSS_k(l_1, l_2)$, the selection of $\hat{i}$ influences the runtime of our divide-and-conquer algorithm. To optimize computational efficiency, we set $\hat{i} = \lfloor \frac{n_1+l_2}{2} \rfloor$. The pseudocode for computing a longest (k)-CSS of S is provided in Algorithm 3.

Algorithm 3. $D\&C(S, k, l_1, l_2)$

Require: $S = \{s_1[1, n_1], s_2[1, n_2]\}$, integers k, l_1 and l_2.
Ensure: The occurrences of a longest (k)-CSS of S in s_1 and s_2 whose members are of length within $[l_1, l_2]$.
1: If $n_1 < 2l_1$ or $n_2 < 2l_1$ or $k = 1$, then return a longest 1-CSS of S whose length is within $[l_1, l_2]$.
2: If $2l_1 \leq n_1 \leq 2l_2$, then invoke Algorithm 2 to return a longest (k)-CSS of S.
3: Let $\hat{i} = \lfloor \frac{n_1+l_2}{2} \rfloor$.
4: Identify $\bar{i}$, $\bar{j}$ and $\bar{t}$ such that $N(\bar{i}, \bar{j}, \bar{t}) = \mathcal{N}(\hat{i})$.
5: Invoke $D\&C(\{s_1[1, \bar{i}], s_2[1, \bar{j}]\}, \bar{t}, l_1, l_2)$ to get Λ_1.
6: Invoke $D\&C(\{s_1[\bar{i}+1, n_1], s_2[\bar{j}+1, n_2]\}, k-\bar{t}, l_1, l_2)$ to get Λ_2.
7: Return $\Lambda_1 \oplus \Lambda_2$.

Since all matrices used in Algorithm 3 contain $O(kl_2n_2)$ cells, the space complexity of the algorithm is $O(n_1 + kl_2n_2)$. It follows by induction on the length of s_1 that the time complexity of the algorithm is $O(kn_1n_2)$ time.

Theorem 2. *Algorithm 3 takes $O(kn_1n_2)$ time and $O(n_1 + kl_2n_2)$ space to return a longest (k)-CSS of S whose members are of lengths within $[l_1, l_2]$.*

5   Subquadratic Time for $LCSS_2(l_1, l_2)$

Given a feasible (2)-CSS of S, there exist indices $i \in [1, n_1]$ and $j \in [1, n_2]$ such that its first member is a regular (1)-CSS of $\{s_1[1, i], s_2[1, j]\}$ and its last member is a (1)-CSS of $\{s_1[i+1, n_1], s_2[j+1, n_2]\}$. To solve $LCSS_2(l_1, l_2)$, we focus on identifying optimal indices $i \in [1, n_1]$ and $j \in [1, n_2]$ that maximize the length of such a (2)-CSS.

For $i \in [1, n_1]$ and $j \in [1, n_2]$, let $Q(i, j)$ denote the length of a (2]-CSS $\lambda_1 \oplus \lambda_2$ of S, where λ_1 is a longest regular (1]-CSS of $\{s_1[1, i], s_2[1, j]\}$ and λ_2 is a longest (1]-CSS of $\{s_1[i + 1, n_1], s_2[j + 1, n_2]\}$. Then $Q(i, j) = W(i, j, 1, 1) + D^*(i + 1, j + 1, 1)$ and $D(n_1, n_2, 2) = \max_{i \in [1, n_1], j \in [1, n_2]} Q(i, j)$.

Let s and λ be two strings on Σ, we refer to $s[i, i + |\lambda| - 1]$ as the *leftmost occurrence* of λ in s if $\lambda = s[i, i + |\lambda| - 1]$ and there is no $i' \in [1, i - 1]$ such that $\lambda = s[i', i' + |\lambda| - 1]$. We refer to (i, j) (resp. $(i + |\lambda| - 1, j + |\lambda| - 1)$) as the *leftmost begin-point pair* (resp. *leftmost endpoint pair*) of λ on S, if $s_1[i, i + |\lambda| - 1]$ and $s_2[j, j + |\lambda| - 1]$ are the leftmost occurrences of λ in s_1 and s_2 respectively.

Let $\lambda_1 \oplus \lambda_2$ denote a longest (2]-CSS of S, then λ_1 is either an empty string or a common substring of S with length in $[l_1, l_2]$. Let $\mathcal{L}$ denote the set of leftmost begin-point pairs of all common substrings of S whose lengths are within $[l_1, l_2]$. Then there exists $(i, j) \in \mathcal{L}$ such that λ_1 is with begin-points at i in s_1 and j in s_2 if $\mathcal{L} \neq \emptyset$. Otherwise, there is no non-empty (2]-CSS of S.

To identify begin-point pair in $\mathcal{L}$ that is of the first member occurrences in s_1 and s_2 of a longest (2]-CSS, we construct suffix tree $\mathcal{T}$ for the concatenated string $s_1 \#_1 s_2 \#_2$ where $\#_1$ and $\#_2$ are not identical to any symbol of s_1 or s_2. By means of $\mathcal{T}$, one can get $\mathcal{L}$ in $O(n_1 + n_2)$ time where $|\mathcal{L}| \leq n_1 + n_2$.

Let $F(i, j)$ denote the longest common prefix length of $s_1[i, n_1]$ and $s_2[j, n_2]$ and $r(i, j) = \min\{F(i, j), l_2\}$. For a longest (2]-CSS of S whose first member is with begin-points at i in s_1 and j in s_2, the corresponding endpoint pair of its first member is in $\{(i + x - 1, j + x - 1) | x \in [l_1, r(i, j)]\}$. Then $D(n_1, n_2, 2) = \max\{Q(i + x - 1, j + x - 1) | (i, j) \in \mathcal{L}, x \in [l_1, r(i, j)]\}$. We now focus on the symbol indices where the first member of a longest (2]-CSS is with endpoints in s_1 and s_2. For $i \in [1, n_1]$ and $j \in [1, n_2]$, let $H(i, j)$ be the length of the longest common substring of $s_1[i, n_1]$ and $s_2[j, n_2]$. Then for $i \in [1, n_1 - 1]$ and $j \in [1, n_2 - 1]$,

$$0 \leq H(i, j) - H(i + 1, j + 1) \leq 1. \tag{8}$$

Assume that (i, j) belongs to $\mathcal{L}$ below. It follows from $W(i + x - 1, j + x - 1, 1, 1) - W(i + x, j + x, 1, 1) = -1$ for $x \in [l_1, r(i, j) - 1]$ and Formula (8) that

$$-1 \leq W(i + x - 1, j + x - 1, 1, 1) + H(i + x, j + x)$$
$$- [W(i + x, j + x, 1, 1) + H(i + x + 1, j + x + 1)] \leq 0 \tag{9}$$

for $x \in [l_1, r(i, j) - 1]$. For $x \in [l_1, r(i, j)]$, if $H(i + x, j + x) < l_1$, then $Q(i + x - 1, j + x - 1) = W(i + x - 1, j + x - 1, 1, 1)$. Otherwise, $Q(i + x - 1, j + x - 1) = W(i + x - 1, j + x - 1, 1, 1) + \min\{H(i + x, j + x), l_2\}$. If $H(i + l_1, j + l_1) < l_1$, then $H(i + x, j + x) < l_1$ for $x \in [l_1, r(i, j)]$, which implies $Q(i + r(i, j) - 1, j + r(i, j) - 1) = W(i + r(i, j) - 1, j + r(i, j) - 1, 1, 1) = G(i - 1, j - 1) + r(i, j) > Q(i + x - 1, j + x - 1)$. Later, assume $H(i + l_1, j + l_1) \geq l_1$.

Lemma 8. *Let* $x^* = max\{x \mid x \in [l_1, r(i, j)], H(i + x, j + x) \geq l_1\}$. *Then* $Q(i + x^* - 1, j + x^* - 1) \geq Q(i + x - 1, j + x - 1)$ *for* $x \in [l_1, r(i, j)]$.

If $H(i + l_1, j + l_1) < l_1$, then there is no x in $[l_1, r(i, j)]$ satisfying $H(i + x, j + x) \geq l_1$. If $H(i + r(i, j), j + r(i, j)) \geq l_1$, then $x^* = r(i, j)$. By Formula (8), the maximal value $x^* \in [l_1, r(i,j)]$ satisfying $H(i + x, j + x) \geq l_1$ can be found via $O(\min\{\log(l_2 - l_1), \log(l_1 - 1)\})$ times of longest common substring queries for $s_1[i + x - 1, n_1]$ and $s_2[j + x - 1, n_2]$. Let $n = n_1 + n_2$ denote the total input length.

Lemma 9. *[2] After $O(n \log^2 n)$-time and $O(n \log n)$-space suffix tree based pre-processing, the longest common substring of any suffix of s_1 and any suffix of s_2 can be obtained in $O(\log n)$ time.*

To find a longest (2]-CSS of S, we start with constructing $\mathcal{L}$, which takes $O(n)$ time. For each pair $(i, j) \in \mathcal{L}$, we need to find the optimal value $x^* \in [l_1, r(i,j)]$ that maximizes the length of a (2]-CSS whose first member is with begin-points at i in s_1 and j in s_2. It takes $O(\log^2 n)$ time to find such a longest (2]-CSS of S because it needs to examine $\log n$ potential values of x^* and each examination takes $O(\log n)$ time according to Lemma 9. Furthermore, through $O(n)$-time preprocessing using suffix tree, the length of the longest common suffix of any prefix of s_1 and any prefix of s_2 can be computed in constant time.

Theorem 3. *An exact solution of $LCSS_2(l_1, l_2)$ can be found in $O(n \log^2 n)$ time and $O(n \log n)$ space, where n is the total input length.*

6   Conclusion

Since $LCSS_2(l_1, l_2)$ is solvable in subquadratic time, it is significant and interesting to bring into consideration of subquadratic time algorithms for $LCSS_k(l_1, l_2)$ with arbitrary k. It is worth investigating the case where each member of a $(k]$-CSS is of length bounded by $[l_1, \infty]$.

Acknowledgments. This work is supported by grants from the NSF of China (NSFC: 62272272 and U24A20257).

Disclosure of Interests. The authors have no competing interests.

References

1. Abboud, A., Backurs, A., Williams, V.V.: Tight hardness results for LCS and other sequence similarity measures. In: Foundations of Computer Science, pp. 59–78. IEEE (2015)
2. Amir, A., Charalampopoulos, P., Pissis, S.P., Radoszewski, J.: Dynamic and internal longest common substring. Algorithmica **82**(12), 3707–3743 (2020)
3. Banerjee, A., Gibney, D., Thankachan, S.V.: Longest common substring with gaps and related problems. In: 32nd Annual European Symposium on Algorithms (ESA 2024), pp. 16:1–16:18. Schloss Dagstuhl–Leibniz-Zentrum für Informatik (2024)
4. Benson, G., Levy, A., Maimoni, S., Noifeld, D., Shalom, B.R.: LCSk: a refined similarity measure. Theoret. Comput. Sci. **638**, 11–26 (2016)

5. Benson, G., Levy, A., Shalom, B.R.: Longest common subsequence in k length substrings. In: Brisaboa, N., Pedreira, O., Zezula, P. (eds.) SISAP 2013. LNCS, vol. 8199, pp. 257–265. Springer, Heidelberg (2013). https://doi.org/10.1007/978-3-642-41062-8_26

6. Fremin, B.J., Bhatt, A.S.: Comparative genomics identifies thousands of candidate structured RNAs in human microbiomes. Genome Biol. **22**(1), 100 (2021)

7. Gusfield, D.: Algorithms on Strings, Trees, and Sequences: Computer Science and Computational Biology. Cambridge University Press (1997)

8. Hirschberg, D.S.: A linear space algorithm for computing maximal common subsequences. Commun. ACM **18**(6), 341–343 (1975)

9. Li, T., Jiang, H., Wang, L., Zhu, D.: Longest (k]-tuple common substrings. In: International Workshop on Frontiers in Algorithmics, pp. 106–114. Springer, Singapore (2024)

10. Li, T., Zhu, D., Jiang, H., Feng, H., Cui, X.: Longest K-tuple common substrings. In: 2022 IEEE International Conference on Bioinformatics and Biomedicine (BIBM), pp. 63–66. IEEE (2022)

11. Masek, W.J., Paterson, M.S.: A faster algorithm computing string edit distances. J. Comput. Syst. Sci. **20**(1), 18–31 (1980)

12. Michael, M., Nicolas, F., Ukkonen, E.: On the complexity of finding gapped motifs. J. Discrete Algorithms **8**(2), 131–142 (2010)

13. Rausch, T., Zichner, T., Schlattl, A., Stütz, A.M., Benes, V., Korbel, J.O.: DELLY: structural variant discovery by integrated paired-end and split-read analysis. Bioinformatics **28**(18), i333–i339 (2012)

14. Wagner, R.A., Fischer, M.J.: The string-to-string correction problem. J. ACM (JACM) **21**(1), 168–173 (1974)

15. Yonemoto, Y., Mieno, T., Inenaga, S., Yoshinaka, R., Shinohara, A.: Subsequence matching and LCS with segment number constraints. arXiv preprint arXiv:2407.19796 (2025)

16. Zhu, D., Wang, L., Wang, T., Wang, X.: A space efficient algorithm for the longest common subsequence in k-length substrings. Theoret. Comput. Sci. **687**, 79–92 (2017)

Finding Cycle Types in Permutation Groups with Few Generators

Markus Lohrey[(✉)] and Andreas Rosowski

Department ETI, Universität Siegen, 57076 Siegen, Germany
{lohrey,rosowski}@eti.uni-siegen.de

Abstract. The problem whether a given permutation group contains a permutation with a given cycle type is studied. This problem is known to be NP-complete. In this paper it is shown that the problem can be solved in logspace for a cyclic permutation group and that it is NP-complete for a 2-generated abelian permutation group. In addition it is shown that it is NP-complete whether a 2-generated abelian permutation group contains a fixpoint-free permutation.

Keywords: Permutation groups · Algorithmic group theory · NP-completeness

1 Introduction

Permutations are ubiquitous objects in combinatorics [4] and group theory [6]. The set of all permutations on a set Ω forms a group $\mathsf{Sym}(\Omega)$ (the *symmetric group* on Ω) under composition. A subgroup of a symmetric group is called a *permutation group*. Cayley's famous theorem states that every group is isomorphic to a permutation group via the right regular representation. Here, we only deal with the case that Ω is finite and write $\mathsf{Sym}(n)$ for $\mathsf{Sym}(\Omega)$ if $|\Omega| = n$.

Having group elements represented as permutations can be often exploited algorithmically. For instance, the subgroup membership problem for symmetric groups (Does a given permutation $\pi \in \mathsf{Sym}(n)$ belong to the subgroup generated by given permutations $\pi_1, \ldots, \pi_k \in \mathsf{Sym}(n)$?) can be solved in polynomial time [10,15,16] and even in NC [3]. Another problem that has an extremely simple algorithm in symmetric groups is the conjugacy problem: given permutations $\pi, \rho \in \mathsf{Sym}(n)$, does there exist $\tau \in \mathsf{Sym}(n)$ such that $\pi = \tau^{-1}\rho\tau$? This is equivalent to say that π and ρ have the same *cycle type*. The cycle type of a permutation $\pi \in \mathsf{Sym}(n)$ specifies for every $\ell \leq n$ the number of cycles of length ℓ when π is written (uniquely) as a product of pairwise disjoint cycles.

In this paper we are interested in the problem whether a given permutation group $G \leq \mathsf{Sym}(n)$ (specified by a list of generators) contains a permutation of a given cycle type. Or equivalently: does G contain an element that is conjugated to a given permutation π? We call this problem CycleType.

Cameron and Wu showed in [8] that CycleType is NP-complete. Moreover, NP-hardness already holds for the case where G is an elementary abelian 2-group

F. V. Fomin and M. Xiao (Eds.): COCOON 2025, LNCS 15984, pp. 367–380, 2026.
https://doi.org/10.1007/978-981-95-0218-9_27

(i.e., an abelian group where every non-identity element has order two). Here we further pinpoint the borderline between tractability and non-tractability: We show that if the input permutation group G is cyclic and given by a single generator then CycleType can be solved in logarithmic space on a deterministic Turing machine (and hence belongs to the complexity class P). On the other hand, we show that CycleType is already NP-complete for the case where G is generated by two commuting permutations, i.e., $G = \langle \pi, \tau \rangle$ with $\pi\tau = \tau\pi$. Moreover, our proof shows that it is already NP-complete whether for two given commuting permutations π and τ the coset $\pi\langle\tau\rangle$ (a coset of a cyclic group) contains a permutation with a given cycle type.

In the last section of the paper, we consider the problem FixpointFree that asks whether a given permutation group contains a fixpoint-free permutation, i.e., a permutation π such that $\pi(a) \neq a$ for all a. It was shown in [5,8] that FixpointFree is NP-complete and as for CycleType, NP-hardness holds already for elementary abelian 2-groups. The restriction of FixpointFree to cyclic permutation groups is not interesting ($\langle\pi\rangle$ contains a fixpoint-free permutation if and only if π is fixpoint-free). We show that the restriction of FixpointFree to 2-generated abelian permutation groups $\langle\pi, \tau\rangle$ is NP-complete. Moreover, it is also NP-complete to check whether a coset $\pi\langle\tau\rangle$ of a cyclic permutation group, where in addition $\pi\tau = \tau\pi$, contains a fixpoint-free permutation.

Related Work. Fixpoint-free permutations are also known as *derangements* and they have received a lot of attention in combinatorics and group theory; see [7] for a survey. Jordan proved in 1872 that every permutation group G that acts transitively on a finite set Ω of size at least two contains a derangement [14]. Arvind proved that in this situation one can compute in polynomial time a derangement in G [2]. In the same paper, Arvind shows that the problem whether a given permutation group G contains a permutation with at least k non-fixpoints is fixed parameter tractable with respect to the parameter k.

2   Preliminaries

2.1   General Notations

For integers $1 \leq i \leq j$ we write $[i, j]$ for the set $\{i, i+1, \ldots, j\}$ and $[j]$ for $[1, j]$. For a prime p and an integer n we denote with $\nu_p(n)$ the largest positive integer d such that $p^d \mid n$ (it is also called the p-adic valuation of n). The greatest common divisor of integers $n_1, \ldots, n_k$ is denoted by $\gcd(n_1, \ldots, n_k)$ and the least common multiple is denoted by $\text{lcm}(n_1, \ldots, n_k)$.

We assume that the reader is familiar with basic concepts of complexity theory; see [1] for more details. With L (also known as *logspace*) we denote the class of all problems that can be solved on a deterministic Turing machine in logarithmic space. It is a subset of P (deterministic polynomial time).

2.2   Permutations

For $n \geq 1$ we denote with $\text{Sym}(n)$ the group of all permutations on $[n]$. The identity permutation is denoted by id. For $\pi \in \text{Sym}(n)$ and $a \in [n]$ we also write $a\pi$ for $\pi(a)$. There are two standard representations for a permutation $\pi \in \mathsf{Sym}(n)$:

- The *pointwise representation* of π is the tuple $[\pi(1), \pi(2), \ldots, \pi(n)]$.
- The *cycle representation* is a list $\gamma_1 \gamma_2 \cdots \gamma_k$ of pairwise disjoint cycles. Every cycle γ_i is written as a list $(a_0, a_1, \ldots, a_{\ell-1})$ (with $a_i \in [n]$) meaning that $a_k \pi = a_{k+1 \bmod \ell}$. Fixpoints (cycles of the form (i)) are usually omitted in the cycle representation, but sometimes we will explicitly list them.

Note that every cycle $(a_0, a_1, \ldots, a_{\ell-1})$ can be replaced by a cyclic rotation. Moreover since disjoint cycles commute, the order of the cycles γ_i is not relevant.

Computing the pointwise representation from the cycle representation is possible in uniform AC^0 (this is a very small circuit complexity class contained in L). On the other hand, the cycle representation can be computed in logspace from the pointwise representation and no better complexity bound is known [9]. Therefore, as long as one works with complexity classes that contain L (which will be the case in this paper), there is no reason to specify which of the above two representations of permutations is chosen.

Let $\mathsf{fpf}(n) = \{\pi \in \mathsf{Sym}(n) \mid a\pi \neq a \text{ for all } a \in [n]\}$ be the set of all *fixpoint-free* permutations. For $\pi_1, \ldots, \pi_k \in \mathsf{Sym}(n)$ we write $\langle \pi_1, \ldots, \pi_k \rangle \leq \mathsf{Sym}(n)$ for the permutation group generated by $\pi_1, \ldots, \pi_k$. The order $\mathrm{ord}(\pi)$ of $\pi \in \mathsf{Sym}(n)$ is the smallest integer $i \geq 1$ such that $\pi^i = \mathrm{id}$. If $\gamma_1 \cdots \gamma_k$ is the cycle representation of π and every cycle γ_i has length ℓ_i then the multiset $\mathrm{ct}(\pi) := \{\!\{\ell_1, \ldots, \ell_k\}\!\}$ is the *cycle type* of π. Note that in this situation we have

$$\mathrm{ord}(\pi) = \mathrm{lcm}(\ell_1, \ldots, \ell_k). \tag{1}$$

The following lemma is well known, see e.g. [6]:

Lemma 1. *For $\pi, \rho \in Sym(n)$ we have $\mathrm{ct}(\pi) = \mathrm{ct}(\rho)$ if and only if there is a $\sigma \in \mathsf{Sym}(n)$ such that $\pi = \sigma^{-1} \rho \sigma$.*

Also the following lemma seems to be folklore. For completeness we give a proof.

Lemma 2. *Let $x \in \mathbb{N}$ and γ be a single cycle of length ℓ. Then the cycle representation of γ^x consists of $\gcd(x, \ell)$ many disjoint cycles of length $\ell / \gcd(x, \ell)$.*

Proof. (Proof of Lemma 2). Let us first consider the case where $\gcd(x, \ell) = 1$. Then there is a $y \in \mathbb{N}$ with $xy \equiv 1 \bmod \ell$. If γ^x consists of at least two cycles of length strictly smaller than ℓ, then the same holds for every power of γ^x. This contradicts $(\gamma^x)^y = \gamma^{xy} = \gamma$. This shows the statement of the lemma for the case $\gcd(x, \ell) = 1$.

For the general case let $m = \gcd(x, \ell), k = \ell/m$ and $z = x/m$. Then we can write the cycle γ as $\gamma = (a_0, \ldots, a_{mk-1})$ for some pairwise different $a_i \in [n]$. For all $i \in [0, m-1]$ and $d \in [0, k-1]$ we have $a_{dm+i}\gamma^m = a_{(d+1)m+i}$ where all arithmetics in the indices is done modulo $\ell = mk$. We obtain

$$\gamma^x = (\gamma^m)^z = \prod_{i=0}^{m-1} (a_i, a_{m+i}, a_{2m+i}, \ldots, a_{(k-1)m+i})^z.$$

Since $\gcd(z, k) = 1$ we obtain from the above case $\gcd(x, \ell) = 1$ that

$$(a_i, a_{m+i}, a_{2\,m+i}, \ldots, a_{(k-1)m+i})^z$$

is a cycle of length k. Hence, γ^x splits into m disjoint cycles of length k.   □

For integers $1 \leq i < j \leq n$ we denote with $([i, j])$ the cycle $(i, i+1, \ldots, j) \in \mathsf{Sym}(n)$. We also use $([i])$ instead of $([1, i])$ for $2 \leq i \leq n$.

We will consider the following two computational problems in this paper:

Problem 1. CycleType is the following problem:

- input: $\pi_1, \ldots, \pi_m, \rho \in \mathsf{Sym}(n)$
- question: Is there an element $\pi \in \langle \pi_1, \ldots, \pi_m \rangle$ such that $\mathsf{ct}(\pi) = \mathsf{ct}(\rho)$?

Problem 2. FixpointFree is the following problem:

- input: $\pi_1, \ldots, \pi_m \in \mathsf{Sym}(n)$
- question: Does $\mathsf{fpf}(n) \cap \langle \pi_1, \ldots, \pi_m \rangle \neq \emptyset$ hold?

Note that the unary encoding of n (from $\mathsf{Sym}(n)$) is implicitly part of the inputs for CycleType and FixpointFree. It is easy to see that CycleType and FixpointFree are in NP: on input $\pi_1, \ldots, \pi_m, \rho \in \mathsf{Sym}(n)$ we guess a permutation $\pi \in \mathsf{Sym}(n)$ and then check in polynomial time whether (i) $\pi \in \langle \pi_1, \ldots, \pi_m \rangle$ [3] and (ii) $\mathsf{ct}(\pi) = \mathsf{ct}(\rho)$ (resp., $\pi \in \mathsf{fpf}(n)$).

For a given number k we denote with CycleType(k) the restriction of Cycle-Type where $m \leq k$ holds. In other words, the input permutation group is generated by k permutations. Moreover, if the input permutations $\pi_1, \ldots, \pi_k$ pairwise commute, then we write CycleType(ab, k) (ab stands for "abelian"). Analogous restrictions are defined for FixpointFree.

3   Cycle Type in Cyclic Permutation Groups

In this section, we study the problem CycleType(1), i.e., CycleType for cyclic permutation groups. Let us fix a symmetric group $\mathsf{Sym}(n)$. We assume that n is given in unary encoding for the following. Note that a brute-force algorithm that iterates over all elements $\pi \in \langle \pi_1 \rangle$ and thereby checks whether $\mathsf{ct}(\pi) = \mathsf{ct}(\rho)$ holds, needs exponential time. In [13, Lemma 2.1] it is shown that for every sufficiently large $n \in \mathbb{N}$, there exists a permutation $\pi_1 \in \mathsf{Sym}(\lfloor 2n^2 \ln n \rfloor)$ such that $\langle \pi_1 \rangle$ has size greater than 2^n.

Let P_n be the set of all primes in $[n]$. One can easily produce a list $p_1 < p_2 < \cdots < p_r$ of all those primes in logspace. For this, one only needs the fact that integer division for unary encoded integers can be done in logspace (actually, integer division of binary encoded integers can be also done in logspace [12] but this is not needed here). We will only consider numbers where all prime divisors are from P_n. For such a number a we denote with $\mathsf{pe}(a)$ (for prime exponents) the tuple $(e_1, \ldots, e_r)$ such that $a = \prod_{i=1}^{r} p_i^{e_i}$ is the prime factorizaton of a. We will represent the exponents e_i in unary notation. From the unary representation of the number $a \in [n]$ one can easily compute in logspace the tuple $\mathsf{pe}(a)$. We need the following fact:

Lemma 3. *From a given permutation $\pi \in \mathsf{Sym}(n)$ one can compute in logspace the tuple $\mathsf{pe}(\mathrm{ord}(\pi))$.*

Proof. Assume that the cycle representation $\pi = \gamma_1 \gamma_2 \cdots \gamma_k$ is given. Let $\ell_i \in [n]$ be the length of the cycle γ_i. We then compute in logspace the tuple $\mathsf{pe}(\ell_i) = (e_{i,1}, \ldots, e_{i,r})$. Since $\mathrm{ord}(\pi) = \mathrm{lcm}(\ell_1, \ell_2, \ldots, \ell_k)$ we have

$$\mathsf{pe}(\mathrm{ord}(\pi)) = (e_1, \ldots, e_r)$$

with $e_i = \max\{e_{1,i}, \ldots, e_{k,i}\}$. Clearly, these exponents e_i can be computed in logspace. □

Lemma 4. *For given permutations $\pi, \rho \in \mathsf{Sym}(n)$ one can check in logspace, whether $\mathrm{ord}(\rho) \mid \mathrm{ord}(\pi)$ holds.*

Proof. Let $\mathsf{pe}(\mathrm{ord}(\rho)) = (e_1, \ldots, e_r)$ and $\mathsf{pe}(\mathrm{ord}(\pi)) = (e_1', \ldots, e_r')$. Then $\mathrm{ord}(\rho) \mid \mathrm{ord}(\pi)$ if and only if $e_i \leq e_i'$ for all $i \in [r]$. Therefore, the statement of the lemma follows from Lemma 3. □

Lemma 5. *There is a logspace algorithm with the following specification:*

- *input: $\pi, \rho \in \mathsf{Sym}(n)$ such that $\mathrm{ord}(\rho) \mid \mathrm{ord}(\pi)$ and $a \in [n]$.*
- *output: $a\pi^d \in [n]$ where $d = \mathrm{ord}(\pi)/\mathrm{ord}(\rho)$*

Proof. By Lemma 3 we can produce in logspace the tuples

$$\mathsf{pe}(\mathrm{ord}(\rho)) = (e_1, \ldots, e_r) \text{ and}$$
$$\mathsf{pe}(\mathrm{ord}(\pi)) = (e_1', \ldots, e_r').$$

Since $\mathrm{ord}(\rho) \mid \mathrm{ord}(\pi)$ we have $e_i \leq e_i'$ for all $i \in [r]$. We then have $\mathsf{pe}(d) = (f_1, \ldots, f_r)$ with $f_i = e_i' - e_i$ and this tuple can be also produced in logspace. Let $\gamma_1 \gamma_2 \cdots \gamma_k$ be the cycle representation of π. We then compute in logspace the length $\ell \in [n]$ of the unique cycle γ_i that contains $a \in [n]$. We have $a\pi^d = a\pi^{d \bmod \ell}$. Since all primes p_i and exponents f_i are given in unary notation, we can compute in logspace the value $d \bmod \ell$ by going over the prime factorization $\prod_{i=1}^r p_i^{f_i}$ and making $\sum_{i=1}^r f_i$ many multiplications modulo ℓ. Once $d \bmod \ell$ is computed, we can finally compute $a\pi^{d \bmod \ell}$ in logspace. □

Lemma 6. *Let $\pi, \rho \in \mathsf{Sym}(n)$. Then the following holds:*

- *If $\mathsf{ct}(\pi) = \mathsf{ct}(\rho)$ then $\mathrm{ord}(\pi) = \mathrm{ord}(\rho)$.*
- *For all $i \in \mathbb{N}$ we have $\mathrm{ord}(\pi) = \mathrm{ord}(\pi^i)$ if and only if $\mathsf{ct}(\pi) = \mathsf{ct}(\pi^i)$.*

Proof. For the first statement note that if $\{\!\{\ell_1, \ell_2, \ldots, \ell_k\}\!\}$ is the common cycle type of π and ρ then $\mathrm{ord}(\pi) = \mathrm{lcm}(\ell_1, \ell_1, \ldots, \ell_k) = \mathrm{ord}(\rho)$ by (1). Therefore we only have to show that if $\mathrm{ord}(\pi) = \mathrm{ord}(\pi^i)$ then $\mathsf{ct}(\pi) = \mathsf{ct}(\pi^i)$. Let $\pi = \gamma_1 \cdots \gamma_k$ be the cycle representation of π. Then we have $\pi^i = \gamma_1^i \cdots \gamma_k^i$. Since $\mathrm{ord}(\pi) = \mathrm{ord}(\pi^i)$ we obtain $\gcd(\mathrm{ord}(\pi), i) = 1$. Because of $\mathrm{ord}(\gamma_j) \mid \mathrm{ord}(\pi)$ we get $\gcd(\mathrm{ord}(\gamma_j), i) = 1$ for all $j \in [k]$. By Lemma 2, γ_j and γ_j^i are cycles of the same length and thus π and π^i have the same cycle type. □

Theorem 1. CycleType(1) *is in* L.

Proof. Let $\pi, \rho \in \mathsf{Sym}(n)$ be the two input permutations of CycleType(1). It is asked whether there is a $q \in \mathbb{N}$ such that $\mathsf{ct}(\pi^q) = \mathsf{ct}(\rho)$. By Lemma 4 we can check in logspace whether $\mathrm{ord}(\rho) \mid \mathrm{ord}(\pi)$ holds. If this is not the case, then by the first statement of Lemma 6 there is no q such that $\mathsf{ct}(\pi^q) = \mathsf{ct}(\rho)$ and we can immediately reject. Let us now assume that $\mathrm{ord}(\rho) \mid \mathrm{ord}(\pi)$ and let $d = \mathrm{ord}(\pi)/\mathrm{ord}(\rho)$ in the following. Note that $\mathrm{ord}(\pi^d) = \mathrm{ord}(\rho)$.

Claim 1. There is a $q \in \mathbb{N}$ such that $\mathsf{ct}(\pi^q) = \mathsf{ct}(\rho)$ if and only if $\mathsf{ct}(\pi^d) = \mathsf{ct}(\rho)$.

Proof of Claim 1. The direction from right to left is trivial. Hence, let us assume that there is a q such that $\mathsf{ct}(\pi^q) = \mathsf{ct}(\rho)$. By Lemma 6, we have $\mathrm{ord}(\pi^q) = \mathrm{ord}(\rho)$. We get $\mathrm{ord}(\pi^d) = \mathrm{ord}(\rho) = \mathrm{ord}(\pi^q)$. Since $\langle \pi \rangle$ has exactly one subgroup of order $\mathrm{ord}(\rho)$ it follows that $\langle \pi^q \rangle = \langle \pi^d \rangle$. Let $\pi^q = (\pi^d)^i$ for $i \in \mathbb{N}$. Since $\mathrm{ord}(\pi^d) = \mathrm{ord}(\pi^q) = \mathrm{ord}((\pi^d)^i)$, the second statement of Lemma 6 implies that π^q and π^d (and hence ρ and π^d) have the same cycle type. This shows Claim 1.

By Claim 1, it suffices to check in logspace whether $\mathsf{ct}(\pi^d) = \mathsf{ct}(\rho)$. By Lemma 5 we can compute in logspace the pointwise representation and hence the cycle representation of π^d. From the cycle representation of a permutation we can of course compute in logspace the cycle type. $\qquad\square$

4   Cycle Type in the 2-Generated Abelian Case

In this section we show that CycleType becomes NP-complete if the input permutation group is abelian and generated by two elements.

Theorem 2. CycleType(ab, 2) *is* NP-*complete.*

Proof. Since CycleType is in NP (see the remark at the end of Sect. 2.2), it remains to show NP-hardness. For this we exhibit a logspace reduction from X3HS (exact 3-hitting set), which is the following problem:

- Input: a finite set S and a set $\mathcal{B} \subseteq 2^S$ of subsets of S all of size 3.
- Question: Is there a subset $T \subseteq S$ such that $|T \cap C| = 1$ for all $C \in \mathcal{B}$?

Note that X3HS is the same problem as positive 1-in-3-SAT, which is a well-known NP-complete problem; see [11] for more details.

Let S be a finite set and $\mathcal{B} \subseteq 2^S$ be a set of subsets of S all of size 3. W.l.o.g. assume that $S = [n]$ and let $\mathcal{B} = \{C_1, \ldots, C_m\}$. Let $p_1 < \cdots < p_{2n}$ be the first $2n$ primes with $p_1 > 3$. Moreover let $q_1 < \cdots < q_m$ be the next m primes with $p_{2n} < q_1$. We associate $i \in S$ with the prime p_i and $C_j \in \mathcal{B}$ with the prime q_j. We will work with the group

$$G = \prod_{i=1}^{n} \mathsf{Sym}(p_i p_{n+i}) \times \prod_{j=1}^{m} \mathsf{Sym}(p_n^3 q_j)^6$$

which naturally embedds into $\mathsf{Sym}(N)$ for

$$N = \sum_{i=1}^{n} p_i p_{n+i} + 6 \sum_{j=1}^{m} p_n^3 q_j.$$

Let $f : G \to \mathsf{Sym}(N)$ be this embedding. When we talk of the cycle type of an element $g \in G$, we always refer to the cycle type of the permutation $f(g) \in \mathsf{Sym}(N)$. If $g = (\pi_1, \ldots, \pi_n, \rho_1, \ldots, \rho_{6m}) \in G$, then this cycle type is obtained by taking the disjoint union (of multisets) of the cycle types of all the π_i and ρ_j.

For $j \in [m]$ we define $r_j = q_j \cdot \prod_{i \in C_j} p_i$. Moreover for $j \in [m]$ and all $d \in [6]$ we define the number $s_{j,d} \in [0, r_j - 1]$ as the smallest positive integer satisfying the following congruences in which we assume $C_j = \{i_1, i_2, i_3\}$ with $i_1 < i_2 < i_3$:

$$
\begin{array}{lll}
s_{j,1} \equiv -1 \bmod p_{i_1} & s_{j,2} \equiv 0 \bmod p_{i_1} & s_{j,3} \equiv 0 \bmod p_{i_1} \\
s_{j,1} \equiv 0 \bmod p_{i_2} & s_{j,2} \equiv -1 \bmod p_{i_2} & s_{j,3} \equiv 0 \bmod p_{i_2} \\
s_{j,1} \equiv 0 \bmod p_{i_3} & s_{j,2} \equiv 0 \bmod p_{i_3} & s_{j,3} \equiv -1 \bmod p_{i_3} \\
s_{j,1} \equiv 1 \bmod q_j & s_{j,2} \equiv 1 \bmod q_j & s_{j,3} \equiv 1 \bmod q_j \\
\\
s_{j,4} \equiv -1 \bmod p_{i_1} & s_{j,5} \equiv -3 \bmod p_{i_1} & s_{j,6} \equiv -2 \bmod p_{i_1} \\
s_{j,4} \equiv -2 \bmod p_{i_2} & s_{j,5} \equiv -1 \bmod p_{i_2} & s_{j,6} \equiv -3 \bmod p_{i_2} \\
s_{j,4} \equiv -3 \bmod p_{i_3} & s_{j,5} \equiv -2 \bmod p_{i_3} & s_{j,6} \equiv -1 \bmod p_{i_3} \\
s_{j,4} \equiv 1 \bmod q_j & s_{j,5} \equiv 1 \bmod q_j & s_{j,6} \equiv 1 \bmod q_j
\end{array}
$$

Moreover, we define the number $t_j \in [0, r_j - 1]$ as the smallest positive integer satisfying

$$t_j \equiv 1 \bmod p_{i_a} \text{ for all } a \in [3] \text{ and } t_j \equiv 0 \bmod q_j.$$

We define the input group elements $\rho, \pi_1, \pi_2 \in G$ as follows, where i ranges over $[n]$, j ranges over $[m]$ and $i_1 < i_2 < i_3$ are the elements of C_j (recall that $([m])$ denotes the cycle $(1, 2, \ldots, m)$):

$$
\begin{aligned}
\rho &= (\zeta_1, \ldots, \zeta_n, \eta_1, \ldots, \eta_m) \\
\zeta_i &= ([p_i p_{n+i}]) \\
\eta_j &= (([r_j])^{p_{i_1} p_{i_2} p_{i_3}}, ([r_j])^{p_{i_1}}, ([r_j])^{p_{i_2}}, ([r_j])^{p_{i_3}}, ([r_j]), ([r_j])) \\
\pi_1 &= (\alpha_1, \ldots, \alpha_n, \beta_1, \ldots, \beta_m) \\
\alpha_i &= ([p_i p_{n+i}]) \\
\beta_j &= (([r_j])^{s_{j,1}}, ([r_j])^{s_{j,2}}, ([r_j])^{s_{j,3}}, ([r_j])^{s_{j,4}}, ([r_j])^{s_{j,5}}, ([r_j])^{s_{j,6}}) \\
\pi_2 &= (\gamma_1, \ldots, \gamma_n, \delta_1, \ldots, \delta_m) \\
\gamma_i &= \mathrm{id} \\
\delta_j &= (([r_j])^{t_j}, ([r_j])^{t_j}, ([r_j])^{t_j}, ([r_j])^{t_j}, ([r_j])^{t_j}, ([r_j])^{t_j})
\end{aligned}
$$

Note that π_1 and π_2 commute.

We will show there are $x_1, x_2 \in \mathbb{N}$ such that $\mathsf{ct}(\rho) = \mathsf{ct}(\pi_1^{x_1} \pi_2^{x_2})$ if and only if there is a subset $T \subseteq S$ such that $|T \cap C_j| = 1$ for all $j \in [m]$.

First suppose that there are $x_1, x_2 \in \mathbb{N}$ with $\mathsf{ct}(\rho) = \mathsf{ct}(\pi_1^{x_1} \pi_2^{x_2})$. We define

$$T = \{ i \in [n] \mid x_2 \not\equiv 0 \bmod p_i \}. \tag{2}$$

Claim 2. For all $i \in [n]$ and $j \in [m]$ we have $x_1 \not\equiv 0 \bmod p_i$, $x_1 \not\equiv 0 \bmod p_{n+i}$ and $x_1 \not\equiv 0 \bmod q_j$.

Proof of Claim 2. The claim follows from Lemma 2 and the following facts:

- ζ_i and α_i are cycles of length $p_i p_{n+i}$.
- π_2 does not contain any cycle whose length is a multiple of p_{n+i}.
- $t_j \equiv 0 \bmod q_j$ and hence π_2 also does not contain any cycle whose length is a multiple of q_j.
- ρ and π_1 both contain 6 pairwise disjoint permutations of the form $([r_j])^z$, where z is not a multiple of q_j. □

Claim 3. For all $C_j = \{i_1, i_2, i_3\} \in \mathcal{B}$ there is a (necessarily unique) $a \in [3]$ such that $x_2 \not\equiv 0 \bmod p_{i_a}$ and $x_2 \equiv 0 \bmod p_{i_b}$ for all $b \in [3] \setminus \{a\}$.

Proof of Claim 3. Let $j \in [m]$ and assume $C_j = \{i_1, i_2, i_3\}$ with $i_1 < i_2 < i_3$. Consider η_j. By Lemma 2 $([r_j])^{p_{i_1} p_{i_2} p_{i_3}}$ consists of $p_{i_1} p_{i_2} p_{i_3}$ cycles of length q_j and these are the only cycles of length q_j in ρ. Hence, $\beta_j^{x_1} \delta_j^{x_2}$ must contain exactly $p_{i_1} p_{i_2} p_{i_3}$ cycles of length q_j. By Lemma 2 this can only be achieved if there is a unique $a \in [6]$ such that

$$\forall c \in [3] : x_1 s_{j,a} + x_2 t_j \equiv 0 \bmod p_{i_c}. \tag{3}$$

Also note that

$$\forall b \in [6] : x_1 s_{j,b} + x_2 t_j \equiv x_1 \not\equiv 0 \bmod q_j$$

by Claim 2 and

$$\forall c \in [3] : x_2 t_j \equiv x_2 \bmod p_{i_c}.$$

We want to show that $a \in [3]$. In order to get a contradiction, suppose that $a \in \{4, 5, 6\}$. The congruence $x_1 s_{j,a} + x_2 t_j \equiv 0 \bmod p_{i_c}$ from (3) gives us

$$\forall c \in [3] : x_2 \equiv -x_1 s_{j,a} \bmod p_{i_c}.$$

Then, for all $b \in [3] \setminus \{a - 3\}$ we have

$$x_1 s_{j,b} + x_2 t_j \equiv x_1 s_{j,b} - x_1 s_{j,a} \equiv x_1(-1 - s_{j,a}) \not\equiv 0 \bmod p_{i_b},$$

where $x_1 \not\equiv 0 \bmod p_{i_b}$ by Claim 2 and $-1 - s_{j,a} \not\equiv 0 \bmod p_{i_b}$ since $s_{j,a} \not\equiv -1 \bmod p_{i_b}$ for $b \neq a - 3$ (also note that $p_{i_b} > 2$). Similarly, for all $b \in [3] \setminus \{a - 3\}$ we get

$$x_1 s_{j,3+b} + x_2 t_j \equiv x_1 s_{j,3+b} - x_1 s_{j,a} \equiv x_1(s_{j,3+b} - s_{j,a}) \not\equiv 0 \bmod p_{i_b},$$

where as above $x_1 \not\equiv 0 \bmod p_{i_b}$ by Claim 2 and $s_{j,3+b} - s_{j,a} \not\equiv 0 \bmod p_{i_b}$ since $a \neq 3 + b$ and $s_{j,a} \not\equiv s_{j,3+b} \bmod p_{i_c}$ for all $c \in [3]$.

Moreover, for all $b \in [3] \setminus \{a - 3\}$ and all $c \in [3] \setminus \{b\}$ we have

$$x_1 s_{j,b} + x_2 t_j \equiv x_1 s_{j,b} - x_1 s_{j,a} \equiv -x_1 s_{j,a} \not\equiv 0 \bmod p_{i_c} \text{ and}$$
$$x_1 s_{j,3+b} + x_2 t_j \equiv x_1 s_{j,3+b} - x_1 s_{j,a} \equiv x_1 (s_{j,3+b} - s_{j,a}) \not\equiv 0 \bmod p_{i_c}.$$

Finally, for all $b \in [6]$ we have $x_1 s_{j,b} + x_2 t_j \not\equiv 0 \bmod q_j$ as pointed out above. Taken together, these congruences yield for all $b \in [3] \setminus \{a - 3\}$:

$$\gcd(x_1 s_{j,b} + x_2 t_j, r_j) = \gcd(x_1 s_{j,3+b} + x_2 t_j, r_j) = 1.$$

Hence, by Lemma 2, $\beta_j^{x_1} \delta_j^{x_2}$ contains at least 4 cycles of length r_j. However η_j contains only 2 cycles of length r_j and ρ does not contain any other cycles of length r_j, which gives us a contradiction. Thus we obtain $a \in [3]$ and by this

$$x_2 \equiv -x_1 s_{j,a} \equiv x_1 \not\equiv 0 \bmod p_{i_a},$$

where $x_1 \not\equiv 0 \bmod p_{i_a}$ holds by Claim 2. Moreover, for all $b \in [3] \setminus \{a\}$ we obtain

$$x_2 \equiv -x_1 s_{j,a} \equiv 0 \bmod p_{i_b}.$$

This shows Claim 3. □

We can now show that $|T \cap C_j| = 1$ for all $j \in [m]$. Let $j \in [m]$. By Claim 3 there is a unique $i \in C_j$ such that $x_2 \not\equiv 0 \bmod p_i$. Thus $i \in T$ by (2). Moreover for all $h \in C_j \setminus \{i\}$ we have $x_2 \equiv 0 \bmod p_h$ by Claim 3 and hence $h \notin T$. Thus, we get $|T \cap C_j| = 1$.

For the other direction, suppose there is a subset $T \subseteq [n]$ such that $|T \cap C_j| = 1$ for all $j \in [m]$. We define $x_1 = 1$ and x_2 as the smallest positive integer satisfying the congruences

$$x_2 \equiv \begin{cases} 1 \bmod p_i & \text{if } i \in T \\ 0 \bmod p_i & \text{if } i \notin T \end{cases}$$

for all $i \in [n]$. Since $x_1 = 1$, ρ and $\pi_1^{x_1} \pi_2^{x_2}$ both contain a unique cycle of length $p_i p_{n+i}$ for all $i \in [n]$. All other cycles in ρ and $\pi_1^{x_1} \pi_2^{x_2}$ result from powers of $([r_j])$ for some $j \in [m]$. Consider a $j \in [m]$ and let $C_j = \{i_1, i_2, i_3\}$ with $i_1 < i_2 < i_3$. By Lemma 2, η_j consists of

(i) $p_{i_1} p_{i_2} p_{i_3}$ cycles of length q_j,
(ii) p_{i_1} cycles of length $p_{i_2} p_{i_3} q_j$,
(iii) p_{i_2} cycles of length $p_{i_1} p_{i_3} q_j$,
(iv) p_{i_3} cycles of length $p_{i_1} p_{i_2} q_j$ and
(v) 2 cycles of length r_j.

We have to show that

$$\beta_j \delta_j^{x_2} = (([r_j])^{s_{j,1}+x_2 t_j}, ([r_j])^{s_{j,2}+x_2 t_j}, ([r_j])^{s_{j,3}+x_2 t_j},$$
$$([r_j])^{s_{j,4}+x_2 t_j}, ([r_j])^{s_{j,5}+x_2 t_j}, ([r_j])^{s_{j,6}+x_2 t_j})$$

contains the same cycle lengths with the same multiplicities as in (i)–(v). Note that $s_{j,d} + x_2 t_j \equiv 1 \bmod q_j$ for all $d \in [6]$. Let $a \in [3]$ be the unique element with $i_a \in T$. Then $x_2 \equiv 1 \bmod p_{i_a}$ and $x_2 \equiv 0 \bmod p_{i_b}$ for all $b \in [3] \setminus \{a\}$. We obtain

$$s_{j,a} + x_2 t_j \equiv -1 + 1 \equiv 0 \bmod p_{i_a} \text{ and}$$
$$s_{j,a} + x_2 t_j \equiv 0 + 0 \equiv 0 \bmod p_{i_b} \text{ for all } b \in [3] \setminus \{a\}.$$

By Lemma 2, $([r_j])^{s_{j,a} + x_2 t_j}$ consists of $p_{i_1} p_{i_2} p_{i_3}$ cycles of length q_j. Moreover

$$s_{j,3+a} + x_2 t_j \equiv -1 + 1 \equiv 0 \bmod p_{i_a} \text{ and}$$
$$s_{j,3+a} + x_2 t_j \equiv s_{j,3+a} + 0 \not\equiv 0 \bmod p_{i_b} \text{ for all } b \in [3] \setminus \{a\}$$

(for the second point we use the fact that all primes p_i are larger than 3). By Lemma 2, $([r_j])^{s_{j,3+a} + x_2 t_j}$ consists of p_{i_a} cycles of length $q_j \prod_{b \in [3] \setminus \{a\}} p_{i_b}$. For all $b \in [3] \setminus \{a\}$ we have

$$s_{j,b} + x_2 t_j \equiv 0 + 1 \equiv 1 \bmod p_{i_a},$$
$$s_{j,b} + x_2 t_j \equiv s_{j,b} + 0 \equiv -1 \bmod p_{i_b} \text{ and}$$
$$s_{j,b} + x_2 t_j \equiv s_{j,b} + 0 \equiv 0 \bmod p_{i_c}, \text{where } \{c\} = [3] \setminus \{a,b\}.$$

By Lemma 2, $([r_j])^{s_{j,b} + x_2 t_j}$ consists of p_{i_c} cycles of length $q_j p_{i_a} p_{i_b}$ with $\{c\} = [3] \setminus \{a,b\}$. Finally, for all $b \in [3] \setminus \{a\}$ we have

$$s_{j,3+b} + x_2 t_j \equiv s_{j,3+b} + 1 \not\equiv 0 \bmod p_{i_a} \text{ and}$$
$$s_{j,3+b} + x_2 t_j \equiv s_{j,3+b} + 0 \not\equiv 0 \bmod p_{i_c} \text{ for all } c \in [3] \setminus \{a\}.$$

Hence, $([r_j])^{s_{j,3+b} + x_2 t_j}$ is a single cycle of length r_j. This shows that $\mathsf{ct}(\eta_j) = \mathsf{ct}(\beta_j \delta_j^{x_2})$ and concludes the proof of the theorem. $\square$

The construction from the previous proof yields the following additional result:

Corollary 1. *The following problem is* NP-*complete:*

- *input: $\rho, \pi_1, \pi_2 \in \mathsf{Sym}(n)$ such that π_1 and π_2 commute*
- *question: Is there is a $\pi \in \pi_1 \langle \pi_2 \rangle$ such that $\mathsf{ct}(\rho) = \mathsf{ct}(\pi)$?*

Proof. The instance ρ, π_1, π_2 of $\mathsf{CycleType}(\mathsf{ab}, 2)$ that we constructed in the proof of Theorem 2 has the property that there are $x_1, x_2 \in \mathbb{N}$ such that ρ and $\pi_1^{x_1} \pi_2^{x_2}$ have the same cycle type if and only if there is $x_2 \in \mathbb{N}$ such that ρ and $\pi_1 \pi_2^{x_2}$ have the same cycle type. This yields the corollary. $\square$

Whereas it can be decided in logspace whether a cyclic permutation group $\langle \pi_1 \rangle$ contains a permutation with a given cycle type (Theorem 1), the same problem for cosets of cyclic permutation groups is NP-complete (Corollary 1).

5 Fixpoint Freeness in the 2-Generated Abelian Case

Our main result for the problem FixpointFree is:

Theorem 3. FixpointFree(ab, 2) *is NP-complete.*

Proof. We give a logspace reduction from 3-SAT (the satisfiability problem for conjunctions of clauses, where every clause consists of exactly three literals and a literal is either a boolean variable x or a negated boolean variable $\bar{x}$). For this take a finite set of variables $X = \{x_1, \ldots, x_n\}$ and a set of clauses $\mathcal{C} = \{C_1, \ldots, C_m\}$. Every $C_j \in \mathcal{C}$ is a set of three literals. When we write C_j as $C_j = \{\tilde{x}_{i_1}, \tilde{x}_{i_2}, \tilde{x}_{i_3}\}$, every $\tilde{x}_{i_k}$ is either x_{i_k} or $\bar{x}_{i_k}$ and we always assume that $i_1 < i_2 < i_3$. A truth assignment $\sigma : X \to \{0,1\}$ is implicitly extended to all literals by setting $\sigma(\bar{x}_i) = 1 - \sigma(x_i)$.

Let $p_1, \ldots, p_n, \bar{p}_1, \ldots, \bar{p}_n$ be the first $2n$ primes. We associate the positive literal x_i with p_i and the negative literal $\bar{x}_i$ with $\bar{p}_i$ and define

$$\tilde{p}_i = \begin{cases} p_i & \text{if } \tilde{x}_i = x_i, \\ \bar{p}_i & \text{if } \tilde{x}_i = \bar{x}_i. \end{cases}$$

For the clause $C_j = \{\tilde{x}_{i_1}, \tilde{x}_{i_2}, \tilde{x}_{i_3}\}$ define $r_j = \tilde{p}_{i_1} \tilde{p}_{i_2} \tilde{p}_{i_3}$. Moreover, for all $i \in [n], l \in [p_i - 1]$ and $k \in [\bar{p}_i - 1]$ let $s_{i,l,k}$ be the unique number in $[p_i \bar{p}_i - 1]$ with

$$s_{i,l,k} \equiv l \bmod p_i \quad \text{and} \quad s_{i,l,k} \equiv k \bmod \bar{p}_i.$$

We will work with the group

$$G = \prod_{i=1}^{n} \left(\mathsf{Sym}(p_i) \times \mathsf{Sym}(\bar{p}_i) \times \mathsf{Sym}(p_i \bar{p}_i)^{(p_i-1)(\bar{p}_i-1)+1} \right) \times \prod_{j=1}^{m} \mathsf{Sym}(r_j).$$

The group G naturally embeds into $\mathsf{Sym}(N)$ for

$$N = \sum_{i=1}^{n} (p_i + \bar{p}_i + p_i \bar{p}_i ((p_i - 1)(\bar{p}_i - 1) + 1)) + \sum_{j=1}^{m} r_j.$$

Now we define the input permutations π_1 and π_2 as follows, where i ranges over $[n]$, l ranges over $[p_i - 1]$, k ranges over $[\bar{p}_i - 1]$ and j ranges over $[m]$:

$$\pi_1 = (\alpha_1, \ldots, \alpha_n, \beta_1, \ldots, \beta_m) \text{ with}$$
$$\alpha_i = (\alpha_{i,1}, \alpha_{i,2}, \alpha_{i,3}, \alpha_{i,1,1}, \ldots, \alpha_{i,p_i-1,\bar{p}_i-1})$$
$$\alpha_{i,1} = ([p_i])$$
$$\alpha_{i,2} = ([\bar{p}_i])$$
$$\alpha_{i,3} = \text{id}$$
$$\alpha_{i,l,k} = ([p_i\bar{p}_i])^{s_{i,l,k}}$$
$$\beta_j = \text{id}$$
$$\pi_2 = (\gamma_1, \ldots, \gamma_n, \delta_1, \ldots, \delta_m) \text{ with}$$
$$\gamma_i = (\gamma_{i,1}, \gamma_{i,2}, \gamma_{i,3}, \gamma_{i,1,1}, \ldots, \gamma_{i,p_i-1,\bar{p}_i-1})$$
$$\gamma_{i,1} = \gamma_{i,2} = \text{id}$$
$$\gamma_{i,3} = \gamma_{i,l,k} = ([p_i\bar{p}_i])$$
$$\delta_j = ([r_j])$$

Note that π_1 and π_2 commute. We will show that $\mathcal{C}$ is satisfiable if and only if there are $z_1, z_2 \in \mathbb{N}$ such that $\pi_1^{z_1}\pi_2^{z_2} \in \text{fpf}(N)$.

First, suppose that there are $z_1, z_2 \in \mathbb{N}$ such that $\pi_1^{z_1}\pi_2^{z_2} \in \text{fpf}(N)$.

Claim 4. For all $i \in [n]$ we have $z_1 \not\equiv 0 \bmod p_i$ and $z_1 \not\equiv 0 \bmod \bar{p}_i$.

We have $\alpha_{i,1}^{z_1}\gamma_{i,1}^{z_2} = \alpha_{i,1}^{z_1} = ([p_i])^{z_1}$ and hence by Lemma 2 we obtain $z_1 \not\equiv 0 \bmod p_i$. Analogously we obtain $z_1 \not\equiv 0 \bmod \bar{p}_i$. □

Claim 5. For all $i \in [n]$ we have $z_2 \equiv 0 \bmod p_i$ if and only if $z_2 \not\equiv 0 \bmod \bar{p}_i$.

Assume that $z_2 \equiv 0 \bmod p_i$ and $z_2 \equiv 0 \bmod \bar{p}_i$. Then we obtain by

$$\alpha_{i,3}^{z_1}\gamma_{i,3}^{z_2} = \gamma_{i,3}^{z_2} = ([p_i\bar{p}_i])^{z_2} = \text{id}$$

a contradiction. Now assume that $z_2 \not\equiv 0 \bmod p_i$ and $z_2 \not\equiv 0 \bmod \bar{p}_i$. Since by Claim 4 we have $z_1 \not\equiv 0 \bmod p_i$ and $z_1 \not\equiv 0 \bmod \bar{p}_i$ we can define $l \in [p_i - 1]$ and $k \in [\bar{p}_i - 1]$ as the smallest positive integers satisfying the congruences

$$l \equiv -z_2 z_1^{-1} \bmod p_i \quad \text{and} \quad k \equiv -z_2 z_1^{-1} \bmod \bar{p}_i.$$

From this we obtain $s_{i,l,k} \equiv -z_2 z_1^{-1} \bmod p_i\bar{p}_i$ and hence

$$\alpha_{i,l,k}^{z_1}\gamma_{i,l,k}^{z_2} = ([p_i\bar{p}_i])^{s_{i,l,k} \cdot z_1}([p_i\bar{p}_i])^{z_2} = ([p_i\bar{p}_i])^{-z_2}([p_i\bar{p}_i])^{z_2} = \text{id},$$

which is again a contradiction. This shows Claim 5 □

Claim 6. For all $j \in [m]$ there is an $a \in [3]$ such that $z_2 \not\equiv 0 \bmod \tilde{p}_{i_a}$, where $C_j = \{\tilde{x}_{i_1}, \tilde{x}_{i_2}, \tilde{x}_{i_3}\}$.

Since we must have $\beta_j^{z_1}\delta_j^{z_2} = \delta_j^{z_2} = ([r_j])^{z_2} \in \mathsf{fpf}(r_j)$ we must have $z_2 \not\equiv 0 \bmod r_j = \tilde{p}_{i_1}\tilde{p}_{i_2}\tilde{p}_{i_3}$. Hence, there is an $a \in [3]$ such that $z_2 \not\equiv 0 \bmod \tilde{p}_{i_a}$. $\qquad\square$

We define the truth assignment $\sigma : X \to \{0,1\}$ by

$$\sigma(x_i) = \begin{cases} 1 & \text{if } z_2 \not\equiv 0 \bmod p_i \\ 0 & \text{if } z_2 \equiv 0 \bmod p_i \end{cases}$$

for all $i \in [n]$ and show that every clause in $\mathcal{C}$ contains a literal that is mapped to 1 by σ. Let $j \in [m]$ and $C_j = \{\tilde{x}_{i_1}, \tilde{x}_{i_2}, \tilde{x}_{i_3}\}$. By Claim 6 there is an $a \in [3]$ such that $z_2 \not\equiv 0 \bmod \tilde{p}_{i_a}$. If $\tilde{x}_{i_a} = x_{i_a}$, then $\tilde{p}_{i_a} = p_{i_a}$ and $1 = \sigma(x_{i_a}) = \sigma(\tilde{x}_{i_a})$. On the other hand, if $\tilde{x}_{i_a} = \bar{x}_{i_a}$, then $\tilde{p}_{i_a} = \bar{p}_{i_a}$ and $z_2 \equiv 0 \bmod p_{i_a}$ by Claim 5. We obtain $1 = 1 - \sigma(x_{i_a}) = \sigma(\bar{x}_{i_a}) = \sigma(\tilde{x}_{i_a})$. Hence, $\sigma(\tilde{x}_{i_a}) = 1$ in both cases.

Vice versa suppose that there is a truth assignment $\sigma : X \to \{0,1\}$ such that every clause in $\mathcal{C}$ contains a literal that is mapped to 1 by σ. We define $z_1 = 1$ and $z_2 \in \mathbb{N}$ as the smallest positive integer satisfying the congruences

$$z_2 \equiv \sigma(x_i) \bmod p_i \quad \text{and} \quad z_2 \equiv 1 - \sigma(x_i) \bmod \bar{p}_i \qquad (4)$$

for all $i \in [n]$. Then $\pi_1^{z_1}\pi_2^{z_2} \in \mathsf{fpf}(N)$ follows from the following points, where $i \in [n]$, $l \in [p_i - 1]$, $k \in [\bar{p}_i - 1]$, and $j \in [m]$ are arbitrary:

- $\alpha_{i,1}^{z_1}\gamma_{i,1}^{z_2} = ([p_i])$, $\alpha_{i,2}^{z_1}\gamma_{i,2}^{z_2} = ([\bar{p}_i])$ and $\alpha_{i,3}^{z_1}\gamma_{i,3}^{z_2} = \gamma_{i,3}^{z_2} = ([p_i\bar{p}_i])^{z_2}$ are fixpoint-free.
- $\alpha_{i,l,k}^{z_1}\gamma_{i,l,k}^{z_2} = ([p_i\bar{p}_i])^{s_{i,l,k}+z_2}$ is fixpoint-free since $s_{i,l,k} + z_2 \equiv l \not\equiv 0 \bmod p_i$ if $\sigma(x_i) = 0$ and $s_{i,l,k} + z_2 \equiv k \not\equiv 0 \bmod \bar{p}_i$ if $\sigma(x_i) = 1$.
- $\beta_j^{z_1}\delta_j^{z_2} = \delta_j^{z_2} = ([r_j])^{z_2}$ is fixpoint-free. To see this let $C_j = \{\tilde{x}_{i_1}, \tilde{x}_{i_2}, \tilde{x}_{i_3}\}$ and $a \in [3]$ be such that $\sigma(\tilde{x}_{i_a}) = 1$. Then (4) yields $z_2 \equiv \sigma(\tilde{x}_{i_a}) \equiv 1 \bmod \tilde{p}_{i_a}$ and hence $z_2 \not\equiv 0 \bmod r_j$. $\qquad\square$

Corollary 2. *It is* NP-*complete to check whether* $\pi_1\langle\pi_2\rangle \cap \mathsf{fpf}(n) \neq \emptyset$ *holds for given* $\pi_1, \pi_2 \in \mathsf{Sym}(n)$ *with* $\pi_1\pi_2 = \pi_2\pi_1$.
Proof. For π_1 and π_2 from the proof of Theorem 3, there are $z_1, z_2 \in \mathbb{N}$ with $\pi_1^{z_1}\pi_2^{z_2} \in \mathsf{fpf}(n)$ if and only if there is $z \in \mathbb{N}$ with $\pi_1\pi_2^z \in \mathsf{fpf}(n)$. $\qquad\square$

6   Conclusion

We proved NP-completeness of the following two problem:

- Does a given 2-generated abelian permutation group contain a permutation with a given cycle type (Theorem 2)?
- Does a given 2-generated abelian permutation group contain a fixpoint-free permutation (Theorem 3)?

One might consider the problems CycleType and FixpointFree also for other classes of permutation groups. Whereas FixpointFree is trivial for transitive permuation groups (by Jordan's theorem [14]), the complexity of CycleType for transitive permuation groups seems to be open.

Acknowledgments. This work has been supported by the DFG research project LO 748/15-1.

References

1. Arora, S., Barak, B.: Computational Complexity - A Modern Approach. Cambridge University Press (2009). https://doi.org/10.1017/CBO9780511804090
2. Arvind, V.: The parameterized complexity of fixpoint free elements and bases in permutation groups. In: Gutin, G., Szeider, S. (eds.) IPEC 2013. LNCS, vol. 8246, pp. 4–15. Springer, Cham (2013). https://doi.org/10.1007/978-3-319-03898-8_2
3. Babai, L., Luks, E.M., Seress, Á.: Permutation groups in NC. In: Proceedings of the 19th Annual ACM Symposium on Theory of Computing, STOC 1987, pp. 409–420. ACM (1987). https://doi.org/10.1145/28395.28439
4. Bóna, M.: Combinatorics of Permutations, 3rd Edn. Discrete Mathematics and Its Applications. CRC Press (2022). https://doi.org/10.1201/9780429274107
5. Buchheim, C., Jünger, M.: Linear optimization over permutation groups. Discret. Optim. **2**(4), 308–319 (2005). https://doi.org/10.1016/J.DISOPT.2005.08.005
6. Cameron, P.J.: Permutation Groups. Cambridge University Press (2010). https://doi.org/10.1017/CBO9780511623677
7. Cameron, P.J.: Lectures on derangements. In: Pretty Structures Conference in Paris, vol. 654 (2011)
8. Cameron, P.J., Wu, T.: The complexity of the weight problem for permutation and matrix groups. Discret. Math. **310**(3), 408–416 (2010). https://doi.org/10.1016/J.DISC.2009.03.005
9. Cook, S.A., McKenzie, P.: Problems complete for deterministic logarithmic space. J. Algorithms **8**(3), 385–394 (1987). https://doi.org/10.1016/0196-6774(87)90018-6
10. Furst, M.L., Hopcroft, J.E., Luks, E.M.: Polynomial-time algorithms for permutation groups. In: Proceedings of the 21st Annual Symposium on Foundations of Computer Science, FOCS 1980, pp. 36–41. IEEE Computer Society (1980). https://doi.org/10.1109/SFCS.1980.34
11. Garey, M.R., Johnson, D.S.: Computers and Intractability: A Guide to the Theory of NP–Completeness. Freeman (1979)
12. Hesse, W., Allender, E., Barrington, D.: Uniform constant-depth threshold circuits for division and iterated multiplication. J. Comput. Syst. Sci. **65**(4), 695–716 (2002). https://doi.org/10.1016/S0022-0000(02)00025-9
13. Jerrum, M.: The complexity of finding minimum-length generator sequences. Theoret. Comput. Sci. **36**, 265–289 (1985). https://doi.org/10.1016/0304-3975(85)90047-7
14. Jordan, C.: Recherches sur les substitutions. J. de Mathématiques Pures et Appliquées **17**, 351–387 (1872). http://eudml.org/doc/234268
15. Sims, C.C.: Computational methods in the study of permutation groups. In: Computational Problems in Abstract Algebra, pp. 169–183. Pergamon (1970). https://doi.org/10.1016/B978-0-08-012975-4.50020-5
16. Sims, C.C.: Computation with permutation groups. In: Proceedings of the 2nd ACM Symposium on Symbolic and Algebraic Manipulation, SYMSAC 1971, pp. 23–28. ACM (1971). https://doi.org/10.1145/800204.806264

Counting Overlapping Pairs of Words

Eric Rivals$^{(\boxtimes)}$ and Pengfei Wang

LIRMM, Université Montpellier, CNRS, Montpellier, France
{rivals,pengfei.wang}@lirmm.fr

Abstract. A correlation is a binary vector that encodes all possible positions of overlaps of two words, where an overlap for an ordered pair of words (u, v) occurs if a suffix of u matches a prefix of v. As multiple pairs can have the same correlation, it is relevant to count how many pairs of words share the same correlation, depending on the alphabet size and word length n. We exhibit recurrences to compute the number of such pairs – which is termed *population size* – for any correlation; for this, we exploit a relationship between overlaps of two words and self-overlap of one word. This theorem allows us to compute the number of pairs with the longest overlap of a given length, solving two open questions Gabric raised in 2022. Finally, we also provide bounds for the asymptotic population ratio of any correlation. Given the importance of word overlaps in areas like combinatorics on words, bioinformatics, and digital communication, our results may ease analyses of algorithms for string processing, code design, or genome assembly.

Keywords: combinatorics · correlation · overlap · border · asymptotic · bounds · expectation · string

1 Introduction

A word u overlaps a word v if a suffix of u equals a prefix of v. The shared suffix-prefix is called a *border* for the ordered pair of words (u, v) (note that other authors call this a *right border*, see [7]). If (u, v) has no border, it is said *unbordered*. The pair (u, v) is said *mutually unbordered* if both (u, v) and (v, u) lack a border. Conversely, if both (u, v) and (v, u) have a border, then the pair is said to be *mutually bordered*. These notions generalize the well-studied concepts of border, bordered, and unbordered words, originally defined for single words, to pairs of words.

Overlapping and unbordered words are central in many applications: bioinformatics, pattern matching, or code design. Computing overlaps between all pairs of sequencing reads is one step of the genome assembly task [10,22]; several algorithms solve it in optimal time [11,17]. The notion of borders is core in combinatorics on word [20,21], the design of pattern matching algorithms [16,35], and in the statistical analysis of pattern finding and discovery [5,24]. For instance, questions in vocabulary statistics deal with the distributions of the number of

F. V. Fomin and M. Xiao (Eds.): COCOON 2025, LNCS 15984, pp. 381–395, 2026.
https://doi.org/10.1007/978-981-95-0218-9_28

missing words or of common words in random texts [26,27], which depend on the overlap structure of words, and find applications in bioinformatics [34] or in the test of random number generators [25]. A set of mutually unbordered words serves as code for synchronization purposes in network communication. A seminal construction algorithm appeared in 1973 [23], and others brought recent improvements in the design of cross-bifix-free codes [1,3] or non-overlapping code [2], a topic of combinatorial interest [4].

The combinatorics of single (not pair) bordered and unbordered words over a q-ary alphabet has been studied in depth. For instance, the recurrence for counting the number of unbordered words of length n over a q-ary alphabet was first given in [23], while the recurrence for counting the number of bordered words (termed "overlapping sequences" in some articles) is proven in [19]. From this, the probability that a random word of length n is unbordered was shown to converge when n tends to infinity in [23], while Holub and Shallit have shown that the expected maximum border length for words of length n over a q-ary alphabet also converges [15]. In a related area (but somehow more distant from our topic), other works have investigated unbordered factors of words [13,14,18], a topic introduced by Ehrenfeucht and Silberger in [6].

Recently, building on ideas similar to those used in [23], Gabric gave three recurrences to count bordered, mutually bordered, mutually unbordered pairs of words of length n over a k-ary alphabet [7]. In his conclusion, he raised two challenging open questions: Q1: Count the number of pairs having the longest border of length j (with j satisfying $0 < j < n$); Q2: What is the expected length of the longest border across all pairs of length-n words. We address and solve both questions in our work (see Sect. 5).

Example: Consider the binary alphabet $\{a, b\}$ and the following three words denoted by u, v, w: abaaa, aaabb, and abbbb. The pairs (u, v) and (v, w) both have the longest border of length 3, but (u, v) has 3 distinct non-empty borders aaa, aa, and a, while (v, w) has only one abb. The pairs (v, u) and (w, v) have no borders, which illustrates the asymmetry of this notion.

First, this example illustrates that the possibilities of overlap of a pair (u, v) depend on the self-overlapping structure of their longest border (compare aaa with abb). Second, it shows that the self-overlap structure of the border limits the number of words having such a shared suffix-prefix, and thus the number of pairs of words to count. Indeed, only words of length 5 having a suffix (resp. prefix) such as aaa or bbb, can participate in a pair having as many and as long borders as (u, v). These observations suggest that, in response to the open question raised by Gabric, one may have to account for the complete overlap structure of a pair of words.

Other authors have proposed to encode the starting position of such overlaps in a binary vector called a *correlation* [8]. In our example, the correlation of the pair (u, v) is 00111, while that of (v, w) is 00100. For any word z, the correlation of (z, z) is called the *autocorrelation* of z. Clearly, multiple pairs can have the same correlation, and hence there are fewer correlations of length n than pairs of words of length n.

Fortunately, one can build on previous studies of the set of autocorrelations, denoted Γ_n, and the set of correlations, denoted Δ_n, for all possible words of length n [8,9,29,30]. It is known that the self-overlap structure of a word [8], as well as the overlap structure of a pair of words [31,32], do not depend on the alphabet size (provided that the alphabet has at least two letters – a unary alphabet makes these questions trivial). Combining a characterization of Δ_n provided in [31,32] and an algorithm for enumerating Γ_n [28], we can enumerate Δ_n to get the list of all correlations of length n.

With the terminology used in [8,27,30], we exhibit two solutions to compute the population size of any correlation, which is the number of pairs of words having the same correlation (in Sect. 3). For this, we exploit two recurrences to compute the population size of autocorrelations [8,29]. With this in hand, in Sect. 5 we derive formulae for the abovementioned open questions (Corollary 22, Eq. 4, Theorem 23). Besides this, we provide bounds for the asymptotic behavior of the population ratio of any correlation (Theorem 20 Sect. 4), which extend the result known for autocorrelations [8]. Finally, we conclude with some open questions (Sect. 6). Additional results and omitted proofs are available in the preprint version of this work on arXiv [33].

2   Preliminaries

Let Σ be a finite *alphabet*, a set of *letters* of cardinality σ. We call a sequence of elements of Σ a *string* or a *word*. The empty word is denoted by ε. We denote by Σ^* the set of all finite words over Σ, and by Σ^n the set of all words of length n over Σ, with $n \in \mathbb{N}$. For a word x, $|x|$ denotes the *length* of x. For two words x, y, we denote their concatenation by xy, and the k-fold concatenation of x with itself by x^k for any $k > 0$. For any $L \subset \Sigma^*$, we define $x.L$ as $\{xy : y \in L\}$.

Let u be a word of Σ^n. We index the letters of u from 0 to $n-1$: $u = u[0]\ldots u[n-1]$. The ith letter of u is denoted by $u[i]$. We also denote by $u[i..j]$ for any $0 \le i \le j < n$ the substring of u starting at position i and ending at position j. A substring is said to be *proper* iff $j - i + 1 < n$. Moreover, $u[0..j]$ is a prefix, $u[i..n-1]$ is a suffix of u.

2.1   Definitions of Borders and Correlation for Pairs of Words

To study overlaps between two words, we consider ordered pairs of words: we denote a pair of words $(u, v) \in \Sigma^n \times \Sigma^m$, which differs from the pair (v, u).

Definition 1 (Border of pair of words). *A border of a pair of words $(u, v) \in \Sigma^n \times \Sigma^m$ is any string that is a non-empty suffix of u, and a non-empty prefix of v. If a border exists, (u, v) is said* bordered, *otherwise it is* unbordered.

A pair may have multiple borders, and in general, the set of borders for (u, v) differs from that of (v, u). In his article, Gabric refers to a border of (u, v) as a right border and to a border of (v, u) as a left border; we use a different terminology.

Guibas & Odlyzko [9] proposed to encode in a binary vector the positions in u at which a border is starting, and they named this notion the *correlation* of a pair of words. From now on, for the sake of simplicity, we focus on pairs of words of equal length, denoted n, although our results can be generalized to the case of unequal lengths.

Definition 2 (Correlation). *Let* $(u, v) \in \Sigma^n \times \Sigma^n$. *The correlation of* (u, v), *denoted by* $c(u, v)$, *is a binary vector of length* n (*i.e.,* $c(u, v) \in \{0, 1\}^n$) *satisfying* $\forall i \in [0, \ldots, n - 1]$

$$c(u, v)[i] = \begin{cases} 1 & \text{if } u[i..n-1] = v[0..n-i-1] \\ 0 & \text{otherwise.} \end{cases}$$

Generally $c(u, v) \neq c(v, u)$. For any length $n \in \mathbb{N}$, we denote the set of all correlations for words of length n by Δ_n and its cardinality by δ_n as in [30].

Definition 3 (Δ_n and δ_n). *Let* $n \in \mathbb{N}$. *The set of all correlations of words of length* n *is:*

$$\Delta_n := \{t \in \{0, 1\}^n : \exists (u, v) \in \Sigma^n \times \Sigma^n : c(u, v) = t\},$$

and its cardinality is denoted by δ_n.

Example 4. Consider the pair of words $(u, v) = (\mathsf{aabbab}, \mathsf{babbaa})$ of length 6 over the binary alphabet $\{\mathsf{a}, \mathsf{b}\}$. The pair (u, v) has a border starting at position 3 in u, and a shorter border starting at position 5. Its correlation is $c(u, v) = 000101$. See Table 1. Of course, a permutation of the alphabet (that is exchanging a with b and vice versa) yields a different pair of words, which has the same correlation as (u, v). Thus, several pairs can share the same correlation.

Table 1. Example of correlations for the words of length 6: $u := \mathsf{aabbab}$ and $v := \mathsf{babbaa}$. Left the table for $c(u, v)$: All possible shifts of v to the right of u are displayed on distinct lines: those at which an overlap exists are colored in blue. The last column shows $c(u, v)$ written top-down, with 1 bits colored in blue corresponding to borders. Right: same table for $c(v, u)$.

pos.	0 1 2 3 4 5	
u	a a b b a b - - - - -	t
v	b a b b a a - - - -	0
	- b a b b a a - - - -	0
	- - b a b b a a - - -	0
	- - - b a b b a a - -	1
	- - - - b a a b a b -	0
	- - - - - b a b b a a	1

pos.	0 1 2 3 4 5	
v	b a b b a a - - - - -	t
u	a a b b a b - - - -	0
	- a a b b a b - - - -	0
	- - a a b b a b - - -	0
	- - - a a b b a b - -	0
	- - - - a a b b a b -	1
	- - - - - a a b b a b	1

A special case arises when u equals v. Then $c(u, u)$ is called the *autocorrelation* of u. We recall definitions of period and some useful known properties of autocorrelations. Their proofs can be found in [8, 12, 31, 32].

Definition 5 (Period). *A word $u = u[0..n-1]$ has period $p \in \{0, 1, \ldots, n-1\}$ if and only if $u[0..n-p-1] = u[p..n-1]$, i.e., for all $0 \leq i \leq n-p-1$, we have $u[i] = u[i+p]$.*

The zero period is called *trivial*. The smallest non-trivial period of u is called its *basic period*.

For all possible words of length $n \in \mathbb{N}$, the set of autocorrelations, denoted by Γ_n, is defined as: $\Gamma_n := \{s \in \{0, 1\}^n : \exists u \in \Sigma^n : c(u, u) = s\}$. We denote by κ_n the cardinality of Γ_n. Clearly, $\Gamma_n \subset \Delta_n$. When $n = 0$ we consider that $\Gamma_n = \{\varepsilon\}$.

Lemma 6. *Let $s \in \Gamma_n$ and $u \in \Sigma^n$ such that $c(u, u) = s$. Let $0 \leq p \leq q < n$ such that $s[p] = 1$. Then, $s[q] = 1$ iff $u[p..n-1]$ has period $(q-p)$ (equivalently the $(q-p)$ bit in $c(u[p..n-1], u[p..n-1])$ equals 1).*

Lemma 7. *Let $s \in \Gamma_n$. For all p satisfying $0 \leq p < n$, and $s[p] = 1$, it follows that $s[kp] = 1$ for all $k \in [2, \ldots, \lfloor \frac{n}{p} \rfloor]$.*

Lemma 8. *Let $\pi(u)$ be the basic period of $u \in \Sigma^n$ and p be a non-trivial period. Then either $p = k \cdot \pi(u)$, with $k \in [1, \ldots, \lfloor \frac{n}{\pi(u)} \rfloor]$, or $p > n - \pi(u)$.*

2.2   Set of All Correlations of Length n and Its Characterization

The first characterization of autocorrelations was given by Guibas and Odlyzko in their seminal paper [8]. They studied the cardinality of Γ_n and provided a lower and an upper bound for $\log(\kappa_n)/\log_2(n)$, and conjectured that their lower bound was also an upper bound. They also proposed an algorithm to compute the number of words in Σ^n that share the same period set, which they termed the *population* of an autocorrelation. A key result of their work is the *alphabet independence* of Γ_n: Any alphabet with $\sigma > 1$ gives rise to the same set of autocorrelations, i.e., to Γ_n.

Rivals et al. [31, 32] have characterized Δ_n and exhibited its relation to the sets Γ_j for $0 \leq j \leq n$, which is stated below.

Lemma 9 (Lemma 21 [31]). *The set of correlations of length n is of the form*

$$\Delta_n = \left\{ 0^{(n-j)}s, \text{ with } s \in \Gamma_j \text{ and } j \in [0, \ldots, n] \right\}.$$

Lemma 9 gives us the **structure of any correlation** for any pair of words (u, v) of length n: it starts with a series of 0, until the leftmost 1, which marks the position in u of the longest border of pair (u, v). Let z denote this border and j denote its length. The above characterization is based on the fact that the suffix of length j of $c(u, v)$ (the one starting with the leftmost 1) must be the autocorrelation of z. Indeed, each border of z is also a border of (u, v). If $j = 0$, then z is empty string and $c(u, v) = 0^n$. Of course, if $u = v$, then the correlation of (u, v) is the autocorrelation of u.

This characterization implies the following **partition** of Δ_n:

Corollary 10. $\Delta_n = \bigcup_{j=0}^n \{0^{n-j}s \mid s \in \Gamma_j\} = \bigcup_{j=0}^n \left(0^{n-j}.\Gamma_j\right).$

Rivals et al. [31,32] studied the cardinalities of Γ_n and Δ_n and proved the asymptotic convergence of ratios involving κ_n and δ_n towards the same limit when n tends to infinity. Precisely, $\frac{\ln \kappa_n}{\ln^2(n)} \to \frac{1}{2\ln(2)}$, and $\frac{\ln \delta_n}{\ln^2(n)} \to \frac{1}{2\ln(2)}$ when $n \to \infty$.

Investigating the algebraic structure of Δ_n, we show that, like Γ_n, Δ_n is a lattice under set inclusion and does not satisfy the Jordan-Dedekind condition. Example 11 and Fig. 1 illustrate the lattice structure of Δ_n for $n = 4$.

Example 11. From Corollary 10, one has $\Delta_4 = \Gamma_4 \cup (0.\Gamma_3) \cup (00.\Gamma_2) \cup (000.\Gamma_1) \cup \{0000\}$. Elements of Γ_4 are shown in green background in Fig. 1.

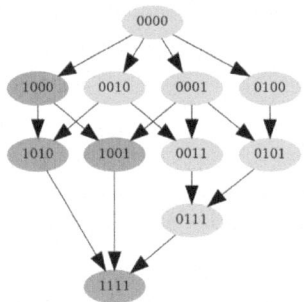

Correlation	Population sizes			
	$\sigma = 2$	$\sigma = 3$	$\sigma = 4$	$\sigma = 5$
0000	74	3678	45132	297020
0001	82	1866	15108	74380
0010	30	480	3060	12480
0011	24	216	960	3000
0100	16	162	768	2500
0101	8	54	192	500
0111	6	24	60	120
1000	6	48	180	480
1001	6	24	60	120
1010	2	6	12	20
1111	2	3	4	5

Fig. 1. The lattice of Δ_4: each node contains a correlation as a binary vector. The elements of Γ_4 are colored in green. Since the chains between 0000 and 1111 differ in length, Δ_4 does not satisfy the Jordan-Dedekind condition.

Fig. 2. Population sizes for correlations of Δ_4 (correlations of words of length $n = 4$) and for alphabet sizes $\sigma = 2, 3, 4$ and 5.

3 Population Size of a Correlation

We define the *population* of a correlation $t \in \Delta_n$ as: $P(t) := \{(u, v) \in \Sigma^n \times \Sigma^n$ such that $c(u, v) = t\}$ and denote its cardinality by $p(t)$. For example, consider the correlation $t := 01010$ from Δ_5: over the alphabet $\Sigma = \{a, b\}$, we have $P(t) = \{(ababa, babaa), (ababa, babab), (bbaba, babab), (bbaba, babaa), (aabab, ababa), (aabab, ababb), (babab, ababa), (babab, ababb)\}$ and $p(t) = 8$.

In the literature, one finds two ways of computing the population size of an autocorrelation (i.e., when $t = c(u, u)$). The first recurrence formula links the population size of t with that of a suffix of t [8][Thm 7.1]; it allows the authors to investigate the asymptotics of the population size [8][Thm 7.2][1]. The second recurrence formula takes advantage of the fact that Γ_n, the set of autocorrelations of length n, forms a lattice with set inclusion [30]. In Sect. 3.1, we review both formulae, and exhibit two recurrence formulae for correlations: the first is based on the suffix (Theorem 16 on page 8), and the other is a lattice-based recurrence (Theorem 18, page 9).

3.1 Computing the Population Size

Before finding a formula to compute the population size $p(t)$ of a correlation t in Δ_n, we show that $p(t)$ is related to the population size of some autocorrelations of words of length $2n$ in Theorem 14. To achieve this, we demonstrate two lemmas linking the borders of a pair (u, v) with the borders of the word vu.

Lemma 12. *Let $(u, v) \in \Sigma^n \times \Sigma^n$. Then $c(u, v)$ is the suffix of length n of $c(xu, vy)$ for any words x, y in Σ^k for some $k \geq 0$.*

Proof. Let $n > 0$, and let u and v be words of Σ^n. Note that for any words x, y in Σ^k for some $k \geq 0$, the suffix of length n of $c(xu, vy)$ encodes the borders between the suffix of length n of xu (which is u) and the prefix of length n of vy (which is v). Thus, $c(u, v)$ is the suffix of length n of $c(xu, vy)$.

Lemma 13. *Let $w \in \Sigma^{2n}$; let u and v be words in Σ^n such that $w = vu$. If w has a border, then the pair of words (u, v) is bordered.*

Proof. Let w, u, and v be as in the lemma. If $u = v$, then u is a border of the pair (u, u). Otherwise, we have $u \neq v$. Let z be a border of w. We distinguish two cases based on $|z|$.

1. Case 1: $|z| \in [1, \ldots, n-1]$. Then, there exist two words x, y of length $n - |z|$ such that $v = zy$ and $u = xz$. Thus, z is a border of (u, v).
2. Case 2: $|z| \in [n+1, \ldots, 2n-1]$. Then, w has a period $p := 2n - |z|$ and $p < n$ (the half $|w|$). According to properties of periods (Lemma 7), the integer $\lfloor \frac{2n}{p} \rfloor p$ is also period of w. Then, if we denote its corresponding border by z', we have $|z'| < n$, and we are back to case 1, with z' being a border of (u, v).

Before stating the theorem on the population size of a correlation, we need a notation. Let $t \in \Delta_n$. We denote by $G(t)$ the set of all words of length $2n$ whose autocorrelation has t as a suffix, and by $g(t)$ its cardinality. Formally, $G(t) := \{w \in \Sigma^{2n} : t \text{ is a suffix of } c(w, w)\}$.

The following theorem shows the relation between the number of pairs of words of length n and the number of specific words of length $2n$. For $t \in \Gamma_n$, $p(t)$ can be directly calculated using Theorem 15. Therefore, we consider $t \in \Delta_n$ but exclude those in Γ_n.

Theorem 14. *Let $t \in \Delta_n \setminus \Gamma_n$. Then, $p(t) = g(t)$.*

---

[1] In their article, the authors use the term "correlation" instead of autocorrelation.

Proof. i/ Let us first prove that $p(t) \leq g(t)$. Let $(u, v) \in P(t)$, that is, $c(u, v) = t$. By Lemma 12, let $x = v$ and $y = u$, we get that $c(vu, vu)$ has $t = c(v, u)$ as a suffix, i.e., $vu \in G(t)$. This implies that $p(t) \leq g(t)$.

ii/ Let us prove that $p(t) \geq g(t)$. Let $w \in G(t)$, and let u and v be words of length n such that $w = vu$. Again, by Lemma 12, we know that $c(u, v) = t$, which implies that $g(t) \leq p(t)$.

Combining both inequalities, we get $p(t) = g(t)$, which concludes the proof.

Now we will calculate the number of pairs of words of length n with the correlation $t = 0^{n-j}s \in \Delta_n$, where $s \in \Gamma_j$, i.e., the population size of t. Thanks to Theorem 14, we provide two different approaches of computing $p(t)$: the first one, based on the recurrence for the population size of autocorrelation using their recursive structure [8, Predicate Ξ], is presented in this section. The second approach based on the recurrence for the population size of autocorrelation that exploits the lattice structure of Γ_n [30], is shown on page 9.

Recurrence Based on the Recursive Structure of Autocorrelations.
We review the recurrence formula given by Guibas & Odlyzko. Let $s \in \Gamma_j$. They define the autocorrelation of length n denoted as $s_n := 10^{n-j-1}s$, and the sequence ψ for $k \in \mathbb{Z}$ depending on s as

$$\psi[k] := \begin{cases} 0 & \text{for } k > j \\ s[j-k] & \text{for } 1 \leq k \leq j \\ \sigma^{-k} & \text{for } k < 1. \end{cases}$$

We will use this definition of s_n in many places. The sequence ψ partitions $\mathbb{N}$ into three distinct ranges. For $k < 1$, $\psi[k]$ equals σ^{-k}. In the interval $1 \leq k \leq j$, $\psi[k]$ equals 1 if $(j - k)$ is a period in s, and 0 otherwise. For any $k > j$, $\psi[k]$ is consistently equal to 0. Let $s \in \Gamma_j$ and assumed fixed. Theorem 15 states their recurrence for $p(s_n)$.

Theorem 15 (Population size of an autocorrelation (Theorem 7.1 [8])).
Let $k \in \mathbb{Z}$. Let $n, j \in \mathbb{N}$ satisfying $0 \leq j < n$. Let $s \in \Gamma_j$ and let $s_n := 10^{n-j-1}s$. Then the number of words of length n that have autocorrelation $s_n \in \Gamma_n$ satisfies the recurrence:

$$p(s_n) + \sum_{k \in \mathbb{Z}} p(s_k)\psi[2k - n] = 2\psi[2j - n]p(s),$$

where $p(s_k) = 0$ for $k < j$, and the sequence ψ is defined as above.

We state our result regarding the population size of a correlation $t = 0^{n-j}s$ with s being fixed. See Fig. 2 for population sizes on different alphabets ($\sigma = 2, 3, 4, 5$) for all correlations in Δ_4. Note that, if $j = n$, then the population size of t is the known population size of s.

Theorem 16 (Population size of a correlation (I)). *Let $j, n \in \mathbb{N}$ satisfying $0 \le j < n$. Let $t := 0^{n-j}s$ be an element of Δ_n with $s \in \Gamma_j$. Then the population size of t satisfies the recurrence*

$$p(t) = \left(\sum_{\lambda=1}^{\lfloor j/2 \rfloor} p(s_{n+\lambda}) \cdot s[j - 2\lambda] \right) + p(s_{2n}).$$

Observe that in Theorem 16, calculating the population size of $t = 0^{n-j}s$ requires to compute $p(s_{(n+\lambda)})$ for all $\lambda \in \{0, \ldots, \lfloor j/2 \rfloor\}$. by Theorem 15. Therefore, we provide a third recurrence on t that calculates $p(t)$ relying only on s; see [33, Theorem 30].

Recurrence Based on the Lattice Structure. As Γ_n equipped with inclusion is a lattice [30, Theorem 3.1], the successor of an autocorrelation s is a more constrained autocorrelation, i.e., one that contains more periods than s. One can use this relationship to compute population sizes. From the proof of Theorem 16, we know the autocorrelation of $w \in G(t)$ satisfies the form: $c(w, w) = (10^{\pi(w)-1})^{\tilde{\lambda}/\pi(w)} s_{(2n-\tilde{\lambda})}$ where $s[j + 2\tilde{\lambda} - 2n] = 1$ for all $\tilde{\lambda} \in [\lceil \frac{2n-j}{2} \rceil, \ldots, n-1] \cup \{0\}$. Clearly $\pi(w)|\tilde{\lambda}$. Thus, we can provide another recurrence by using the notion of *number of free characters (nfc for short)* introduced in [30]. The *nfc* of an autocorrelation $s \in \Gamma_n$ is the maximum number of positions in a word u with $c(u, u) = s$ that are not determined by the periods. For instance, the *nfc* of $100001001 \in \Gamma_9$ is 4 since a word u with $c(u, u) = 100001001$ must satisfy *character equations*: $u[0] = u[3] = u[5] = u[8], u[1] = u[6]$, and $u[2] = u[7]$. Thus $u = u[0]u[1]u[2]u[0]u[4]u[0]u[1]u[2]u[0]$ where $u[0], u[1], u[2], u[4] \in \Sigma$. Theorem 17 states the recurrence on population sizes for autocorrelations.

Theorem 17 (Population size of an autocorrelation (Theorem 6.1 [30])). *Let $n \in \mathbb{N}$ and v_k be the kth ($k = 1, \ldots, \kappa_n$) autocorrelation of Γ_n. Let ρ_k denote the number of free characters of v_k. The population size $p(v_k)$ satisfies the recurrence*

$$p(v_k) = \sigma^{\rho_k} - \sum_{j:v_k \subset v_j} p(v_j).$$

The proof of Theorem 17 relies on the *nfc* of a given autocorrelation, on the lattice structure of Γ_n, and on the following idea. Consider the set $\mathcal{A}$ of words that satisfy the *character equations* imposed by autocorrelation v_k. As a word in $\mathcal{A}$ can satisfy additional character equations, $\mathcal{A}$ contains all words whose autocorrelation is v_k, but also words whose autocorrelations are $s_{(j)}$ with $j : v_k \subset v_j$. We reuse this idea to compute $p(t)$.

Let $\tilde{\lambda}_1, \ldots, \tilde{\lambda}_m$ be the proper positive divisors of $\tilde{\lambda}$. From the proof of Theorem 16, the autocorrelation of $w \in G(t)$ could be decomposed based on $\tilde{\lambda}_i$ for $i \le m$, and $\tilde{\lambda}$ for $\tilde{\lambda} \in \{\lceil \frac{2n-j}{2} \rceil, \ldots, n-1\}$ as follows:

$$c(w, w) = s_{(2n, \tilde{\lambda}, \tilde{\lambda}_i)} = (10^{\frac{\tilde{\lambda}}{\tilde{\lambda}_i} - 1})^{\tilde{\lambda}_i} 10^{2n - \tilde{\lambda} - j - 1} s.$$

Let $s \in \Gamma_j$ and consider correlation $t := 0^{n-j}s$ as fixed. Recall that $g(t)$ is the cardinality of the set $G(t)$ of all words of length $2n$ whose autocorrelation has t as the suffix. To state our second formula for the population size of t, we assume that all autocorrelations of Γ_{2n} have been calculated. For consistency, we use the notation ρ. to refer to the number of free characters of an autocorrelation (as in Theorem 17).

Theorem 18. *Let $n, i \in \mathbb{N}$. Let $\rho_{(2n,\tilde{\lambda},\tilde{\lambda}_i)}$ denote the number of free characters of $s_{(2n,\tilde{\lambda},\tilde{\lambda}_i)}$ and ρ be the number of free characters of s_{2n}. The population size $p(t)$ satisfies:*

$$
p(t) = \left(\sum_{\tilde{\lambda}=\lceil \frac{2n-j}{2} \rceil}^{n-1} \sum_{\tilde{\lambda}_i} (\sigma^{\rho_{(2n,\tilde{\lambda},\tilde{\lambda}_i)}} - \sum_{v \in \Gamma_{2n} : s_{(2n,\tilde{\lambda},\tilde{\lambda}_i)} \subset v} p(v)) \right) + \sigma^\rho - \sum_{y \in \Gamma_{2n} : s_{2n} \subset y} p(y).
$$

4   Asymptotics on the Population Ratios

The population ratio of a correlation $t \in \Delta_n$ is $p(t)/\sigma^{2n}$. Here, we study the asymptotic lower and upper bounds for this ratio. Before stating our result, we recall some definitions introduced by Guibas & Odlyzko [8]. Recall that Theorem 15 on the population size of an autocorrelation s_n relies on a sequence $\psi[k]$. They define three generating functions (with dummy variable z) two for $p(s_n)$ and $\psi[k]$, and introduce $\tilde{h}(z)$, which is the normalization of $h(z)$ by $p(s)$. Their definitions are as follows:

$$
h(z) = \sum_{n=0}^{\infty} p(s_n)z^{-n}; \quad \psi(z) = \sum_{n=0}^{\infty} \psi[k]z^{-n}; \quad \tilde{h}(z) = \frac{h(z)}{p(s)}.
$$

Thus, Theorem 15 can be rewritten as:

$$
\tilde{h}(z) + \psi(z)\tilde{h}(z^2) = 2\psi(z)z^{-2j}. \tag{1}
$$

Hence, the asymptotics of $p(s_n)$ as $n \to \infty$ with s being fixed follows.

Theorem 19 (Asymptotics on the population sizes [8]). *Let μ be any small positive complex number. Let $j \in \mathbb{N}$ satisfying $0 \le j < n$. Let $s \in \Gamma_j$ and let $s_n := 10^{n-j-1}s$. The population size of $s_n \in \Gamma_n$ divided by the population size of s over an alphabet of cardinality $\sigma \ge 2$ satisfies*

$$
\frac{p(s_n)}{p(s)} = \left(\frac{2}{\sigma^{2j}} - \tilde{h}(\sigma^2) \right) \sigma^n + O((\sigma + \mu)^{\frac{n}{2}}),
$$

where $\tilde{h}(\sigma^2)$ satisfies the Functional Eq. (1).

Denote $c = \frac{2}{\sigma^{2j}} - \tilde{h}(\sigma^2)$. Note that c is the asymptotic limit of $p(s_n)/(p(s)\sigma^n)$; thus $c \cdot p(s)$ provides the limiting value of $p(s_n)/\sigma^n$. Here, we state our result on the population size of correlation $t \in \Delta_n$ with s being assumed fixed. In Table 2 we show, for some interesting cases, the limiting values of $p(s_n)/\sigma^n$ and the asymptotic bounds on $p(t)/\sigma^{2n}$.

Theorem 20 (Asymptotics on the population ratios). *Let μ be any small positive complex number. Let $t := 0^{n-j}s \in \Delta_n$ with $j \in [0, \ldots, n-1]$. Over an alphabet of cardinality $\sigma \geq 2$, the ratio $p(t)/p(s)$ satisfies the asymptotic inequality:*

$$c \cdot \sigma^{2n} + O((\sigma + \mu)^n) \leq \frac{p(t)}{p(s)} < \frac{c \cdot \sigma}{\sigma - 1} \cdot \sigma^{2n} + O(n(\sigma + \mu)^n). \qquad (2)$$

In particular, we have the asymptotic bounds on the population ratio $p(t)/\sigma^{2n}$

$$c \cdot p(s) \leq \lim_{n \to \infty} \frac{p(t)}{\sigma^{2n}} < \frac{c \cdot \sigma}{\sigma - 1} \cdot p(s). \qquad (3)$$

Table 2. Population ratios for alphabet sizes $\sigma = 2, 3$, and 24. Columns 3 and 4 give resp. the limiting values of $p(s_n)/\sigma^n$ and the asymptotic bounds for $p(t)/\sigma^{2n}$ for the correlations s in column 2. The lower bound in col. 4 matches the value of col. 3 (taken from [8] for a given s and σ). The correlations ε and $0^{n-1}1$ are the most populated ones for $\sigma = 2$. For $\sigma \geq 3$, the correlation 0^n becomes the most populated one.

Alphabet Size σ	Autocorrelation s	$p(s_n)/\sigma^n$	$p(t)/\sigma^{2n}$
2	ε	0.268	[0.268, 0.536)
	1	0.300	[0.300, 0.600)
	10	0.110	[0.110, 0.220)
	11	0.089	[0.089, 0.178)
3	ε	0.557	[0.557, 0.836)
	1	0.283	[0.283, 0.424)
	10	0.072	[0.072, 0.108)
	11	0.032	[0.032, 0.048)
24	ε	0.957	[0.957, 0.999)
	1	0.042	[0.042, 0.044)

5   Solutions to Gabric's Open Questions

5.1   Counting Pairs of Words with a Longest Border in a Fixed Range

In an article about bordered and unbordered pairs of words [7], the author raises a challenging question Q1: *How many pairs of length-n words have the longest border of fixed length j?* Note that with his terminology, a border is either a right border or a left border, depending on the order of words in the pair. As the words play symmetrical roles in the definition of the border, the counts for the question are equal.

For the question, we answer a more complex question than the one asked by Gabric: *How many pairs of length-n words have the longest border within the fixed length range [i..k]*. From the characterization of the set of correlations (Lemma 9), we know that correlations are partitioned by their longest border (Corollary 10). To consider pairs with longest border of length in the range $[i..k]$, we must count pairs having a correlation t in the subset $\left\{ \bigcup_{j=i}^{k}(0^{n-j}.\Gamma_j) \right\}$ of Δ_n. With the recurrence that computes the population size for any correlation t (Theorem 16), it suffices to sum up $p(t)$ overall t in this subset to answer our question, which yields Theorem 21. By shrinking the range to a single value, we exactly answer Gabric's question, as addressed in Corollary 22.

Theorem 21. *Let $L_{[i..k]}$ be the number of pairs of words of length n that have a longest border within the fixed length range $[i..k]$ where $i \le k \in \{0, \ldots, n-1\}$. Let $j \in \{i, \ldots, k\}$. Let s be any autocorrelation of Γ_j. Let $t := 0^{n-j}s \in (0^{n-j}.\Gamma_j)$. Let $s_{(n+\lambda)} = 10^{n+\lambda-j-1}s \in \Gamma_{(n+\lambda)}$ where $\lambda \in \{0, \ldots, \lfloor \frac{j}{2} \rfloor\}$. Then*

$$L_{[i..k]} = \sum_{t \in (\cup_{j=i}^{k}(0^{n-j}.\Gamma_j))} p(t) = \sum_{\lambda = \lceil \frac{2n-j}{2} \rceil}^{n-1} \sum_{s \in (\cup_{j=i}^{k}\Gamma_j)} p(s_{(2n-\lambda)}) \cdot s[j+2\lambda-2n] + \sum_{s \in (\cup_{j=i}^{k}\Gamma_j)} p(s_{2n}).$$

In particular, $L_{[0..k]}$ represents the number of pairs of words of length n that have the longest border of length at most k, and $L_{[i..n-1]}$ counts the number of pairs of (distinct) words of length n that have the longest border of length at least i. By restricting the length range to a single value j, we get the following corollary that answers Gabric's first question.

Corollary 22. *Let L_j be the number of pairs of words of length n that have the longest border of length j. Let s be any autocorrelation of Γ_j. Let $t := 0^{n-j}s \in (0^{n-j}.\Gamma_j)$. Let $s_{(n+\lambda)} = 10^{n+\lambda-j-1}s \in \Gamma_{(n+\lambda)}$ where $\lambda \in \{0, \ldots, \lfloor \frac{j}{2} \rfloor\}$. Then*

$$L_j = \sum_{t \in (0^{n-j}.\Gamma_j)} p(t) = \sum_{\lambda=1}^{\lfloor \frac{j}{2} \rfloor} \sum_{s \in \Gamma_j} p(s_{(n+\lambda)}) \cdot s[j - 2\lambda] + \sum_{s \in \Gamma_j} p(s_{2n}).$$

5.2   Expected Value of the Longest Border of a Pair of Words

In [7], Gabric considers a fixed alphabet size σ and a Bernoulli i.i.d model for random words. In this model, the probability that a character occurs at any position is independent of other positions and equals $1/\sigma$. For a fixed word length n, the probability of any pair of words (u, v) both of length n is $1/\sigma^{2n}$. Gabric shows that the expected length of the **shortest border** of a pair of words converges to a constant. In this section, we show that the expected length of the **longest border** of a pair of words also converges, and thereby answer Q2.

Define X to be the length of the longest border of a pair of words (u, v). Then, the expectation of X is

$$E(X) = \sum_{j=0}^{n-1} j \cdot Pr(X = j) = \sum_{j=1}^{n-1} j \cdot \frac{L_j}{\sigma^{2n}} = \sum_{j=1}^{n-1} j \cdot \frac{\sum_{t \in (0^{n-j}.\Gamma_j)} p(t)}{\sigma^{2n}}. \qquad (4)$$

Theorem 23. *The asymptotic expected length of the longest border of a pair of words $(u, v) \in \Sigma^n \times \Sigma^n$ converges. Furthermore, we have that*

$$\sum_{j=1}^{J-1} \sum_{s \in \Gamma_j} \frac{j \cdot p(s_{2n})}{\sigma^{2n}} + O(\frac{1}{\sigma^J}) \leq E_\infty(X) \leq \frac{\sigma}{(\sigma - 1)^2},$$

where $J \geq 2$ and J is any j that satisfies Theorem 19.

6   Conclusion

Our work focuses on counting ordered pairs of words (u, v) that satisfy a given correlation, which is a binary vector that encodes all overlaps of u over v. The set of such pairs is called the population of the correlation, and their number the population size. The main results on population size are stated in Theorems 16 and 18. With these at hands, one can count the number of pairs of length-n words with the longest border of length j as asked by Gabric (open question Q1) [7], or the expected length of this longest border across all pairs of length-n words (open question Q2), since the longest border is encoded in the correlation. Thus, the answer to Q1 is, for instance, a corollary of Theorem 21, which answers a more complex question. This emphasizes the importance of accounting for the complete overlap structure of a pair of words when investigating such questions. Another result illustrating this is the asymptotic convergence of the expected length of the longest border (open question Q2–see Eq. 4, Theorem 23), which is in line with the case of single words of length n [15]. We conclude our work by proposing one conjecture and one open question:

1. We conjecture that the population ratio $p(t)/\sigma^{2n}$ converges, and its asymptotic behavior equals the limiting value of $p(s_n)/\sigma^n$: $\lim_{n \to \infty} p(t)/\sigma^{2n} = \lim_{n \to \infty} p(s_n)/\sigma^n$.
2. What is the variance or distribution of the length of the longest border of a pair of words?

Acknowledgments. This study was funded by the European Union's Horizon 2020 research and innovation programme under the Marie Skłodowska-Curie grant agreement No 956229. We thank the anonymous reviewers for a suggestion on the proof of Lemma 12. The authors have no competing interests relevant to this work.

References

1. Bajic, D., Loncar-Turukalo, T.: A simple suboptimal construction of cross-bifix-free codes. Cryptogr. Commun. **6**(6), 27–37 (2014)
2. Barcucci, E., Bernini, A., Pinzani, R.: A strong non-overlapping Dyck code. In: Moreira, N., Reis, R. (eds.) DLT 2021. LNCS, vol. 12811, pp. 43–53. Springer, Cham (2021). https://doi.org/10.1007/978-3-030-81508-0_4

3. Bilotta, S., Pergola, E., Pinzani, R.: A new approach to cross-bifix-free sets. IEEE Trans. Inf. Theory **58**(6), 4058–4063 (2012)
4. Blackburn, S.R., Esfahani, N.N., Kreher, D.L., Stinson, D.R.: Constructions and bounds for codes with restricted overlaps. IEEE Trans. Inf. Theory **70**(4), 2479–2490 (2024)
5. Cakir, I., Chryssaphinou, O., Månsson, M.: On a conjecture by Eriksson concerning overlap in strings. Comb. Probab. Comput. **8**(5), 429–440 (1999)
6. Ehrenfeucht, A., Silberger, D.: Periodicity and unbordered segments of words. Discret. Math. **26**(2), 101–109 (1979)
7. Gabric, D.: Mutual borders and overlaps. IEEE Trans. Inf. Theory **68**(10), 6888–6893 (2022)
8. Guibas, L.J., Odlyzko, A.M.: Periods in strings. J. Comb. Theory Ser. A **30**, 19–42 (1981)
9. Guibas, L.J., Odlyzko, A.M.: String overlaps, pattern matching, and nontransitive games. J. Comb. Theory Ser. A **30**(2), 183–208 (1981)
10. Gusfield, D.: Algorithms on Strings, Trees, and Sequences - Computer Science and Computational Biology. Cambridge University Press (1997)
11. Gusfield, D., Landau, G.M., Schieber, B.: An efficient algorithm for the all pairs suffix-prefix problem. Inf. Proc. Lett. **41**(4), 181–185 (1992)
12. Halava, V., Harju, T., Ilie, L.: Periods and binary words. J. Comb. Theory Ser. A **89**(2), 298–303 (2000)
13. Harju, T., Nowotka, D.: Periodicity and unbordered words. In: Diekert, V., Habib, M. (eds.) STACS 2004. LNCS, vol. 2996, pp. 294–304. Springer, Heidelberg (2004). https://doi.org/10.1007/978-3-540-24749-4_26
14. Holub, S., Nowotka, D.: The Ehrenfeucht-Silberger problem. J. Comb. Theory Ser. A **119**(3), 668–682 (2012)
15. Holub, S., Shallit, J.O.: Periods and borders of random words. In: STACS 2016. LIPIcs, vol. 47, pp. 44:1–44:10 (2016)
16. Knuth, D., Morris, J., Pratt, V.: Fast pattern matching in strings. SIAM J. Comput. **6**, 323–350 (1977)
17. Lim, J., Park, K.: A fast algorithm for the all-pairs suffix-prefix problem. Theoret. Comput. Sci. **698**, 14–24 (2017)
18. Loptev, A., Kucherov, G., Starikovskaya, T.: On maximal unbordered factors. In: Cicalese, F., Porat, E., Vaccaro, U. (eds.) CPM 2015. LNCS, vol. 9133, pp. 343–354. Springer, Cham (2015). https://doi.org/10.1007/978-3-319-19929-0_29
19. Lossers, O.P.: Overlapping binary sequences. SIAM Rev. **37**(4), 619–620 (1995)
20. Lothaire, M. (ed.): Algebraic Combinatorics on Words. Cambridge University Press (1997)
21. Lothaire, M. (ed.): Combinatorics on Words. Cambridge University Press (1997)
22. Mäkinen, V., Belazzougui, D., Cunial, F., Tomescu, A.I.: Genome-Scale Algorithm Design. Cambridge University Press (2015)
23. Nielsen, P.T.: A note on bifix-free sequences (Corresp.). IEEE Trans. Inf. Theory **19**(5), 704–706 (1973)
24. Nielsen, P.T.: On the expected duration of a search for a fixed pattern in random data (Corresp.). IEEE Trans. Inf. Theory **19**(5), 702–704 (1973)
25. Percus, O.E., Whitlock, P.A.: Theory and application of Marsaglia's monkey test for pseudorandom number generators. ACM Trans. Model. Comp. Simul. **5**(2), 87–100 (1995)

26. Rahmann, S., Rivals, E.: Exact and efficient computation of the expected number of missing and common words in random texts. In: Giancarlo, R., Sankoff, D. (eds.) CPM 2000. LNCS, vol. 1848, pp. 375–387. Springer, Heidelberg (2000). https://doi.org/10.1007/3-540-45123-4_31

27. Rahmann, S., Rivals, E.: On the distribution of the number of missing words in random texts. Comb. Probab. Comput. **12**(01) (2003)

28. Rivals, E.: Incremental computation of the set of period sets. In: SOFSEM 2025: Theory and Practice of Computer Science. LNCS, vol. 15539, pp. 254–268 (2025)

29. Rivals, E., Rahmann, S.: Combinatorics of periods in strings. In: Orejas, F., Spirakis, P.G., van Leeuwen, J. (eds.) ICALP 2001. LNCS, vol. 2076, pp. 615–626. Springer, Heidelberg (2001). https://doi.org/10.1007/3-540-48224-5_51

30. Rivals, E., Rahmann, S.: Combinatorics of periods in strings. J. Comb. Theory Ser. A **104**(1), 95–113 (2003)

31. Rivals, E., Sweering, M., Wang, P.: Convergence of the number of period sets in strings. In: ICALP 2023. LIPIcs, vol. 261, pp. 100:1–100:14 (2023)

32. Rivals, E., Sweering, M., Wang, P.: Convergence of the number of period sets in strings. Algorithmica **87**, 690–711 (2025)

33. Rivals, E., Wang, P.: Counting overlapping pairs of strings (2024). https://doi.org/10.48550/arXiv.2405.09393. arXiv:2405.09393

34. Robin, S., Rodolphe, F., Schbath, S.: DNA. Words and Models. Cambrigde University Press (2005)

35. Smyth, W.F.: Computating Pattern in Strings. Pearson Addison Wesley (2003)

Author Index

© The Editor(s) (if applicable) and The Author(s), under exclusive license
to Springer Nature Singapore Pte Ltd. 2026
F. V. Fomin and M. Xiao (Eds.): COCOON 2025, LNCS 15984, pp. 397–399, 2026.
https://doi.org/10.1007/978-981-95-0218-9

The manufacturer's authorised representative in the EU is Springer
Nature Customer Service Centre GmbH, Europaplatz 3, 69115 Heidelberg,
Germany. If you have any concerns regarding our products, please
contact ProductSafety@springernature.com

Printed and bound by CPI Group (UK) Ltd, Croydon, CR0 4YY
29/04/2026
02099551-0003